Algorithmic and Architectural Gaming Design:

Implementation and Development

Ashok Kumar
University of Louisiana at Lafayette, USA

Jim Etheredge
University of Louisiana at Lafayette, USA

Aaron Boudreaux
University of Louisiana at Lafayette, USA

Managing Director: Lindsay Johnston
Senior Editorial Director: Heather A. Probst
Book Production Manager: Sean Woznicki
Development Manager: Joel Gamon
Development Editor: Hannah Abelbeck
Acquisitions Editor: Erika Gallagher
Typesetters: Russell A. Spangler, Deanna Jo Zombro
Cover Design: Nick Newcomer, Lisandro Gonzalez

Published in the United States of America by
Information Science Reference (an imprint of IGI Global)
701 E. Chocolate Avenue
Hershey PA 17033
Tel: 717-533-8845
Fax: 717-533-8661
E-mail: cust@igi-global.com
Web site: http://www.igi-global.com

Library of Congress Cataloging-in-Publication Data

Algorithmic and architectural gaming design: implementation and development / Ashok Kumar, Jim Etheredge and Aaron Boudreaux, editors.
p. cm.
Includes bibliographical references and index.
Summary: "This book discusses the most recent advances in the field of video game design, with particular emphasis on practical examples of gaming development as well as the design, implementation, and testing of actual games"--Provided by publisher.
ISBN 978-1-4666-1634-9 (hardcover) -- ISBN 978-1-4666-1635-6 (ebook) -- ISBN 978-1-4666-1636-3 (print & perpetual access) 1. Video games--Design. 2. Computer games--Design. 3. Computer games--Programming. I. Kumar, Ashok, 1968- II. Etheredge, Jim, 1949- III. Boudreaux, Aaron, 1982-
GV1469.3.A37 2012
794.8'1--dc23

2012003348

British Cataloguing in Publication Data
A Cataloguing in Publication record for this book is available from the British Library.

Table of Contents

Section 4
Game Models and Implementation

Section 5
Serious Games

Detailed Table of Contents

This chapter takes a scientific view of gameplay as a combination of challenges and methods. Good gameplay is viewed as the creation of enjoyable tension between pressure from a challenge and understanding how to deal with it using the methods at a player's disposal. A collection of challenges and methods are presented for synthesizing as well as analyzing gameplay in video games. Using the proposed model, two sample games are designed for illustration.

This chapter creates a model that can then be used to guide the game's implementation. The authors use the game Othello as an example. The simple story of Othello, where two players alternate placing either black or white pieces on an 8x8 grid, is used to develop a finite state machine. This machine is used to guide the step-by-step implementation of the game code. From there, the authors make two alterations to the game to include both single player and networked versions.

Section 5
Serious Games

Serious games are great tools for instruction and learning. The topics covered under the Serious Games section include development of a music tutor game using tower defense strategy, and the development of a game for training navy personnel about movement of ships.

Vivace, the game presented in this chapter, is a version of the classic tower defense game, where the player must protect a growing tree sapling by using flowers. Each type of flower represents a different musical note, and this note is played when the flower attacks an enemy. Chords containing multiple notes can also be created. Learning the correct chords to defeat the enemy attack makes the game significantly easier. Algorithms discussed include procedural generation of the tree, game balancing, and the design of a finite state machine to control the boss enemy.

Preface

The title of this book is " Algorithmic and Architectural Gaming Design: Implementation and Development" and it does an admirable job of conveying both its theme and its focus. The theme is the investigation of the basic components of the computer programs that are created to implement video games. The general term for these systems is "game engine". As is the case with virtually all computer applications, algorithms of one sort or another form the heart and soul of a game engine's functionality. Most game engines are complex systems because they incorporate all of the components necessary to develop complete video games. To the casual user video games may seem relatively simple, or at least their preoccupation with the game itself leaves them no time to ponder the astonishing amount of computation required (graphics, game play updating for every animate object in the game, sound, physics, artificial intelligence, data storage and retrieval, etc.) to provide realistic game play at speeds of around thirty frames a second. The only way to manage this combination of complexity and speed is through the use of algorithms to provide the functionality and game engine architecture to lend some kind of order to the design of the game engine used to develop the games.

An algorithm is defined in literature as a "step-by-step procedure for calculations. More precisely, it is an effective method expressed as a finite list of well-defined instructions for calculating a function." In a practical sense, algorithms are ways of accomplishing a task that have been found to be effective and refined over many uses to improve efficiency and adapt to variations of a specific task. The tricky part is deciding which algorithm to use (there can be several established algorithms applicable to the same kind of task) and deciding exactly how they need to be adapted.

Within the context of this book, the term architecture refers to the software architecture of the game engine used to develop video games. A game engine's architecture is a description of the conceptual model upon which it is based, its components along with their function and interaction, and usually some connection to lower level software systems (i.e. graphics libraries) or even hardware components (i.e. video cards).

While video game algorithms and software architecture are the theme of the book, their implementation and development are its focus. It has been our experience that many game development issues are discussed in general in the literature but seldom are sufficient details included to allow the interested reader to actually build the component in software. It is the goal of this book to, whenever possible, provide the reader with implementation details of the topics covered. To avoid problems caused by different languages, the implementations presented throughout the book are written in pseudo-code. The book also contains chapters dedicated to the architecture of game engines for those readers interested in understanding or perhaps even implementing a game engine of their own.

With the commercialization of video games in the late 1970s and early 1980s the race was on to provide users with the most realistic and satisfying gaming experience possible. In the relatively short time span since then the design and development of video games and game development engines has become a major entertainment industry worldwide with revenues expected to exceed $112 billion dollars in 2015. Algorithms and game engines to speed development quickly became a hot topic and the potential source of a competitive edge in a growing video game market. Recently, there has been a resurgence of individuals and small programming teams developing casual games with shorter production times and less investment in both time and money. This, coupled with the availability of open source game engines has led to a newly revived interest in both game algorithms and game engine architectures. A book that focuses on the implementation details of developing video games is especially useful to a wide spectrum of people with an interest in this rapidly expanding field.

The target audience for this book consists of undergraduate and graduate students, researchers, and game professionals working in the area of video game design and implementation. Readers will benefit by learning about many core concepts and how to implement them. Some of the latest research and implementation work is covered in the book and will provide both researchers and practitioners an opportunity to learn and apply the new-found knowledge. The book also provides a rich source of bibliographical information for each chapter for further study which graduate students and researchers will find very useful.

The book is organized into five major sections labeled as Artificial Intelligence in Games, Game Physics, Collision Detection in Games, Game Models and Implementation, and Serious Games. The topics covered under the Artificial Intelligence in Games section include group formation and steering behavior for navigation, pathfinding, behavior trees, collective decision making, intelligent and dynamic adaptation of difficulty, application of AI algorithms, and evolving bots. The topics covered under the Game Physics section include a rigid body linear complementary (LCP) solver and implementation of rocket jump mechanics for side scrolling platform games. The topics covered under the Collision Detection section includes two detailed chapters explaining how algorithms can be implemented efficiently for detecting collisions in a game. The topics covered under Game Models and Implementation covers development of a software framework for designing multiplayer online games, illustration of modular design of a game engine, an algorithmic perspective on game play model, and an example of a game's development. The topics covered under the Serious Games section include development of a music tutor game using tower defense strategy, and the development of a game for training navy personnel about movement of ships.

Chapter 1, entitled "*Co-ordinating Formations: A Comparison of Methods*", focuses on group behavior in computer games, which is directly applicable to real-time strategy games as well as any situations that involve control and coordination of a group of characters or entities. The task of coordinating formations is divided into defining a control structure for the formation, finding a path for the formation, and steering individual entities within the formation. Modeling of the formation is done with six different control structures. Several strategies are discussed for moving the characters into a formation, and three decision making methods for coordinating a formation are discussed and implemented. For a rigorous testing of the approach, obstacles of different types and sizes are considered for pathfinding while maintaining a formation. Tests are run on the three proposed methods, called behavior-based steering, fuzzy logic control, and mass-spring systems and results are reported. It is found that the behavior based steering is easy to implement, efficient, and yields the best results.

Chapter 2, entitled "*Adapting Pathfinding with Potential Energy*", extends traditional pathfinding by allowing the non-player characters (NPCs) to learn from the experience of traversing generated paths. This is accomplished by allowing waypoints to have either positive or negative energy based on events that have happened to NPCs around that node. As a result, NPCs can adapt to the environment and can avoid or be attracted to parts of the game world as appropriate.

Chapter 3, entitled "*Behavior Trees: Introduction and Memory-Compact Implementation*", addresses the decision making and control of NPCs through behavior trees. Behavior trees are a method of organizing behaviors in a tree structure such that appropriate actions are taken in a cost-effective way and possibly without incurring huge memory footprints. The chapter presents, in detail, the concept of behavior trees including behavior execution states, actions, conditions, deciders, priority and probability selectors, decorators, and concurrency and speed up issues. After introducing the concept, the authors also give a sample implementation that uses a small memory footprint.

In Chapter 4, entitled "*Nonmanipulable Collective Decision-Making for Games*", the authors introduce a novel method for allowing human and/or non-player characters to find a group consensus. The power of their approach is that no participant can manipulate the results in his or her favor by voting insincerely. The chapter gives an algorithm and example for several types of scenarios. The approach is applicable to situations that require cooperation and consensus among players or NPCs. The issue of asynchronous decision making has also been addressed, which makes the approach useful for a variety of game scenarios.

In Chapter 5, entitled "*Understanding and Implementing Adaptive Difficulty Adjustment in Video Games*", the authors address the issue of providing the game an opportunity to dynamically adjust the difficulty of the game as it is played. The work proposes a way to model, build, and validate a system for dynamic decision adjustment, and it provides a deep insight into how to create a video game with adaptive difficulty.

In Chapter 6, entitled "*Application and Evaluation of Artificial Intelligence Algorithms for StarCraft*", the author shows how to build intelligent NPCs for StarCraft. First, the installation and setup of the Brood War API (BWAPI) is discussed. Then, algorithms to automate essential game tasks, such as producing workers and assigning them to harvest resources, are covered. Also included are gathering algorithms, applications of swarm intelligence, expert systems, and hill climbing AI, with examples of situations where each method can be useful.

In Chapter 7, entitled "*Evolving Bots' AI in Unreal*™", the authors propose and implement a system to control the behavior of bots by changing the default AI of the bot in Unreal™. A genetic algorithm based approach is used to change the default, hardcoded values that determine bots' parameters, and genetic programming is applied to obtain new sets of rules that govern bots' behavior.

Chapter 8, entitled "*Practical Introduction to Rigid Body Linear Complementary Problem (LCP) Constraint Solvers*", is useful for implementing real-time physics in games. It explains the principles for LCP constraint solvers with practical examples and implementation details. It provides code for modeling a rigid body class, an iterative LCP solver, and adaptation details. Examples of constraints, how they are handled, and a variety of implementation methods are discussed. Also, stability and reliability issues are addressed.

Chapter 9, entitled "*Rocket Jump Mechanics for Side Scrolling Platform Games*", takes a popular concept from first person shooter games and incorporates it into a two dimensional platform game. It discusses the mechanics of rocket launching, static and timed obstacles, and rhythm based implementation. Next, it compares the results of their rocket jump implementation with that of conventional jump, and it also notes the importance of player perception, experience, and rhythm.

Chapter 10, entitled "*Collision Detection in Video Games*", explains collision detection in 3D environments. After a brief review of vector math, the chapter goes into the details of three collision detection methods: axially-aligned bounding boxes (AABBs), Object-Oriented Bounding Boxes (OOBs), and the Gilbert-Johnson-Keerthi distance algorithm (GJK). Each section gives a sample implementation of the algorithm, as well as methods for optimization by removing unnecessary intersection tests.

Chapter 11, entitled "*Collision Detection Using the GJK Algorithm*", provides an in-depth tutorial of the famous algorithm. The chapter begins with an introduction to the basic principles needed to understand GJK, determining if a given shape is convex and calculating the Minowski sum and difference of two shapes. Each section of the algorithm is further explained with easy-to-follow examples. After the explanation of GJK, the author provides examples where the algorithm is useful. Finally, an extension to GJK, known as the Expanding Polytope Algorithm, that provides additional information used to resolve collisions is explained in detail.

Chapter 12, entitled "*Designing Multiplayer Online Games Using the Real-Time Framework*", describes a framework for designing and executing online computer games using a new software platform developed by the authors. Major design issues including scalability and single server vs. multi-server applications are considered. The usefulness of the approach is shown with the design of a new multiplayer online game and the extension of the single server commercial game Quake 3 to multiple servers.

Chapter 13, entitled "*Modular Game Engine Design*", provides a tutorial on using software engineering principles in creating a game engine. Specifically, the use of managers as classes that oversee objects with similar functions is emphasized. The authors use a simple tower defense game to demonstrate the implementation of managers to handle game states, game objects, and other game components.

Chapter 14, entitled "*A Gameplay Model for Understanding and Designing Games*", takes a scientific view of gameplay as a combination of challenges and methods. Good gameplay is viewed as the creation of enjoyable tension between pressure from a challenge and understanding how to deal with it using the methods at a player's disposal. A collection of challenges and methods are presented for synthesizing as well as analyzing gameplay in video games. Using the proposed model, two sample games are designed for illustration.

In Chapter 15, entitled "*From a Game Story to a Real 2D Game*", the authors create a model that can then be used to guide the game's implementation. The authors use the game Othello as an example. The simple story of Othello, where two players alternate placing either black or white pieces on an 8x8 grid, is used to develop a finite state machine. This machine is used to guide the step-by-step implementation of the game code. From there, the authors make two alterations to the game to include both single player and networked versions.

Chapter 16 is entitled "*Music Tutor Using Tower Defense Strategy*". Vivace is a version of the classic tower defense game, where the player must protect a growing tree sapling by using flowers. Each type of flower represents a different musical note, and this note is played when the flower attacks an enemy. Chords containing multiple notes can also be created. Learning the correct chords to defeat the enemy attack makes the game significantly easier. Algorithms discussed include procedural generation of the tree, game balancing, and the design of a finite state machine to control the boss enemy.

In Chapter 17, entitled "*Low Cost Immersive VR Solutions for Serious Gaming*", the authors discuss the implementation of a real-time ship training simulator. The life cycle and architecture of the project are described followed by physics based modeling of the ship motion and waves. Implementation of real time motion detection algorithm is discussed followed by rendering details and the validation process. By using off the shelf components and free and open source software, the cost of the simulator was kept to a minimum.

Few books focus on a practical approach to game development concepts, focusing instead on theoretical concepts. It has been shown many times that practical, hands-on learning contributes to a greater understanding of the material, and it has been our experience that when students know how an aspect of a popular game works, they are much more excited about the topic. This book gives readers of all levels insight into practical implementation details for a variety of game development topics.

Instructional chapters, such as the physics chapter using LCP solvers, collision detection using AABBs, OBBs, and the GJK algorithm, and the modular game engine chapter give readers at all levels an opportunity to use hands-on learning to achieve a greater understanding of the concept. With this knowledge, the reader can then go on to extend the given implementation or create one of their own.

Two chapters focus on the importance of designing a game before starting implementation. The defining aspect of a game is its gameplay, or how a player interacts with the game. This knowledge is important for students and anyone starting to create video games. By breaking games down into gameplay elements, the implementation of the game is separated from aspects that are largely interchangeable, such as the art style. The developer can then focus on the implementation details.

Several chapters feature new contributions to the field. For example, the Real-Time Framework provides a high-level method of developing online games such as MMOG (Massively Multiplayer Online Games) and first-person shooters. Scalability, distribution of game data, and server communication are just some of the important issues in the field of online games. The Real-Time Framework addresses these issues and more. Another chapter discusses the implementation of a model for adaptive difficulty in video games. Currently, some players are under-challenged and others are over-challenged, with neither scenario being enjoyable for the player. The goal is to have the difficulty of a game change based on the capabilities of the player to provide an appropriate level of challenge. The authors have implemented a new type of adaptive difficulty model that determines the skill level of the player and changes the difficulty accordingly.

Finally, games can be used for more than just entertainment purposes. The use of simulators is important in many fields for training purposes since using actual equipment for training can be extremely costly and dangerous. Using simulators allows for training scenarios that are expensive, difficult, or impossible to reproduce in any other way. The immersive training simulator shows that effective simulators can be made cheaply.

While not every game is developed using a commercial, open source, or user developed game engine, the material in this book can offer insight into some of the issues critical to the successful development of video games and assist the reader in understanding and implementing game engine components common to virtually all game development projects. It is our sincere hope that readers will take away a deeper understanding of some topic of particular interest to them or a clearer understanding of how to go about the task of actually implementing a common video game algorithm or building a game engine.

Sincerely yours,

Ashok Kumar
University of Louisiana at Lafayette, USA

Jim Etheredge
University of Louisiana at Lafayette, USA

Aaron Boudreaux
University of Louisiana at Lafayette, USA

Acknowledgment

This book would not have been possible without the help of family, friends, and colleagues. The editors would like to thank everyone that contributed to this project in some way. In particular, the editors thank the chapter authors for their contributions to the field of computer science. With their assistance, we have provided a book that focuses on practical implementation of computer science concepts that is accessible to readers of all levels.

We also thank the members of the Editorial Advisory Board for their guidance, insights, and invaluable contributions to the success of this book. The members of the EAB are Remi Arnaud, Jessica Bayliss, Al Biles, Joel Gonzalez, Klaus Jantke, Mick Mancuso, Lee Mendoza, Ian Parberry, Tim Roden, and Jouni Smed.

Those that participated in the manuscript review process also deserve thanks. Our reviewers were Remi Arnaud, Golam Ashraf, Jessica Bayliss, Al Biles, William Bittle, Luke Deshotels, Blake Edler, Joel Gonzalez, Sergei Gorlatch, Thomas Hartley, Klaus Jantke, Ben Kenwright, Bjoern Knafla, Nihal Kodikara, Jussi Laasonen, Jennifer Lavergne, Kurt Lavergne, Joachim LeBlanc, Rob LeGrand, Mick Mancuso, Lee Mendoza, Ian Parberry, Brandon Primeaux, Tim Roden, Benjamin Rodrigue, Damitha Sandaruwan, Nick Sangchompuphen, Jouni Smed, Jonathan Tremblay, and Chong-wei Xu. Their efforts were instrumental in assuring the highest possible quality in the material included.

Finally, the editors want to thank the IGI Global staff for their efforts in getting this book published. Ms. Hannah Abelbeck in particular has provided immeasurable support and guidance throughout the entire process. She was truly a pleasure to work with.

Ashok Kumar
University of Louisiana at Lafayette, USA

Jim Etheredge
University of Louisiana at Lafayette, USA

Aaron Boudreaux
University of Louisiana at Lafayette, USA

Section 1
Artificial Intelligence in Games

Chapter 1
Co-ordinating Formations:
A Comparison of Methods

Jussi Laasonen
University of Turku, Finland

Jouni Smed
University of Turku, Finland

ABSTRACT

Moving in a formation is a basic group behaviour needed in computer games. This chapter presents different methods for co-ordinating formations in real-time game environments. To compare the methods in different situations, how well the formations stay organized and how fast they are able to navigate through test courses is measured. Additionally, the authors analyse whether the methods can cope with pathological situations such as passing through a narrow canyon. The effect of different formation types and study the scalability when the formations get larger are also compared.

INTRODUCTION

Computer-controlled entities often have to move in a formation. For example, a real-time strategy game can include hundreds of computer-controlled entities on a battlefield, which are not to be controlled individually but as groups. If the entities are commanded on this higher level, the player can control even larger units with less effort. Therefore, formations can improve the visual appeal and realism of a game, because the units seem to move in a natural looking way rather than in unorganized groups.

The task of co-ordinating formations can be divided into three subtasks:

1. Define a control structure for the formation.
2. Find a path for the whole formation.
3. Steer the individual entities.

Control structure is used to model the formation, and it defines how to calculate the places of the agents (or entities) in a formation (for basic formation types, see Figure 1). To realize this we

DOI: 10.4018/978-1-4666-1634-9.ch001

Figure 1. Basic formations types: (a) column, (b) line, (c) wedge, (d) vee, (e) box, (f) right flank, (g) left flank, and (h) diamond

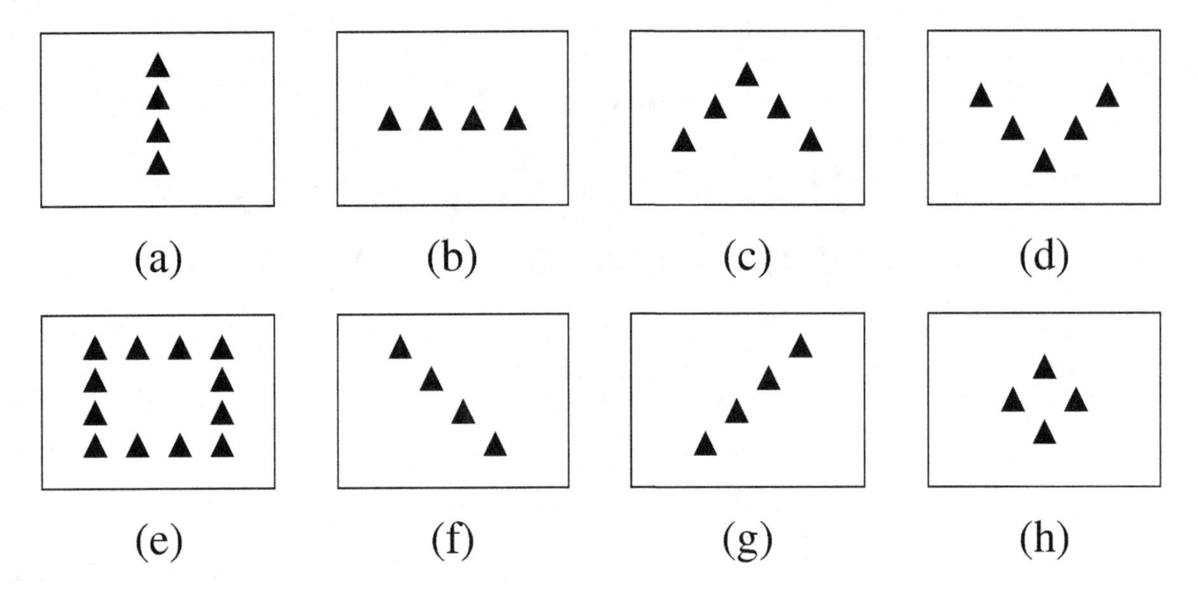

can choose from several different control structures, which can be based on following a leader or the average position of the neighbours.

In path finding, we search for the shortest viable path from a start position to a goal position. For this subtask, the most commonly used path finding algorithm is A*, which has many game-dependent variants to make it faster. The path found by the algorithm can look unrealistic, but it can be improved by path realization methods to make it look more natural.

When we have found a path, the agents are guided through the game world using a steering method. It tries to keep the agents as close to their intended place in the formation as possible. Moreover, it should allow the agents to deviate if the environment does not allow them to maintain a tight formation.

In general, decision-making can be divided into three levels: strategic, tactical, operative (Smed & Hakonen, 2006). Strategic level contains long-term decisions based on large amounts of data. Tactical level handles group-level decisions aiming at fulfilling a plan made at the strategic level. Operational level focuses on short-term decisions from a limited set of alternatives. In this chapter, we will focus on operational level decision-making in the domain of co-ordinating formations. We assume that the path for the formation has already been found and we must now steer the group using a control structure. For this reason, we have implemented three different formation methods: a steering behaviour-based method, a fuzzy logic controller, and a mass-spring system. We test the performance of these methods with different metrics, measuring the level of organization of a formation and the speed at which they solve test tracks. We analyse closely pathological cases such as passing through a narrow canyon. Moreover, we compare the results for different types of formations (e.g., square, line and column) to see their effect on the methods. Finally, we present results from massive formations that can include up to 100 computer-controlled entities.

FORMATIONS

In this section, we present methods for modelling and controlling formations. First we look at different approaches to model formations, which are classified according to how the point-of-reference

Figure 2. Formation models: (a) leader referenced, (b) unit-centre referenced, (c) neighbour referenced, (d) social potentials, (e) virtual structure, and (f) virtual leaders

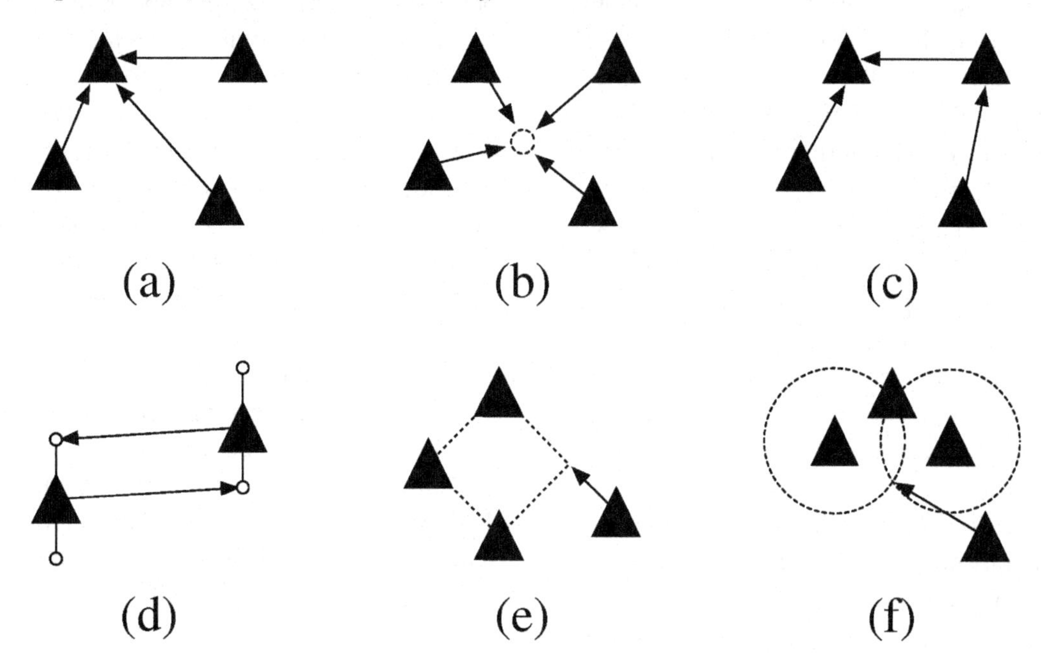

of an agent in a formation is calculated. After that, we study obstacle avoidance and, finally, how to maintain the correct position in a formation. We do not focus on path finding but assume that the path to follow has been given (e.g., from a human player or a higher-level decision-making system).

Most of the research work on formation has been carried out in the field of robotics, whereas for computer games, the research has been scarce: Dawson (2002) gives a short introduction to the use of formations in computer games, Verth *et al.* (2000) present a solution used in the game *Force 21*, and Pottinger (1999a; 1999b) proposes ideas albeit without a deeper analysis. Heiden *et al.* (2008) discuss a dynamic approach on organizing formations in a real-time strategy game.

Modelling Formations

We can model a formation with different control structures, which can be classified according to how the position of an agent in a formation is defined (see Figure 2).

(a) In a *leader-referenced* formation (Figure 2a), the position of an agent is defined in relation to the position of the leader (Balch & Arkin, 1998). Moreover, the movement of the formation is based on the movement of the leader. The main problem of this approach is that if the leader does not receive enough updates from the other agents, some agents may lag behind without the leader slowing down, which can cause the formation to disperse.
(b) In a *unit-centre-referenced* formation, the position of an agent depends on the average position of the agents (Balch & Arkin, 1998). The main difficulty of this approach is to get the formation move purposefully, which requires that all the agents move towards to the destination.
(c) In a *neighbour-referenced* formation, the position of an agent is defined in relation to a pre-selected neighbour agent (Balch & Arkin, 1998; Fredslund & Mataric, 2002; Naffin & Sukhatme, 2004).

(d) In a formation using *social potentials*, each agent has a given number of attachment sites, which attract other agents (Balch & Hybinette, 2000). By changing the configuration of the attachment sites we can change the shape of the formation. The benefit of social potentials is that they scale up well even to larger formations. The drawback is that with some attachment point configurations the shape of the resulting formation is not unambiguous. Moreover, if the formation gets divided (e.g., to avoid an obstacle) and returns back together, the shape of the formation can be different than before the division
(e) In a *virtual structure* formation, the position of an agent is bound to a geometric form (Lewis, 1997; Young *et al.*, 2001). When a virtual structure moves, we try to adapt first the structure to the current positions of the agents as well as possible, and then move the structure towards to the goal. After that, all the agents are moved towards their new positions in the formation.
(f) In a *virtual leaders* formation, the agents try to maintain a given distance to one another and to one or many virtual leaders (Ögren *et al.*, 2002). The formation is, therefore, not necessarily ambiguous but it can, for example, revolve around a virtual leader so that the formation stays together. To define the positions of the agents more closely we can add more virtual leaders.

Obstacle Avoidance

We can divide the obstacles encountered along the route into three groups according to their size:

1. Equal to or larger than the formation
2. Smaller than the formation
3. Canyons

Obstacles that belong to the first group can be omitted here, because it is enough that the path finding can solve a route around them.

The second group is more problematic, because a formation as a whole can travel through an area with smaller obstacles but the individual agents must be able to move around them. We can solve this by applying local path finding or obstacle avoidance methods. Local path finding guarantees that an individual agent has an optimal path around an obstacle, but it does not guarantee that the formation stays cohesive. In obstacle avoidance, the agents try to keep in their designated positions, but move so that they do not collide with the obstacles. The easiest way to realize this is to set repulsion which is directed away from the centre of an obstacle. If we observe a collision is about to occur, the agent is steered away from the obstacle (Reynolds, 1999). Pottinger (1999a; 1999b) proposes that when a formation encounters an obstacle, it is divided into two groups both of which avoid the obstacle from different sides. After the obstacle, the groups are merged in one formation again. However, this division and merging can become problematic if there are many obstacles.

Canyons and narrow passages often mean that the formation has to disassemble or change its size or type (e.g., into a column) in order to pass through. For example, Pottinger (1999a) suggests scaling down the formation so that it fits into the canyon, or, if this cannot be realized, finding a new path that goes around the canyon. Verth *et al.* (2000) recount that obstacle avoidance and canyons were solved in similar fashion in the game *Force 21*: If the formation is located so that some agents' positions are inside an obstacle, these positions are relocated along the route from the centre of the formation so that each agent finds a free position. Derenick and Spletzer (2007) present a method for co-ordinating large-scale formations in polygonal environments based on convex optimization.

Maintaining the Formation

The assignment of optimal positions for *n* agents in a formation has a complexity of *O*(*n!*). Because of this, we need in practice an approximation such as sorting the agents according to their distance to the closest available position and assigning the positions in this order (Dawson, 2002). We can simplify the approximation even further by using the distance to the centre of the formation as a sorting criterion.

When we have assigned the positions for all the agents, the movement into the formation can be realized using three different strategies (Dawson, 2002):

1. The formation is located in the centre of the agents, and the agents move first into the formation before they start moving towards the destination as a formation (see Figure 3a).
2. Agents move individually to a rally-point located nearby the destination and from there continue as a formation (see Figure 3b). Here, a possible problem is that the agents closer to the destination arrive at the rally-point earlier and have to stop then to wait for the ones coming from afar.
3. Agents use path finding and move individually to the correct positions in a formation at the destination (see Figure 3c). This approach guarantee that the agents will be in a formation at the destination, but movement of the agents can appear senseless and the benefits of moving in a formation are lost.

If the agent moves into all directions, it is easy to correct the position its position in a formation by simply moving it in the right direction. However, if the agent's mobility is limited (e.g., a vehicle can move forward and turn within a given radius but cannot make 180 degree turn on the spot), maintaining a correct position is harder. In this case, we have to correct it by accelerating or decelerating or turning the direction (Balch & Arkin, 1998).

Figure 3. Strategies for realizing the movement: (a) agents move first into a formation and to the destination, (b) agents move into a formation rally-point along the way to the destination and move there to the destination, and (c) agents move into a formation at the destination

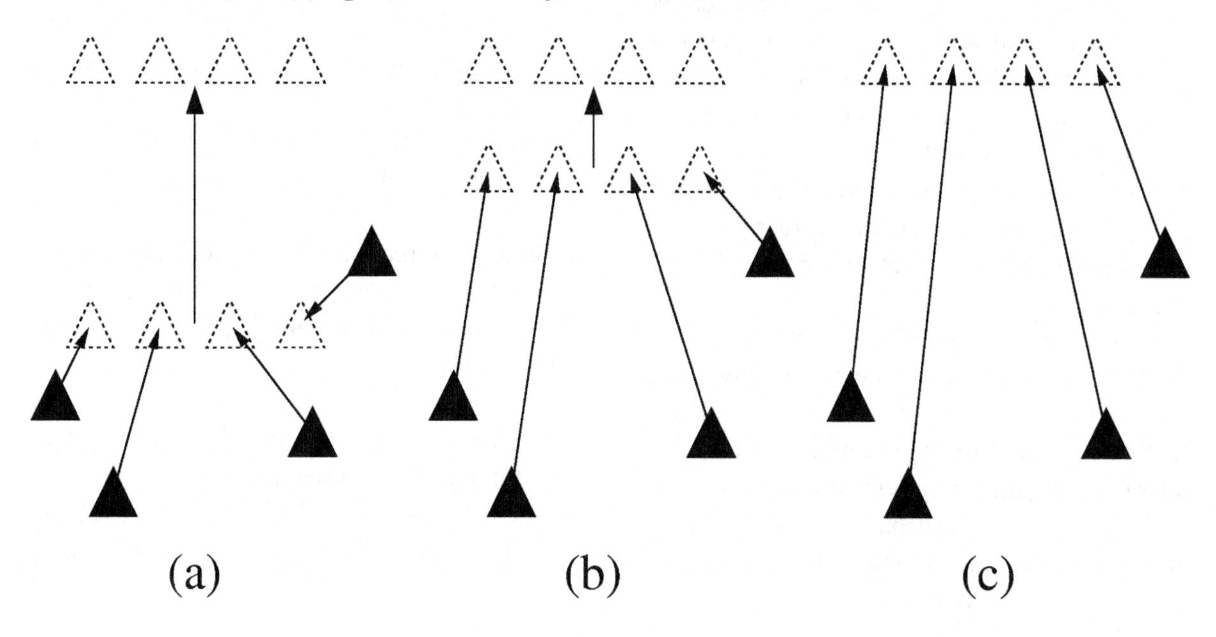

DECISION-MAKING

This section presents different decision-making methods for moving formations. We begin with behaviour-based steering, where the agents' behaviours are represented as vectors and these behaviour vectors are combined to get acceleration for the agent. After that, we examine how fuzzy logic control can be used to control formations. Finally, we present the concepts behind mass-spring systems which allow us to model formations.

Behaviour-Based Steering

In behaviour-based steering, behaviours are represented with vectors modelling the agent's urge to move in certain directions (Reynolds, 1987; Reynolds, 1999). In addition to the direction, the length of the vector's length indicates how strong the urge is. An agent makes the steering decision by combining the vectors and using the end result as the acceleration of the agent. The most important behaviours are

- *Seek* or *flee*: In seeking, the agent's velocity is directed towards a goal, whereas in fleeing, the steering vector is the inverse of seeking.
- *Pursuit* or *evasion*: In this case the goal is moving, and we calculate a future position for it towards which the agent is directed.
- *Arrival*: Resembles seek but the agent slows down whilst nearing the goal.
- *Obstacle avoidance*: Finds the most imminent obstacle along the current path and assigns a steering vector away from it.

Behaviour-based steering is a stateless algorithm that does not need to maintain any extra information. Moreover, there is no centralized control for the formation but the behaviour emerges from simple individual rules. Balch & Arkin (1998) show that behaviour-based steering suits well controlling formations both in simulations and with real vehicles.

Fuzzy Logic Control

A fuzzy set is generalization of a Boolean (or crisp) set, where each element can belong partially to a set (Zadeh, 1965). We can define a fuzzy set using a membership function $\mu: U \rightarrow [0, 1]$, where U is the set of all elements. Using membership functions we define all the classical set operations for fuzzy sets. For example, the intersection of fuzzy sets A and B $\mu_{A \cap B}(x) = \min(\mu_A(x), \mu_B(x))$, the union $\mu_{A \cup B}(x) = \max(\mu_A(x), \mu_B(x))$, and the complement of the fuzzy set A $\mu_{A'}(x) = 1 - \mu_A(x)$.

Fuzzy logic control comprise IF…THEN rules using fuzzy predicates (Yager & Filev, 1994). Assume that the rule-base has the form

IF U_1 is B_{11} AND U_2 is B_{12} AND … AND U_n is B_{1n} THEN V is D_1;

…

IF U_1 is B_{m1} AND U_2 is B_{m2} AND … AND U_n is B_{mn} THEN V is D_m

where $U_1,\ldots, U_n$ are parameters and $B_{11},\ldots, B_{mn}$ and $D_1,\ldots, D_m$ are fuzzy sets. The result of the output V can be computed as follows:

1. For each rule, compute the degree τ in which it gets fired.
2. Compute the result of the rule.
3. Aggregate the results as an output.

In order to realize a fuzzy decision, we have to defuzzify it to get a crisp result. To realize this we can use different methods (Yager & Filev, 1994):

- Maximum grade: Select the element value with the highest membership.
- Centre of area: Calculate the centre of the area indicated by the membership function.

Figure 4. An example of a mass-spring system. The circles represent point masses, the dashed lines the structural springs, solid lines the shear springs, and the bold lines the flexion springs

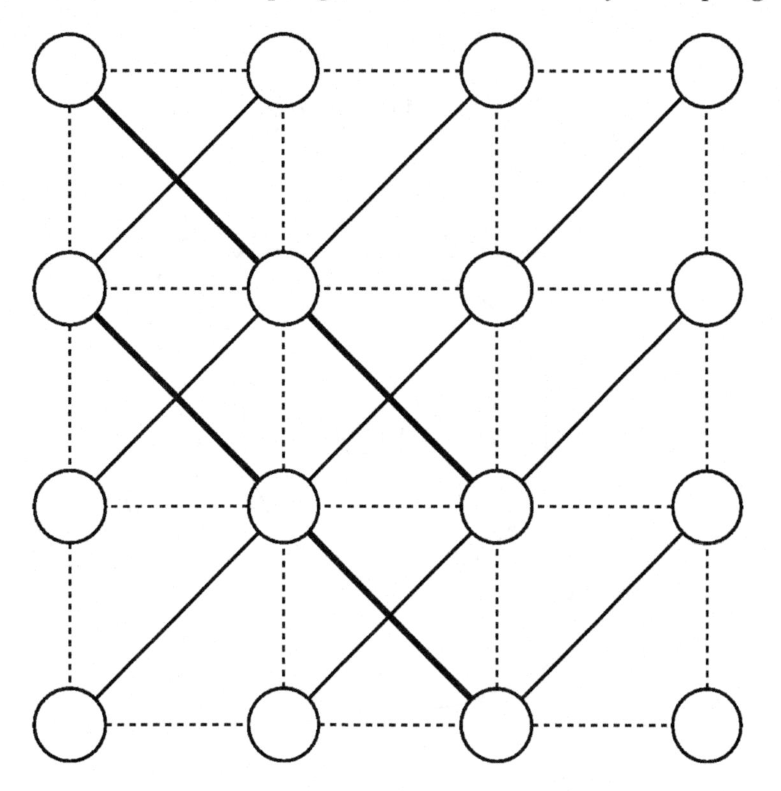

- Mean of maxima: Select the elements that have the highest membership and calculate their mean.
- Random generation: The fuzzy set is transformed into probability distribution, and the crisp value is chosen randomly using this probability distribution.

If we want that some elements will not be selected, we can assign a constraint that sets their membership value to zero.

Mass-Spring Systems

Spring models are used in computer games to simulate soft objects, but they can be used also in formations. A mass-spring system comprises point masses and springs connecting them. Each spring causes the force $F = -kx$, where k is a spring coefficient and x the distance from the static state. To build a structure we need three kinds of springs (Wang & Devarajan, 2004):

- Structural springs model the interaction between different parts of the object.
- Shear springs model the resistance against bending.
- Flexion springs model the shearing resistance.

To use a mass-spring system for a formation, we define a set of springs between the agents. The simplest way would be to create a spring between all agent pairs, but we would end up having excessive springs and the updating would be too slow. Also, excess springs increase the rigidity of the formation. When modelling a two-dimensional object, it is enough to have horizontal, vertical and diagonal springs between neighbouring agents (Wang & Devarajan, 2004).

Figure 5. Zones for the magnitude of maintaining the formation: the control vector is maximum in the ballistic zone, zero in the dead zone, and decreases linearly in the controlled zone the closer the agent gets the dead zone

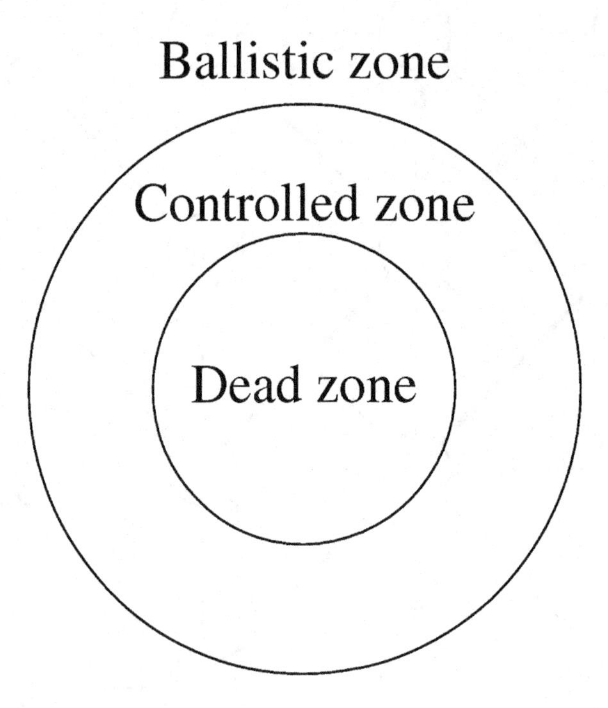

One of the agents is selected as a leader of the formation, and it acts as a fixed point in the formation unaffected by the spring forces. The formation moves when we apply a constant force to the leader pulling it towards the destination.

IMPLEMENTATION

To test different methods we use a test-bench system, which allows us to analyse the methods in two views: A basic test view is intended for visual assessment and for trying out different methods. The batch view is intended for the automatic evaluation of multiple test runs. The implemented methods include Reynolds' behaviour-based steering (BBS), fuzzy logic control (FLC) and mass-spring system (MSS). For detailed information about implementation (e.g., test cases and metrics), see Laasonen (2008).

Behaviour-Based Steering

In BBS, the formation is modelled using a unit-centre-referenced control structure. We use the same behaviours as Balch and Arkin (1998) except for avoidance, which is based on Reynolds (1999):

Maintaining formation: Depending on the distance the agents can be in one of three zones (see Figure 5): Ballistic zone, where control vector has the maximal length. Controlled zone, where the length of the control vector changes linearly from maximum to zero. Dead zone where the control vector always has the length of zero. The dead zone provides a more stable target for the agents than a single point. This can be useful, for example, if there is uncertainty, when we can set the dead zone larger than the error due to uncertainty. In our test, we set the radius of the dead zone to 0 unit, because there is not uncertainty. The controlled zone has the radius of 10 units. The behaviour has the weight 0.75.

Figure 6. In collision avoidance, the collision cylinder has the width of the agent and the length is set according to the agent's speed and manoeuvrability. If there are obstacles intersecting the collision cylinder, the agent chooses the most threatening one and steers away from it

Avoiding agents: We calculate a control vector for all the other agents. The vector is directed to the opposite direction and has the length w / d^2, where d is the distance between the agents and w is the weight of the behaviour (in our tests, w = 1.0).

Avoiding obstacles: Balch and Arkin (1998) proposed that the control vector for obstacle avoidance is calculated only if the obstacle is close enough to the agent. In our tests, this approach turned out to work not so well, especially if there are several obstacles nearby. For this reason, we use collision cylinders (see Figure 6). The length of the cylinder depends only on the speed of the agent, has the length of 10 times the speed of the agent. The weight of the behaviour is 2.0.

Noise: Depending solely on reactive behaviour can lead the agents to get stuck on local maxima or minima or reverting to cyclic behaviour. We can avoid this by adding noise in the agent's control. Here, the length of control vector is the weight of the behaviour and it has a random direction. Noise has a predefined duration when it remains constant. In our test, the weight of the noise behaviour is 0.1 and its duration is 7 turns.

Seeking the target: Seeking behaviour is realized with a control vector that is directed to the agent's correct position in the formation. This behaviour has the weight 0.2.

Fuzzy Logic Control

FLC is implemented using two fuzzy logic controllers, one for steering and another for throttle. Simply put, the rules for the controllers are:

- If the agent is far from the correct formation place and the sensors detect no obstacles, then the steering controller tries to steer towards the formation place.
- If the sensors detect no obstacles, then the steering controller steers towards the next

Figure 7. The membership functions for the FLC: (a) the quantitative fuzzy sets, (b) the steering fuzzy sets, and (c) the distance fuzzy sets

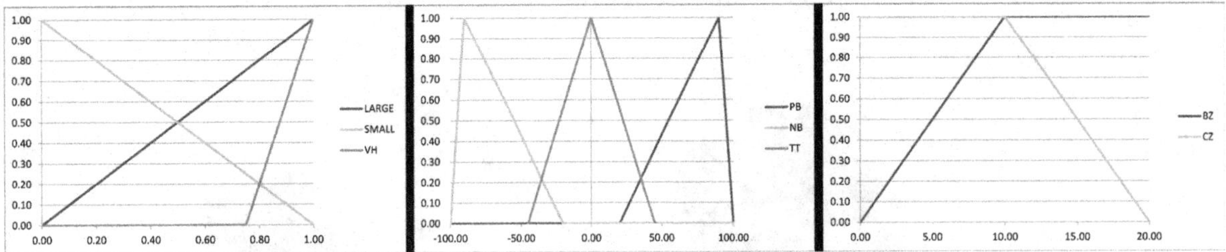

waypoint. If a sensor detects an obstacle, the steering controller tries to steer in the opposite direction of the sensor and the throttle controller slows down the speed.

We use the collision detection scheme proposed by Teiner *et al.* (2003) where only one resultant vector representing the direction and distance to the obstacles is input to the controller. To minimize the effects of different collision avoidance methods we used a similar collision cylinder than in the behaviour-based method. The controller uses two inputs from the collision detection: the direction and the magnitude of the collision. The direction is represented in radians and the magnitude has a value between 0.0 and 1.0, where 1.0 means that the obstacles are very close and 0.0 means that there are no obstacles. If the obstacles intersect the cylinder, the collision direction is perpendicular to the agent's heading and its magnitude is set to 1.0.

The inputs to the fuzzy logic controller are:

- CM: the magnitude of the collision
- CD: the direction of the collision
- AM: the agent's magnitude
- AD: the agent's direction
- FCV: is the centre of the formation visible
- FLV: is the agent's correct location in the formation visible
- D: the distance to the correct location in the formation

There is no controller for the throttle but we use always the maximum acceleration. For steering, we define the fuzzy sets, TT_D (towards the destination), TT_F (towards the formation), TT_{AAD} (away from the agent's direction), NB (steep right) and PB (steep left). To approximate the distance we define the fuzzy sets BZ (ballistic zone) and CZ (controlled zone). We also define the quantitative fuzzy sets VH (very high), LARGE and SMALL. The membership functions for these fuzzy sets are illustrated in Figure 7.

The steering controller has four types of rules:

- Seeking the target:
    - **IF** D is CZ **AND** CM is SMALL **AND** AM is SMALL, **THEN** steering is TT_D
    - **IF** FLV is SMALL **AND** CM is SMALL, **THEN** steering is TT_D
    - **IF** FCV is SMALL **AND** CM is SMALL, **THEN** steering is TT_D
- Staying in the formation:
    - **IF** FCV is LARGE **AND** FLV is LARGE **AND** D is BZ **AND** CM is SMALL **AND** AM is SMALL, **THEN** steering is TT_F
- Obstacle avoidance:
    - **IF** CD is PB **AND** CM is VH, **THEN** steering is NB
    - **IF** CD is NB **AND** CM is VH, **THEN** steering is PB
- Avoid other agents:

- ◦ **IF** AM is LARGE, **THEN** steering is TT_{AAD}

The resulting fuzzy sets are defuzzified into crisp values using the mean-of-maxima method.

Mass-Spring System

The way to model formations in MSS resembles the social potentials and the neighbour-referenced control structure described earlier. The basic idea of MSS is to create a spring between agent pairs. The number of springs required depends on the formation configuration (Wang & Devarajan, 2004). We tested two different spring configurations:

- *Dense*: A spring is created between all agent pairs, which ensures that we create all springs required to keep the formation stable regardless of its shape.
- *Sparse*: Only structural and shear springs are created. This approach works well with rank and file formations used in our tests, but it does not suit, for example, a box formation.

Adding more springs makes the formation more rigid and more resistant to changes. A sparser configuration gives more room for avoiding collision but the agents are less eager to maintain the correct place in the formation.

The value of the spring coefficient is set to 0.2 and the spring force is calculated using normal equation for the spring force $F = -kx$. A simple approach is to create a spring between the agent and the centre of the obstacle if the distance between them is small enough and calculate the spring force using $F = -k \cdot \text{sgn}(x)\, x^2$. As we are not trying to simulate physical springs, we can use arbitrary functions for spring force.

The repulsion-based obstacle avoidance turned out not to work well enough in environments with a high number of obstacles or with non-symmetrical obstacles. Instead, we implemented collision cylinder avoidance resembling the one used in the behaviour-based method. Because the spring forces can get arbitrarily large if one of the agents gets left behind an obstacle, the formation can easily get stuck. There are several ways to overcome this:

1. Make sure that the obstacle avoidance force is sufficiently large to overcome the spring forces in all situations.
2. Limit the spring force magnitude.
3. Use a repulsion-based force when the agent is very close to an obstacle.

For the target seeking, a force pulling towards the destination is applied to all the agents. For the leader, the strength of the force is 8.0 and for the other agents 0.2. The reason for this is that if the force is applied to the leader only, it is difficult to get the formation to move fast enough the without leader getting too far ahead. Moreover, if an equal force is applied to all the agents, the formation does not turn properly towards the waypoints.

RESULTS

The decision-making methods are compared with two test cases: A method comparison test with a track containing turns and small obstacles and a canyon test where the formation has to pass a narrow canyon. The best performing decision-making method is also tested using different formation types on the basic test track and its scalability is tested with a large formation. Examples of the different test cases can be found in Figure 8.

All of the tests were performed on a computer with Intel Core 2 Duo CPU at 2.66 GHz and 4 GB of memory running the Windows 7 operating system. Graphics rendering was disabled during the test runs. The test application was implemented with Java 1.6 and the size of the compiled application including libraries was 1.6 MB.

Figure 8. Examples of the initial situation for each test type. Solid circles represent the agents, open circles indicate the waypoints, lines the path through the waypoints, and grey areas are obstacles that agents should avoid

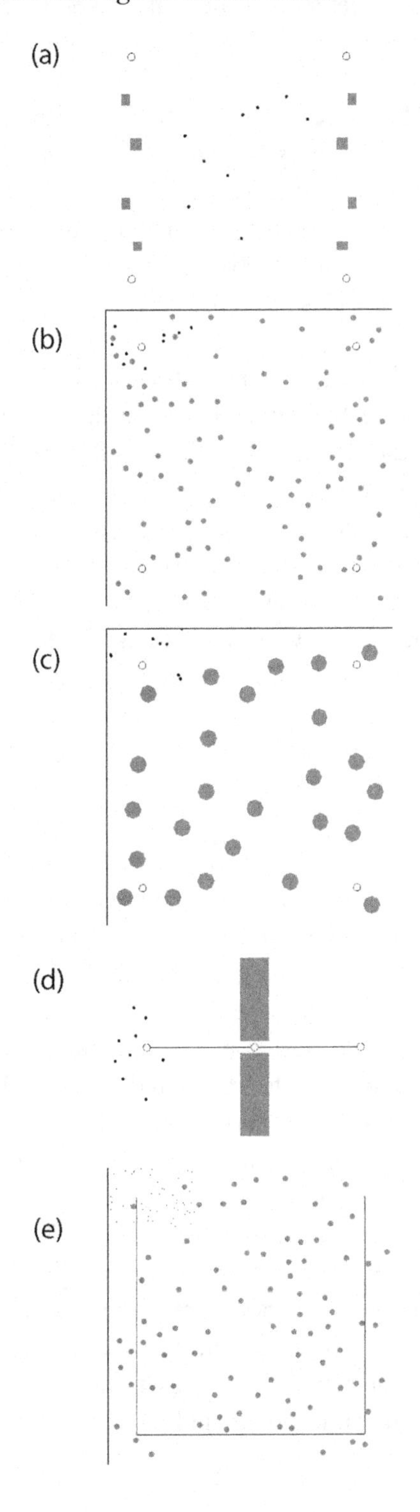

Method Comparison

In the method comparison test, the decision-making methods were tested with three different set-ups shown in Figure 8a-c. The test area has a size of 400×400 units, and nine agents are placed randomly in the centre of the test area. We make 100 test runs for each decision-making method, which are evaluated using the metrics based on the metrics proposed by Balch and Arkin (1998).

Let us assume that

- N is the total number of turns in the test run,
- A is the number of agents,
- p_{ij} is the position of the jth agent in the ith turn,
- l_{ij} is the correct formation position of the jth agent in the ith turn, and
- $e(i, j) = |p_{ij} - l_{ij}|$ is the jth agent's distance to the correct formation position in the ith turn.

In an ideal case, the formation would get organized in every turn. In a reality, however, some agents occasionally fall out of the formation due to turning and obstacle avoidance. This situation is nearly impossible to avoid completely, but a good method should maximize the *number of turns in a formation*

$$\sum_{i=1}^{N} f(i) / N,$$

where

$$f(i) = \begin{cases} 0, \exists j : e(i, j) > 10 \\ 1, \forall j : e(i, j) \leq 10 \end{cases}.$$

This, however, does not take into account how many of the agents are out of the formation. To get a more detailed view of the level of organization we also measure the *proportion of agents in a formation*:

$$\sum_{i=1}^{N}\sum_{j=1}^{A} f_2(i,j) \,/\, \left(AN\right),$$

where

$$f_2\left(i,j\right) = \begin{cases} 0, e(i,j) > 10 \\ 1, e(i,j) \le 10 \end{cases}$$

and the *mean distance to the correct formation location*:

$$\sum_{i=1}^{N}\sum_{j=1}^{A} e(i,j) \,/\, \left(AN\right).$$

Keeping the agents organized in a formation is one thing, but a good method for co-ordinating formations should also be able to move the formation efficiently. By measuring the *total cumulative distance travelled by all agents*

$$\sum_{i=2}^{N}\sum_{j=1}^{A} p_{ji} - p_{j(i-1)}$$

we can compare how efficiently the tested methods pass the test track. We also compare the *total number of turns* in which the methods pass the track.

Finally, to compare the computational efficiency of the methods we measure the *time* used to pass the track.

The results are collected in Figures 9-11.

Passing the track: All the tested methods were able to pass the test track on all of the 100 test runs. In our initial experiments, MSS was not able to pass the track on all test runs due to too rigid spring model and poor collision avoidance.

Maintaining the formation: Both BBS and FLC were able to keep in average 80 per cent of the agents in formation, whereas MSS managed to keep only 60 per cent..The spring model propagates displacement of one agent to others more efficiently than the unit centric formation used in BBS and FLC. When MSS moves past or around an obstacle, the formation tends to get deformed. In contrast, BBS and FLC forces only one or two agents nearest to the obstacle to move out of their places while the rest remain in correct positions. The performance of all of the methods decreases as the number or the size of the obstacles increases.

Staying in the formation: BBS was able to keep all agents in a formation 40-80 per cent of the turns. This is significantly better than FLC, which managed to keep all agents in a formation 20-70 per cent of the turns, and MSS, which managed to keep all agents in a formation only under 20 per cent of the turns. These results are consistent with the average distance to the agent's correct formation position: The average distance was 3-5 for BBS, 4-6 for FLC, and 13-16 for MSS. Again, the performance of all methods decreases when the number or the size of obstacles increases, and the effect is larger than in the agents in the formation metric.

Efficiency: In the basic track there is not much difference between the methods in terms of the travelled distance or the number of used turns. When obstacles are added to the test, MSS becomes the most efficient method. Computationally FLC has the worst performance: it is magnitude slower than BBS or MSS. This poor performance can be partly attributed to design decisions made during the development, because the implementation aims at making it easy to create and modify rules and controllers at the cost of computational complexity.

Appearance: The tested methods have noticeable visual differences. When using BBS or FLC the movement of the formation looks very natural, and the individual agents seem to act intelligently, for example, when avoiding obstacles. In contrast, MSS provides natural looking movement when there are no obstacles. When encountering

Figure 9. Comparison of the methods in the basic track with 100 test runs

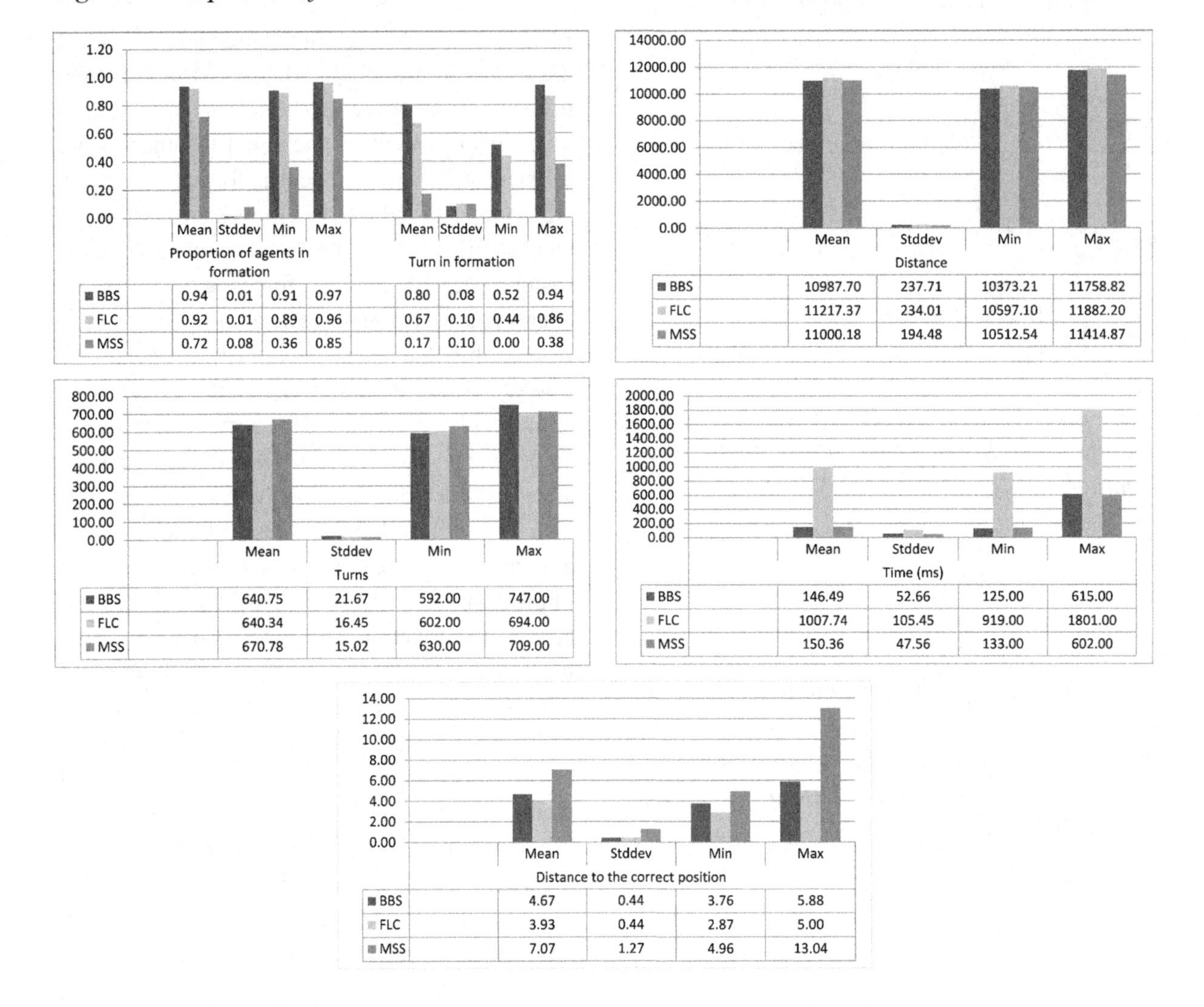

	Proportion of agents in formation				Turn in formation			
	Mean	Stddev	Min	Max	Mean	Stddev	Min	Max
BBS	0.94	0.01	0.91	0.97	0.80	0.08	0.52	0.94
FLC	0.92	0.01	0.89	0.96	0.67	0.10	0.44	0.86
MSS	0.72	0.08	0.36	0.85	0.17	0.10	0.00	0.38

Distance	Mean	Stddev	Min	Max
BBS	10987.70	237.71	10373.21	11758.82
FLC	11217.37	234.01	10597.10	11882.20
MSS	11000.18	194.48	10512.54	11414.87

Turns	Mean	Stddev	Min	Max
BBS	640.75	21.67	592.00	747.00
FLC	640.34	16.45	602.00	694.00
MSS	670.78	15.02	630.00	709.00

Time (ms)	Mean	Stddev	Min	Max
BBS	146.49	52.66	125.00	615.00
FLC	1007.74	105.45	919.00	1801.00
MSS	150.36	47.56	133.00	602.00

Distance to the correct position	Mean	Stddev	Min	Max
BBS	4.67	0.44	3.76	5.88
FLC	3.93	0.44	2.87	5.00
MSS	7.07	1.27	4.96	13.04

obstacles, MSS formation with large number of springs seems too rigid and the agents do not have so much freedom to move individually, which causes the formation to deform in a way that does not look like natural formation behaviour. When using only structural springs formation is more flexible and moves naturally around obstacles.

Canyon Track

In the canyon test the formation has to move from the left side to the right side of an area of the size 400 × 400 units. In the centre of the area there are two obstacles forming a canyon (Figure 5d). Nine agents are placed randomly near the first waypoint.

In our initial experiments BBS was the only method able to pass the canyon test. The FLC formations jammed inside the canyon and MSS formations were not even able to get close to the canyon's entrance. In general, the reason for the poor performance was that FLC and MSS tried to maintain to tight a control of the formation – even to the extreme that the formation cannot disassemble in order to move forward. In particular MSS suffered from collision avoidance based on repulsive forces, because repulsion from both sides of the canyon prevents the formation from entering the canyon.

Figure 10. Comparison of the methods in the random small obstacles track with 100 test runs

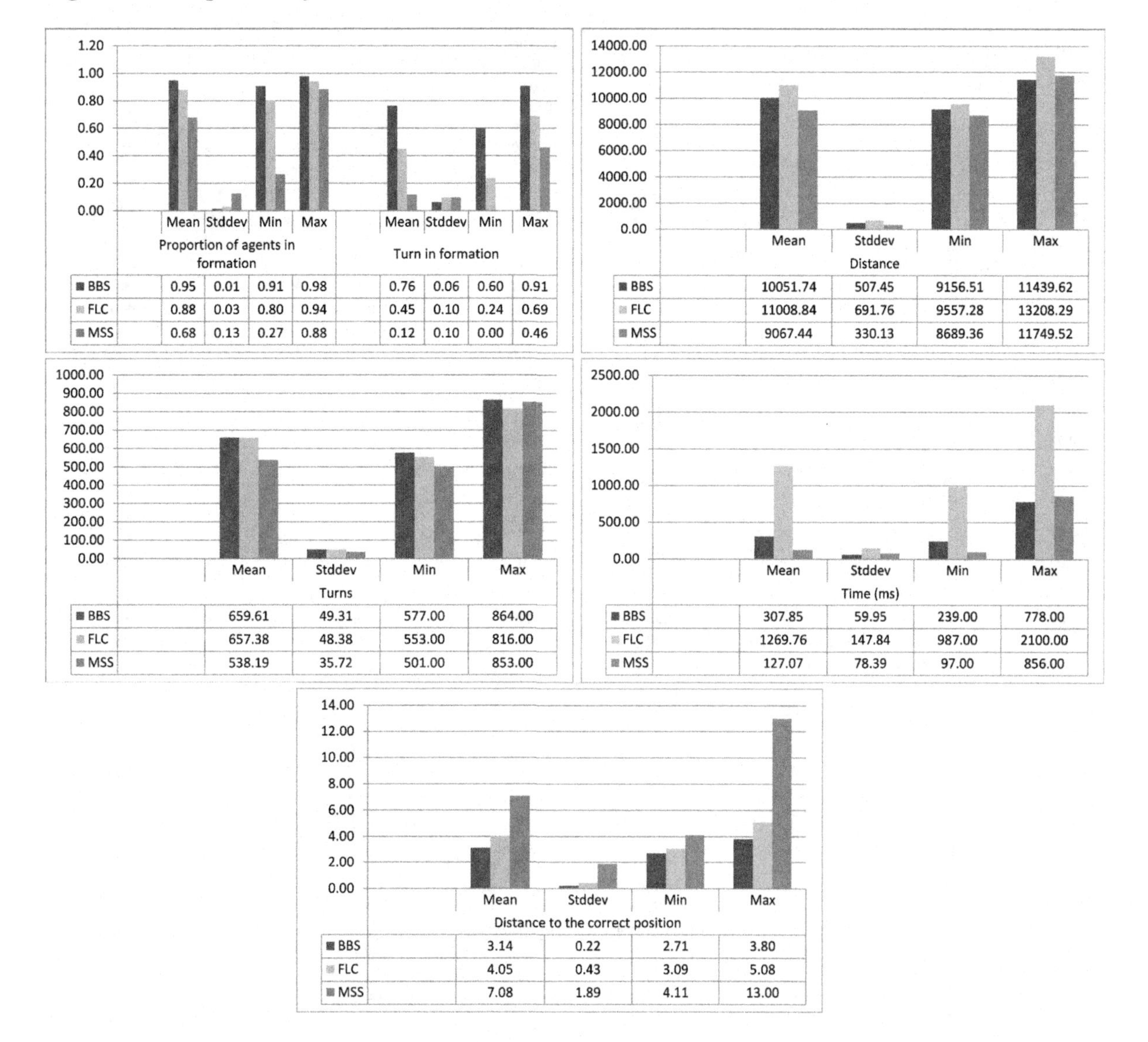

	Proportion of agents in formation: Mean	Stddev	Min	Max	Turn in formation: Mean	Stddev	Min	Max
BBS	0.95	0.01	0.91	0.98	0.76	0.06	0.60	0.91
FLC	0.88	0.03	0.80	0.94	0.45	0.10	0.24	0.69
MSS	0.68	0.13	0.27	0.88	0.12	0.10	0.00	0.46

Distance	Mean	Stddev	Min	Max
BBS	10051.74	507.45	9156.51	11439.62
FLC	11008.84	691.76	9557.28	13208.29
MSS	9067.44	330.13	8689.36	11749.52

Turns	Mean	Stddev	Min	Max
BBS	659.61	49.31	577.00	864.00
FLC	657.38	48.38	553.00	816.00
MSS	538.19	35.72	501.00	853.00

Time (ms)	Mean	Stddev	Min	Max
BBS	307.85	59.95	239.00	778.00
FLC	1269.76	147.84	987.00	2100.00
MSS	127.07	78.39	97.00	856.00

Distance to the correct position	Mean	Stddev	Min	Max
BBS	3.14	0.22	2.71	3.80
FLC	4.05	0.43	3.09	5.08
MSS	7.08	1.89	4.11	13.00

The updated versions of FLC and MSS were able to pass the canyons tests. For example, to make FLC to be able to pass the canyon, we had to the following rules allowing the agents to give up formation behaviour in difficult situations:

- **IF** FLV is SMALL **AND** CM is SMALL, **THEN** steering is TD
- **IF** FCV is SMALL **AND** CM is SMALL, **THEN** steering is TD

Formation Type Comparison

Because BBS turned out to be the most promising method, we test it further using square, line and column formation types (see Figure 1) to evaluate their effect on the performance. The formation comparison test has a setup similar to the method comparison test but it uses only 4 agents (Figure 5c). We make 100 test runs for each formation using BBS method.

Overall, the results do not indicate any large differences between different formation types. The

Figure 11. Comparison of the methods in the random large obstacles track with 100 test runs

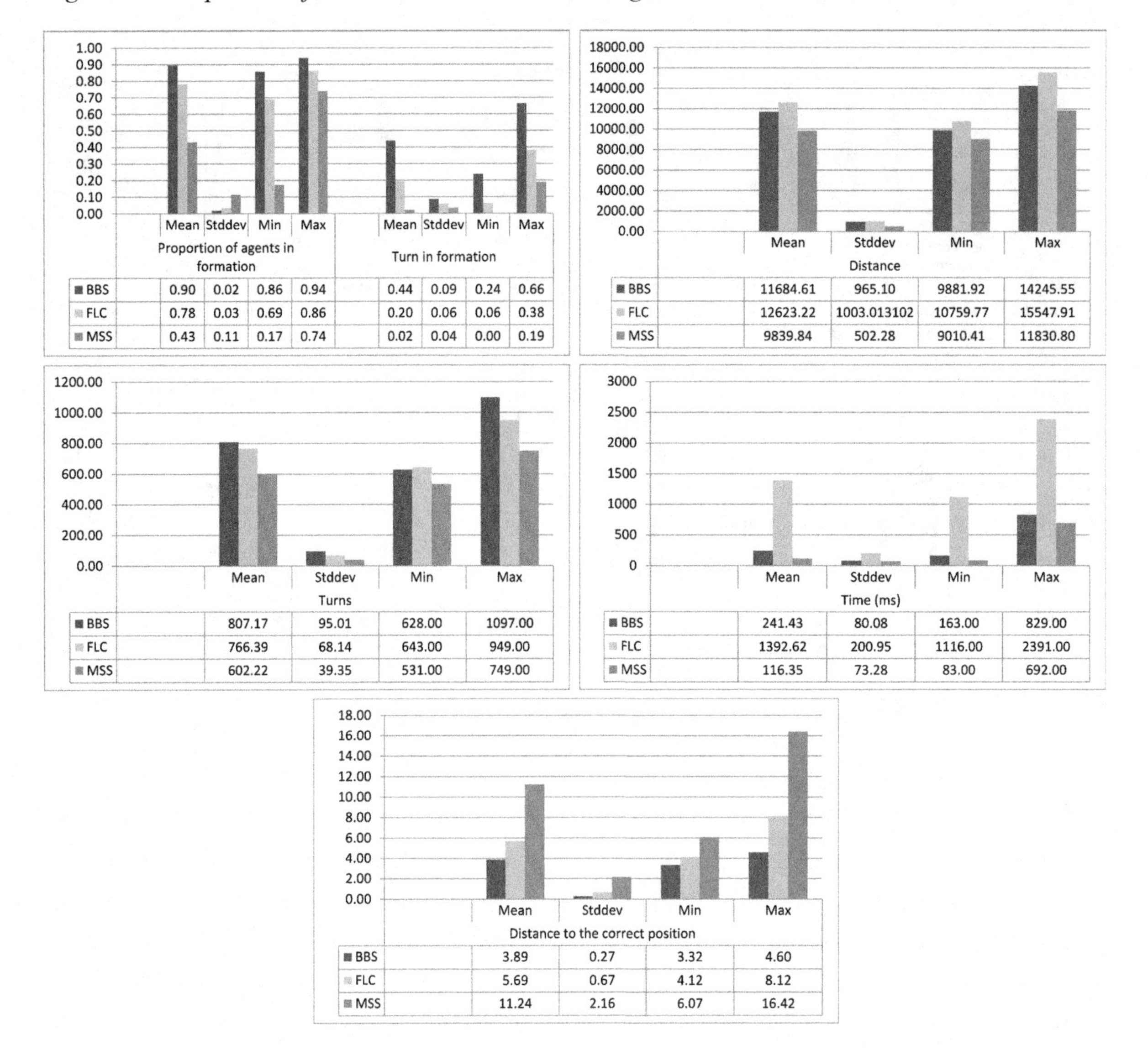

square formation is slightly better in turning, because the agents are closer to the formation centre than in a line or column formation. Moreover, the agents in the square formation return faster to their correct place and turning causes smaller increase in the distance to formation places. The overall results are consistent with Balch & Arkin (1998).

Large Formation

To test the scalability of the BBS method we scale up the basic test area to the size of 2000 × 2000 units including obstacles (Figure 5b). Also, the number of agents is increased so that 100 agents are placed randomly in the centre of the test area.

In general, BBS and FLC scale up quite well to handle a formation of 100 agents. Visually, the repulsion between the agents

Figure 12. Comparison of the methods in the canyon with 100 test runs

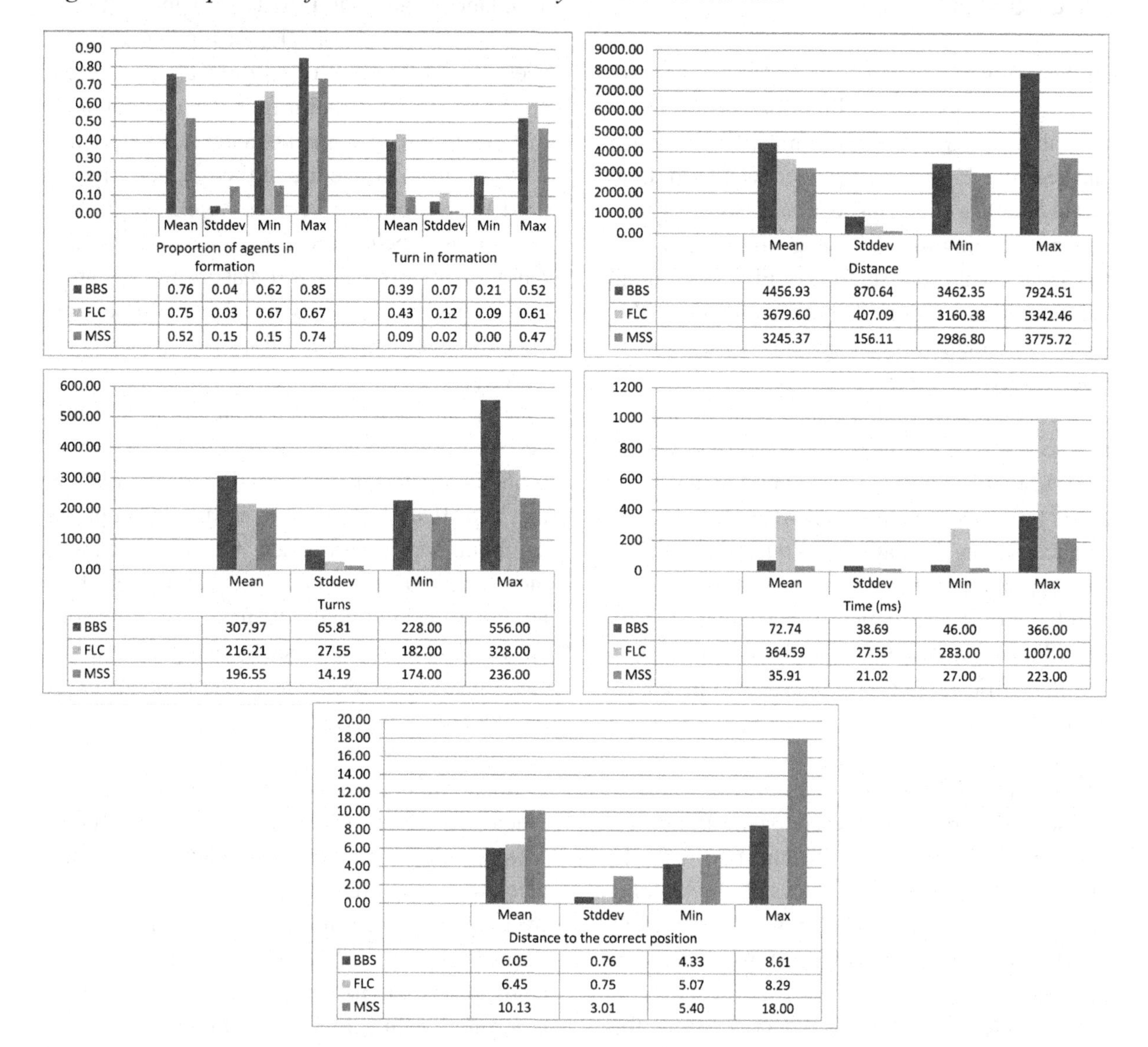

can cause some problems in viscosity on the edges of the BBS formation. The agents on the front side of the formation tend to get pushed forwards out of the formation, whereas the agents on the backside are lagging behind. On the left and right hand sides this effect is not that large and the agents are able to maintain correct places. FLC and MSS do not suffer from the repulsion problem but the formation gets much distorted.

Obstacle avoidance works well also with a large formation and the agents move fluidly around the obstacles. Because in our test the obstacles are larger than the spacing between the agents, many agents are forced out of their places but they return quickly to the correct places after going around the obstacle.

DISCUSSION

Of the test methods, BBS and FLC turned out to be the best ones for controlling formations. They were able to pass the test track on all of the test runs and keep the formation organized while avoiding obstacles. On a closer inspection, BBS is slightly better at keeping formation organized than FLC. MSS was not as good as BBS or FLC in keeping the formation organized, but compensated the lack of organization with speed and the shortest total travelled distance.

The weakness of the dense MSS is that it tries to keep the formation too organized, which leads to that it could not pass the track on some of the test runs. It works well when there are no obstacles along the path, but – due to its rigidity – even a small obstacle cause most of the agents to fall out of the formation..When using MSS to co-ordinate formations the number of springs and spring coefficients need to be selected based on the task in hand.

There is not much difference in the time at which methods pass the basic track, whereas sparse MSS formations turned to take around 10 to 20 per cent less time than the other two methods in the random obstacle test tracks. When looking at the computation requirements, however, FLC requires much more processing power than BBS or MSS.

The canyon test turned to be a touchstone for the methods, and in our initial test BBS was only one to pass it. For instance, dense MSS keeps the agents too tightly under control. Granted, we can reduce the number of springs to make MSS more flexible, but it quickly lead to that the formation loses it overall shape. After improving the collision avoidance, adding specific rules to break the formation in FLC, and reducing number of springs in MSS, all of the methods are able to pass the canyon test.

One of the overall findings is that collision avoidance has a significant impact on the formation performance. Reynolds' cylinder-based avoidance behaviour turned out to perform very well in different set-ups. However, because it is uses the centre point of an obstacle as a reference, it is possible that it fails when encountering an arbitrarily shaped obstacle. If we do not have a priori knowledge of the obstacle shapes or if the shapes are very complex, Reynolds' containment behaviour could be used instead. Another possibility is to measure the distance to the obstacle's surface in different angles and base the avoidance behaviour on the resulting vectors.

The rule base of a fuzzy logic system can get very large on complex problems. The main problem we faced when implementing FLC is that is difficult to fine-tune the rules to improve computational performance. Obviously, adding more detailed rules makes the agents steer better, but each additional rule makes the controller slower. This can be avoided by using a hierarchical controller so that each single controller can be relatively simple.

Another problem in FLC is that it is difficult to make the obstacle avoidance rules dominant when needed. This requires adding a term to lower the degree of firing all the other rules when a collision is imminent. This could also be fixed by implementing a hierarchical controller proposed by Teiner *et al.* (2003), where the top-level controller would use only input from avoidance controller when collision is imminent.

In many test cases MSS passed the track using the smallest number of turns and the agents travelled the shortest total distance but had still the worst organization. Therefore, MSS seems to trade formation organization for speed and adaptability. This indicates the principal problem of formation control: how to maintain global order while allowing local liberty

Because our algorithms use local steering and the agents make decisions independently from one another, the presented methods can be parallelized easily. A naïve approach would be just to use an own thread for every agent running the whole algorithm. Nevertheless, some of the data is the same for all the agents (e.g., the formation location and spring forces), which is why a better approach would be to calculate this common information first and then make the agents' decisions in parallel. In the MSS method, the spring forces can be calculated also in parallel. Similarly, assigning every rule to an own computing unit could parallelize the FLC method even further. Implementation and analysis of the parallelized versions, however, fall beyond the scope of this chapter.

CONCLUDING REMARKS

In this chapter, we studied different methods for co-ordinating formations in computer games. Three different decision-making methods were implemented and tested. The methods were behaviour-based steering, fuzzy decision-making and mass-spring system. We presented test results, which indicate that the best method for co-ordinating formations is the behaviour-based steering, because it is efficient, simple to implement and yields good overall results.

REFERENCES

Balch, T., & Arkin, R. C. (1998). Behavior-based formation control for multi-robot teams. *IEEE Transactions on Robotics and Automation, 14*(6), 926–939. doi:10.1109/70.736776

Balch, T., & Hybinette, M. (2000). Social potentials for scalable multi-robot formations. In *Proceedings of the IEEE International Conference on Robotics and Automation* (pp. 73-80). Piscataway, NJ: IEEE.

Dawson, C. (2002). Formations. In Rabin, S. (Ed.), *AI game programming wisdom*. Hingham, MA: Charles River Media.

Derenick, J. C., & Spletzer, J. R. (2007). Convex optimization strategies for coordinating large-scale robot formations. *IEEE Transactions on Robotics, 23*(6), 1252–1259. doi:10.1109/TRO.2007.909833

Fredslund, J., & Mataric, M. J. (2002). A general algorithm for robot formations using local sensing and minimal communication. *IEEE Transactions on Robotics and Automation, 18*(5), 837–846. doi:10.1109/TRA.2002.803458

Heijden, M., van der Bakkes, S., & Spronck, P. (2008). Dynamic formations in real-time strategy games. In *2008 IEEE Symposium On Computational Intelligence and Games* (pp. 47-54). doi: 10.1109/CIG.2008.5035620

Laasonen, J. (2008). *Muodostelmien hallinta reaaliaikastrategiapeleissä*. Master's thesis, University of Turku, Finland.

Lewis, A. M. (1997). High precision formation control of mobile robots using virtual structures. *Autonomous Robots, 4*(4), 387–403. doi:10.1023/A:1008814708459

Naffin, D. J., & Sukhatme, G. S. (2004). Negotiated formations. In *Proceedings of the International Conference on Intelligent Autonomous Systems (IAS)*. In F. Groen, N. Amato, A. Bonarini, E. Yoshida, & B. Kröse (Eds.), *Proceedings of the 8th International Conference on Intelligent Autonomous Systems* (pp. 181-190).

Ögren, P., Fiorelli, E., & Leonard, N. E. (2002). Formations with a mission: Stable coordination of vehicle group maneuvers. In D.S. Gilliam & J. Rosenthal (Eds.), *Electronic Proceedings of the 15th International Symposium on Mathematical Theory of Networks and Systems*. Retrieved May 27, 2011, from http://www.nd.edu/~mtns/papers/4615_3.pdf

Pottinger, D. C. (1999a). Coordinated unit movement. *Gamasutra*, January 22, 1999. Retrieved May 27, 2011, from http://www.gamasutra.com/features/19990122/movement_01.htm

Pottinger, D. C. (1999b). Implementing coordinated movement. Gamasutra, January 29, 1999. Retrieved May 27, 2011, from http://www.gamasutra.com/features/19990129/implementing_01.htm

Puustinen, I., & Pasanen, T. A. (2006). Game theoretic methods for action games. In T. Honkela, T. Raiko, J. Kortela & H. Valpola (Eds.), *Proceedings of the Ninth Scandinavian Conference on Artificial Intelligence (SCAI 2006)* (pp. 183-188). Espoo, Finland: Finnish Artificial Intelligence Society.

Reynolds, C. W. (1987). Flocks, herds, and schools: A distributed behavioral model. *Computer Graphics*, *21*(4), 25–34. doi:10.1145/37402.37406

Reynolds, C. W. (1999). Steering behaviors for autonomous characters. In *Proceedings of the Game Developers Conference* (pp. 763-782). San Francisco, CA: Miller Freeman Game Group.

Smed, J., & Hakonen, H. (2006). *Algorithms and networking for computer games*. Chichester, UK: John Wiley & Sons. doi:10.1002/0470029757

Teiner, M., Rojas, I., Goser, K., & Valenzuela, O. (2003). A hierarchical fuzzy steering controller for mobile robots. In *2003 International Conference Physics and Control. Proceedings (Cat. No.03EX708)* (pp. 7-12). doi: 10.1109/CIMSA.2003.1227193

Verth, J. V., Brueggemann, V., Owen, J., & McMurry, P. (2000). Formation-based pathfinding with real-world vehicles. In *Proceedings of the Game Developers Conference*. Retrieved May 27, 2011, from http://citeseerx.ist.psu.edu/viewdoc/summary?doi=10.1.1.17.2031

Wang, X., & Devarajan, V. (2004). 2d structured mass-spring system parameter optimization based on axisymmetric bending for rigid cloth simulation. In *Proceedings of the 2004 ACM SIGGRAPH International Conference on Virtual Reality Continuum and its Applications in Industry Vrcai 04* (pp. 317-323). New York, NY: ACM Press.

Yager, R. R., & Filev, D. P. (1994). *Essentials of Fuzzy Modeling and Control*. New York, NY: John Wiley & Sons.

Young, B. J., Beard, R. W., & Kelsey, J. M. (2001). A control scheme for improving multi-vehicle formation maneuvers. In *Proceedings of the American Control Conference* (pp. 704-709).

Zadeh, L. A. (1965). Fuzzy sets. *Information and Control*, *8*(3), 338–353. doi:10.1016/S0019-9958(65)90241-X

ADDITIONAL READING

Bellman, R. E., & Zadeh, L. A. (1970). Decision-making in a fuzzy environment. *Management Science*, *17*(4), 141–164. doi:10.1287/mnsc.17.4.B141

Dubois, D., Fargier, H., & Prade, H. (1996). Possibility theory in constraint satisfaction problems: Handling priority, preference and uncertainty. *Applied Intelligence*, *6*, 287–309. doi:10.1007/BF00132735

Evans, R. (2002). Varieties of learning. In Rabin, S. (Ed.), *AI Game Programming Wisdom* (pp. 567–578). Hingham, MA: Charles River Media.

Fullér, R., & Carlsson, C. (1996). Fuzzy multiple criteria decision making: Recent developments. *Fuzzy Sets and Systems*, *78*, 139–153. doi:10.1016/0165-0114(95)00165-4

Guesgen, H. W. (1994). A formal framework for weak constraint satisfaction based on fuzzy sets. *Proceedings of, ANZIIS-94*, 199–203.

Herrera, F., & Verdegay, J. L. (1997). Fuzzy sets and operations research. Perspectives. *Fuzzy Sets and Systems, 90*, 207–218. doi:10.1016/S0165-0114(97)00088-2

Higgins, D. (2002). Pathfinding design architecture. In Rabin, S. (Ed.), *AI Game Programming Wisdom* (pp. 122–132). Hingham, MA: Charles River Media.

Johnson, G. (2003). Avoiding dynamic obstacles and hazards. In Rabin, S. (Ed.), *AI Game Programming Wisdom 2* (pp. 161–170). Hingham, MA: Charles River Media.

Kaukoranta, T., Smed, J., & Hakonen, H. (2003). Understanding pattern recognition methods. In Rabin, S. (Ed.), *AI Game Programming Wisdom 2* (pp. 579–589). Hingham, MA: Charles River Media.

Kennedy, J., Eberhart, R. C., & Shi, Y. (2001). *Swarm Intelligence*. San Francisco, CA: Morgan Kaufmann.

Novak, J. (2007). *Game Development Essentials: An Introduction* (2nd ed.). Clifton Park, NY: Delmar Cengage Learning.

Pearl, J. (1986). Fusion, propagation, and structuring in belief networks. *Artificial Intelligence, 29*(3), 241–364. doi:10.1016/0004-3702(86)90072-X

Reynolds, C. W. (1987). Flocks, herds, and schools: A distributed behavioral model. *Computer Graphics, 21*(4), 25–34. doi:10.1145/37402.37406

Salen, K., & Zimmerman, E. (2004). *Rules of Play: Game Design Fundamentals*. Cambridge, MA: MIT Press.

Shafer, G. (1990). Perspectives on the theory and practice of belief functions. *International Journal of Approximate Reasoning, 4*(5–6), 323–362. doi:10.1016/0888-613X(90)90012-Q

Slany, W. (1995). Comparing partial constraint satisfaction models. *Workshop Notes of the CP '95 Work-shop on Over-Constraint Systems*, (pp. 151–159). Cassis, France.

Smed, J., & Hakonen, H. (2005). Synthetic players: A quest for artificial intelligence in computer games. *Human IT, 7*(3), 57–77.

Smed, J., & Hakonen, H. (2006). *Algorithms and Networking for Computer Games*. Chichester, UK: John Wiley & Sons. doi:10.1002/0470029757

Snook, G. (2000). Simplified 3D movement and pathfinding using navigation meshes. In DeLoura, M. (Ed.), *Game Programming Gems* (pp. 288–304). Hingham, MA: Charles River Media.

van der Sterren, W. (2003). Path look-up tables – small is beautiful. In Rabin, S. (Ed.), *AI Game Programming Wisdom 2* (pp. 115–129). Hingham, MA, USA: Charles River Media.

Woodcock, S. (2002). Recognizing strategic dispositions: Engaging the enemy. In Rabin, S. (Ed.), *AI Game Programming Wisdom* (pp. 221–232). Hingham, MA: Charles River Media.

Yager, R. R. (1981). A new methodology for ordinal multiobjective decisions based on fuzzy sets. *Decision Sciences, 12*, 589–600. doi:10.1111/j.1540-5915.1981.tb00111.x

Yager, R. R. (1988). On ordered weighted averaging aggregation operators in multicriteria decision making. *IEEE Transactions on Systems, Man, and Cybernetics, 18*(1), 183–190. doi:10.1109/21.87068

Yager, R. R., & Filev, D. P. (1994). *Essentials of Fuzzy Modeling and Control*. New York, NY: John Wiley & Sons.

Zadeh, L. A. (1965). Fuzzy sets. *Information and Control*, *8*(3), 338–353. doi:10.1016/S0019-9958(65)90241-X

KEY TERMS AND DEFINITIONS

Agent: An autonomous entity that aims at a goal by observing and acting in an environment.

Behaviour-Based Steering: A control system where the agents are independent and reactive and have a relative small model of the environment.

Control Structure: Models a formation and defines the correct places of the agents in it.

Formation: The arrangement of moving agents so that they stay in a pre-defined order.

Fuzzy Logic Control: A control system based on defining the decision criteria as fuzzy sets (i.e., allowing partial fulfilment of the criteria).

Mass-Spring System: A control system that connects the agents with springs and applies spring forces to keep them in order.

Path Finding: Finding the shortest viable path from a start position to an end position. Commonly solved using by discretizing the problem into a graph and applying the A* algorithm.

Chapter 2
Adapting Pathfinding with Potential Energy

Thomas Hartley
University of Wolverhampton, UK

ABSTRACT

Movement through a computer game environment is an essential requirement of non-player characters (NPCs) in today's computer games. Local movement is typically reactive and based on the current state of the game and the NPC. Long-range movement is concerned with determining a short and appropriate route from one location in the game environment to another. A desired destination is typically not known in advance. Therefore, techniques are needed to determine a route while a game is being played. This problem is known as pathfinding or path planning. Traditionally, pathfinding systems have focused on determining the shortest path between locations; however, many computer games are beginning to incorporate terrain and strategic reasoning (also known as tactical location analysis) into the pathfinding process. This chapter describes an approach to strategic and tactical pathfinding that learns in-game from an NPC's experience of executing previously generated paths. The experience is used to adapt future pathfinding and therefore allows NPCs to avoid (or be attracted to) areas of the game world. Hence NPCs can improve their chance of success and encourage the human player to adapt their behavior.

INTRODUCTION

Virtual or animated characters in computer games have been around for nearly as long as computer games themselves. They are more commonly known as non-player characters (NPCs) or computer controlled characters and are typically the most visible manifestation of game artificial intelligence (AI). The term "game AI" is used to refer to many aspects of a computer game, such as decision making, path-finding, collision detection, animation systems and audio cues. Game AI can be organized according to a number of different tasks, namely low-level movement,

DOI: 10.4018/978-1-4666-1634-9.ch002

decision making, tactical decision making and strategic decision making (Millington & Funge, 2009; Thurau, 2006; Tozour, 2002a).

The most essential requirement of an NPC is to move through the game environment. Early computer games primarily made use of only movement algorithms and had very little, if any, high-level decision making abilities; for instance, the first arcade videogame to include an animated computer opponent was 1974's Shark Jaws (Byl, 2004). In this game the NPC took the form of a shark which preyed on the human player's avatar, a scuba diver, by swimming directly towards it. In addition to limited high-level decision making, many early games took place in small and relatively simple environments that were designed specifically for their zombie-like NPC inhabitants. For example, the Pac-man game world had no dead ends, so the movement decisions of the ghosts only needed to be taken when they reached a junction. As computer games have increased in complexity and changed from 2D to 3D environments a need for more sophisticated abilities has developed. NPCs in today's computer games should not get trapped behind in-game objects such as vehicles or buildings; they should take terrain into account and take suitable routes to desired locations.

NPC behavior in modern commercial computer games is frequently achieved through a combination of fundamental techniques: movement algorithms, finite state machines (FSMs), pre-programmed scripts and pathfinding. These techniques have produced remarkable character behavior; however, they all need to be programmed to include every situation an NPC may encounter. They lack the ability to adapt and generalize to unforeseen situations; this means that they tend to produce lifeless characters that are prone to repetitive and predictable behavior. Furthermore, this results in diminished enjoyment for the human player as they are able to exploit repetitive behavior, poor decision making and NPC errors to achieve game objectives. If games are to maintain high-levels of entertainment for the human player they need overcome these deficiencies through NPCs that are more believable and responsive to the player.

The work presented in this chapter pursues approaches to enhancing NPC behavior in computer games that will produce characters that are more believable, challenging and responsive to the human player. This aim can be achieved in a variety of ways, such as through offline learning (Funge, 2004). Offline learning occurs before a game is release or while a game is not being played. It can produce more believable behavior for NPCs; however, it still encounters similar limitations to current FSM and scripting techniques. This is illustrated by McGlinchey (2003), who states that humans may act differently in identical situations, but his offline learning system is always completely deterministic. In addition, Laird (2001) found that games developers were impressed by his offline behavior modeling system, but would invariably ask, "Does it anticipate the human player's actions?" (Laird, 2001). Consequently, NPCs that learn offline are susceptible to repetitive behavior and are not responsive to the human player's actions.

In-game or online learning occurs when a computer game is being played by a human player and is typically achieved through unsupervised learning techniques. Practical and efficient in-game learning and adaptation are key capabilities of human game players that are not typically employed by NPCs in commercial computer games. Primarily this is because online learning through interactions with a human user can result in poor NPC behavior and can be computationally expensive (Manslow, 2006), both of which are undesirable in processor-intensive computer games. However, games which include online learning offer a number of benefits to game players and game developers, such as being able to adapt to human player behavior, generating new behavior according to past experiences, overcoming poorly developed "conventional AI" and scaling their difficulty according to a human player's ability (Spronck & Postma, 2003; Andrade *et al.*, 2004).

This chapter presents an approach to strategic and tactical pathfinding from the perspective of in-game learning and adaptation. The technique allows NPCs to adapt their behavior based on experiences gained from interacting in the game world. The experiences are used to adapt future pathfinding, this allows NPCs to avoid (or be attracted to) areas of the environment; hence game characters can improve their chance of success and encourage the human player to adapt their behavior. The approach will be described in terms of a multi-round first person shooter (FPS) computer game scenario. FPS games are a genre of commercial computer games that take place in a simulated 3D environment which mimics the real world. For example, they simulate a virtual space that computer controlled characters and human players can move through.

BACKGROUND

This section serves as background to the work presented in this chapter. It provides a brief overview of pathfinding techniques used in computer games, with a particular focus on NPCs in first person shooter (FPS) games. The section highlights key pathfinding algorithms, game world representations for pathfinding and approaches to strategic and tactical pathfinding.

Pathfinding and Game World Representations

A key ability of NPCs is to be able to move around the environment created by the game developer. Pathfinding (also known as path planning) is concerned with determining a short and appropriate route from one location in the game environment to another. The basic pathfinding problem falls between low-level movement and decision making (Millington and Funge, 2009). Movement is typically reactive and based on the current state of the game. For example, avoiding and steering around obstacles are typical movements. Decision making focuses on determining what an NPC should do. Once a decision has been made (e.g., a goal location has be selected) the pathfinding system can determine a suitable route.

As a desired destination is not known in advance, techniques are required to determine a route while a game is being played. In order to achieve this two key components are needed, namely a world representation and an algorithm to find an appropriate path. Pathfinding algorithms cannot work with the complex 3D game environment data used to render the game world, as it contains an enormous amount of detail that is required for presenting the game world to the human player (Millington and Funge, 2009). This detailed information is not necessary for pathfinding; therefore, a simpler, more easily searched representation is used.

Graphs form the bases for efficient representation of the game world. A graph is built from a set of vertices and edges. The vertices represent accessible locations and the edges represent reachable locations or paths between vertices. The graph provides the search space for the pathfinding algorithms. Navigation graphs can represent the game environment using a number of different constructs; these include waypoint, navigation mesh and grid based graphs. Waypoint maps have traditionally been the most popular approach to representing game worlds (Tozour, 2002b). A waypoint map (Waveren and Rothkrantz, 2001), as illustrated in Figure 1, is a collection of nodes placed on top of the game world to represent accessible locations.

Once the waypoint map has been created, search algorithms can be used to find paths from one node to another; essentially, a waypoint map can be treated as a directed weighted graph (Byl, 2004). A variety of algorithms can be used to search the map (Namee, 2004), such as Dijkstra's algorithm or Floyd's algorithm (Lorch, 2000). However, the technique of choice for most games developers is the A* algorithm, which is often

Figure 1. An illustration of a waypoint map used for path-planning in computer games

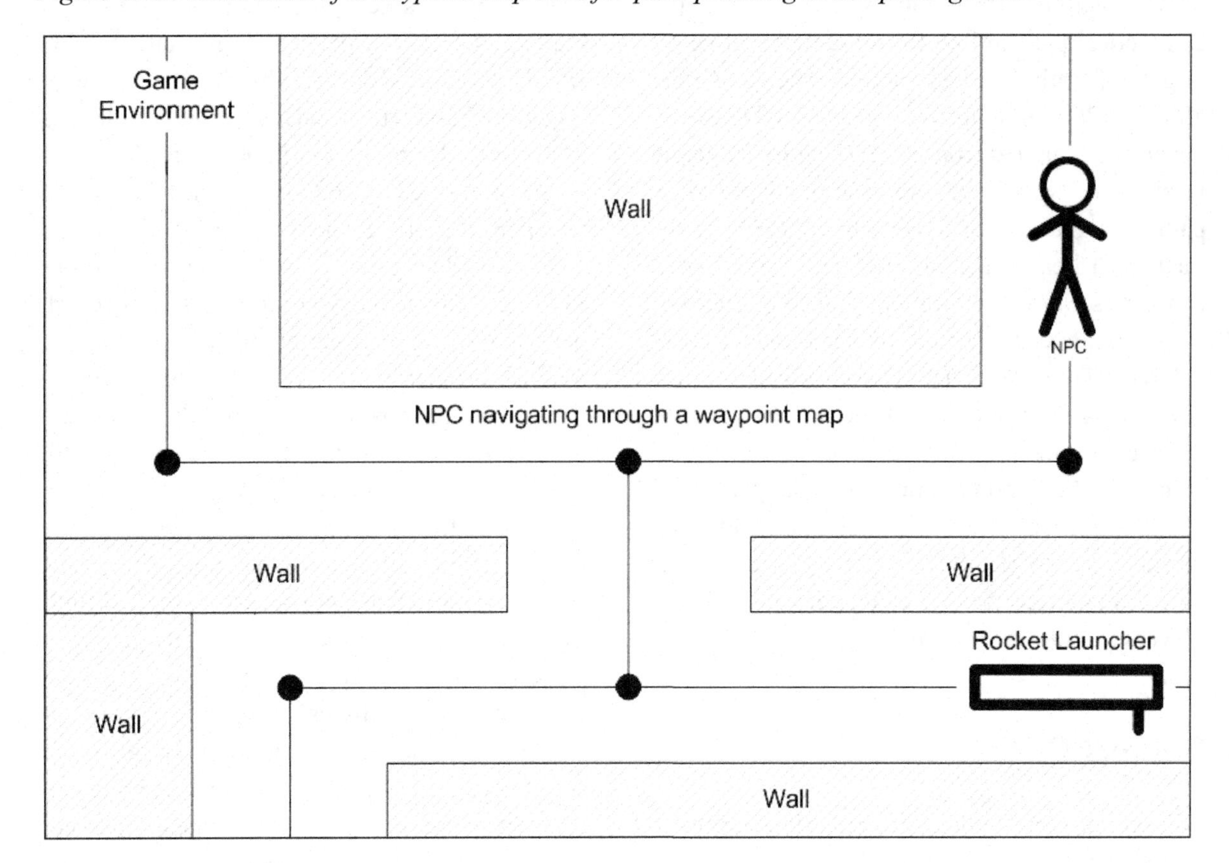

combined with enhancements to improve its computational efficiency (Higgins, 2002a; Higgins, 2002b) and realism (Pinter, 2002). The algorithm's popularity is due largely to its being a tried and tested approach that is relatively simple to understand and is computationally efficient.

Figure 2 outlines the pseudo-code for the basic A* algorithm. The input for the algorithm is a start node, end node and a waypoint graph. The function returns the path from the start node to the end node that has the lowest cost. A* examines one waypoint node at a time, beginning with the start node. The node being examined by the algorithm is referred to as the current node. The algorithm selects the current node using a heuristic that incorporates the cost of reaching the current node and the estimated cost to the goal node. A* looks at the paths of the current node and examines the connected nodes. The algorithm records how it reaches the connected nodes and the cost of reaching the node from the start. It also records an estimate of the total cost from the start node to the end node through the connected node. This is called the estimated total cost and is calculated using a heuristic. The heuristic typically uses the cost to the current node and an estimated cost to the goal node.

In order to keep track of path nodes that have been explored the algorithm maintains an open and closed list. The open list contains path nodes that have been seen, but not had their connected nodes explored. The closed list contains path nodes that have had their connected nodes explored. The algorithm may reach a connected node that has already been explored and is on the closed list. If this occurs the A* algorithm compares the new cost-so-far with the current cost-so-far. If the new cost is lower, A* updates the costs for the

Figure 2. Pseudo-Code for the A algorithm*

```
Inputs:
      startNode                    // The start node.
      endNode                      // The end node.
      waypointGraph                // The waypoint graph.
aStar (startNode, goalNode, waypointGraph) {
  // Initialise the open and closed lists.
  openList = new aStarlist();
  closedList = new aStarlist();

  // Add the start node to the open list.
  aListNode.node = startNode;
  aListNode.cameFrom = None;
  aListNode.costSoFar = 0;
  aListNode.estimatedTotalCost = heuristicEstimate(startNode, goalNode);
  openList.add(aListNode);

  // Loop through each node in the open list until it is empty or
  // the goal has been found.
  While openList is not empty {
    currentNode = openList.getNodeWithLowest EstimatedTotalCost();
    if (currentNode == goalNode)
      break while loop;

    currentNodePaths = currentNode.getWaypointMapPaths();
    for each path in currentNodePaths {
      // Get the node at the end of the currently examined path.
      pathEndNode = path.getToNode();
      // Get the cost of reaching the node at the end of the currently
      // examined path.
      pathEndNodeCost = currentNode.costSoFar + path.getCostToEndNode();

      // If the end path node is on the closed list.
      if(closedList.contains(pathEndNode)) {
        pathEndNodeListRecord = closedList.get(pathEndNode);
        if(pathEndNodeCost < pathEndNodeListRecord.costSoFar){
          // We have a new path node that is lower than the current cost.
          closedList.remove(pathEndNode);
          // Determine the end node's heuristic value.
          pathEndNodeHeuristic = heuristicEstimate(pathEndNode, goalNode);
        } else {
          continue for;
        }
      } else if(openList.contains(pathEndNode)) {
        // If the end path node is on the open list.
        pathEndNodeListRecord = openList.get(pathEndNode);
        if(pathEndNodeCost < pathEndNodeListRecord.costSoFar){
          // We have a new path node that is lower than the current cost.
          // Determine the end node's heuristic value.
          pathEndNodeHeuristic = heuristicEstimate(pathEndNode, goalNode);
        } else {
          continue for;
        }
      } else {
        // If the end path node is not on any list we need to add it to the open list.
        pathEndNodeListRecord.node = pathEndNode
        pathEndNodeHeuristic = heuristicEstimate(pathEndNode, goalNode);
      }

      // Update the path node.
      pathEndNodeListRecord.costSoFar = pathEndNodeCost;
      pathEndNodeListRecord.cameFrom = path;
      pathEndNodeListRecord.estimatedTotalCost = pathEndNodeCost + pathEndNodeHeuristic;
      if(closedList.contains(pathEndNode) == false) {
        openList.add(pathEndNode);
      }
    }

    openList.remove(currentNode);
    closedList.add(currentNode);
  }

  if(currentNode is not goalNode) {
    return null;
  } else {
    // Rebuild the path using the cameFrom node property.
    pathList = new pathList();
    while currentNode is not startNode {
      pathList.add(currentNode.cameFrom);
      currentNode = currentNode.cameFrom.getFromPathNode();
    }
    return reverseOrder(pathList);
  }
}
```

connected node, removes it from the closed list and adds it to the open list so that the new lower cost can be propagated to the nodes connections. When the algorithm reaches the end node, the path from the goal to start node is reconstructed using the recorded information. Due to the approach A* uses to examine the nodes the reconstructed path is the shortest route from the start to the end node.

Strategic and Tactical Pathfinding

Behaviors in computer games, such as FPS games, may initially appear to be reactive and very similar to early shooter games such as Space Invaders. In actual fact, a large amount of tactical and strategic reasoning is also required (Thurau, 2006; Tozour, 2002a). For example, during an encounter with an opponent, a player would need to make reactive decisions, such as avoiding a projectile; tactical decisions, such as getting to a protected location or a good location to attack; and strategic decisions, such as whether to continue the attack or retreat to a new area of the environment to find a better weapon. Strategic behaviors refer to actions an NPC takes in order to achieve its long-term goals (Thurau, 2006). This typically means that an NPC performs behaviors that attempt to increase its competitive advantage in relation to its opponents before they are encountered. Tactical behaviors are used to achieve an objective, which usually assists in the completion of a goal or strategic behavior. For example, an attacking tactic could be used to gain control of an area of the environment. Tactical behaviors are more localized than strategic behavior and they attempt to deal with the current situation in a smart way that is not purely reactive (Thurau *et al.*, 2004).

Traditionally, game navigation systems have focused on determining the shortest path between locations (Sterren, 2001) and have not made use of any reasoning or analysis of world terrain or game play data. However, this is beginning to change (Straatman *et al.*, 2006; Darken & Paull, 2006; Paanakker, 2008). Terrain and strategic reasoning (also known as tactical location analysis) incorporates knowledge of the game world, such as current influence, cover, sniping, light levels and death rates, into high-level decision making and planning (Straatman *et al.*, 2006; Sterren, 2001). Therefore, in addition to navigation information, it is also possible for waypoint nodes to store a vast array of statistical data on their location, the surrounding waypoints and the game play performance history. For example, Sterren (2001) highlights a number of waypoint properties that can be evaluated, including the light level, the terrain type, the line of sight to other nodes and the focus to other nodes (i.e., the lines of sight to other nodes are all in one direction). The waypoint properties can then be used to categorize behavior and game play statistics in order to determine desirable and undesirable locations. Tactical pathfinding uses these techniques to include tactical information in the pathfinding process (Straatman *et al.*, 2005; Millington & Funge, 2009). This results in routes that are tactically and strategically aware, such as avoiding lines of sight and keeping to the shadows, rather than simply the shortest route.

PATHFINDING WITH POTENTIAL ENERGY

The principles of electric potential energy (Weisstein, 2007) offer an elegant approach to adapting strategic and tactical behavior. They can be used to attract or repulse locations during the pathfinding process. For example, a tactical waypoint (or navigational mesh cell) could become repulsive (i.e., in potential energy terms have a positive charge) due to an NPC being ambushed and killed a high number of times around the location. Nearby waypoints would therefore have high potential energy due to their proximity to the repulsive source. The potential energy of a tactical waypoint is incorporated into pathfinding algorithms to represent the virtual cost needed to reach the waypoint. The cost is considered to be "virtual" because the force exerted by the source does not physically exist in the environment; it is internal to the NPC.

Figure 3 illustrates a high-level overview of the proposed strategic and tactical adaptation system. It makes use of repulsive and attractive charges of different intensities to adapt the NPC's pathfinding. Waypoint maps are used to represent the game world topology and A* is used to search the map. However, as previously discussed, other representations and search algorithms could be used. The "Game Environment" section of Figure 3 shows a simplified game world overlaid with a waypoint map. The NPC in Figure 3 is performing a "collect item" strategy by navigating to the nearest rocket launcher (goal location), whilst taking into account learnt repulsive and attractive charges. Pathfinding through the navigation mesh is achieved by an A* algorithm that incorporates the potential energy of a navigation mesh in its cost function.

The NPC has learnt a repulsive charge for the lower right tactical waypoint. The adaptation system incorporates the learnt charge values into the pathfinding process.

The repulsive and attractive charges used in the pathfinding process are selected by the NPC when the path is requested. The charge values are learnt in-game or pre-computed offline (e.g. in-between game levels or before the game is released). To enable an NPC to learn and adapt online the game AI programmer must first define the data or properties the game AI or an NPC should monitor at a waypoint node. For instance, an NPC could observe its death rate around a waypoint node or specific groups of waypoint nodes and store the information as properties that have repulsive charges. The charge value of the "death rate" property can then be applied to a pathfinding algorithm.

Instances of a property type (e.g., "death rate" or "kill rate") can be grouped according to game conditions, such as an NPC's current strategy, state or preferences. This combining of property types and game conditions enables the proposed system to represent more accurately property charge values and their relationship to game strategies, states or preferences. For example, the "kill rate" property can be associated to the NPC state of "using rocket launcher". This allows an NPC to accurately monitor its rocket launcher kill rate at each waypoint node. The combined property and game state can then be applied to pathfinding when an NPC is using a rocket launcher. If a rocket launcher is not being used a general "kill rate" node property can be selected, which represents their "kill rate" using any weapon. The following subsections contain a detailed discussion of how node properties, game conditions and charge values are organized in the system.

The approach and techniques used in the proposed system are inspired by research in the areas of tactical pathfinding, influence maps and artificial potential fields. Tactical pathfinding (Millington & Funge, 2009) is discussed below. Influence maps (Schwab, 2004) are 2D arrays of data overlaid on a game world. Each array element stores data about a specific world position, such as a numerical value representing terrain information or a player's influence on the world position. Artificial potential fields (Arkin, 1998) are vectors that represent force which are generated by objects or target positions. The fields can be repulsive or attractive. For example, obstacles would a have repulsive force, while target positions would have an attractive force.

In addition, previous strategy navigation research has influenced this approach. A number of offline strategic navigation techniques have been developed by researchers in the area. For instance, Thurau (2006) makes use of artificial potential fields to control NPC behavior and Gorman and Humphrys (2005) present an approach based on reinforcement learning. However, both these works focus on offline strategic navigation and do not address online learning and adaptation.

Figure 3 .A high-level overview of the strategic and tactical behavior adaptation system

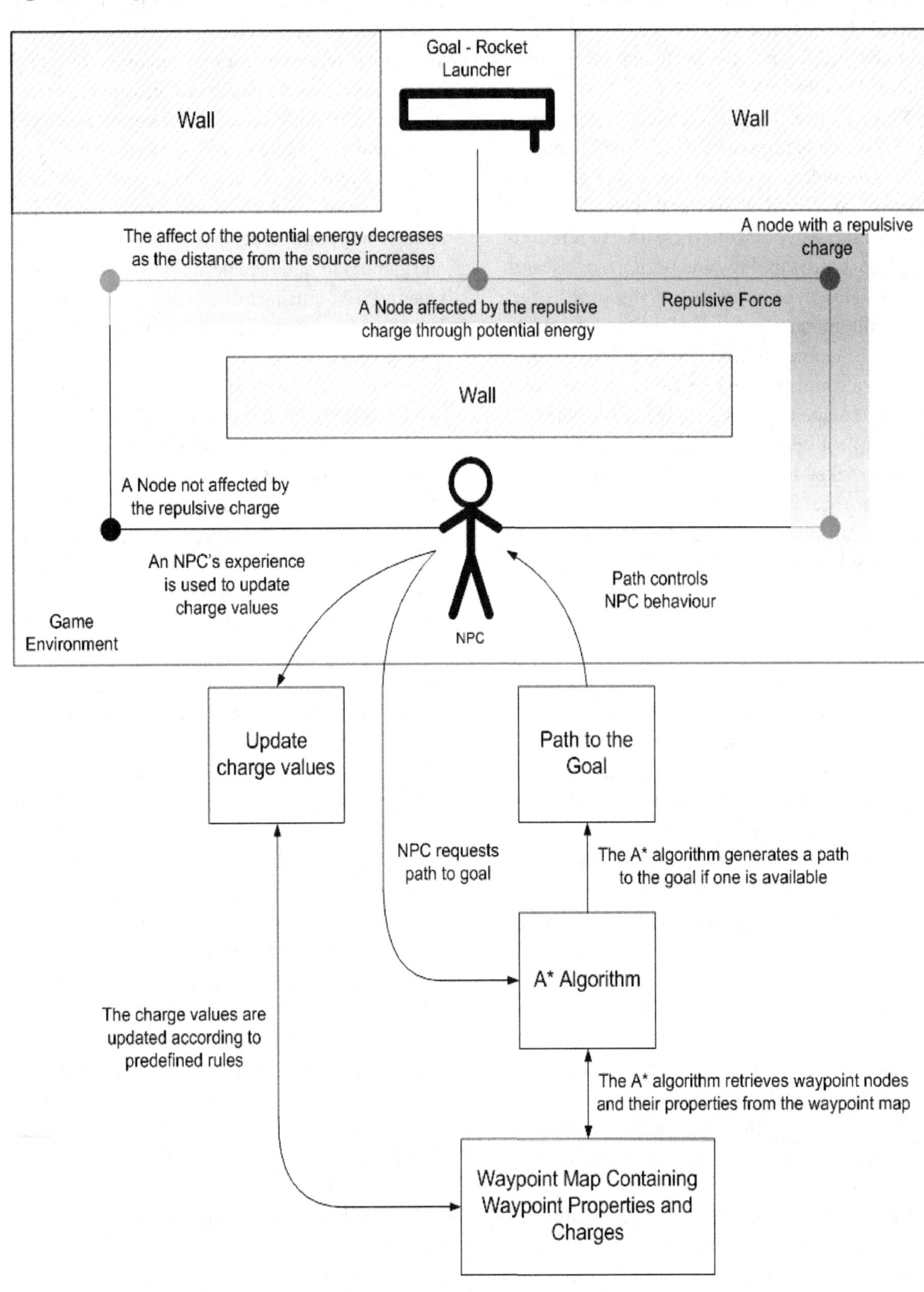

As discussed in the Background section, pathfinding and its related techniques has been used in conjunction with tactical and strategic game AI through techniques such as waypoint tactical information (Straatman *et al.*, 2006; Straatman *et al.*, 2005; Sterren, 2001) and tactical analysis or influence maps (Millington and Funge, 2009). Tactical pathfinding incorporates additional cost into the process; however, the approach does not propagate the cost to visible waypoint nodes. Influence maps provide many different ways to use game location data, a number of which have similarities to the proposed approach. For example, games developers can use both techniques to monitor game information and adapt NPC pathfinding. However, influence maps typically require large data structures, iterative processing and a lot of memory (Schwab, 2004). In contrast, the proposed approach only calculates potential energy when a node is accessed (e.g., during the search algorithm); therefore, less memory and processing are needed. Influence maps have also been combined with search based path algorithms; for example, Paanakker (2008) uses an influence-map-based approach in conjunction with the A* algorithm to create risk-adverse path planning. In the approach game entities such as towers have circles of influence, referred to as influence discs that indicate the entities influence on the surrounding world. The circles of influence are used to effect the path returned by the A* algorithm. The approach is successful; however, it focuses on RTS games and the circles of influence do not take line of sight into account.

Artificial potential fields have achieved some success in robotics (Sørensen, 2003) and computer games (Schwab, 2004; Thurau, 2006) for tasks such as path planning and obstacle avoidance. The techniques used to generate the artificial potential fields have inspired the use of potential energy in the developed system. However, instead of using the artificial potential field equation for force, the architecture uses the equation for potential energy itself. This approach is taken because the architecture does not need to calculate forces and the mechanics of movement. Pathfinding focuses on the potential for movement; therefore, the simpler and more computationally efficient potential energy approach can be taken.

Representing the Game Environment

The online strategic adaptation architecture utilizes a waypoint representation of the game world and the A* path-planning algorithm when adapting strategic and tactical behavior. However, it is important to note that alternative representations, such as navigation meshes and other graph based search techniques (e.g., Dijkstra's algorithm) can be used.

When determining the potential energy experienced at a waypoint node it is important to take into account the topology of the game environment. That is to say, the energy from a node should not travel through walls or game objects. Possible solutions to this issue include pre-computing the distance between each waypoint node and every other node or propagating the potential energy through iterative processes, such as those used in Influence Maps (Schwab, 2004) and Game Value Iteration (Hartley *et al.*, 2004). Game Value Iteration (GVI) is an adaptation to Value Iteration. In a computer games context Value Iteration would overlay a 2D array on a game world, in a similar way to influence maps. Each array element would store utility value that represents the usefulness of an array element. The Value Iteration (VI) algorithm uses an iterative process to calculate the usefulness of each element. The utility value is then used by a game character to select an optimal action. GVI adapts the algorithm by apply a new stopping criteria that reduces the number of iterations needed.

Neither of the approaches highlighted above are particularly desirable as they require significant processing and memory requirements. Therefore, the repulsive or attractive charge exerted by a waypoint is felt by nodes with a shared edge or line of sight. This approach is computationally efficient as line of sight information can be easily determined (Straatman *et al.*, 2006) or pre-computed and stored relatively efficiently in waypoint nodes (Linden, 2002). In addition, waypoint nodes can be manually added to pre-computed lists, allowing the charge's effects to be more accurately represented or customized to a level designer's needs.

Reasoning with Potential Energy

The proposed approach allows games developers to take behavior, terrain and game play experience into account when pathfinding to game world locations; therefore, an NPC can adapt its strategic and tactical behavior (e.g., produce more sophisticated routes). Properties of the game that evaluate terrain and game play statistics are used in the approach to generate repulsive or attractive charges that reflect the observed experience at a navigation waypoint node. The charges are integrated into the waypoint map so that a node can reference many properties (i.e., charge values) and the potential energy from surrounding waypoints can easily be calculated.

Each property (e.g., the behavior, terrain or game play data) being monitored by the system has a property ID, condition rule ID and a charge value. The property ID identifies the property and the condition rule ID identifies a particular instance of the property. An NPC uses these IDs to specify which charge values should be used during pathfinding to calculate the potential energy of a waypoint. The properties defined for a game can be applied to all of a waypoint map or a subset of programmer defined waypoints, which cover a game area or multiple areas, such as the main combat zones. For example, the monitor "death rate" property could be applied to cells in predefined combat areas.

There are many different properties that can be learnt as highlighted above. In addition, properties can be organized into different categories, namely static and dynamic (Millington and Funge, 2009). Properties such as light level at waypoint nodes and terrain are likely to remain static throughout the game; therefore, they can be pre-computed before a game is released. However, a number of games, such as Battlefield: Bad Company 2 (EA Games, 2010) offer large and expansive environments that are almost fully destructible, this means that many static properties are becoming dynamic.

Properties in the proposed system include a condition ID that identifies the game conditions, such as an NPC's current strategy (e.g., Attacking), state (e.g., Health less than 50%) or preferences (e.g., cautious), to which the property would be applied. This allows multiple instances of a property to be added to a waypoint node and be updated independently. For example, a sniping NPC would rate a location as highly undesirable when it has been spotted and killed once or twice, but a more reckless mid-range NPC would require a number of deaths to rate a location undesirable.

The selection of waypoint node properties during path planning will vary according to game conditions. Furthermore, the game conditions alter the desired effect of a waypoint property. For example, instances of a waypoint "occupancy level" property (e.g., the number of enemies previously observed at waypoint nodes), would attract an NPC that is looking for a fight and has high health, but would repulse an NPC with low health.

Instances of node properties are generated and updated according to observations of the game world. As with the selection of waypoint properties the updating of the charge values varies according to game conditions; therefore, the exact implementation of an update algorithm is impossible to specify. However, each approach will have common characteristics that can be identified. For example, an update algorithm for a "death rate" waypoint property could have two game conditions associated to it, namely 'grunt', an NPC type used in general ranged combat and 'ninja', a specialized NPC type which stealthily attacks enemies at close quarters. Each of these conditions needs an instance of the "death rate" waypoint property to be updated in a different manner. The death rate for a 'grunt' NPC increases slowly and it takes many kills for a location to become repulsive. The death rate for a 'ninja' updates quickly, forcing it to find new paths to surprise a human player.

In addition to learning property charge values they also need to be forgotten. For example, a poor location for an NPC (e.g., a place where it gets killed a high number of times) may change as the human player's strategy changes or as game objectives change. Therefore, it is desirable for the NPC to forget charges over time. This can be achieved through a variety of techniques that vary in complexity. The most computationally efficient approach is to simply delete learnt charge values. However, this may result in poor behavior if not properly managed. For example, the resetting of death rate charges to zero could result in NPCs returning to poor quality navigation paths. However, this could be avoided by careful use or by resetting charges to a base value that reflects the quality of the location. For example, each property charge value could be scaled according to a factor so that its reset value is proportional to its previously learnt value. This is accomplished using the following equation, where b is the base value, cv is the current charge value and 0.2 is the scaling factor.

$$b = cv \times 0.2$$

The frequency of forgetting charge values depends on the scenario in which the approach has been implemented. The charge values should be forgotten in order to prevent the human player from learning the adapting NPCs tactics and for the pathfinding algorithm to respond more quickly to the human player. For example, a FPS game is fast paced, this means that the charge values should be reset a number of times during the game. This could be coordinated with game events, such as the opening up of a new area in the game world or success rates. If an NPC is losing repeatedly then it may be due to poor charge values. The use of a scaling factor, such as 0.2, means that some knowledge regarding waypoint charge values can be maintained. However, the scaling factor should be relatively large in order to significantly reduce the charge values and allow the pathfinding algorithm to be responsive to a human player's behavior.

Incorporating Potential Energy into Pathfinding

The proposed system learns to associate world locations with repulsive or attractive charges of different intensities. These charges influence future pathfinding and therefore adapt navigation. Any waypoint map node in the game environment can contain attractive or repulsive charges that results in potential energy being exerted on nearby cells. The effect of the charge decreases as the distance between the source and a cell increases. The potential energy at a cell or tactical waypoint is determined using the following equation, where PE is the potential energy at cell i, G is the charge, k is all the cells with charges, d is the distance to the charge and e is a distance scaling factor. In addition if cell i contains a charge it is added to PE.

$$PE_i = \sum_k \frac{G_k}{ed_k}$$

In order to incorporate potential energy into the A* search algorithm the path cost function is adapted. Traditionally the cost in A* is determined by the cheapest path from the start to the current node. The cost function therefore determines the distance from the current node to the next node, and this is added to the current cost. In the system the potential energy of a location is incorporated into the path cost function along with scaled distance. Consequently the returned path from the

Figure 4. The adapted A cost algorithm, which incorporates potential energy*

```
Inputs:
      n                           // The current node.
      n'                          // The successor node.
cost (n, n') {
      cost = 0                    // The cost of moving from node n to n'.
      e_DistanceScaling = 0.1     // A distance scaling factor.
      k_n = getNodes(n)           // A list of the nodes with charges affecting n.
      k_n' = getNodes(n')         // A list of the nodes with charges affecting n'.
      tempPE = 0                  // Store a temporary potential energy value.

      // First determine the potential energy of n.
      // Determine the potential energy of the current node. This is taken to be its charge.
      tempPE = n.charge()

      //Determine the potential energy effecting the current node.
      for each node ki in k_n {
            if lineOfSight(n, ki) then
                  tempPE = tempPE + ki.charge() / (Math.log10(EuclidianDistance(n, ki)))
      }
      Pe_n = tempPE

      // Next determine the potential energy of n'.
      // Determine the potential energy of the current node. This is taken to be its charge.
      tempPE = n'.charge()

      //Determine the potential energy effecting the current node.
      for each node ki in k_n' {
            if lineOfSight(n, ki) then
                  tempPE = tempPE + ki.charge() / (Math.log10(EuclidianDistance(n', ki)))
      }
      Pe_n' = tempPE

      // Determine the Euclidian distance between n and n'. Then add it to the difference
      // in potential energy to determine the cost of moving from n to n'.
      cost = Pe_n' - Pe_n + (e_DistanceScaling * EuclidianDistance(n, n'))

      return cost
}
```

A* algorithm is the route that combines the lowest difference in potential energy and the shortest distance. The algorithm developed for determining the cost of moving between two waypoint nodes is illustrated in Figure 4.

In the algorithm $e \in [0, \text{maxValue}]$ is a distance scaling factor. The function 'getNodes' determines a list of nodes with charges affecting current node. The function 'lineOfSight' determines if there is line of sight between two nodes. The calls to this function can be removed if waypoint nodes store line of sight information. In this case the 'getNodes' function would return nodes that have line of sight to a node and contain a charge. Finally, the 'EuclidianDistance' function calculates the Euclidean distance between two nodes. The remainder of the A* algorithm follows the standard implementation, but the heuristic h is determined using a scaling factor. This means that in this A* implementation $h = ed$ where e is a scaling factor and d is the distance to the goal waypoint.

Experimental Results

The table below summarizes results from a test scenario implementation in the first person shooter game Unreal Tournament (UT) (Unreal Tournament, 2004). The case study examines scaling factors and charge value increments in the UT game map "dm-1on1-Crash". A case study was developed in which an adapting NPC performs the strategic behavior "collect item". The adapting NPC uses the pathfinding adaptation system to plan a path to goal locations (e.g., item pickups) that incorporates its past experience. The adapting NPC monitors the "death rate" waypoint node property, which is initially set to zero. When the adapting NPC dies, it updates the "death rate" property of the closest waypoint. The "death rate" waypoint property was selected for this case study because avoiding locations with high death rates increases the NPC's chance of success and encourages the human player to adapt their behavior; therefore, it is a key waypoint property to monitor. The adapting NPC's start location is set at a predefined spawning point. Initially the case study is run without a scripted opponent to determine the shortest route between the start and goal location. A scripted NPC is then placed at a waypoint node along the shortest route, near to the goal location. The adapting NPC is tasked with reaching the desired goal by adjusting its route to avoid the scripted opponent. The experiment is repeatedly run to determine the number of encounters before the adapting NPC alters its route and successfully reaches the goal position.

During the experiment the charge values of waypoint nodes were initially set to zero and increased by predefined amounts if the "death rate" property is triggered. The distance scaling factor in the A* algorithm was also varied, ranging from 1 to 0.001. A scaling factor of 1.0 results in distances not being scaled, while a scaling factor of 0.001 results in distances being 1000th their original length. The UT game uses its own measurement system for distance called Unreal Units (UU). It is an abstract system that is only meaningful to the Unreal game engine; however, in the UT games 1 meter is roughly equivalent to 52.5 UU (Unreal Wiki, 2007). This means that distances in UT can have large values that will engulf a single charge range of ± 200; consequently distances may need to be scaled. In the results below the charge value increment is set to 50.

The results in Figure 5 show that a distance scaling factor of 1 or 0.1 result in the adaptation of behavior within a few encounters. Larger scaling factors result in behavior adaptation after 1 encounter. This rate of adaptation is too quick for FPS games and is likely to lead to reasonable locations being unnecessarily avoided by the adapting NPC. The results also show that a distance scaling factor of 1 or 0.1 leads to the adaptation of behavior in the UT game map. This is due to the alternative routes being less costly. These results indicate that this rate of scaling factor is more appropriate for the commercial UT game levels.

FUTURE RESEARCH DIRECTIONS

"Genuinely Adaptive AIs will change the way in which games are played by forcing the player to continually search for new strategies to defeat the AI, rather than perfecting a single technique." (Manslow, 2002). The previous quote highlights the potential effect of online learning and adaptation on commercial computer games. Human players are currently faced with largely static game AI that is predictable and can be defeated through the use of a few, quickly learnt strategies. Learning and adaptation in games is considered by the games industry to be an important advance that is highly desirable for computer games in order to maintain their entertainment value (Manslow, 2006; Baekkelund, 2006).

The approach presented in this chapter is designed to support this desirable endeavor from a strategic and tactical standpoint. It allows game characters to robustly learn in-game from past experience and consequently adjust their pathfinding. NPCs which incorporate in-game learning techniques have the potential to adapt to a human player's actions, generate new behavior according to past experiences and overcome poorly developed scripts. Consequently game developers who apply these techniques can improve the challenge of their games and maintain an entertainment level for the human player.

Figure 5. A summary of results from an investigation into scaling factors and charge value increments for the UT game map "dm-1on1-Crash"

Unreal Tournament Map: dm-1on1-Crash				
	Experiment #01	**Experiment #02**	**Experiment #03**	**Experiment #04**
Adapting NPC Start Position:	PathNode11 (496.0,1840.0,-81.0)	PathNode69 (-1880.0,2016.0,47.0)	PlayerStart2 (154.11,-1498.78,-145.0)	PlayerStart2 (154.11,-1498.78,-145.0)
Goal Location:	InventorySpot60 (-1282.0,-522.0,-257.0)	InventorySpot3 (-643.59,-1333.73,-153.0)	InventorySpot11 (496.0,1840.0,-81.0)	InventorySpot6 (-1471.29,1629.38,-209.0)
Static NPC Location:	PathNode20 (-464.0,-568.0,-81.0)	PathNode5 (-1338.58,-703.51,47.0)	PathNode72 (-1960.0,1184.0,47.0)	PathNode50 (-1320.0,608.0,-241.0)
Charge Value Increment:	50	50	50	50
Distance Scaling Factor	**Number of encounters before the adapting NPC alters its route:**			
1	3	More than 10	More than 10	9
0.1	1	6	More than 10	1
0.01	1	1	2	1
0.001	1	1	1	1

The experimental results outlined here indicate that this approach could be a good technique for adapting pathfinding in fast paced action games, such as FPSs. However, there are a number of areas that would benefit from future research. Firstly, the work presented here focuses on a single-player multi-round FPS computer game scenario. Future research will evaluate and extend the system to operate in a multi-player (partner and opponent) environment. Secondly, more complex game scenarios will be explored. It is important to determine how well the system scales to typical commercial computer game environments and scenarios.

CONCLUSION

This chapter has outlined an approach to strategic and tactical pathfinding from the perspective of in-game learning and adaptation. The technique enables NPCs in games such as FPSs to adapt their routes through a game environment based on previous experience. A potential energy based approach to augmenting navigation was developed. Instead of simply path planning the shortest route to a goal, the approach combines waypoint properties and potential energy so that previous experience can be incorporated into paths. Therefore, an NPC can be steered away from potentially dangerous locations and guided toward potentially beneficial locations.

The developed system showed that the approach can successfully adapt behavior based on previous encounters. The system allows NPCs to exhibit self-correction and creativity through the adaptation and creation of pathfinding routes. The system is effective because waypoint properties are created by the AI programmer; therefore, any path adaptations are based on their domain knowledge of game factors. The strategy system is robust because updates to waypoint properties do not immediately alter NPC paths. Multiple updates are typically required before the system alters routes. In addition, the waypoint properties can be reset in order for any poorly learnt waypoint property values to be removed.

REFERENCES

Andrade, G. D., Santana, H. P., Furtado, A. W. B., Leitão, A. R. G. A., & Ramalho, G. L. (2004) Online adaptation of computer games agents: A reinforcement learning approach. *1st Brazilian Symposium on Computer Games and Digital Entertainment (SBGames2004).*

Arkin, R. (1998). *Behavior-based robotics*. Cambridge, MA: MIT Press.

Baekkelund, C. (2006). A new look at learning and games. In Rabin, S. (Ed.), *AI Game Programming Wisdom 3* (pp. 687–692). Hingham, MA: Charles River Media.

Byl, P. B. (2004). *Programming believable characters for computer games. Game Development* (1st ed.). Hingham, MA: Charles River Media.

Funge, J. (2004). *Artificial intelligence for computer games.* Wellesley, MA: A K Peters.

Games, E. A. (2010). Battlefield: Bad Company 2. Retrieved October 05, 2011, from http://www.battlefieldbadcompany2.com.

Gorman, B., & Humphrys, M. (2005). Towards integrated imitation of strategic planning and motion modelling in interactive computer games. In *Proc. 3rd ACM Annual International Conference in Computer Game Design and Technology (GDTW 05)*(pp 92-99).

Hartley, T., Mehdi, Q., & Gough, N. (2004). Applying Markov decision processes to 2d real time games. In Medhi, Q., Gough, N., Natkin, S., & Al-Dabass, D. (Eds.)*CGAIDE 2004 5th International Conference on Intelligent Games and Simulation* (pp. 55-59).

Higgins, D. (2002a). Pathfinding design architecture. In Rabin, S. (Ed.), *AI Game Programming Wisdom* (pp. 122–132). Hingham, MA: Charles River Media.

Higgins, D. (2002b). How to achieve lightning fast a. In Rabin, S. (Ed.), *AI Game Programming Wisdom* (pp. 133–145). Hingham, MA: Charles River Media.

Johnson, G. (2006). Smoothing a navigation mesh path. In Rabin, S. (Ed.), *AI Game Programming Wisdom 3* (pp. 129–139). Charles River Media.

Laird, J. (2001). It knows what you're going to do: Adding anticipation to a quakebot. Proceedings from *Fifth International Conference on Autonomous Agents* (pp. 385-392). Retrieved November 21, 2010 from http://ai.eecs.umich.edu/people / laird/ papers/Agents01.pdf.

Linden, L. (2002). Strategic and tactical reasoning with waypoints. In Rabin, S. (Ed.), *AI Game Programming Wisdom* (pp. 211–220). Hingham, MA: Charles River Media.

Lorch, W. (2000). An introduction to graph algorithms. Retrieved September 10, 2010, from http://www.cs.auckland.ac.nz/~ute/220ft/graphalg/node21.html.

Manslow, J. (2006). Practical Algorithms for In-Game Learning. In Rabin, S. (Ed.), *AI Game Programming Wisdom 3* (pp. 599–616). Hingham, MA: Charles River Media.

Manslow, P. (2002). Learning and adaptation. In Rabin, S. (Ed.), *AI Game Programming Wisdom* (pp. 557–566). Hingham, MA: Charles River Media.

McGlinchey, S. (2003). Learning of ai players from game observation data. In Medhi, Q., Gough, N., & Natikin, S. (Eds.), *GAME-ON 2003 4th International Conference on Intelligent Games and Simulation* (pp. 106-110).

Millington, I., & Funge, J. (2009). *Artificial intelligence for games* (2nd ed.). San Francisco, CA: Morgan Kauffman.

Namee, B. M. (2004). *Proactive persistent agents: Using situational intelligence to create support characters in character-centric computer games*. Doctoral Dissertation. Trinity College of Dublin, Ireland.

Paanakker, F. (2008). Risk-adverse pathfinding using influence maps. In Rabin, S. (Ed.), *AI Game Programming Wisdom 4* (pp. 173–178). Hingham, MA: Charles River Media.

Pinter, M. (2002). Realistic turning between waypoints. In Rabin, S. (Ed.), *AI Game Programming Wisdom* (pp. 186–192). Hingham, MA: Charles River Media.

Rabin, S. (2000). A* aesthetic optimizations. In DeLoura, M. (Ed.), *Game programing gems*. Hingham, MA: Charles River Media.

Schwab, B. (2004). *AI game engine programming. Game Development Series*. Hingham, MA: Charles River Media.

Sørensen, M. J. (2003). Artificial potential field approach to path tracking for a nonholonomic mobile robot. In *Proceedings of the 11th Mediterranean Conference on Control and Automation*. Rhodes, Greece.

Spronck, P., Sprinkhuizen-Kuyper, I., & Postma, E. (2003). Online adaptation of game opponent ai in simulation and in practice. In Medhi, Q., Gough, N., & Natikin, S. (Eds.), *GAME-ON 2003 4th International Conference on Intelligent Games and Simulation* (pp. 93 - 100).

Sterren, W. (2001). Terrain Reasoning for 3D Action Games. Retrieved from http://www.gamasutra.com/features/20010912/sterren_01.htm

Straatman, R., Beij, A., & Sterren, W. (2006). Dynamic tactical position evaluation. In Rabin, S. (Ed.), *AI Game Programming Wisdom 3* (pp. 389–403). Hingham, MA: Charles River Media.

Straatman, R., Sterren, W., & Beij, A. (2005). Killzone's AI: Dynamic procedural combat tactics. Retrieved from http://www.cgf-ai.com/docs/straatman_remco_killzone_ai.pdf

Thurau, C. (2006). Behavior Acquisition in Artificial Agents. Doctoral Dissertation. Bielefeld (Germany): Bielefeld University.

Thurau, C., Bauckhage, C., & Sagerer, G. (2004). Imitation learning at all levels of game-ai. In *Proc. Int. Conf. on Computer Games, Artificial Intelligence, Design and Education* (pp. 402–408).

Tournament, U. (2004). *Ut2004*. Retrieved from http://www.unrealtournament2004.com/ut2004/modes.html.

Tozour, P. (2002a). First-person shooter ai architecture. In Rabin, S. (Ed.), *AI Game Programming Wisdom* (pp. 387–396). Hingham, MA: Charles River Media.

Tozour, P. (2002b). The evolution of game ai. In Rabin, S. (Ed.), *AI Game Programming Wisdom* (pp. 3–15). Hingham, MA: Charles River Media.

Waveren, J., & Rothkrantz, L. J. M. (2001). Artificial Player For Quake III Arena. In Medhi, Q., Gough, N., Natkin, S., & Al-Dabass, D. (Eds.) *GAME-ON 2001 2nd International Conference on Intelligent Games and Simulation* (pp 48 – 55).

Weisstein, E. (2007). *Gravitational potential energy*. Retrieved from http://scienceworld.wolfram.com/physics/GravitationalPotentialEnergy.html.

Wiki, U. (2007). *Unreal unit*. Retrieved from http://wiki.beyondunreal.com/Legacy:Unreal_Unit.

ADDITIONAL READING

Orkin, J. (2006). 3 states and a plan: The ai of f.e.a.r. *Game Developers Conference*. San Jose, CA.

Reed, C., & Geisler, B. (2004). Jumping, climbing, and tactical reasoning: How to get more out of a navigation system. In Rabin, S. (Ed.), *AI Game Programming Wisdom 2* (pp. 141–150). Hingham, MA: Charles River Media.

Russell, S., & Norvig, P. (1995). *Artificial intelligence a modern approach*. New York, NY: Prentice-Hall.

Spronck, P. (2005). *Adaptive game ai*. Universiteit Maastricht, The Netherlands: Doctoral Dissertaion.

Strout, B. (1999). Smart move: Intelligent Path-Finding. Retrieved October 15, 2011, from http://www.gamasutra.com/view/feature/3317/smart_move_intelligent_.php.

Tournament, U. (2004). UT2004. Retrieved from http://www.unrealtournament2004.com/ut2004/modes.html.

Tozour, P. (2002). Building a bear-optimal navigation mesh. In Rabin, S. (Ed.), *AI Game Programming Wisdom* (pp. 171–185). Hingham, MA: Charles River Media.

van Waveren, J. P. (2001). The Quake III Arena Bot. Master's Dissertation, Delft University of Technology, The Netherlands.

KEY TERMS AND DEFINITIONS

First Person Shooter Computer Games: A genre of commercial computer games that take place in a simulated 3D environment which mimics the real world. For example, they simulate a virtual space that computer controlled characters and human players can move through.

Game Artificial Intelligence (AI): The techniques, algorithms, etc. that control a computers decision making with a game. The most visible manifestation of game AI is NPC behavior.

Influence Maps: 2D arrays of data overlaid on a game world. Each array element stores data about a specific world position, such as a numerical value representing terrain information or a player's influence on the world position.

In-Game or Online Learning: Occurs when a computer game is being played by a human player and is typically achieved through unsupervised learning techniques.

Non Player Character (NPC): A virtual character that is controlled by the computer.

Pathfinding: Also known as path planning. It is concerned with determining a short and appropriate route from one location in a game environment to another.

Chapter 3
Behavior Trees:
Introduction and Memory-Compact Implementation

Björn Knafla
Bjoern Knafla Parallelization + AI + Gamedev, Germany

Alex J. Champandard
AiGameDev.com, Austria

ABSTRACT

Behavior trees (BTs) are increasingly deployed in the games industry for decision making and control of non-player characters (NPCs, also named agents or actors) and gameplay. Behavior trees are incrementally created from reactive and goal-oriented behaviors by hierarchically compositing purposeful sub-behaviors. Composite behaviors (branches in the tree) decide which child behavior to run when and react to changing behavior execution states. Leaf nodes of the behavior tree check conditions, or control the execution of game state affecting actions. This chapter introduces a basic behavior tree concept, describes how behavior tree traversal drives NPC decision making via an example, and sketches a memory-compact implementation.

INTRODUCTION

Games are about entertainment and a great player experience. According to Chris Butcher from Bungie (Valdes, 2004), non-player characters (NPCs), often called agents or actors, should behave in ways that the player believes them to be intelligent. Artificial intelligence in games is more concerned with believability of the computed behavior than with simulating intelligence itself. Big and rich behavior repertoires enable agents to react to many situations in a variety of ways so they seem to have common sense as Isla puts it (2005).

Creating such huge sets of behaviors, organizing the decision-making process when to behave

DOI: 10.4018/978-1-4666-1634-9.ch003

in which way, and maintaining and iterating on the collection of decisions and actions to tune them for fun, is challenging. Part of the challenge stems from the balancing act between authoring autonomous and reactive agents while not losing control over the gameplay experience, which should adhere to the intended game design.

Behavior trees (BTs) are increasingly employed throughout the games industry to meet these challenges (Champandard, 2011). An agent's behavior is modularly assembled and structured by hierarchically organizing sub-behaviors - decisions, actions, or (sub-)trees formed out of decisions and actions - which are successively connected to form larger, more expressive behavior trees. These connections between decision nodes and their children are the sole way to model relationships between behaviors.

To update or tick the behavior of an agent during a simulation step, its behavior tree is searched best-first to find the most appropriate action(s) to run in the current game world situation which is (are) then executed. Traversal is steered by the semantics of decider nodes and the behaviors execution states.

The following topic background is briefly mentioned before explaining a concept of behavior trees based on a small example. Then a memory-compact implementation approach is described. Afterwards, future study and research directions are noted and the chapter is concluded.

BACKGROUND

Scripting and finite-state machines (FSMs) have been the primary tools to model decision making, and action selection and execution for computational behavior of agents in games (Champandard, 2007).

Scripting offers the most direct control over the gameplay in a scene. However, it also complicates reuse, analyzation, and maintenance of behaviors, especially if the controlled entities should remain reactive to the actions of the player (Champandard, 2007).

Finite-state machines provide a simple and intuitive mechanism to model reactive behavior (Champandard, 2007). In games, finite-state machines aren't used to accept languages, but to process and react to continuous streams of events or inputs (Fu & Houlette, 2004). A (finite) set of states corresponds to behaviors while (agent external or internal) events trigger transitions between states.

Diagrams of finite-state machines look conceptually simple but they hide many details necessary to implement them as Isla (2005) observes, e.g., when to follow a transition? Some transitions are voluntary, e.g., when a guard decides to idle instead of searching on for an intruder, while others are forced by an event, e.g., when a trespasser enters a guard's viewing field and the guard transitions from idle to the attack state.

The more states and transitions are used, the more labor intensive it gets to maintain and understand the flexible transition wiring. Hierarchical finite-state machines (HFSMs) lower this complexity by enabling reuse of smaller HFSMs inside of states of higher-level HFSMs and by allowing transitions between states in the different levels of the hierarchy (Millington, 2006). Though even with hierarchical finite-state machines Isla's critique and the complexity due to the ad-hoc transition wiring remains, e.g., there is no way to express sequences or explicit selection of behaviors in a straightforward manner with HFSMs.

Isla's article from 2005 about the artificial intelligence (AI) in Halo 2 describes how behavior trees can be used to handle the complexity of behavior modeling in games. Since then, different flavors of behavior trees have been deployed in addition to finite-state machines by increasing numbers of game developers (Champandard, 2011; Hecker, 2009; Pillosu, 2009).

Champandard broke down the different concepts and extended them with his own and thereby developed a behavior tree language (Champandard, 2007, 2008). His work builds the foundation for this chapter's behavior tree concept.

When programming for today's gaming platforms and the prospective future hardware and processor architectures, the widening gap between computational and memory access speeds and memory bandwidth (data rate) (Albrecht, 2009a; Borkar & Chien, 2011; Patterson, 2004; Hennessey & Patterson, 2007) becomes an important concern. To keep processor cores busy instead of wasting precious cycles waiting on data to arrive from main memory, computations need to utilize memory hierarchies - storage nearer to cores is faster but also smaller than memory farther away - by focusing on accessing and reusing data local to processor cores.

Homogeneous and heterogeneous parallel processor architectures, introduced to keep power consumption and heat generation at bay, contain many computing cores and elaborate memory hierarchies. Care is needed to exploit but not over-saturate their memory systems.

Where performance matters, many game developers turn to so called data-oriented design and programming (DOD and DOP) (Albrecht, 2009b; Llopis, 2009, 2011a, 2011b). Not the code but the required output data to transform from input data takes precedence during programming. A trend that is mirrored by the different data-oriented behavior tree implementations presented at the Paris Game AI Conference and Shooter Symposium 2011. The behavior tree implementation sketch of this chapter also adheres to data-oriented design.

BEHAVIOR TREE CONCEPT

Behaviors can be understood as tasks fulfilling certain purposes (Champandard, 2008), e.g., a monster might exhibit a behavior to spot intruders and then attack them with the goal of defending its dungeon.

Actions ("attack intruder") and deciders (decide which action to run next) are basic behaviors which are hierarchically aggregated into behaviors with actions as leafs and deciders as inner nodes (a.k.a. branches). These (sub-)behavior trees can be recursively assembled to form even larger behavior trees. As behaviors can be modularly reused at multiple places in the hierarchy it would be more accurate to speak of directed acyclic graphs (DAGs) but the term behavior tree is already well established. Often, behaviors are shared and reused while editing behavior graphs but at runtime each occurrence of a shared behavior gets its own instance, resulting in real trees.

To update an agent's decision making process and execute its behaviors during a simulation step, the agent's behavior tree is traversed via a best-first search for the action(s) to run in the current game world state. Entering a behavior during traversal means to call or execute it. When traversal leaves a behavior and goes back up to its parent decider, an execution state is passed up. The decider reacts to the execution state by guiding the ongoing traversal according to its traversal semantics (see the decider node descriptions below), e.g., by selecting another child behavior to visit or by handing its own execution state and therewith traversal control up to its own parent node.

Execution states and their meaning and the different behavior node types and how they affect traversal are investigated below. The next section looks at a behavior tree example to make the interplay between behaviors and how they control an agent more concrete.

Behavior Execution States

Behaviors can be in different execution states:

- Ready, a ready behavior is activated when traversed and starts execution.
- Running, is returned when (direct or indirect child-)actions of the behavior aren't

completed yet and the behavior should be ticked again during the next traversal.

- Successful completion signals that the behavior accomplished its objective and finished execution.
- Failed completion shows that an error occurred and/or that the behavior couldn't accomplish its goal.

Success and failure are termination states while ready and running are scheduling states. After successful or failed behavior completion, behaviors are assumed to be in the ready execution state again. The ready state is never returned to a calling parent behavior - a behavior has to be in a running state or completed when traversal leaves it.

Event-driven behavior tree implementations additionally have a suspended scheduling execution state. Basic behaviors are implemented as tasks which are managed and executed by a scheduler (Champandard, 2008). A suspended behavior can be thought to be running but it isn't actively ticked. Instead, it waits for a notification or an event to change its state. Such a state change, from suspended to success or fail, triggers the scheduler to actively tick the behavior again.

Basic Behavior (Node) Types

Behavior nodes are largely self-contained. Relationships between behaviors are only established by explicitly adding them as children to composite a.k.a. decider nodes. This rigid structure reduces coupling and therefore increases modularity and reuse of behaviors.

Actions and Conditions

Action nodes, or actions for short, are the interface between behavior trees and the game world representation. Actions are the leaf nodes of behavior trees. Action execution might change the behavior tree controlled agent and/or the game world state.

When a ready action is visited during traversal it is activated and ticked. Some actions need multiple simulation ticks until they terminate successfully or fail, e.g., a locomotion action following a path arrives at its target position and succeeds, or determines that a dynamic environment change makes the target location unreachable and returns a fail state. On completion, an action deactivates automatically. If a higher-level decision abandons a sub-tree, for example to run to a higher-priority behavior, the running actions of the abandoned sub-tree need to be explicitly deactivated/cancelled.

Actions that check if the actor or game is in a certain state, e.g., if a guard can perceive an intruder, are called conditions. Typically they finish during the same update tick in which they are activated without any side effects on the game state. Conditions query the game world or agent state and return a success execution state if their condition check is fulfilled or fail otherwise.

Deciders

Decision nodes, a.k.a. deciders, are branches/inner nodes of the behavior tree and manage one or multiple child behaviors. When entered by the best-first search through the behavior tree to update the agent, a decider chooses which of its child behaviors to traverse. Once traversal leaves this sub-behavior the decision node evaluates the returned execution state to decide whether to traverse down into another child node or if to return traversal control to its own parent decider, passing up its own execution state. As deciders react to the result execution states of their children and steer traversal accordingly, they can be understood to model the control flow of a behavior tree and schedule actions to tick.

When decision nodes are left and return a run execution state they might store a decider state to remember which child behavior to jump to if they are traversed again during the next behavior tree update. If a higher-up decider abandons child behaviors which are in a running state from last traversal, then it recursively tells them to deactivate and to reset to their initial or default state.

The most common decider types and their traversal semantics are listed below.

Sequence Deciders

Sequence deciders (sequences) run their child behaviors linearly, one after the other, as long as they complete successfully. When all children succeed, the sequence succeeds, too. If a child fails, then the whole sequence fails. In both cases, success and failure, the sequence passes the execution state up to its parent node and the next sequence traversal restarts afresh from the first sub-behavior.

If a child passes the running execution state to the sequence decider it stores this child as the next one to visit and returns with a running state, too.

A sequence could be used to model a "detect intruder" behavior of a guard whose purpose is to react to noises by checking if they are caused by an intruder. For example, the following actions could be sequence sub-behaviors:

1. Heard noise condition.
2. Move to origin of noise.
3. Spot intruder.
4. Raise the alarm.

The first child is a pre-condition. Only if a noise was heard will the "detect intruder" objective be pursued further by traversing the other behaviors. If a noise was heard, then the next behavior finds a path to the sound location and walks the actor there. On successful arrival the guard looks around if it can find a trespasser. If an intruder is detected, the alarm is raised and the "detect intruder" sequence completes, it has fulfilled its objective.

In practice it makes sense to remove the "Heard noise" condition from the sequence and put it below a higher-up decider. The sequence becomes an "investigate disturbance" behavior which can be applied to resolve different "disturbance-detected" conditions.

Priority Selectors

Priority selectors, a.k.a. priority deciders, order child behaviors from highest to lowest priority and run through them starting from the highest-priority child until one sub-behavior is found that does not return a fail state. If a child returns running, then the priority selector stores it in its decider state and returns running, too. If no child is found that completes successfully or is running, then the priority decider fails.

One variant of the priority selector, the so called static priority selector, always returns to a child which returned running during the previous behavior tree tick to run it till it terminates.

In comparison, a dynamic priority selector always starts from the highest priority behavior for each new traversal. If a sub-behavior succeeds before reaching the child behavior with a running state from the previous behavior tree update, then the previously running child behavior is abandoned and deactivated/cancelled. This traversal semantic allows quick reactions to game world state changes and triggers more important behaviors to run. Some game developers use priority selectors as the only behavior tree decision node type to simplify and streamline their behavior tree implementation. An example priority selector at the behavior tree root of a dungeon guarding monster could have the following sub-behaviors:

1. Die theatrically.
2. Fight enemy.
3. Detect intruder.
4. Idle.

The highest priority child behavior checks if the monster has been slain by the intruding player to prevent other behaviors from running in that case. Mikael Hedberg from DICE talked at the Paris Game AI Conference 2010 about the strong impact death animations had on players, especially as non-player characters only possess seconds of lifetime in first-person shooter games.

If a guard is alive, its next most important behavior is to fight the enemy intruder. As long as no trespassers have been spotted the monster tries to detect them. If it can't find any troublemakers the priority selector falls back to the always succeeding "Idle" behavior - which might trigger actions to guide the player's attention to the guarded dungeon entry so she/he can't miss it.

Probability Selectors

Probability selectors randomly pick one of their children to execute and also return its execution state to their own parent node. Static probability deciders maintain a formerly running child stored in their decider state while a dynamic probability selector always reselects which child to run anew and deactivates children which are in a running state from the behavior tree update before. One form of probability selectors assigns each sub-behavior a different probability sub-range from 0.0-1.0 whereas all sub-ranges together cover the full probability range.

Decorators

Decorator decider nodes only have a single child behavior. Some decorators transform the execution state from their sub-behavior, e.g., to always return a success state to their parent behavior. The filter variant of decorator deciders maintains a counter or stores a timer value in a persistent decider state to restrict how often and when their sub-behavior can be entered, e.g., to prevent repetition. A decorator could also use a counter to loop its child for a given number of cycles. It is up to the game specific decorator implementation if a decorators decider state is preserved or reset to an initial state if a higher-up decider abandons and deactivates the sub-behavior containing the decorator.

Parallel / Concurrent Deciders

Parallel or concurrent deciders tick all their children from left to right during each traversal. Their sub-behaviors appear to execute concurrently. Some parallel decider variants fail if at least one child fails and then might not execute the following sub-behaviors to the right of the failing one. All children that are in a running state need to be deactivated explicitly before the parallel decider bails out and returns a fail execution state to its own parent node. Other parallel variants need a higher child fail count until they fail, too. Some parallel deciders return a success state only if all of their children succeed, while other parallel decider types might succeed the moment a smaller predefined number of sub-behaviors succeeded. If no sub-behavior of a parallel fails and at least one returns a running execution state, then the following child behaviors are still traversed, however the parallel decider returns a running state on finishing traversal. A typical use case for a concurrent decider is to monitor assumptions that need to hold for other sub-behaviors to keep executing, e.g., by placing a condition as the first child. This condition functions like an invariant check which needs to succeed to execute the remaining - monitored - sub-behaviors. If the invariant fails, then the parallel decision node bails out, too - the invariant for the following sub-behaviors doesn't hold true anymore.

Semaphore Decorators

With parallel deciders a behavior tree can contain more than one sub-behavior which is in a running execution state after an update tick. Therefore multiple actions might be active, too. These actions might use the same resource, e.g., try to animate the same limb of an actor's animation skeleton. To prevent conflicting resource accesses so called semaphore deciders can be deployed. With them, only one behavior succeeds in obtaining ownership of a semaphore and the ownership is guarded for

Figure 1. Legend of graphical behavior type symbols

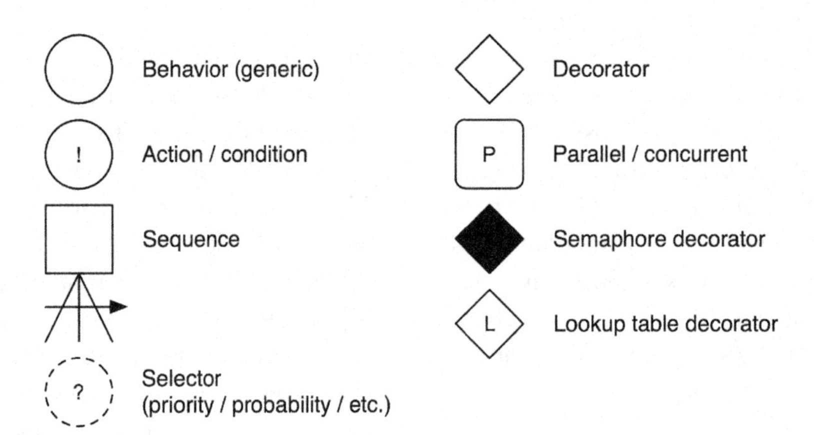

as long as its sub-behavior keeps running. Other behaviors fail when their semaphore rejects resource access and the behavior tree traversal falls back and searches for a non-conflicting behavior option to run. There is no known case of semaphore decider deployment in the games industry yet.

Lookup Tables

Lookup tables and the lookup table decorators accessing them improve the modularity of behavior trees. A lookup table has multiple slots, each associated with a different objective, e.g., "die theatrically", "fight", "detect intruders", and "idle", and each referencing a sub-behavior (which can be a full sub-tree). Lookup decorators are set up to access a specific objective in the lookup table assigned to them. On traversal they access the behavior stored in the objective slot and execute it.

By defining all necessary table slots and then iteratively adding more and more behaviors to them, behavior trees can be build step by step, from a simple and small up to a big and behavior rich tree, while being tested for correctness and gameplay fun all along the way.

Lookup tables can be chained in an inheritance hierarchy. Generic base-tables model generic default behaviors while sub-tables override and/or add a few slots with more specific sub-behaviors according to the character type of the agent to control, e.g., all non-player characters have a given set of idle behaviors but monsters with big noses carry an additional behavior to blow their nose if they caught a cold.

Some implementations might use lookup tables to dynamically change an actor's behavior at runtime, e.g., by adding a special sub-table to the inheritance hierarchy the moment an actor leveled up or found a magic item. Though, oftentimes lookup tables and their decorator deciders are mainly used during development and the described behavior tree remains static for the lifetime of the actor it controls.

Figure 1 shows graphical representations for the different behavior node types.

A BEHAVIOR TREE TRAVERSAL EXAMPLE

Following, an example for a behavior tree (Figure 2) to control a monster guarding the entry to a dungeon is described. Figure 3 annotates (sub-) behaviors with their purposes.

At the root of the example behavior tree is a priority selector. Its highest-priority sub-behavior

Figure 2. Example behavior tree to control a guard agent

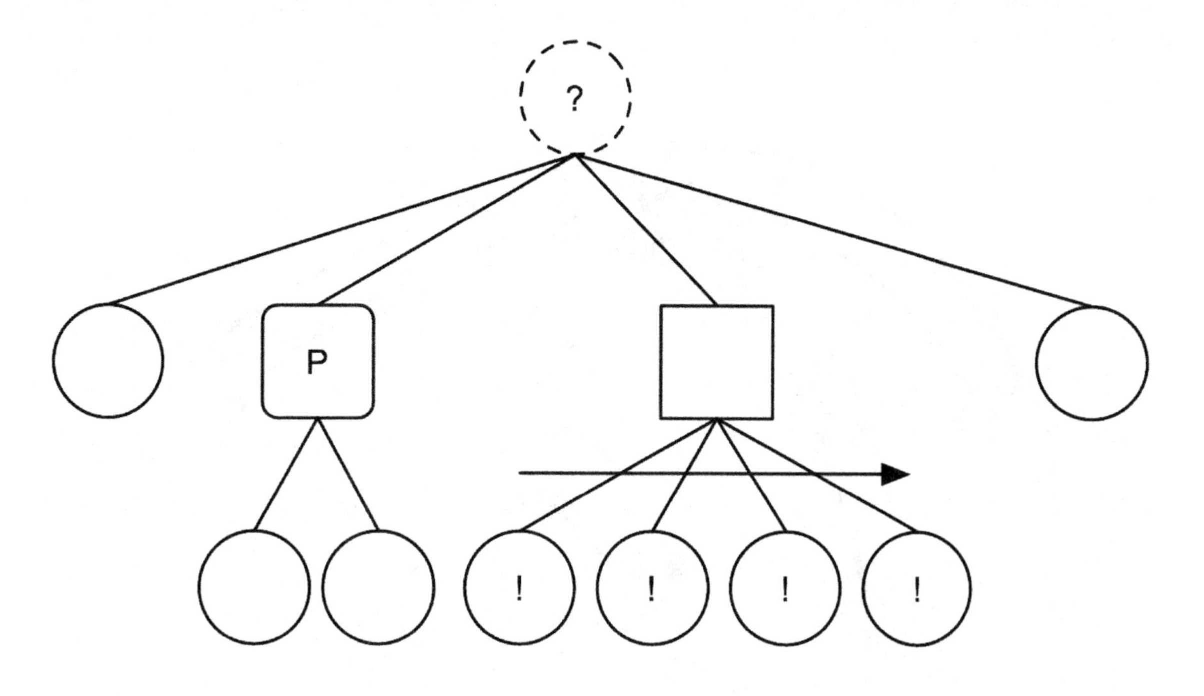

Figure 3. Same behavior tree as in Figure 1 with objectives ascribed to (sub-behaviors)

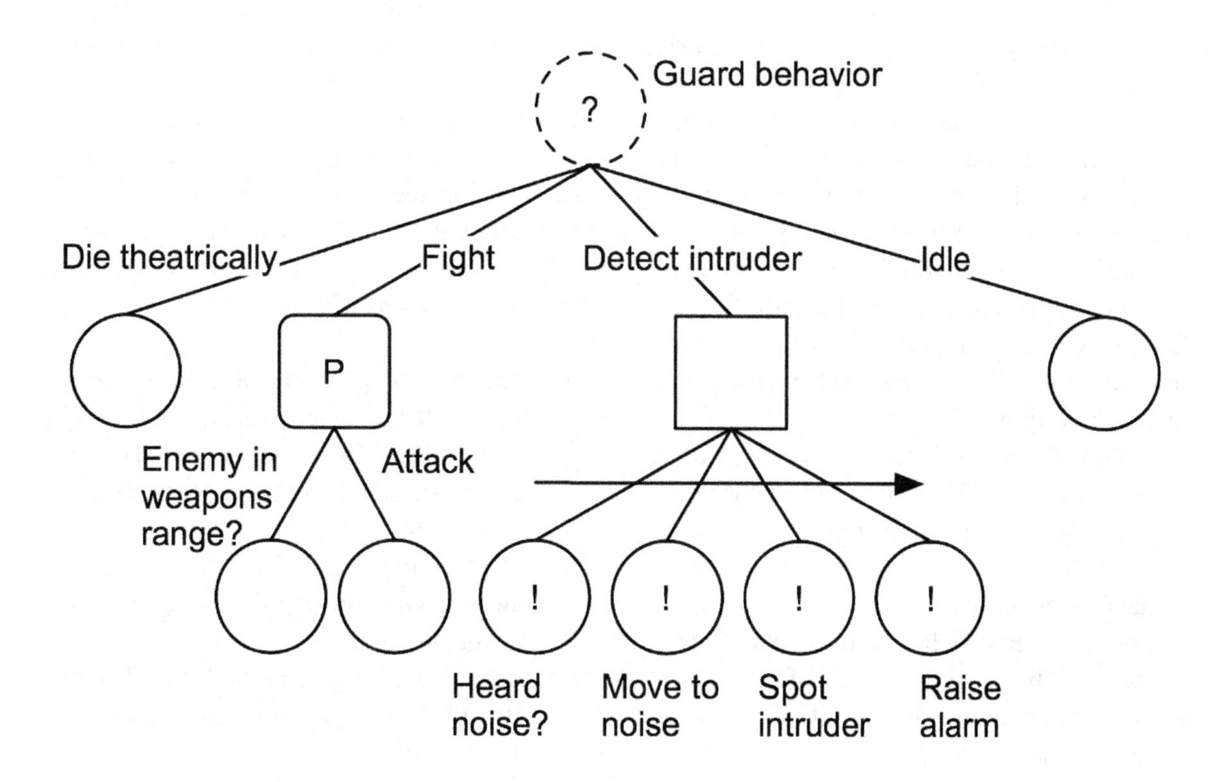

Figure 4. To update the decision making process of an agent and execute the game world state appropriate actions, its behavior tree is traversed in a best-first search fashion to find a behavior which succeeds or runs

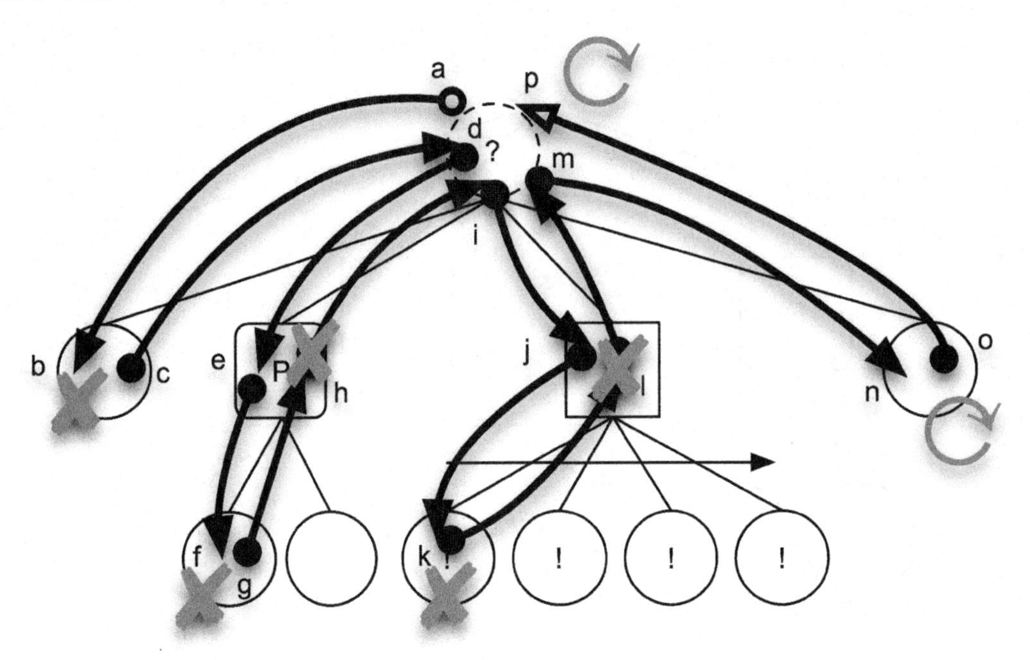

is on the left, the lowest-priority sub-behavior is on the right.

The leftmost sub-behavior, "Die theatrically", is a placeholder for a sub-tree, probably a lookup table decorator. It checks if the controlled agent is dead. In that case it selects and runs a theatrical death animation and returns a success state, otherwise it returns a fail state. A counting decorator might be used to only show the death animation once but still succeed afterwards so no other behavior can take over the dead actor until it is removed from the game world simulation.

Second, the "Fight" behavior consists of a parallel decider. Its first child is a behavior that tests if an enemy is inside the range of the available weapons. The other child behavior selects an attack behavior to execute.

Right next to the "Fight" behavior is the "Detect intruder" sub-behavior. It consists of a sequence decider with children as described in the earlier sequence decider example.

The lowest-priority child of the root priority selector is a sub-behavior that always succeeds in picking an "Idle" behavior. It serves as a fallback if none of the higher-priority behaviors returns with a success or run execution state.

Figure 4 visualizes a traversal path through the behavior tree to update the guard agent. At traversal start, all deciders are in their initial or default state. Based on the game world state, traversal unfolds as follows:

a) The decision process and action execution update of the agent begins at the priority root node of its behavior tree. Traversal descents into the highest-priority child behavior first.
b) The "Die theatrically" behavior is traversed (not shown in the path) to find an action to run or succeed.
c) As the monster guard is well off, "Die theatrically" fails and traversal returns to its parent node.

Figure 5. An intruder has sneaked up since the last update and the behavior tree traversal reacts accordingly

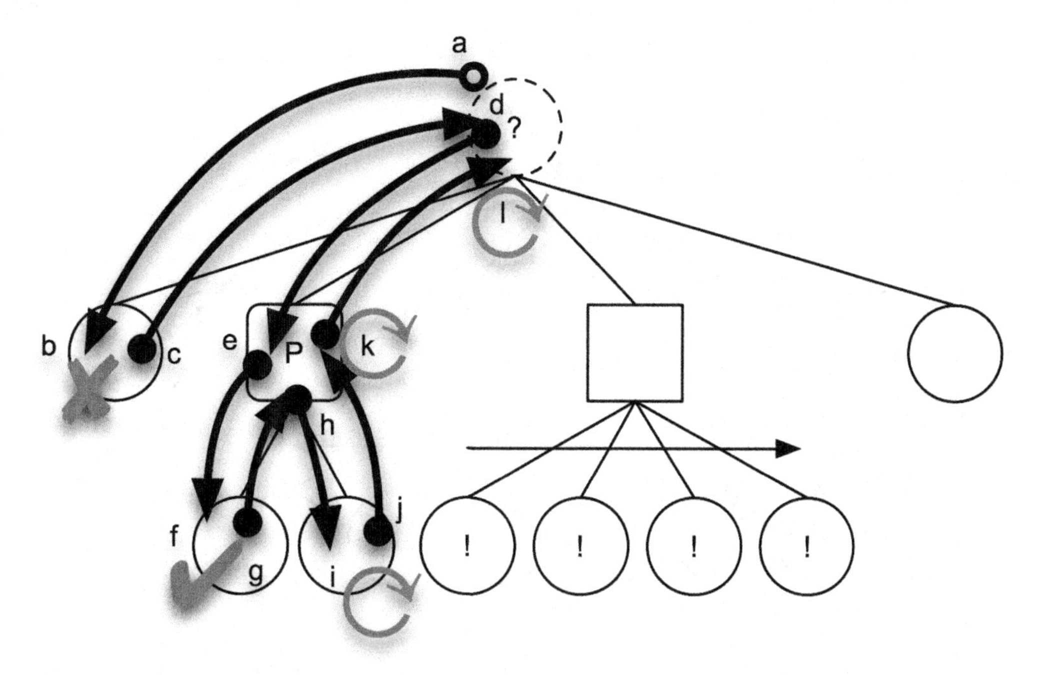

d) Based on its decider semantics the priority selector executes its next child.
e) A parallel decider roots the "Fight" sub-behavior. Traversal is steered into the first child behavior.
f) "Enemy in weapons range?" sub-behavior is traversed (not shown in the example).
g) No intruder or attacker is found so the behavior fails.
h) With a failing child behavior the parallel monitor doesn't execute its other children and fails, too.
i) The root priority decider searches in its next child if it wants to run or if it might even succeed.
j) "Detect intruder" is a sequence. Visited in its default state it goes on to execute its first child.
k) However, as no noise has been heard by the monster, the condition fails and returns to its parent sequence.
l) With a failing child the sequence fails, too.
m) Time for the priority selector to fall back to its lowest-priority sub-behavior: "Idle".
n) The idle behavior is configured to always find an action to run.
o) Consequently it returns with a running state.
p) Finally, the priority decider found a non-failing child and returns with a running state itself.

Herewith the agent update completes. All activated actions have been ticked or have been dispatched to their associated game systems, e.g., the locomotion or animation system, to execute them.

After the last agent update the game world state has changed an the player has sneaked up undetected and assaults the guard monster. During the next behavior tree update the agent receives the chance to react to the ambush, see Figure 5:

a) Again, the behavior tree of the guard is traversed beginning at the root priority selector.

Figure 6. To finish traversal, which selected a higher-priority behavior than during the previous update, the previously running idle sub-behavior is traversed to deactivate it and all of its associated actions

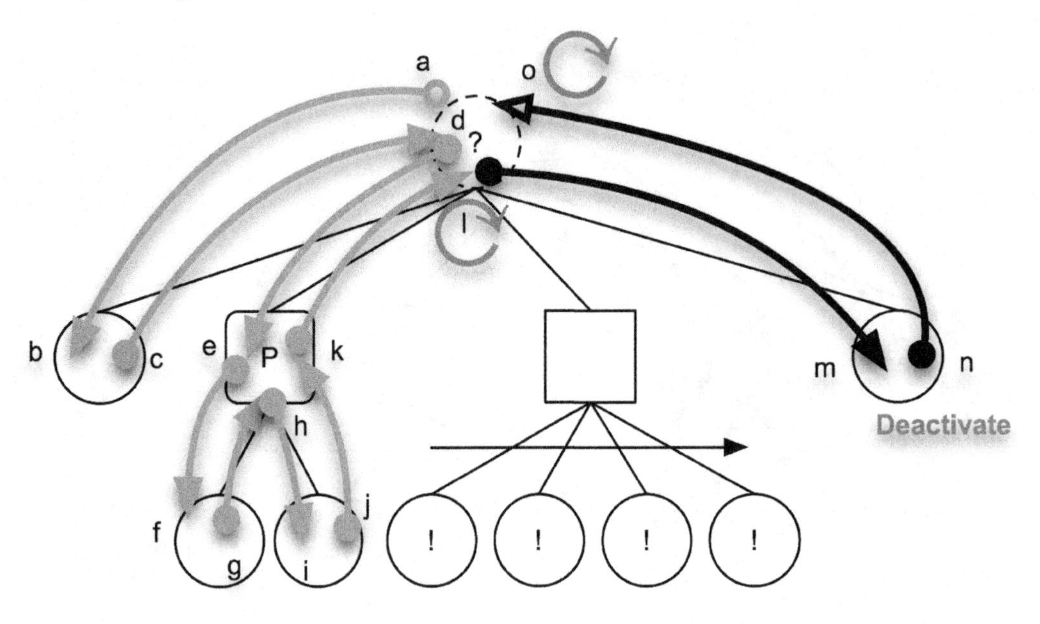

b) At first the "Die theatrically" sub-behavior is probed.
c) However the monster is alive and kicking, so the behavior fails.
d) Upon a fail the next priority selector sub-child is executed: "Fight"
e) "Fight" is a parallel monitor decider which descents into its first child behavior.
f) The "Enemy in weapons range?" behavior is traversed.
g) Strike! The attacking player can be attacked by the guard. A success state is returned.
h) No child failed so the parallel decider executes its next child.
i) "Attack" is traversed, and...
j) the best-first search found a combat action that returns a running execution state. Attack actions and their animations aren't instantaneous but need a few ticks to play out.
k) With a success and a running state returned by its child behaviors the parallel node passes a running state up to its own parent node.
l) Bingo, the priority decider selected a running child and doesn't need to search on for a lower-priority child to run during this update tick.

Although the priority node found a running (or succeeding) child it notices that in the update before a lower-priority behavior was running.

Figure 6 shows how the priority node traverses its abandoned running child behavior from the last update tick to deactivate it:

m) As the "Idle" sub-behavior is still in a running execution state it is visited to be deactivated recursively.
n) Upon deactivation the behavior returns to its parent node.
o) That's it. Traversal finishes with a running state from the "Fight" behavior.

Alternatively, deciders could be used that never switch away from a running sub-behavior. To keep the actor reactive to game world changes no action should run for more than a few ticks at max.

Behind the scenes it is up to the locomotion and animation system to handle the transition from the idle animation to the attack animations to play. Note that low-level action executing systems might model different action states and transitions between them as small finite-state or action state machines (Gregory, 2009) with limited number of states and a strong focus on modeling the transitions between them.

It is interesting to note in the given example that the "Detect intruder" sequence might not finish if the "Move to noise" action returns with a running state, and, if during the next update, the "Enemy in weapons range?" behavior of the "Fight" sub-tree succeeds, so the sequence sub-behaviors are deactivated. It might be sensible to design the "Fight" conditions to need a special "intruder spotted" flag to run. The "Spot intruder" action becomes responsible for setting this flag if it succeeds. Consequently, the sequence will trigger the "Raise alarm" action during the same update traversal, directly after the "Spot intruder" action succeeded. Furthermore, an action should be added to the "Fight" behavior which also sets the intruder flag if the agent is under a direct attack so it reacts immediately instead of going through the "Detect intruder" sequence and therefore delaying fighting back to the next update.

Another option would be to extract "Heard noise?" and "Move to noise" and combine them into an "Investigate possible intruder" sub-behavior with a lower priority than "Detect intruder". This would lead to a stronger separation of concerns.

MEMORY COMPACT IMPLEMENTATION SKETCH

The implementation approach sketched here is based on Knafla's blog series about data-oriented behavior trees (2011a, 2011b, 2011c, 2011d). It strives for the following guiding principles:

- Minimize the memory needed to store the static structure of behavior trees and its traversal state.
- Minimize the memory bandwidth necessary to move behavior tree data from main memory to local caches or local memory.
- Minimize random memory accesses when traversing a behavior tree where possible.
- Enable streaming of chunks of behavior tree data from main memory to local caches/memory.
- Don't use pointers in behavior tree data structures that should be serialized to simplify serialization and the movement of data between non-unified memory areas, e.g., from main memory to the local storage of a Cell Broadband Engine SPU.

Accordingly the implementation relies completely on plain old data structures (PODs) to represent behavior tree nodes, arrays to store them, array indices instead of pointers to express the hierarchy, and functions operating on sub-arrays.

The way actions are handled is introduced first. Afterwards the three core behavior tree representations are described: the shape of a behavior tree that is shared by multiple agents, the actor data structure which stores action and decider states for an agent, and the interpreter state. How all these data structures are used in the behavior update or interpretation process is discussed last.

Immediate and Deferred Actions

Execution of arbitrary actions might crunch through huge amounts of data. To prevent actions from trashing the cache that should stay filled with behavior tree specific data during traversal, two action types with clear data-usage constraints are introduced:

- Immediate actions run immediately during behavior tree traversal. They can only access agent-private data stored in a blackboard which should be resident in local storage/cache.

Figure 7. Stages per main loop cycle necessary to update an actor via its behavior tree

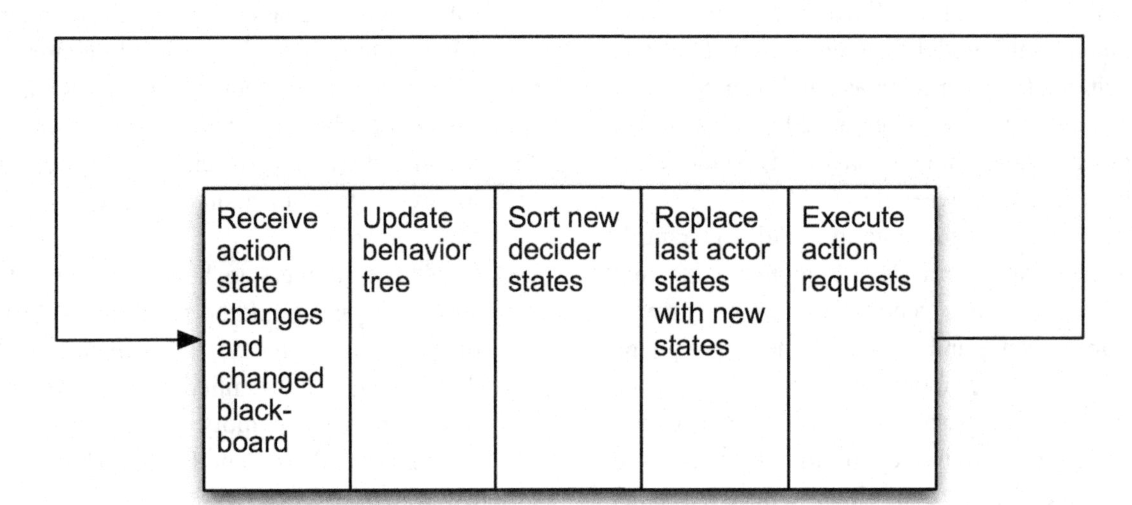

- Deferred actions don't execute their action immediately. When ready to run they emit an action request and return a running execution state. After updating all agents behavior trees and collecting their action requests, other game systems get the chance to batch process the received requests in data-optimal ways. Until the requested actions are completed by their game systems the deferred action nodes maintain and return a running state. Once a game system completes execution of a deferred action, the termination status is sent back to the actor (see below). During the following update tick the associated deferred action node reads and returns its received termination state.

An agent's blackboard is used to aggregate all action-specific game world knowledge in it. Immediate actions get all agent-relevant data from its blackboard (Isla, 2005).

For an effective use of deferred actions the game main loop cycle is divided into multiple stages that are executed one after the other. Figure 7 shows the order of these stages from the perspective of processing behavior trees. The functional aspect of an agent, covered by the deferred action handled by another system, is sometimes called an aspect or component (Cohen, 2010).

When a running deferred action is explicitly deactivated by a higher-up decider, a deactivation or cancellation request is emitted for the associated game system.

Behavior Tree Shape

The static hierarchical structure or shape of a behavior tree can be flattened and effectively turned into an array or stream by enumerating its nodes via preorder traversal (see Figure 8) and storing items associated with the nodes in increasing order (Knafla, 2011a, 2011b, 2011c, 2011d) as shown in Figure 9. Sub-trees rooted at branches are flattened into a contiguous sub-stream with the inner node's item as the first item of the sub-stream. To update/traverse/execute a behavior means to iterate over its behavior (sub-)stream and to interpret the contained shape item(s). Shape item interpretation affects further iteration, e.g., the shape item sub-stream of untraversed sub-behaviors is skipped and iteration proceeds from an item behind the avoided sub-stream.

Figure 8. Guard behavior tree nodes with numbers assigned according to their preorder traversal order. Sub-nodes of generic behaviors are omitted

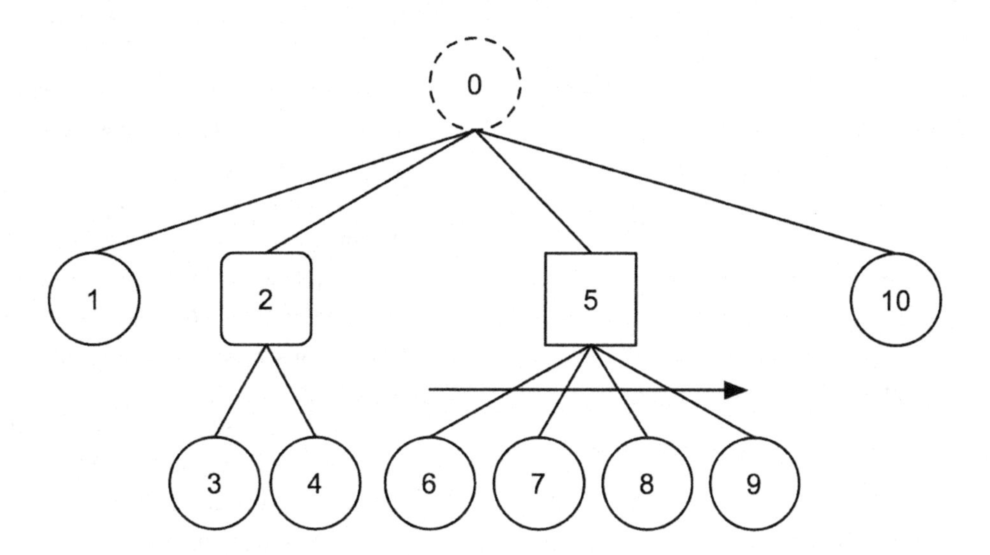

Shape items always have the same size in bits regardless of their type. Each shape item data structure contains an item type identifier and further type related data:

- Immediate action items contain an index into a buffer of user supplied functions to run when traversing them.
- Deferred action items include an action handle which identifies their associated game system and the action it should execute.
- All decider type items contain the index of the shape item marking the end of their sub-stream.

In the C programming language a shape item is represented by a 64 bit-sized struct containing a union of structs for all the different shape item types and a type item field. Fields are ordered from the largest type to the smallest to enable tight packing and alignment according to the largest field type by the compiler:

Figure 9. Guard behavior tree flattened into a shape item stream based on preorder traversal ordering of tree nodes

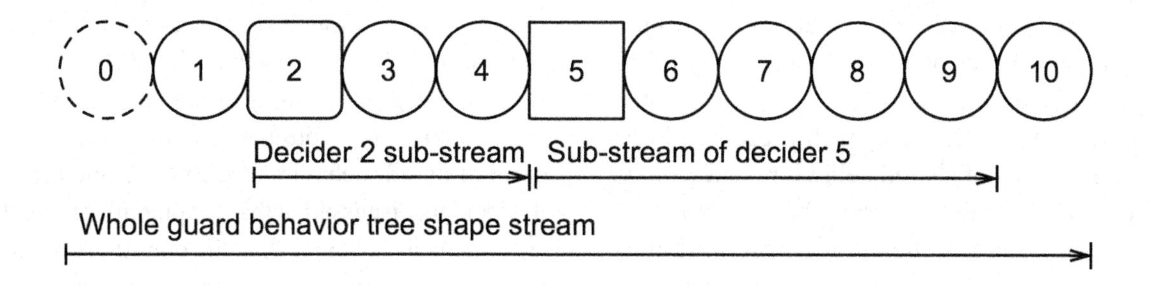

Code 1. Shape item struct

```
struct bt_shape_item_s {
    union {
        struct {
            bt_uint32_t action_id;
            bt_uint16_t padding_dum-
my0;
            bt_uint8_t padding_dum-
my1;
        } immediate_action;
        struct {
            bt_uint32_t action_id;
        } deferred_action;
        struct {
            bt_uint16_t end_item_in-
dex;
        } sequence_decider;
        struct {
            bt_uint16_t end_item_in-
dex;
        } concurrent_decider;
        struct {
            bt_uint16_t end_item_in-
dex;
        } dynamic_priority_decider;
    } data;

    bt_uint8_t type;
};
```

Shape item indices are encoded via 16 bit integral types to constrain the shape item type size to 64 bit while still supporting reasonably large arrays to represent the behavior tree shape.

While the concept of a behavior tree might modularly share some (sub-)behaviors at different places in the directed acyclic graph, the implementation approach presented here creates a copy of shared behaviors where they occur. Consequently, every behavior, every behavior node and the item representing it is unique - the directed acyclic graph is an editing construct which is turned into a tree for the game's runtime.

A behavior tree shape data structure refers to:

- Its shape stream consisting of shape items.
- An array of functions accessed by immediate action items.
- An ordered array of shape item indices. Agents controlled by this behavior tree have equally sized persistent state arrays. A shape's item index and an agent's persistent state stored at the same position of their respective arrays belong together and establish an association of the state with the indexed shape item (see below).
- Sizes of the different arrays.
- Capacity requirements, e.g., how many action and decider states are generated and used during a behavior tree traversal at max.

At runtime a shape item C struct is created which points to the associated arrays (see Code 2).

Capacity requirement for shapes are calculated by traversing a behavior tree bottom up and successively aggregating the maximum number of states that can be active per update. Persistent state slots of decorators are always summed. Most deciders, e.g., sequences or priority selectors, only need to maintain the state for the currently executed sub-behavior, therefore they select the maximal action request slots, decider states, and action states from their children. Each maximum is selected independently of the others, probably from different sub-behaviors. The same is done for the maximum number of explicit persistent state changes, although sequences sum them up because it might happen that all sub-behaviors succeed and emit persistent state changes. Only the parallel decider adds the necessary state count of all of its sub-behaviors together. As the priority deciders discussed in this chapter might select an earlier sub-behavior than the one still in a running state from previous traversal, buffer space

Code 2. Shape item struct

```
struct bt_shape_s {
    bt_shape_item_t *shape_items;
    bt_immediate_action_func_t *action_functions;
    bt_uint16_t *persistent_state_shape_item_indices;

    bt_uint16_t shape_item_count;
    bt_uint16_t action_function_count;
    bt_uint16_t persistent_state_count;

    bt_uint16_t required_decider_guard_capacity;
    bt_uint16_t required_decider_state_capacity;
    bt_uint16_t required_action_state_capacity;
    bt_uint16_t required_action_request_capacity;
    bt_uint16_t required_persistent_state_change_capacity;
};
```

needs to be reserved to request the newly activated actions and for deactivation requests for the previously running ones. Additionally, the maximum depth of the behavior tree is determined as it dictates the size of the stack maintaining the current location in the tree during traversal (see decider guards below). By extracting the capacity requirements from a behavior tree's shape, the associated buffers for states and action requests can be pre-allocated per agent so no time is wasted on memory management during update ticks.

Behavior Tree Actor

Each agent has a unique view of the game world. Its surrounding and situation differs from that of other agents, e.g., every agent is located at a different position and has a unique internal state. Nonetheless, multiple agents are controlled by the same behavior tree shape. Therefore, each agent needs its own set of action and decider states to maintain its individual traversal state. This text calls the data structure which holds an agent's behavior tree states a behavior tree actor, or actor.

Such an actor needs to reference and access the following data:

- Non-default decider states, e.g., to maintain which sub-behavior of a decider previously returned a running execution state.
- Action states for immediate and deferred actions which ran and did not terminate during the last behavior tree update.
- A reference to the agent's blackboard.
- A buffer to hold the persistent states for an actor's decorator deciders.

Here's the C struct for such an actor:

Code 3. Actor struct

```
struct bt_actor_s {
    bt_decider_state_buffer_t decid-
er_states;
    bt_action_state_buffer_t action_
states;
    bt_persistent_t *persistent_
states;
```

```
    void *blackboard;
};
```

Buffers use an array to store their data and carry the array capacity and the count of contained data items. Persistent states aren't stored in a buffer construct because the actor's shape already defines the exact number of persistent states needed so a straightforward array created at runtime suffices. Decider and action states are only stored if they differ from default states or if a deferred action termination state is needed for the next update tick. This minimizes memory requirements.

An actor's action and decider states are associated with the specific shape item they provide the state for by storing the state data and shape item index in the same data structure. They are sorted in increasing shape item index order so traversal along the shape stream and along the state buffers is aligned.

Agent blackboards are assumed to be one blob of data which can be moved as a whole or streamed chunk-wise from main memory to the local memory or cache of a processor core. In its simplest form a blackboard could be a structure with developer defined fields, or it could be a dictionary or key-value store.

Persistent states are used to maintain counter or timer information for an agent's decorator deciders even when they aren't traversed regularly. An actor's persistent states can be set directly when interpreting a decorator if the change shouldn't be cancelable, or persistent state changes are emitted to a stack of the interpreter (see below). These explicitly represented changes are applied to the persistent states when finishing an agent's traversal. Explicitly represented persistent state changes are cancelled by simply not applying them.

The Interpreter

To update an agent, its actor, blackboard, and the behavior shape stream controlling it are handed to an interpreter. When looking at the behavior shape stream as byte code and at the actor state as a kind of variable-value binding construct, the interpreter can be thought of as a virtual machine or a compute kernel processing the shape and state streams to generate the next decider and action states. Furthermore, action launch and cancellation requests are emitted during interpretation and collected and applied by the interpreter.

During interpretation, an interpreter data structure is used to buffer the following data:

- Action states for immediate action shape items whose interpretation returned a running execution state.
- Running deferred action states are preserved so the agent can't forget which are active even if the action processing game system doesn't return a running state for every behavior tree update tick.
- Enqueued deferred action activation and deactivation requests.
- Non-default decider states are collected the moment traversal leaves a decider behavior upwards to its parent.
- Changes to persistent states, which might be cancelled by higher-up decider decisions, are buffered.
- A stack to keep track of the deciders above the currently traversed shape item in the behavior tree shape. These stack's items are called decider guards and the stack is called a decider guard stack.
- The interpreter state data structure also contains indices into the shape stream and the actor and interpreter buffers to maintain which shape item and which state items it accessed last.
- Last but not least, a cancellation token stores the shape item range of actions to deactivate.

In C code, the data structure for the interpreter might look as Code 4.

Code 4. Interpreter struct

```
struct bt_interpreter_s {
    bt_decider_state_stack_t next_decider_states;
    bt_action_state_stack_t next_action_states;
    bt_double_sided_action_request_stack_t next_action_requests;
    bt_persistent_state_change_stack_t persistent_state_changes;

    bt_decider_guard_stack_t decider_guard_stack;
    bt_cancellation_range_t cancellation_range;

    int shape_item_index;
    int actor_decider_state_index;
    int actor_action_state_index;
    int actor_persistent_state_index;

    bt_execution_state_t execution_state;
};
```

Emitted states are pushed onto stacks. The current top index of such a stack can be queried and stored to allow a quick rollback to that position should cancellation require abandonment of the states.

Deferred action activation and cancellation requests are pushed to the different ends of a two-sided stack. This saves the memory two separate stacks would require because an action that emits an activation request doesn't emit a deactivation request during the same update tick, and the other way around. To cancel an activation request emitted during the same execution tick, its side of the stack is simply rolled back so no request will be passed to the associated action handling game system.

Integers are used to store the current indices into the different arrays and buffers instead of smaller integral types for best computational performance.

When interpreting a decider shape item, an associated decider guard is pushed onto the stack. Decider guards are used to react to the execution states returned by "guarded" sub-behaviors according to the decider's traversal semantics. Decider guard data structures carry:

- Their decider's shape item index and type.
- The index for the first shape item of the last reached child's sub-stream. If a behavior contained within the sub-stream triggers a return to the guarded decider and passes up a running execution state, then the last reached child item index marks the entry to return to and tick the previously running behavior during the next update.
- An index marking the end of the decider's sub-stream.
- Indices to roll the interpreter's buffers for decider states, persistent state changes, and deferred action launch requests back to their configuration before the decider has been traversed. A buffer rollback effectively cancels the traversal of the decider sub-stream by getting rid of the generated traversal states.
- The execution state of a parallel decider depends upon the execution state of all of its children. A single failing sub-behavior trumps all running or succeeding children. A single running child trumps all succeed-

Code 5. Decider guard struct

```
struct bt_decider_guard_s {
    bt_uint16_t shape_item_index;
    bt_uint16_t reached_child_item_index;
    bt_uint16_t end_item_index;
    bt_uint16_t next_decider_state_rollback_index;
    bt_uint16_t action_launch_request_rollback_index;
    bt_uint16_t persistent_state_change_rollback_index;
    bt_uint8_t type;
    bt_uint8_t aggregated_state;
};
```

ing sub-behaviors. The aggregated decider state is maintained in another data structure field. Its default state is the ready execution state.

The decider guard C data structure in Code 5 - for optimal packing by the compiler the fields are ordered in decreasing order of their type's size in bits:

After an actor update, the new states generated by the interpretation process are sorted for their shape item indices and then replace the old states of the actor from the previous update. An interpreter's buffers must offer enough capacity to hold all possible next actor states as defined by the restrictions of the behavior shapes to work on.

Interpretation Process

All the described data structures are involved in the decision making and action scheduling process of a behavior tree controlled agent. Their interplay during an update tick is shown in Figure 10 and unfolds in detail as follows.

During game main loop cycles in which an agent is updated via its behavior tree, deferred action state changes are integrated into their actor's action state buffer (Figure 7). Afterwards, an interpreter function iterates over the behavior shape stream and actor state. For each traversed shape item the interpretation process goes through the following steps:

1. Determine item (action or decider) state.
2. Interpret the shape item based on its item state.
3. Decider guard traversal handling:
    - 3.1. If the decider guard stack is empty go to step 4.
    - 3.2. Interpret the top decider guard on the stack to determine the next shape item index to visit.
    - 3.3. If the next item index is inside the top decider guard sub-stream goto step 4, otherwise pop the decider guard stack and loop to 3.1.
4. Handle explicit action deactivation if necessary.
5. If the next shape item index is inside the shape item stream, then select the indexed shape item and loop to step 1, otherwise go to step 6.
6. Finish the actor's update.

After all actors are updated the collected deferred action requests are sorted and batched to their respective game systems. Deferred action deactivation requests are submitted before action launch requests so the game systems can first free

Figure 10. Read and write data accesses of an interpreter while updating a behavior tree controlled agent

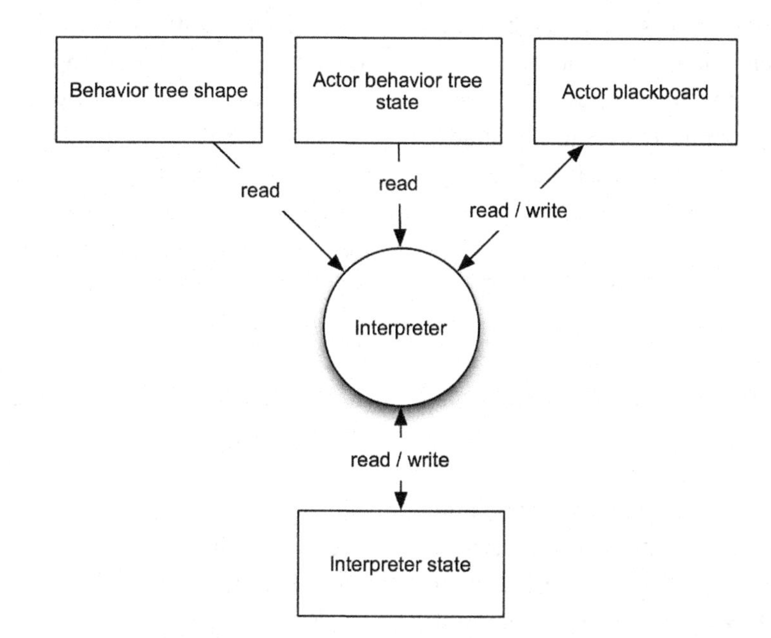

game world resources in use by cancelled actions which therefore become available for the newly activated actions.

Next, some of the interpretation process steps are inspected and explained in greater detail.

1. Determine Item State

Based on the item's type the interpreter searches forward over the applicable state buffers of the actor to find the state associated with the shape item. As state buffer (streams) are ordered according to the states shape item indices, arriving at a state with a greater index than searched shape item index indicates that no state for the current shape item is contained and consequently a default state is created, e.g., a ready state for non-running immediate and deferred actions. The interpreter picks up from the current position in the state stream for the next search. States with lower indices than the one searched for are skipped which mirrors how their associated behaviors are not traversed by higher level deciders.

2. Interpret Shape Item

An item type specific function is called to interpret the shape item based on the item state from step one. Immediate actions also get access to the actor's blackboard. Deferred actions emit an action activation request if they aren't running yet. Interpretation of a decider shape item pushes an associated decider guard onto the interpreter's decider guard stack.

If an immediate action returns a running execution state, then the state is immediately pushed onto the interpreter's next action state buffer. If the last state of a deferred action is a running execution state, then it is pushed onto the next action state buffer even before interpreting the deferred action shape item - it will simply return the buffered state.

When an emitted deferred action request isn't subsequently cancelled but transmitted to its associated game system, the behavior tree interpreter can assume that the deferred action is active and puts a running action state onto the interpreter's next action state buffer in step 6.

After handling an action shape item the interpreter's shape item index is increased by one. Deciders set the shape item index according to their state - according to which child of them was running during the previous traversal. The index value is only a suggestion which is refined by decider guards within step 3.

3. Decider Guard Traversal Handling

After interpreting a shape item, the top decider guard from the interpreter's decider guard stack is used to determine which shape item to interpret next based on the decider guard's traversal semantics, the suggested shape item index, and the last returned execution state. This top decider guard is like the decider node in the non-flattened behavior tree which controls traversal. The first decider guard pushed onto the stack represents the root decider of the behavior tree. When it is finally popped and the guard stack becomes empty then traversal and the interpretation process for the current agent end.

Tables 1-3 show how the decider guards for sequences, priority selectors, and parallel deciders react to the different combinations of shape item index suggestions and sub-behavior execution states. A ready execution state should only be returned by interpreting a decider shape item as its true execution state is not known before its decider guard is popped. If it is not explicitly mentioned, then the execution state of the last shape item is taken over as the decider guard's execution state to pass to its own parent decider.

When the suggested next shape item index stays inside the current top decide guard's sub-stream, then the index references the first item of the next sub-behavior's sub-stream. An index outside of the guarded sub-stream identifies a sub-behavior belonging to one of the decider guard's parents in the behavior tree hierarchy.

By setting the suggested shape item index to the first item outside the guarded decider sub-stream, the interpreter will pop the decider guard stack automatically. Traversal returns from the decider associated with the popped guard to the parent's decider guard - the new top guard on the stack.

Both, (dynamic) priority selectors and parallel deciders are always traversed starting from their first child, while sequences start from their first child or from the sub-behavior that previously returned a running state. Therefore a sequence state is only emitted for the next update if its decider guard detects a running child.

A sequence only proceeds from sub-behavior to sub-behavior if they complete execution successfully. The children handle their states and the need to cancel actions independently from the sequence. Even if a sub-behavior fails, the sequence relies on it to set up the interpreter's cancellation token by itself.

Priority deciders run over their child behaviors until one does not fail and instead returns a success or running execution state. If all child behaviors fail, the priority decider fails, too. As a lower-priority child might have been running during the previous update, the priority decider guard always sets the range of the interpreter's cancellation token to include all of the yet untraversed decider sub-stream.

When the guard for a parallel decider receives a fail execution state from a child, some earlier sub-behaviors might already have been traversed and have emitted deferred action activation requests, next decider states, already running immediate and deferred action states, and persistent state changes to the interpreter's buffers. Therefore, the whole shape item index range of the guarded sub-stream is added to the cancellation token for later cancellation. As decider states, deferred action requests, and persistent state changes don't require explicit deactivation, their interpreter buffers are rolled back to the state they had on entering the guards decider item. This guarantees that the preallocated buffers have enough free capacity for states subsequently emitted during the ongoing traversal process.

Table 1. Sequence decider guard interpretation

Last shape item execution state	Shape item index suggestion inside of decider guard sub-stream	Shape item index suggestion outside of decider guard sub-stream
Ready	Set reached child index to shape item index suggestion.	Nothing to do.
Running	1. Emit decider state for last reached child with running state. 2. Set shape item index suggestion to first item outside the decider sub-stream.	Emit decider state for last reached child with running state.
Success	Set reached child index to shape item index suggestion.	Nothing to do.
Fail	Set shape item index suggestion to first item outside the decider sub-stream.	Nothing to do.

Table 2. Priority selector decider guard interpretation

Last shape item execution state	Shape item index suggestion inside of decider guard sub-stream	Shape item index suggestion outside of decider guard sub-stream
Ready	Nothing to do.	Nothing to do.
Running	1. Adapt interpreter cancellation token to include the sub-stream index range starting from the suggested shape item index to the end of the decider sub-stream. 2. Set shape item index suggestion to first item outside the decider sub-stream.	Nothing to do.
Success	1. Adapt interpreter cancellation token to include the sub-stream index range starting from the suggested shape item index to the end of the decider sub-stream. 2. Set shape item index suggestion to first item outside the decider sub-stream.	Nothing to do.
Fail	Nothing to do.	Nothing to do.

Table 3. Parallel decider guard interpretation

Last shape item execution state	Shape item index suggestion inside of decider guard sub-stream	Shape item index suggestion outside of decider guard sub-stream
Ready	Nothing to do.	Set execution state to decider guard aggregated execution state.
Running	Set decider guard aggregated execution state to running.	Do nothing.
Success	Set the decider guard aggregated execution state to success if it isn't set to running already.	1. Set the decider guard aggregated execution state to success if it isn't set to running already. 2. Set execution state to decider guard aggregated execution state.
Fail	1. Adapt interpreter cancellation token to include the sub-stream index range of the complete guarded decider sub-stream. 2. Roll the following interpreter buffers back to the positions stored in the decider guard: next decider states, next deferred action requests, persistent state changes. 3. Set shape item index suggestion to first item outside the decider sub-stream.	1. Adapt interpreter cancellation token to include the sub-stream index range of the complete guarded decider sub-stream. 2. Roll the following interpreter buffers back to the positions stored in the decider guard: next decider states, next deferred action requests, persistent state changes.

4. Handle Explicit Action Deactivation If Necessary

Decider guard handling ends if the stack is empty or if the next shape item to interpret is inside the guarded decider sub-stream, which means that traversal doesn't go up but down into a decider's sub-behavior. If the interpreter's cancellation token contains a non-empty shape stream index range, then the running immediate and deferred actions covered by that range need to be explicitly deactivated before new states are buffered.

Explicit deactivation, alias cancellation, happens in two phases:

1. Iterate from the last state on the interpreter's next action state buffer backward until its beginning is reached or until accessing the first state outside of the cancellation range. If the shape item associated with an iterated state is an immediate action, then call a cancellation function so it gets a chance to cleanup blackboard manipulations it started. If the shape item type identifies a deferred action, then emit a deactivation request. At the end, roll back the next action state buffer to the first state outside the cancellation range.
2. Iterate forward over the actor's last action state stream not yet accessed by the interpreter until its end or until hitting a state with an associated shape item index outside of the cancellation range. Explicitly cancel running immediate actions and enqueue cancellation requests for running deferred actions inside the cancellation range.

The first cancellation phase clears off states generated by actions interpreted by parallel deciders during the current update tick, while the second phase handles deactivation of unfinished actions from the previous update.

It might be problematic that previously running actions of a priority selector are deactivated only after higher-priority ones succeeded because the higher-priority immediate actions access an actor blackboard that might contain side effects from currently active actions. Therefore, immediate actions need to be designed to cope with potentially uncleaned blackboard data, or, a more drastic solution, immediate actions might be prohibited from entering a running execution state that spans multiple update ticks.

5. Finish the Actor's Update

Finally, to finish an actor's update, the interpretation process:

1. Sorts the interpreter's decider states in increasing order of their associated shape item indices, and replaces the actor's decider states with them.
2. Sorts and applies the collected persistent state changes to the actor's persistent states.
3. Generates running action states for the collected action launch requests and adds them to the interpreter's action states.
4. Sorts the interpreter's action states in increasing order of their associated shape item indices, and replaces the actor's action states with them.
5. Merges the collected action launch and cancellation requests with the requests from other actors.

FUTURE RESEARCH DIRECTIONS

Quick iteration times are very important for rapid development and tuning of computational behavior for a fun gameplay experience. The runtime-centric implementation of this chapter should be extended with methods to allow interactive behavior tree manipulations while the game runs, e.g., via lookup table support. Furthermore, mechanisms to trace update traversal to help behavior selection debugging is missing. Both features, life editing and runtime monitoring can benefit from the way behavior streams and states are stored in flat arrays, as it eases copying, data streaming, and simplifies taking a snapshot of an agent's traversal state.

Many different behavior tree concepts exist. Based on a game's needs, a simple concept can lead to an equally simple and efficient implementation, e.g., many behavior tree approaches don't have parallel deciders and therefore don't need to maintain multiple running behaviors.

Straightforward object-oriented behavior tree implementations might be fast enough for a game's performance requirements, and, combined with a scheduler, enable event-driven traversal.

A first step towards an event-driven update of the presented design could be to let immediate actions only quickly check flags in their agent's blackboard. Expensive condition check computations for determining the flags states are handled externally and are only triggered due to monitored game state mutations.

Another approach to combine data-oriented behavior trees with event-driven traversal might be derived from the observation that a bottom-up traversal from running and terminated action states can recreate the associated decider states and therefore the full traversal state of a behavior tree. Consequently, an actor doesn't need to store the last decider states, and, by only triggering decider state recreation and the subsequent behavior tree traversal once action states change, even a data-oriented behavior tree might be driven by events.

CONCLUSION

Behavior trees are a flexible tool for the development of computational behavior for games. The behavior tree concept introduced in this chapter forms a behavior language which imposes a structure for how to think about behaviors and therefore guides purposeful, hierarchical behavior modeling.

It is good practice to create many small trees and reuse them to build bigger behavior trees. Start with a simple behavior model and incrementally flesh it out based on the behavior observed in-game.

An actor might be controlled by multiple behavior trees on different levels, e.g., for strategic, tactical, and purely reactive decision making. A behavior tree instance might even be used to coordinate the cooperation of a group of actors.

Behavior trees can be deployed to not only control game characters but also to orchestrate cut scenes or other areas of the game which benefit from conditions and sequences (Van Dongen, 2010, 2011).

Behavior trees can also be used when game designers ask for direct script control over a game scene and want to suppress autonomous and reactive computational behavior. One way to achieve script control is to modify the game state accessible to an actor's behavior tree conditions, e.g., by overriding sensor data in the actor's blackboard, so it behaves in a specific way. An alternative method is to add high-priority sub-behaviors to the tree which hand over control to an embedded script.

Always use the most applicable tool to solve a problem. If the set of needed states or behaviors is small and won't change much, or if the transition between these states is the key to solve the problem, then a finite-state machine might be preferable to a behavior tree. If the number and organization of behaviors is unpredictably large and/or will change a lot during development, then a behavior tree might be the right tool for the job.

REFERENCES

Albrecht, T. (2009a). The latency elephant. *Seven Degrees of Freedom*. Retrieved from http://seven-degrees-of-freedom.blogspot.com/2009/10/latency-elephant.html

Albrecht, T. (2009b). Pitfalls of Object Oriented Programming. *Seven Degrees of Freedom*. Retrieved from http://seven- degrees-of-freedom.blogspot.com/2009/12/pitfalls-of-object-oriented-programming.html

Borkar, S., & Chien, A. A. (2011). The future of microprocessors. *Communications of the ACM*, *54*(5), 67–77. doi:10.1145/1941487.1941507

Champandard, A. J. (2007). Behavior trees for next-gen game ai. *Game Developers Conference 2007*. Retrieved from http://aigamedev.com/insiders/presentation/behavior-trees/

Champandard, A. J. (2008). Getting started with decision making and control systems. In Rabin, S. (Ed.), *AI Game Programming Wisdom 4* (pp. 257–264). Boston, MA: Course Technology.

Champandard, A. J. (2011). This year in game ai: Analysis, trends from 2010 and predictions for 2011. *AiGameDev.com*. Retrieved from http://aigamedev.com/open/editorial/2010-retrospective/

Cohen, T. (2010). A Dynamic Component Architecture for High Performance Gameplay. *Game Developers Conference Canada*. Retrieved from http://www.insomniacgames.com/a-dynamic-component-architecture-for-high-performance-gameplay/

Fu, D., & Houlette, R. (2004). The ultimate guide to fsms in games. In Rabin, S. (Ed.), *AI Game Programming Wisdom 2* (pp. 283–302). Hingham, MA: Charles River Media.

Game, A. I. Conference (2010). *Game/AI Conference*. Retrieved from http://gameaiconf.com/

Gregory, J. (2009). *Game Engine Architecture*. Wellelsley, MA: AK Peters.

Hecker, C. (2009). My liner notes for spore/spore behavior tree docs. *Chris Hecker's Website*. Retrieved July from http://www.chrishecker.com/My_Liner_Notes_for_Spore/Spore_Behavior_Tree_Docs

Hennessy, J. L., & Patterson, D. A. (2007). *Computer architecture: A Quantitative Approach* (4th ed.). San Francisco, CA: Morgan Kaufmann.

Isla, D. (2005). Handling complexity in the halo 2 ai. *Proceedings of the Game Developers Conference 2005*. Retrieved from http://www.gamasutra.com/gdc2005/features/20050311/isla_01.shtml

Knafla, B. (2011a). Introduction to behavior trees. *Altdevblogaday Series About Data-Oriented Behavior Trees*. Retrieved from http://altdevblogaday.com/2011/02/24/introduction-to-behavior-trees/

Knafla, B. (2011b). Shocker: Naive object-oriented behavior tree isn't data-oriented. *AltDevBlogADay Series About Data-Oriented Behavior Trees*. Retrieved from http://altdevblogaday.com/2011/03/10/shocker-naive-object-oriented-behavior-tree-isnt-data-oriented/

Knafla, B. (2011c). Data-oriented streams spring behavior trees. *AltDevBlogADay Series About Data-Oriented Behavior Trees*. Retrieved from http://altdevblogaday.com/2011/04/24/data-oriented-streams-spring-behavior-trees/

Knafla, B. (2011d). Data-Oriented Behavior Tree Overview. *AltDevBlogADay Series About Data-Oriented Behavior Trees*. Retrieved from http://altdevblogaday.com/2011/07/09/data-oriented-behavior-tree-overview/

Llopis, N. (2009). Data-oriented design (Or why you might be shooting yourself in the foot with oop). *Games From Within*. Retrieved from http://gamesfromwithin.com/data-oriented-design

Llopis, N. (2011a). Data-oriented design now and in the future. *Games From Within*. Retrieved from http://gamesfromwithin.com/data-oriented-design-now-and-in-the-future

Llopis, N. (2011b). High-performance programming with data-oriented design. In Lengyel, E. (Ed.), *Game Engine Gems 2* (pp. 251-261). Natick, MA: A K Peters.

Millington, I. (2006). *Artificial Intelligence for Games*. San Francisco, CA: Morgan Kaufmann.

Patterson, D. A. (2004). Latency lags bandwidth. *Communications of the ACM, 47*(10), 71–75. doi:10.1145/1022594.1022596

Pillosu, R., Jack, M., Kirst, K., Mohr, M., & Zielinski, M. (2009). Coordinating agents with behavior trees. *Paris Game AI Conference 2009*. Retrieved from http://staff.science.uva.nl/~aldersho/GameProgramming/Papers/Coordinating_Agents_with_Behaviour_Trees.pdf

Valdes, R. (2004). The artificial intelligence of halo 2. *HowStuffWorks.com*. Retrieved from http://electronics.howstuffworks.com/halo2-ai.htm

Van Dongen, J. (2010). AI in swords and soldiers (part 1). *Joost's Dev Blog*. Retrieved from http://joostdevblog.blogspot.com/2010/12/ai-in-swords-soldiers-part-1.html

Van Dongen, J. (2011). AI in swords and soldiers (part 2). *Joost's Dev Blog*. Retrieved from http://joostdevblog.blogspot.com/2011/01/ai-in-swords-soldiers-part-2.html

ADDITIONAL READING

Dyckhoff, M. (2008). Decision making and knowledge representation in halo 3. *Bungie Publications*. Retrieved from http://www.bungie.net/images/Inside/publications/presentations/publications-des/engineering/nips07.pdf

Jacobsson, J. (2009-ongoing). Calltree open source behavior tree editor, compiler, and runtime project. *Calltree*. Retrieved from https://github.com/jjacobsson/calltree

Johansen, E. (2008-ongoing). Behave behavior tree editor, compiler, and runtime for Unity project. *Behave*. Retrieved July 27, 2011, from http://angryant.com/behave

Mateas, M., & Stern, A. (2002). A behavior language for story-based believable agents. *AAAI Spring Symposium on Artificial Intelligence and Interactive Digital Entertainment*, 68-75. Retrieved from http://users.soe.ucsc.edu/~michaelm/publications/mateas-aaai-symp-aiide-2002.pdf

KEY TERMS AND DEFINITIONS

Action: Basic behavior at the leafs of a behavior tree and interface between the behavior tree and the game world state.

Actor or Agent: Game entity whose decision making and action scheduling process is controlled by a behavior tree, often a non-player character (NPC).

Behavior: Hierarchical composition of action and decider nodes to fulfill a certain purpose.

Behavior Tree: Hierarchical composition of behaviors to control an agent's decision making and action scheduling.

Blackboard: Data structure holding agent specific game world knowledge. Accessed by actions and conditions.

Condition: Action which succeeds if it detects a certain game world state configuration and otherwise fails.

Decider: Inner behavior tree node which controls update traversal by deciding which sub-behaviors to run and when to pass traversal back to its own parent decider.

Decorator: Behavior with a single child. Controls access to its child via counters or timers or just transforms its sub-behavior's execution state.

Deferred Action: Behavior tree node which emits an action request to an associated game system when activated during behavior tree update traversal.

Execution Status: Scheduling and/or termination status of a behavior.

Immediate Action: Action which only accesses an agent's blackboard and is executed immediately when visited during a behavior tree traversal.

Parallel/Concurrent Decider: Concurrently runs all child behaviors. Succeeds if a pre-defined number of sub-behaviors succeed, fails if a pre-defined number of sub-behaviors fail.

Priority Selector: Runs from its highest- to lowest-priority sub-behavior until one succeeds. Fails if all children fail.

Probability Selector: Randomly selects one sub-behavior to execute. Succeeds and fails with the chosen child behavior execution state.

Semaphore Decorator: Limits concurrent access to game world resources accessed by decorated actions.

Sequence Decider: Sequentially executes its child behaviors as long as they return success execution states. Succeeds if all of its sub-behaviors succeed. Fails with the first failing child.

Shape: Static structure or shape of the behavior tree.

Chapter 4
Nonmanipulable Collective Decision-Making for Games

Rob LeGrand
Angelo State University, USA

Timothy Roden
Angelo State University, SUSA

Ron K. Cytron
Washington University in St. Louis, USA

ABSTRACT

This chapter explores a new approach that may be used in game development to help human players and/or non-player characters make collective decisions. The chapter describes how previous work can be applied to allow game players to form a consensus from a simple range of possible outcomes in such a way that no player can manipulate it at the expense of the other players. Then, the text extends that result and shows how nonmanipulable consensus can be found in higher-dimensional outcome spaces. The results may be useful when developing artificial intelligence for non-player characters or constructing frameworks to aid cooperation among human players.

INTRODUCTION

Teamwork is important in many games. Whether they are human players or non-player characters (NPCs) or both, game entities must often work together to achieve goals. However, those goals do not always coincide perfectly, and, even when they do, players will not always agree on the best next course of action to take.

DOI: 10.4018/978-1-4666-1634-9.ch004

Much research (see especially Rabin, 2002, sec. 7) has explored effective decision-making for individual game agents, even in a multiagent context. By contrast, in this work, we assume that all agents have already individually decided which of the available outcomes (which are usually actions) they prefer over others, and we assume the agents desire to use those preferences to reach consensus for the group.

For an example game situation, imagine a team of wargame players with a common goal: band together to attack the western coast of a continent held by a common enemy. They could attack the coast's northernmost point, the southernmost point or anywhere in between, and each player has a different favorite attack point. If the players can be trusted to express sincere preferences, their preferred points could simply be averaged to give the consensus point. Averaging, however, can allow some players to gain a better outcome from their point of view by exaggerating their preferences, whereas other aggregation mechanisms may never reward such insincerity.

When a group of players aims to benefit all members of the group by coordinating their actions, a method of combining their preferences into a single outcome is useful, but the usefulness may disappear if individual players can manipulate the outcome by expressing insincere preferences. Here we present a set of nonmanipulable collective-decision-making methods that apply to a wide range of game situations.

In the sections below we review previous work that informs ours, look at several game situations that motivate our approach and present the ideas that provide an innovative solution.

BACKGROUND IDEAS

The core ideas of this chapter, while new, are based in extant work from fields such as computer science, mathematics, political science and economics.

Mechanism Design

Returning to the above wargame example, if a team of players is trying to agree on a coastal attack point, their preferred points could simply be averaged to give the consensus point, but doing so sometimes rewards insincerity on the parts of the players. The field of *mechanism design* (Nisan, 2007) has evolved to find decision-making mechanisms that satisfy particular properties, often some kind of immunity or resistance to strategic manipulation.

Strategic manipulation is a common problem in collective decision-making. It is well known that voters can gain advantage under most voting systems by voting insincerely (Gibbard, 1973; Satterthwaite, 1975). Examples include voting for an alternative that is not a voter's first choice and ranking alternatives untruthfully. Traditionally, this problem is discussed in political science, but more recently the techniques of computer science have been applied with success (Bartholdi, Tovey & Trick 1989, Conitzer & Sandholm, 2003; Elkind & Lipmaa, 2005; Procaccia & Rosenschein, 2006). In this chapter we explore a particular approach to creating manipulation-resistant mechanisms.

The Declared-Strategy Voting Framework

Declared-Strategy Voting (DSV) is a computationally-based response to manipulable voting systems (Cranor & Cytron, 1996; Cranor, 1996). Under DSV, each voter submits preferences over the available outcomes. The DSV system then uses those preferences to vote optimally (and, perhaps, insincerely) on each voter's behalf in a simulated election using some underlying voting method. It continues to cast optimal ballots on behalf of each voter until an equilibrium is found or some other stopping criterion is reached. The outcome at equilibrium is then taken as the DSV outcome, or the results of the voting rounds could be used by a policy-maker to reach a justifiable decision.

The hope is that, since the DSV system is attempting to vote strategically on each voter's behalf, no voter will have a reason to mislead the system by expressing insincere preferences—in fact, an attempt to mislead the system may easily backfire. DSV has been shown to be effective in transforming some manipulable voting systems into manipulation-immune ones with the same available outcomes.

A Previous Result: AAR DSV

In past work, we have successfully applied DSV to an average-approval-rating (AAR) system, which takes voters' ratings between 0% and 100% as input and outputs their average (LeGrand, 2008; LeGrand & Cytron, 2008). Such systems are widely used for rating movies, music, and buyer/seller reliability on the Internet on websites such as Amazon, Metacritic and Rotten Tomatoes; similar systems are used for many diverse applications. We assume only that each voter has an ideal outcome between 0% and 100% and prefers the outcome to be as close to that ideal as possible. So, for example, a voter whose ideal outcome is 20% cannot prefer 40% to 30%. We found that the resulting system had some surprising and attractive properties: Given reasonable assumptions, optimal voting strategy is unique and can be fully characterized, and, given any input ratings, the outcome at equilibrium is unique. We discuss these results in more detail below.

Most importantly, we were able to prove that no AAR DSV voter can achieve an outcome closer to ideal than by voting sincerely. As an example, imagine a group of three game players who are deciding how much of a collective resource to use towards an immediate goal, such as how much of their magic store to employ against the next major enemy, or how much of a healing pack to use before a fight. Say that the three players have sincere preferences [25%, 40%, 70%] and express them sincerely to an AAR DSV system. After DSV iteratively applies optimal strategies on behalf of each player, the equilibrium becomes [0%, 20%, 100%], giving the outcome 40%. Now, if the first player, somewhat dissatisfied with this outcome, had expressed the preference 0% instead of 20% in an effort to pull the outcome closer to 20%, the DSV equilibrium and outcome would be the same as above; in essence, the DSV system is already manipulating on behalf of each player, so misleading the DSV system is unnecessary and fruitless. For this AAR voting system, the DSV framework perfectly internalizes voting strategy—players can never gain advantage by exaggerating their position.

If this meta-voting system is to be used to find consensus in real games, it will be important to be able to calculate the outcome quickly. Fortunately, it turns out that there is an efficient algorithm to calculate the AAR DSV outcome, one that is not significantly slower than simply sorting the numerical inputs of the players.

AAR DSV offers a way for agents to find consensus within a numerical range (a line segment) without the possibility of successful manipulation by selfish agents. Many game situations, however, require finding consensus within other outcome spaces. We will see that the DSV framework is flexible enough to work well with other useful outcome spaces.

A NEED: FINDING COLLECTIVE CONSENSUS IN GAMES

Some games are purely competitive and have no room for cooperation and thus no need to find consensus among a group of players. But more common are games that feature multiple agents who are, at least under some circumstances, motivated to cooperate to the advantage of all, and the rise of online and social gaming is making these situations more common. Agents in games include both human players and computer-controlled NPCs, and opportunities for cooperation can emerge among human players, among NPCs, and even between humans and NPCs. Frequently, to cooperate most efficiently, these groups of agents need to come to some sort of explicit consensus. We will give three example situations from real games that illustrate benefits of our approach, each with a different outcome space.

Finding Consensus in One Dimension

Often game agents need to come to a numerical consensus within the space of a line segment, the endpoints of which may be thought of as 0% and 100%. Many times, agents may need to agree on how much, from 0% to 100%, of a common resource to use at a given time. As another, more concrete, example, imagine a game like Sid Meier's Colonization, in which a single colony may have many cities, each individually controlled by either a human player or an NPC. One of the most important facets of the game is deciding when your colony should revolt against its parent country, and a colony's players may disagree on the best time to revolt. Triggering a successful revolt is based on the average level of discontent for the entire colony, so it is sometimes desirable to build up a measure of discontent in individual cities. At any given time, then, each player may have a different ideal target for the colony-wide average level of discontent, and the player controlling a given city has exclusive control over the level of discontent associated with that player's city.

As an example, consider a colony with three cities, each controlled by a different player. The three players prefer that the colony-wide average level of discontent be [25%, 40%, 70%]. The players could manipulate their own cities' discontent levels to move the colony-wide average towards their ideals, struggling against one another, but doing so would require sacrifices to be made in other ways. Alternatively, they could "vote" their ideal colony-wide averages and agree to abide by the results, thus requiring smaller changes to each city's discontent level. If the system that resolves input ratings into an outcome is chosen carefully, it will never reward insincerity and the players will be able to find consensus with confidence. The AAR DSV system, introduced above and explored in more detail below, fits the bill: It takes a vector of ratings between 0% and 100% as input, outputs a consensus rating, and is impossible for a voter to manipulate through insincere voting. If the input vector were [25%, 40%, 70%] as in the example above, the DSV equilibrium would be [0%, 20%, 100%], giving the outcome 40%, which is the ideal outcome for one of the players.

Finding Consensus within a Hypercube

In other game situations, agents need to find consensus inside a hypercube of some dimensionality; each dimension can be seen as ranging between 0% and 100%. One motivating example comes from Age of Empires, a real-time strategy game that can accommodate many human players and many NPCs. Several human players working together may need to decide collectively the location for their newest base. Every point in the game's square map has two coordinates, one ranging between the west edge, 0%, and the east edge, 100%, and one between the south edge, 0%, and the north edge, 100%. Each player has a most-preferred point for the base and would like to see the base built as close to that ideal point as possible. Players may be assumed to prefer points closer by Euclidean distance to their ideal point to points farther from it, so a player with the ideal point (60%, 10%) can be assumed to prefer that the base be built at (65%, 20%) than at (50%, 30%). When a nonmanipulable system is used to aggregate input points into an outcome point, it will not be possible for the player with ideal (60%, 10%) to change the outcome from (50%, 30%) to (65%, 20%) by changing their vote point from the sincere (60% 10%) to any other point.

Finding Consensus within a Simplex

Another common kind of game decision to make collectively involves the allocation of a finite resource among several uses. Essentially, the agents must agree on a point within a simplex of some dimensionality, where each dimension ranges between 0% and 100% and the values chosen for each dimension sum to exactly 100%. For example, in a space war game such as Star Trek Online, the officers on a starship must often decide how to deploy its limited energy resources; shields, weapons and maneuvering all require energy. The ship's officers may be controlled by separate AI programs, which may sincerely disagree on the policy best for the ship. At a given time, one AI officer may estimate (30%, 60%, 10%) as the best policy, giving 30% of the available energy to shield systems, 60% to weapons and 10% to the helm, while another may prefer (50%, 50%, 0%). Each officer may be assumed to prefer points nearer to ideal in each dimension to points farther in each dimension; for example, we may assume that an officer that most prefers (30%, 60%, 10%) will prefer the outcome (40%, 40%, 20%) to (50%, 30%, 20%). Even when all relevant agents are controlled by a computer, it may be that a nonmanipulable consensus-finding system is preferred; for the sake of realism, AI agents may be programmed to take advantage of manipulable systems whenever possible (unless they are part of a hive collective such as the Borg).

The above examples are inspired by currently available games, but we believe social cooperative games are potentially the best-suited games for our method. Most of these games are still in development as designers and game programmers figure out how to implement effective cooperative strategies among players that play non-concurrently; our techniques do not require players to indicate their preferences simultaneously, and so they may prove especially useful for this kind of game.

USING DSV TO FIND STABLE CONSENSUS IN ONE DIMENSION

First we consider the simplest outcome space that players may need, which is a line segment; we will assume that the available outcomes range between 0 (0%) and 1 (100%). We further assume:

- There are n players that want to find a consensus in the inclusive interval from 0 to 1.
- Each player i submits a value v_i in the inclusive interval between 0 and 1; the resultant vector $\vec{v}$ is used to determine the consensus outcome.
- Each player i has an ideal outcome r_i and prefers that the outcome be as near to r_i as possible.

This last assumption, that of *single-peakedness* (Moulin 1980), is required for the conclusions reached below, but it is important to point out that it rules out certain preference orderings. We are assuming, for example, that an agent that prefers an outcome of 0.2 to 0.3 cannot also prefer 0.4 to 0.3. Still, it is an imminently reasonable assumption for many game situations.

Average Aggregation

Perhaps the most natural outcome function to use is the average of the inputs, which minimizes the sum of squared distances between the outcome and the inputs. While the Average aggregation function is sensitive to each voter's input, it has an important disadvantage: Voters can often gain by voting insincerely. For example, if $n = 3$, $\vec{r} = [0.25, 0.4, 0.7]$ and all three players express their sincere preferences, then $\vec{v} = [0.25, 0.4, 0.7]$ and the Average outcome is 0.45. Consider player 3, whose ideal outcome is $r_3 = 0.7$. That player could achieve a better outcome by not expressing the sincere preference $v_3 = 0.7$ and instead choosing $v_3 = 1$. The resulting Average aggregation yields the outcome 0.55, which, being closer to 0.7, is preferred by player 3 to 0.45.

Rationally Optimal Strategy for Average Aggregation

Using the Average outcome opens the door for manipulation, but investigating the nature of that manipulation further will prove fruitful. If we assume that all players want only to optimize the outcome from their own points of view, we can characterize rational voting. If all other players have expressed their preferences and player *i* is deciding how to choose v_i, the ideal outcome r_i could be achieved by choosing $v_i = r_i n - \sum_{j \neq i} v_j$, but this choice for v_i is allowed only if it is between 0 and 1. In general, player *i* can move the Average outcome as near to r_i as is possible by choosing $v_i = \min\left(\max\left(r_i n - \sum_{j \neq i} v_j, 0\right), 1\right)$, which is the rationally optimal strategy for any player *i*.

An example will illustrate that, if all players use this strategy iteratively, an equilibrium may be reached from which no player would change. Imagine that the three players from above with sincere preferences $\vec{r} = [0.25, 0.4, 0.7]$ begin by expressing $\vec{v} = [0,0,0]$. If they then rationally adjust their strategies in order of descending ideal preferences, they would calculate as follows.

Calculate: $$\begin{aligned} v_3 &= \min\left(\max\left(r_3 n - \sum_{j \neq 3} v_j, 0\right), 1\right) \\ &= \min\left(\max\left(2.1 - 0, 0\right), 1\right) = 1 \end{aligned}$$

Update: $\vec{v} = [0, 0, 1]$

New outcome: $\overline{v} \approx 0.333$

Calculate: $$\begin{aligned} v_2 &= \min\left(\max\left(r_2 n - \sum_{j \neq 2} v_j, 0\right), 1\right) \\ &= \min\left(\max\left(1.2 - 1, 0\right), 1\right) = 0.2 \end{aligned}$$

Update: $\vec{v} = [0, 0.2, 1]$

New outcome: $\overline{v} = 0.4$

Calculate: $$\begin{aligned} v_1 &= \min\left(\max\left(r_1 n - \sum_{j \neq 1} v_j, 0\right), 1\right) \\ &= \min\left(\max\left(0.75 - 1.2, 0\right), 1\right) = 0 \end{aligned}$$

Update: $\vec{v} = [0, 0.2, 1]$

New outcome: $\overline{v} = 0.4$

After one pass through the players, an equilibrium from which no player would change has been found, and the Average outcome at this equilibrium is $\overline{v} = 0.4$. Player 1 would prefer a smaller outcome, but v_1 is already as small as is allowed; player 3 would prefer a larger outcome, but v_3 is already as large as is allowed; player 2 has set v_2 exactly where it must be to achieve the outcome $\overline{v} = r_2 = 0.4$. Thus Average aggregation is manipulable by strategic players willing to submit insincere preferences.

But strategic manipulation may not be so undesirable if an equilibrium can always be found as rapidly as in the above example whatever the players' sincere preferences. Given a set of *n* players and their sincere preferences $\vec{r}$, LeGrand (2008) shows, by counting the number of players that must be strategizing at each of the two extremes, that any average $\overline{v}$ at equilibrium must satisfy two inequalities:

$$\left|\left\{i : \overline{v} < r_i\right\}\right| \leq \overline{v} n$$

$$\overline{v} n \leq \left|\left\{i : \overline{v} \leq r_i\right\}\right|$$

LeGrand (2008) then proves that:

- At least one equilibrium $\vec{v}$, at which $(\forall i)\ v_i = \min\left(\max\left(r_i n - \sum_{j \neq i} v_j, 0\right), 1\right)$ (so that no player i would be motivated to change v_i unilaterally), must exist. (The proof shows that the same algorithmic approach as seen in the example above will always find an equilibrium in one pass, which requires showing that no player *i*, after choosing v_i, would want to change its value later in the pass.)
- Multiple different such equilibria may exist, but all such equilibria must have the same Average outcome. (The proof shows by contradiction that any two averages at equilibrium, which must satisfy the above two inequalities, must be equal.)

It follows that, given the vector of sincere preferences $\vec{r}$ as input, the equilibrium Average outcome is unique and can be defined as a mathematical function.

Average-Approval-Rating DSV

The DSV framework allows players to express their ideal outcomes, then "votes" for them iteratively until a stable outcome emerges. So, applied to the Average system discussed above, players would input their preferences and the DSV system would simulate the same rationally strategic voting illustrated above, reliably giving a unique outcome. Doing so explicitly provides an effective AAR DSV algorithm, but LeGrand (2008) proves that, given a vector of expressed preferences $\vec{v}$, the AAR DSV outcome can be calculated yet more efficiently:

sort $\vec{v}$ so that $(\forall i \leq j)\ v_i \geq v_j$

$w \leftarrow 0$

for $i = 1$ to n do

$$w \leftarrow w + \min\left(\max\left(v_i n - w, 0\right), 1\right)$$

return $\frac{w}{n}$

This algorithm runs in $O(n \log n)$ time if a $O(n \log n)$-time sort is used.

The most important property of this new AAR DSV system is that it cannot be manipulated by strategic players who are willing to submit insincere preferences, as the Average system can. LeGrand (2008) proves that AAR DSV never rewards insincerity: No player *i* can move the AAR DSV outcome closer to the ideal r_i by expressing a preference other than $v_i = r_i$; the proof entails first proving that:

- An AAR DSV outcome cannot be increased (respectively, decreased) without increasing (decreasing) at least one of the inputs. (The proof is by contradiction and uses the two inequalities that any average at equilibrium must satisfy.)
- If an AAR DSV outcome is smaller (respectively, larger) than one of the inputs, the outcome does not change when that input is increased (decreased). (The proof relies on the uniqueness of the average at equilibrium.)

Once these two points are proved, it is straightforward to show that no player *i* can gain (but may lose) from moving v_i away from sincerity. This nonmanipulability result is satisfying because it shows that the DSV framework, by using the inputs to "vote" on each player's behalf, perfectly internalizes all required strategy, allowing players to focus on more important matters than attempts to manipulate the consensus.

Therefore, if a collective decision is to be made inside a line segment, the AAR DSV approach provides a way to find a consensus in $O(n \log n)$ time without the possibility of any advantage gained by insincere players. Above we imagined a team of wargame players banding together to attack the western coast of a continent held by a common enemy. If they used AAR DSV to decide how far north or south to attack, none of the players would have any reason to express an insincere preference, and the players would be able to forget about outfoxing each other and concentrate on the attack itself.

USING DSV TO FIND STABLE CONSENSUS IN A HYPERCUBE

As discussed above, another potentially important outcome space for game players is a hypercube, essentially a cross product of line segments. For example, players might want to agree on a point inside the two- or three-dimensional game map to attack or at which to meet, or they may be making several independent 0%-to-100% decisions at once. This problem is essentially equivalent to making two, three or more 0%-to-100% decisions as a package; 0% might mean the west edge of a map and 100% might mean the east edge, etc. We will assume that the available outcomes in each of the *d* dimensions range between 0 (0%) and 1 (100%). We further assume:

- There are *n* players that want to find a consensus inside the hypercube of dimension *d*.
- Each player *i* submits a *d*-dimensional vector v_i, each scalar coordinate of which is in the inclusive interval between 0 and 1; the resultant vector $\vec{v}$ is used to determine the consensus outcome.
- Each player *i* has an ideal outcome r_i and prefers that, in each of the *d* dimensions individually, the outcome be as near to r_i as possible.

Instead of this last assumption, that of intradimensional single-peakedness, we could assume simply that each player *i* would like to minimize the Euclidean distance between the outcome and r_i, but that is actually a stronger assumption than we require. Fortunately, if each player aims simply to minimize the Euclidean distance between the actual outcome and their ideal outcome, as arguably would often be the case, then this notion of dimension-independence holds.

However, our assumption does rule out certain preference orderings that might conceivably occur in a real game. For example, consider a game map over which a group of RPG players is deciding collectively where to meet. It may be that one player, player *i*, prefers to meet near a river that flows northeast to southwest, so if the consensus is to meet somewhere in the west, player *i* would prefer meeting points farther to the south, whereas if the consensus is to meet somewhere in the east, player *i* would prefer meeting points farther to the north. In other words, player *i* might prefer both (0, 0) to (1, 0) and (1, 1) to (0, 1). Such a player would not be able to isolate one ideal point r_i that satisfies our assumptions; only a richer input space would allow expressing such preferences. But our assumption is reasonable for many common game situations, including any in which the dimensions are completely independent.

Average aggregation is easily generalized to a higher-dimension hypercube by taking the average of each coordinate separately, effectively calculating the *centroid*, the center of mass given a set of unit masses. Alternatively, one can imagine attaching Hookean springs of equal spring constants to each fixed input point, then gluing the other ends of the springs together; the glue point will come to rest at the centroid. This generalization is equivalent to finding the point that minimizes the sum of squared distances between that point and all of the input points. The resulting system is rotationally invariant and is equivalent to conducting *d* separate and independent Average elections

(LeGrand & Cytron, 2008). Thus, the results above for strategic behavior under the one-dimensional Average system apply to the "election" for each coordinate. In particular, if each voter has *separable* preferences (Border & Jordan, 1983), so that preferences in one dimension are independent of preferences in all other dimensions, conducting a d-dimensional AAR DSV election is equivalent to conducting d parallel one-dimensional AAR DSV elections, and so gives a nonmanipulable system. (Such a preference-function space is not *abundant* by Zhou's (1991) definition.)

To illustrate briefly, consider two players with ideal preferences $\vec{r} = \left[\left(0.2, 0.3\right), \left(0.6, 0.1\right)\right]$. If they used Average aggregation and applied rational strategy iteratively, they would soon reach the equilibrium $\vec{v} = \left[\left(0, 0.6\right), \left(1, 0\right)\right]$, giving the outcome $\left(0.5, 0.3\right)$. At this equilibrium they are pulling each coordinate of the outcome toward their ideal outcomes to the greatest extent possible, which in this case requires insincerity. If the players decided to use AAR DSV instead of Average, they could submit their sincere preferences, $\vec{v} = \vec{r} = \left[\left(0.2, 0.3\right), \left(0.6, 0.1\right)\right]$, giving the outcome $\left(0.5, 0.3\right)$ directly, and the players would have no incentive to be insincere.

Therefore, if a collective decision is to be made inside a hypercube, the AAR DSV approach provides a way to find a consensus without the possibility of any advantage gained by insincere players: Simply use the one-dimensional AAR DSV in each of the d dimensions. Using the AAR DSV algorithm given above, a nonmanipulable consensus can thus be found in $O\left(dn \log n\right)$ time.

USING DSV TO FIND STABLE CONSENSUS IN A SIMPLEX

The AAR DSV approach gives satisfying results for the line-segment and hypercube outcome spaces, but there are other outcome spaces that may be useful in game situations. Another useful way to generalize the one-dimensional space between 0 and 1 into higher dimensions is into a simplex, as discussed above. For example, consider another decision situation, in which players decide how a fixed amount of a resource should be allocated among several uses, such as (30%, 60%, 10%). An AAR-style method would have each player suggest an allocation and then average them to give the outcome. We will assume that the available outcomes in each of the d dimensions range between 0 (0%) and 1 (100%); these coordinates of any one input point or outcome must sum to 1 (100%). We further assume:

- There are n players that want to find a consensus inside the simplex of d dimensions. (Because the coordinates must sum to 1, this simplex is mathematically a $(d-1)$ -dimensional space, but we will refer to d dimensions for convenience.)
- Each player i submits a d-dimensional vector v_i, each scalar coordinate of which is in the inclusive interval between 0 and 1 and all of which sum to 1; the resultant vector $\vec{v}$ is used to determine the consensus outcome.
- Each player i has an ideal outcome r_i and prefers outcome a to outcome b whenever a is nearer to r_i than b is in at least one of the d dimensions and a is farther from r_i than b is in none of them.

Again, instead of this last assumption, we could assume simply that each player i would like to minimize the Euclidean distance between the outcome and r_i, but it would be a stronger assumption than we require.

In the hypercube space, it was possible to use one-dimensional AAR DSV in each dimension to arrive at a multidimensional consensus. Effectively, each dimension was completely independent of the others. But in the simplex space, the coordinates of a point restrict the allowed values of the other coordinates; in a sense, the dimensions fail to be independent by virtue of the shape of the space. For example, if one coordinate of a point has the value 0.4, then no other coordinate can have a value higher than 0.6.

Still, as before, we can characterize optimally strategic "voting" for Average aggregation within a simplex. As in the one-dimensional space, a rational player would express $v_i = r_i n - \sum_{j \neq i} v_j$ if it were inside the allowed simplex space. Otherwise, it must be projected to the (always uniquely) nearest point in the simplex, which may be on a vertex, an edge, a face, etc. For example, if $d = 3$, then the simplex is an equilateral triangle with vertices $(1,0,0)$, $(0,1,0)$ and $(0,0,1)$, and rational strategy can be characterized as follows: If we calculate $r_i n - \sum_{j \neq i} v_j$ and call its three coordinates *x*, *y* and *z*, then player *i* should express

$$v_i = \begin{cases} (1,0,0) & \text{if } x \geq y+1 \text{ and } x \geq z+1 \\ (0,1,0) & \text{if } y \geq x+1 \text{ and } y \geq z+1 \\ (0,0,1) & \text{if } z \geq x+1 \text{ and } z \geq y+1 \\ \left(x + \frac{z}{2}, y + \frac{z}{2}, 0\right) & \\ \quad \text{if } x \leq y+1 \text{ and } y \leq x+1 \text{ and } z \leq 0 & \\ \left(x + \frac{y}{2}, 0, z + \frac{y}{2}\right) & \\ \quad \text{if } x \leq z+1 \text{ and } y \leq 0 \text{ and } z \leq x+1 & \\ \left(0, y + \frac{x}{2}, z + \frac{x}{2}\right) & \\ \quad \text{if } x \leq 0 \text{ and } y \leq z+1 \text{ and } z \leq y+1 & \\ (x,y,z) & \text{if } x \geq 0 \text{ and } y \geq 0 \text{ and } z \geq 0 \end{cases}$$

The rational strategy functions for higher-dimensional simplex spaces will have more cases but follow a similar pattern.

As an example of rationally strategic players finding consensus in a simplex, consider $n = 2$ players trying to decide on the best way to allocate an amount of magic points among $d = 3$ uses, such as to attack the current enemy, to heal the players and to save for future situations. The two players have sincere preferences $\vec{r} = \left[(0.5, 0.5, 0), (0.25, 0.25, 0.5)\right]$ and use Average aggregation. After repeatedly applying the rational strategy detailed above, they would soon reach the unique equilibrium $\vec{v} = \left[(0.5, 0.5, 0), (0, 0, 1)\right]$, giving the outcome $(0.25, 0.25, 0.5)$, which is player 2's ideal outcome. From this equilibrium, neither player can move the Average outcome nearer to ideal by Euclidean distance.

Using Average aggregation in the one-dimensional space, it was possible to prove the following properties (LeGrand, 2008):

- A strategic equilibrium always exists.
- The equilibrium outcome is unique.
- The resulting DSV function is nonmanipulable by insincere agents.

The equivalent properties may yet be proved in the simplex space; until then, it will be useful to look for counterexamples via Monte Carlo simulations.

Experiments and Results in a Simplex

We wrote software using the C programming language to simulate many random "elections", checking each one to see whether the properties held. The setup was as follows:

- For each combination of *d* and *n* from $d \in \{3,4,5,6\}$ and $n \in \{2,3,5,8,13\}$, run one million simulated elections in *d* dimensions with *n* agents.

- For each election, do the following:
  - Randomly generate a vector $\vec{r}$ of n d-dimensional points, each coordinate of which has exactly p_g decimal places, from a uniform distribution in the simplex.
  - Taking the points in $\vec{r}$ as the agents' sincere preferences, iteratively vote strategically on behalf of each agent, using a randomized ordering of agents, until an equilibrium $\vec{v}$ is reached. Stop only when all coordinates of the points in $\vec{v}$ are stable to p_e decimal places.
  - Repeat the iterative strategy from scratch, again using a randomized ordering of agents, until an equilibrium is reached. Take note of an *outcome-uniqueness violation* if the Average outcomes at the two found equilibria differ to p_d decimal places in any coordinate.
  - Construct another vector $\vec{r}'$ of points such that r_1' is regenerated randomly as before and $r_i' = r_i$ for all $i > 1$, representing an attempt on the part of agent 1 to deceive the DSV system with insincere strategy, and find an equilibrium with the new $\vec{r}'$. Take note of a *successful Euclidean-distance manipulation* if the equilibrium Average outcome using $\vec{r}'$ is nearer, to p_d decimal places, in Euclidean distance to r_1 than is the equilibrium Average outcome using $\vec{r}$. Also take note of a *successful dimension-domination manipulation* if the equilibrium Average outcome using $\vec{r}'$ is nearer, to p_d decimal places, to r_1 than is the equilibrium Average outcome using $\vec{r}$ in at least one of the d dimensions and farther, to p_d decimal places, in none of them.

We ran the above simulations with the values $p_g = 4$, $p_e = 13$ and $p_d = 7$. We chose p_d larger than p_g so as to detect much smaller differences among outcomes than the differences among input points, and we chose p_e larger than p_d so as to ensure that we did not detect a difference among outcomes that was due only to the inaccuracy of the DSV outcomes; the choice of $p_e = 13$ was small enough for the DSV function to converge to an equilibrium quickly in practice. None of the simulated elections failed to find an equilibrium, and none found two equilibria with different Average outcomes. While by no means a proof, this result leads us to suspect simplex AAR DSV to be a mathematical function of its inputs, meaning that, given the same inputs, it will always give the same outcome.

On the other hand, we found very quickly that simplex AAR DSV *is* Euclidean-distance-manipulable by insincere agents; Table 1 shows how often the attempted manipulations were successful in our simulations. In fact, a manipulation opportunity exists for the simplex example given above. If both players submit sincere preferences, the DSV system will apply rational strategy on the part of both players, find the unique equilibrium $[(0.5, 0.5, 0), (0, 0, 1)]$ and give the outcome $(0.25, 0.25, 0.5)$, player 2's ideal outcome. But if player 1, instead of the sincere $(0.5, 0.5, 0)$, insincerely expresses the preference $(0.2, 0.6, 0.2)$, the DSV system would instead find the unique equilibrium $[(0, 1, 0), (0.25, 0, 0.75)]$. The resulting outcome, $(0.125, 0.5, 0.375)$, is only 0.28125 away from player 1's ideal in Euclidean distance, an improvement from the sincere outcome, which was 0.375 away. Player 1 has thus successfully manipulated simplex AAR DSV, at least by Euclidean distance.

Table 1. Incidence of successful Euclidean-distance manipulation in simplex simulation

d \ n	2	3	5	8	13
3	6.4109%	4.2370%	4.0829%	3.6739%	3.4174%
4	4.7949%	7.0336%	5.7254%	5.3783%	5.0183%
5	3.1722%	6.3082%	6.6661%	6.1147%	5.7822%
6	2.0793%	4.9974%	7.0774%	6.5012%	6.2127%

However, while this manipulated outcome is closer to player 1's ideal outcome in the second and third dimensions (0 *vs.* 0.25 and 0.375 *vs.* 0.5, respectively), it is farther from ideal in the first dimension (0.375 *vs.* 0.25). In fact, our simulated elections found no example of a successful dimension-domination manipulation (Table 2). This result, while again no proof, suggests that the simplex AAR DSV system may be immune to strategic manipulation at least in a weak sense: It may be impossible for an agent to use insincerity to move a simplex AAR DSV outcome to one which is strictly better, *i.e.*, closer to ideal on at least one dimension and farther on no dimension. In other words, from any one agent's point of view, insincerity that improves the outcome in one dimension must make it worse in another dimension.

This narrower notion of immunity to strategic manipulation, which essentially makes strong assumptions about preferences voters may have, avoids previous impossibility results (Moulin 1980, Zhou 1991). Even if this assumption about players' preferences would not always hold in real game situations, this simplex AAR DSV method does seem to be at least quite resistant to strategic manipulation.

MAKING ASYNCHRONOUS DECISIONS

In many real games, collective decisions must be made quickly and players may need to express their preferences simultaneously. In such situations, while a nonmanipulable method of aggregating preferences removes any chance of successful manipulation, a manipulable method like Average aggregation may work well enough in practice if the players have too little insight into the others' preferences to take advantage of them. A player may not know enough to inform a strategically insincere "vote".

On the other hand, in many social games, players are often playing at different times and on different schedules; Sid Meier's Civilization World (under development; originally titled Civilization Network) may prove to be an excellent example. If a cooperating group of players needs to make a collective decision, the players may often need to register their preferences asynchronously. Our AAR-DSV-based methods adapt to this environment gracefully. In fact, even if expressed preferences are made public before all of them have been expressed, players cannot take advantage of

Table 2. Incidence of successful dimension-domination manipulation in simplex simulation

d \ n	2	3	5	8	13
3	0.0000%	0.0000%	0.0000%	0.0000%	0.0000%
4	0.0000%	0.0000%	0.0000%	0.0000%	0.0000%
5	0.0000%	0.0000%	0.0000%	0.0000%	0.0000%
6	0.0000%	0.0000%	0.0000%	0.0000%	0.0000%

them. Players can never do better (in the senses discussed above) than to express sincere preferences, and so no unfair advantage is generally gained by expressing preferences sooner or later than other players.

Besides Civilization World, there are many other games currently being developed for Facebook and other social networking websites. Our approach could potentially find a use in many games in which players cooperate by casting "votes" for collective actions or other kinds of outcomes.

FUTURE RESEARCH DIRECTIONS

There are several potentially interesting and useful directions for future research. First, in light of the results of our simulations, it may be possible to prove for the simplex outcome space the equivalent results already proved for the line-segment and hypercube spaces. Also worth consideration may be collective-decision-making methods with other outcome spaces, whether continuous spaces of different shapes or discrete spaces.

Discrete outcome spaces certainly occur in real games. They occur whenever players are faced with a collective decision among finitely many mutually exclusive choices, such as whether to deploy phasers or photon torpedoes against an enemy, or which of several computed paths to follow. A continuous outcome space may even be effectively discretized when players' preferences are multiple-peaked, such as when all players would prefer to attack either of the two flanks of an opposing army than to attack the center. Previous impossibility results (Arrow, 1951; Gibbard, 1973; Satterthwaite, 1975) may preclude the possibility of finding perfectly nonmanipulable decision-making protocols with (effectively) discrete outcome spaces, depending on assumptions made about players' preferences, but the DSV framework may prove successful in minimizing opportunities for insincere strategy.

Perhaps more immediately important than future theoretical directions, however, is deployment of these techniques in real games. It will be instructive to note how useful and intuitive human players find them when cooperating with other human players and NPCs. These techniques could also be used when groups of NPCs make collective decisions, and such use may make a game's AI become more effective and/or seem more realistic—they may prove to increase the apparent independence of NPC agents. Whether used with human players or NPCs, these techniques offer a flexible alternative to hierarchical group organization. Eventually, they may inform future designs for multiagent AI frameworks in real games.

CONCLUSION

In this chapter we have presented new methods for collective decision-making in games and explored their immunity and/or resistance to manipulation by insincere players. We specifically applied our approach to the line-segment, hypercube and simplex outcome spaces, giving examples of each.

We believe that our AAR DSV approach has many advantages:

- It can contribute to both the AI and the player-to-player frameworks.
- It allows agents to concentrate on determining optimal policies instead of on deceiving other agents with whom they are ostensibly cooperating.
- It places a relatively small burden on players: They need only indicate their ideal outcome; no complex ranking or rating of outcomes is needed.
- It allows cooperating agents to find a compromise immediately, which is painless, rather than by fighting it out through the game, which may be costly to all.

- It allows important decisions to be made whether players indicate their preferences simultaneously in real time or at different times.
- Its outcome functions are decisive, efficient and easily implemented.
- It is sufficiently general to be applied to decision situations in almost any kind of game.

Certain assumptions, concerning the players' preferences and continuity of the outcome space, are required for AAR DSV to work perfectly, but these assumptions are at least approximately true in many game situations. We conclude that our approach is likely to be useful when designing real games.

REFERENCES

Arrow, K. J. (1951). *Social Choice and Individual Values*. New York, NY: John Wiley & Sons.

Bartholdi, J. J. III, Tovey, C. A., & Trick, M. A. (1989). The computational difficulty of manipulating an election. *Social Choice and Welfare*, *6*, 227–241. doi:10.1007/BF00295861

Border, K. C., & Jordan, J. S. (1983). Straightforward elections, unanimity and phantom voters. *The Review of Economic Studies*, *50*(1), 153–170. doi:10.2307/2296962

Conitzer, V., & Sandholm, T. (2003). *Universal voting protocol tweaks to make manipulation hard.* Paper presented at the 18th International Joint Conference on Artificial Intelligence, Acapulco, Mexico.

Cranor, L. F. (1996). Declared-strategy voting: An instrument for group decision-making. Doctoral Dissertation. Washington University, St. Louis.

Cranor, L. F., & Cytron, R. K. (1996). *Towards an information-neutral voting scheme that does not leave too much to chance.* Paper presented at the Midwest Political Science Association Annual Meeting, Chicago, Illinois.

Elkind, E., & Lipmaa, H. (2005). *Hybrid voting protocols and hardness of manipulation.* Paper presented at the 16th Annual International Symposium on Algorithms and Computation (ISAAC), Sanya, Hainan, China.

Gibbard, A. (1973). Manipulation of voting schemes: A general result. *Econometrica: Journal of the Econometric Society*, *41*(3), 587–601. doi:10.2307/1914083

LeGrand, R. (2008). Computational aspects of approval voting and declared-strategy voting. Doctoral Dissertation. Washington University, St. Louis.

LeGrand, R., & Cytron, R. K. (2008). *Approval-rating systems that never reward insincerity*. Paper presented at the 2nd International Workshop on Computational Social Choice (COMSOC), Liverpool, England.

Moulin, H. (1980, January). On strategy-proofness and single peakedness. *Public Choice*, *35*(4), 437–455. doi:10.1007/BF00128122

Nisan, N. (2007). Introduction to Mechanism Design (for Computer Scientists). In Nisan, N., Roughgarden, T., Tardos, E., & Vazirani, V. V. (Eds.), *Algorithmic Game Theory* (pp. 209–241). Cambridge, UK: Cambridge University Press. doi:10.1017/CBO9780511800481.011

Procaccia, A. D., & Rosenschein, J. S. (2007). Junta distributions and the average-case complexity of manipulating elections. *Journal of Artificial Intelligence Research*, *28*, 157–181.

Rabin, S. (2002). *AI Game Programming Wisdom*. Hingham, Massachusetts: Charles River Media.

Satterthwaite, M. A. (1975). Strategy-proofness and arrow's conditions: Existence and correspondence theorems for voting procedures and social welfare functions. *Journal of Economic Theory*, *10*(2), 187–217. doi:10.1016/0022-0531(75)90050-2

Zhou, L. (1991). Impossibility of strategy-proof mechanisms in economies with pure public goods. *The Review of Economic Studies*, *58*(1), 107–119. doi:10.2307/2298048

ADDITIONAL READING

Black, D. (1948). On the rationale of group decision-making. *The Journal of Political Economy*, *56*(1), 23–34. doi:10.1086/256633

Conitzer, V., Lang, J., & Sandholm, T. (2003). How many candidates are needed to make elections hard to manipulate? In *Proceedings of the 9th Conference on Theoretical Aspects of Rationality and Knowledge (TARK-03)* (pp. 201–214). Bloomington, Indiana.

Conitzer, V., & Sandholm, T. (2002). Complexity of manipulating elections with few candidates. In *Proceedings of the Eighteenth National Conference on Artificial Intelligence (AAAI '02)* (pp. 314–319). Edmonton, Canada.

Downs, A. (1957). *An Economic Theory of Democracy*. New York, NY: Harper.

Ephrati, E., & Rosenschein, J. S. (1993). Multiagent planning as a dynamic search for social consensus. In *Proceedings of the Thirteenth International Joint Conference on Artificial Intelligence (IJCAI '93)* (pp. 423–429). Chambéry, France.

Merrill, S. III. (1988). *Making Multicandidate Elections More Democratic*. Princeton, NJ: Princeton University Press.

Pennock, D. M., Horvitz, E., & Giles, C. L. (2000). Social choice theory and recommender systems: Analysis of the axiomatic foundations of collaborative filtering. In *Proceedings of the Seventeenth National Conference on Artificial Intelligence (AAAI '00)* (pp. 729–734). Austin, Texas.

Procaccia, A. D., Rosenschein, J. S., & Zohar, A. (2007, January). Multi-winner elections: Complexity of manipulation, control and winner-determination. In *Proceedings of the Twentieth International Joint Conference on Artificial Intelligence (IJCAI '07)* (pp. 1476–1481). Hyderabad, India.

Vazirani, V. V., Nisan, N., Roughgarden, T., & Tardos, É. (2007). *Algorithmic Game Theory*. Cambridge, UK: Cambridge University Press.

Von Neumann, J., & Morgenstern, O. (1953). *Theory of Games and Economic Behavior*. Princeton, NJ: Princeton University Press.

KEY TERMS AND DEFINITIONS

Average-Approval-Rating (AAR) DSV: A specific application of the DSV framework to a continuous outcome space that uses the Average aggregation procedure (simply averaging the inputs) as the internal voting method for the simulated election.

Declared-Strategy Voting (DSV): A framework for collective decision-making that uses agents' preferences to vote on their behalf in a simulated election, the result of which becomes the DSV outcome.

Hypercube: A generalization of a cube into any dimension. For example, $\left\{(x,y,z) : 0 \le x \le 1 \text{ and } 0 \le y \le 1 \text{ and } 0 \le z \le 1\right\}$ is a 3-hypercube (cube).

Mechanism Design: A subfield of game theory that aims to design mechanisms that output a decisive outcome given agents' input and are relatively robust to rationally strategic agents.

Outcome: The output of a collective-decision-making mechanism.

Outcome Space: The set of allowed outcomes over which agents have preferences.

Simplex: A generalization of a triangle into any dimension. For example, $\left\{(x,y,z) : 0 \le x \le 1 \text{ and } 0 \le y \le 1 \text{ and } 0 \le z \le 1 \text{ and } x+y+z=1\right\}$ is a 2-simplex (triangle).

Single-Peaked Preferences: Preferences for which it is never true that *a* and *c* are each preferred to *b* when $a \le b \le c$.

Chapter 5
Understanding and Implementing Adaptive Difficulty Adjustment in Video Games

Tremblay Jonathan
LIARA, Canada

Bouchard Bruno
LIARA, Canada

Bouzouane Abdenour
LIARA, Canada

ABSTRACT

This chapter begins with an introduction to different concepts evolving around the adaptive difficulty in video games (i.e. problematic definition, existing models of dynamic difficulty adjustment, evaluating the player's experience, transposing the player's skills into numerical values, using these numerical values as seeds for the difficulty level, etc.). Further on, this chapter covers the implementation of a novel adaptive model and the validation of such a model. This model uses a normal distribution system (ELO ranking) to determine the player's skill level and then adapt the difficulty to their needs. In order to validate this model, 42 players play-tested two versions of the game, one with adaptive difficulty and one without any difficulty adaptation.

INTRODUCTION

The great video game companies such as Nintendo, Microsoft and Sony have succeeded in opening up the video game market to a wider audience of players (Pelland, 2009). This audience is composed of non-experienced and experienced players that share the same games. This reality has brought new challenges to game companies and game designers, such as making the game stimulating and interesting for this wider range of players. To reach all of them, the game has to provide the right amount of challenge for their skill level, as a game that is too hard or too easy

DOI: 10.4018/978-1-4666-1634-9.ch005

is frustrating for them (Csikszentmihalyi, 1990). Also, one faculty of game, mainly used by serious games, is the capability of games to teach the player using game mechanics (the player's actions) (Koster 2004). In this serious context, it is crucial that the game is able to reach a wider audience, for example in the classroom the game would be played by experienced players and neophytes alike. Therefore, it is important for the player's experience that the challenge meets their expectations, as their enjoyment and learning results are closely related to the challenge. A classic approach to this conundrum is to let the player choose their difficulty level (easy, normal, hard, etc.) at the beginning of the game, without even trying it. Without any clear standard definition for difficulty, the player could easily choose the wrong level expecting the game to be something that it is not. This would lead the player to leave the game as they are experiencing frustration in the form of boredom and/or anxiety (Schell, 2008). Also, giving one unique level of difficulty is not really a proper solution, though it is a widespread one. Unique challenges are mostly interesting for a narrow range of players depending on their skill levels. Nevertheless, a well developed learning curve can bring many players, but the majority to the same point (Rolling & Adams, 2003).

To address this issue, some researchers and commercial video games such as *Max Payne* from *Remedy Entertainment* or *Left 4 Dead* one and two from *Valve* have built different models to dynamically adapt the difficulty of the game based on the player's performance or intensity level. One of the main challenges regarding difficulty is the subjective factor that stems from the interaction between the player and the challenge. The perceived difficulty is also not a static property: it changes with time as the player learns the game skills (Goetschalckx, Missura *et al.*, 2010). One of the major issues in this particular field is that the proposed approaches or games often do not offer any implementation explanations or guidelines for the dynamic difficulty model used. For example, the work of (Lindley & Sennersten, 2008) where they present an adaptive approach using gameplay schema that recognizes the player's actions and based on them adapts the gameplay to the player's preferences. In this particular example, the model is essentially theoretical and experienced in a non-realistic context. From this context it is nearly impossible to determine its actual effectiveness and bring to light the problems that had not been taken into account in the design. Nevertheless, experimentation and testing remain the best way to assure the robustness of a dynamic system.

Other works propose models that have been implemented and tested such as (Hunicke & Chapman. 2004), where they present a computational model using normalization distribution to evaluate the player's skill level. A different approach proposed by (Chen. 2006) introduces a game design process leaving the player with the control of the difficulty of the game during gameplay. These types of work are essential for the comprehension and the elaboration of dynamic difficulty in games, for they give us an indication of how to implement and/or design their models. These works also provide discussion on their implementation and testing. They focus on how their models interact with the game but do not include deep explanations on the impact their model might have on the player.

The goal of this chapter is precisely to address this important issue that has been raised. This chapter has two distinct parts. At first the different concepts revolving around an adaptive model will be presented; such as the provenance of pleasure for the player in video games and their understanding of the difficulty and in game challenge. From there, a novel adaptive model for dynamically adjusting, in real-time, the difficulty level of a game according to the players' performance in order to enhance their gaming experience will be presented. The foundation of this model relies on a computational approach for Dynamic Difficulty Adjustment (DDA) (Hunicke & Chapman, 2004) and an extension of our previous model proposed in (Tremblay, Bouchard *et al.,* 2010). The model is composed of two distinct layers. The first is called the *tactical layer*, which is unstable in the long-term but is useful to evaluate the players' skills for quick actions. The second part is called the *strategic layer*, which is used to adjust the difficulty in the long run by exploiting the theory behind the well-known ranking system ELO (Coulom 2010). The calculated rank is used to understand the player's performance over a long period of time and to adjust the game according to their performance. In other words, the model feeds important parts of the game (AI, generated content, etc.) based on the ranking value. Using the *tactical* and the *strategic* layers together in a video game, it is possible to build a flexible game experience that better suits different types of players. This model is meant to be easily deployed and inexpensively implemented, allowing it to be an asset in any small or large production. Hence, answers are proposed to the following conundrums:

- How to understand the player's expectation about challenge in video games
- How to visualize the difficulty in game
- How to transpose the player's actions into mathematical values
- How to implement a simple DDA system
- How to use those values to feed the difficulty in the game
- How to validate the DDA model using play testing

This chapter is organized as follows: section 2 will cover the player's expectations and pleasure regarding video games based on scientific research. This will help to understand the different design choices that were taken into account in the other sections. Section 3 will present the different approaches researchers have developed in past years to address this issue regarding DDA. The manner in which these approaches influence the presented model will also be discussed. Section 4 specifies the model's use to determine the player's difficulty level. This model is based on a simple ranking system that evolves with the game. Section 4 will also cover the implementation of such a model and how the player interacts with it and how it interacts with the game. Section 5 presents our testing phase, details the experimental protocol, shows the obtained results, and discusses the various problems the model confronted. The final section concludes the chapter, and opens up on different perspectives that may be interesting to study in future works.

PLAYER AND DIFFICULTY

In Game Player's Enjoyment and Experience

The in-game experience for the player is unstable; it moves quickly and depends on a wide range of elements such as: challenge, interactions, aesthetics, storytelling, etc. It is important to understand the underlying structure of fun in game to clearly design any adaptive system for it. The game structure promises much greater pleasure from the game than the immediate impulse gratification (Vygotsky, 1976). In other words, performing an action within a gaming environment for a player is more gratifying than the action itself without any game structure. For instance, it is more fun and gratifying to score during a hockey game than scoring on an empty net during practice. The player's pleasure is an effect of submitting themselves to the rules of the game (Salen & Zimmerman, 2003). The gamer has to respect the game in order to experience fun.

The game exists inside a frame, a specially delimited time and space. When the player is entering that space, they are ready to respect the game and let it be their leader. This particular moment is defined as the magic circle (Schell, 2008). In the magic circle, the player accepts the rules and the limitations of a game because of its emergent pleasure. The magic circle is responsible for keeping players from cheating, i.e. in a track race, every racer stays in their lane and does not try to change lanes or block any of the competitors.

Though for a gamer, entering the magic circle is straight forward; the controller (game pad) restricts the player's freedom in the game which in itself introduces the possibility of pleasure in the game. When trying to describe feelings and fun in video games, it becomes hard to put a gaming experience into words.

Koster points out that fun is all about the brain feeling good (Koster, 2004); it is the moment a player masters a task or understands something. In games, this particular moment arrives often and quite easily. Just by solving conflict, the player learns and rewards their intellect. The act of solving a puzzle itself makes a game fun. In other words, with games, learning is like a drug. The puzzle is a particular conflict that has to be solved by the player (Salen & Zimmerman, 2003). Any conflicts in game are indeed the challenges (difficulty level) that the player must overcome. When the game is no longer challenging, the player will most likely stop playing as it is becoming uninteresting.

Although conflict is not everything in games, a lot of gamers expect games to provide a background story inside the game as a form of pleasure. This form of pleasure is called the narrative play and its structure can be embedded in the game. These elements are pre-generated narrative components such as video clips which are scripted scenes and/or emergent scenes. Narrative elements are created on-the-fly as the player interacts with the game, arising from the operation of the game system (LeBlanc, 1999). When playing a game a narrative play emerges from the underlying structure of the game, this structure is composed of all the elements that participate in the game representation: instructional text, in-game cinematic, interface elements, game objects and other visual and audio elements (Wark, 2007). It is important that these elements are carefully crafted with a narrative experience in mind. The accumulation of the elements emerges as a narrative play that is quite sensitive to the player's point of view and can easily be disturbed.

Aesthetics is closely related to the factor of fun in game, although it is more of a form of visual enjoyment than one of pleasure derived from accomplishments (Koster, 2004). Koster states that beauty is found in the tension between our expectations and reality. Although this form of visual enjoyment does not last forever, the prettiest game can only be pretty for a certain period of time. The quality of the game relies on the caliber and type of interactions and the underlying structure. Marc Leblanc (LeBlanc, 2000) describes fun or enjoyment in game as merely a stand-in term for a more complex phenomenon that no one really understands. He developed a typology that lists eight categories that describe the kind of experimental pleasure that arises from playing games.

- Sensation: Game as sense-pleasure
- Fantasy: Game as make-believe
- Narrative: Game as drama
- Challenge: Game as obstacle course
- Fellowship: Game as social framework
- Discovery: Game as uncharted territory
- Expression: Game as self-discovery
- Submission: Game as masochism

This model is intended to assist game designers in understanding different ranges of fun and also to provide a common language (nomenclature) to define games. For example, playing a role playing game such as *Dragon Age Origin* from *Bioware* might be high on Sensation, Fantasy, Narrative and Discovery and low in Expression, Submission and Challenge. Using such nomenclature helps understand what kind of fun the player may experience.

One of the qualities of any game is to give a challenging and intense experience to the player; it is possible to look at games as a matter of challenge (Salen & Zimmerman, 2003; Koster, 2004). Csikszentmihalyi (Csikszentmihalyi, 1990) defines the state of mind that a participant achieves with a high degree of focus and enjoyment during any activity as flow. The definition of flow is easily transposable to video games. In order for a player to experience flow, a game should provide tasks that can be completed, allow the player to focus on these tasks, define clear goals, and provide constant feedback concerning the results of the player's actions (Sweetser & Wyeth, 2005). In return, the player experiences a sense of control over the game, disconnection with everyday life, the sense of self disappearing and of time being altered. Experiencing flow for the player is so gratifying that the player is ready to perform all sorts of activities even though they could be in danger in real life (Csikszentmihalyi, 1990). Flow also requires that the challenge should not be too hard or too easy for the player.

Based on that last qualification we can easily represent how a player would react to a certain difficulty level according to their skill level, see Figure 1. In order to let the player experience a constant flow state, the game has to be balanced between skill level and challenge. If the challenge is too hard for the player's skill level, the player gets anxious (Koster, 2004). On the other hand, if the player is too skilled for the in-game challenge, the player will get bored because the challenge is too easy (Tremblay, Bouchard *et al.*, 2010; Chanel, Rebetez, *et al.,* 2008). As the player experiences flow, they are experiencing pleasure in the game.

Figure 1. Flow model for challenge and skills

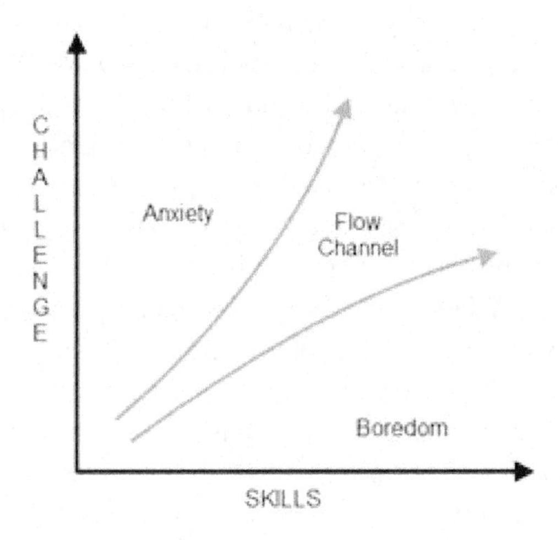

Dynamic Difficulty Adjustment

As explained, fun is intrinsically linked to flow theory. Most DDA systems are based on that assumption; give the player a challenge that suits their skill level. There are two main approaches to dynamically change the difficulty of the game; the design and the computational approaches. The first approach gives control of the difficulty of the game to the player with imbedded game mechanics. The second approach attempts to understand the player's state in order to adapt the content of the game to their expectations. In both situations, it is crucial to make it impossible for the player to abuse of the system. If they do find a way to abuse it, they will look for ways to cheat instead of trying to master the game (Baker, Corbett *et al.*, 2006).

Design Approach

In general, the design approach consists of designing the game in such a way that the player has the control over the difficulty content. It does not need complex algorithms or mathematic systems. The application of positive and negative feedback and adaptive flow are the most common approaches (Salen & Zimmerman, 2003; Chen, 2006).

Positive and negative feedback is created using a cybernetic system that monitors the in game environment and modifies it according to its state or value (Littlejohn, 1989). A common example of a cybernetic system is the thermostat. The system (thermostat) can interact with the environment by changing the temperature using a heater. In the cold, if the temperature is below a certain value, the system will activate the heater until the temperature reaches that certain value. This is a negative feedback loop example; negative systems maintain a certain equilibrium. On the other hand, a positive feedback creates an exponential growth or decline of the environment. For example, in the summer time imagine the system is still equipped with a heater and when the temperature is over a certain value the system starts the heater, the environment would get hotter and hotter as the heater increases the temperature.

In video games, positive and negative feedback can be used to control the difficulty of the game when well designed. Imagine the game of basketball using its classic rules but adding a particular one: for every N points of difference in the two teams' scores, the losing team may have an extra player in play (LeBlanc, 2000). Every time a team is winning with a great difference of points, more players would be added to the loosing team and the loosing team would have a better chance to score. The game would come back to a stable state; the score difference tends towards zero. A negative feedback would dramatically affect the outcome of the game.

The use of that particular system is also know by players as the "rubber band" system (Saltsman, 2009). In the game Mario Kart from Nintendo, the DDA system puts great effort into restricting the maximum distance between first and last place. The game gives less useful objects to the first player. The last player gets very useful items that can completely reverse the outcome of the race. With this kind of design, experienced as well as inexperienced players can still have a competitive race together.

Adaptive Flow

Adaptive flow is based on the assumption that the player's flow experience is subjective (Chen, 2006). Adaptive flow proposes giving the control of the game difficulty to the player using imbedded game mechanics. Letting the player navigate their own flow experience leads to wider control over the game. The game designer has to provide game mechanics that will give a wider spectrum of difficulty to the player. Once a network of choices is applied, the flow experience is easily customized by the player. If they start feeling bored, they can choose to play harder, and vice versa.

An example of such a system is the game *flOw* from *ThatGameCompany*. The player uses the mouse cursor to navigate an organism through a surreal biosphere where it can consume other organisms, evolve, and advance into the abyss. The game mechanics are quite simple: the only actions the player can perform are to swim around and eat other organisms that are in front of their mouths. When eating an organism, the player's avatar gets bigger and then they can eat bigger organisms. Instead of forcing the player to have to complete the level in order to progress to the next one which is harder than before, *flOw* offers the power to the player to control their progression. By choosing different food to eat, the player can advance to a more difficult level and return to the easier level at any time. Also, the player can choose to avoid the challenge, skip the level, and come back later when their avatar is bigger (stronger) or they feel more confident. This particular model allows players to play the game at their own pace.

Computational Approach

Another approach to DDA in games is to let a computational model change, in real time, the difficulty of the game (Gilleade & Dix, 2004; Hunicke, 2005; Tremblay, Bouchard, *et al.*, 2010; Um, Kim, *et al.*, 2007). A computational system of DDA is based on the presupposition that the player produces data when playing the game such as health level, velocity level, game states, etc. Figure 2 expresses the idea of such a model. The player produces data and the DDA system evaluates it. The system determines what kind of changes should be made in order to enhance the player's experience. This system is part of the game, it could impact multiple aspects of the game such as velocity, AI, ammo, level design, etc.

A computational DDA model is useful only if the algorithms behind the model are well devel-

Figure 2. General computational DDA system (Hunicke 2005)

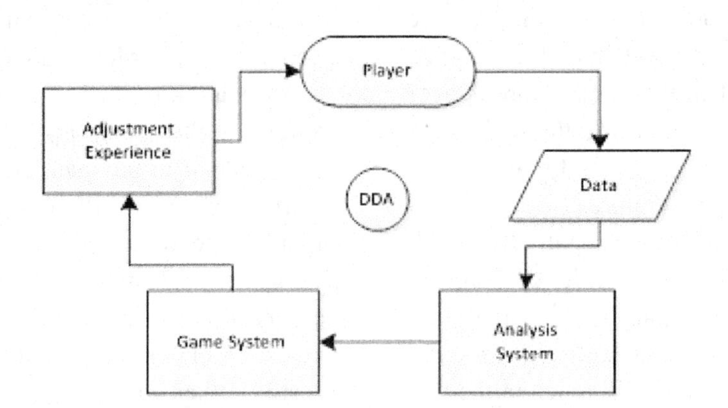

oped and can truly understand the state of the player. For example, imagine a First Person Shooter (FPS) such as Half Life 2; where the same enemies have killed the player at the same spot many times. The model would make some changes in the game such as changing the accuracy of the AI or making more ammo available for the player. These changes could make the game just a little bit easier for the player and they could experience a better gaming session. On the other hand, if the player is really experienced and they go through the game quite easily, the system could change the AI accuracy and make the ammo scarcer.

DESIGNING DDA

From the player's perspective, a game relies on the fact that it is fun and relaxing. Is it possible to enhance this sentiment of relaxation and fun using a DDA system? This proposal is inspired by the previous works presented in the last section and consists of redefining a new computational DDA model while introducing different layers. These layers can be threaded independently in order to enhance the different control the players have over the game, as (Chen, 2006) points out regarding the design of flow in games. To represent the player's experience, the player's tactics and strategies in the game were taken into account. The word tactics refers, in military science, to a short term calculated decision, for instance an action, a move or a manoeuvre against an enemy. On the other hand, the word strategy refers to a particular long-term plan for success. Using these words to define the two model layers demonstrates an effort to represent and respond to either the players' short term behavior or actions (tactics), and to the overall gaming experience (strategy). To portray these two specifications, we designed two algorithmic layers that can be used independently and/or congruently. The first is called the tactical layer and the second the strategic layer. The tactical layer uses small feedback variables that are easy to adjust and that follow the players' fast actions with ease. It allows us to quickly respond to the players' immediate situation. The strategic layer is used to understand the player's skills in the long run. Therefore, putting these two layers together allows us to have a better understanding of the players' experience over both a short and a long period of time in the game, which provides us with a dynamic adjustment approach that works on two different layers.

Tactical Layer

The tactical layer is used to understand a small and/or short action that the player performs in the game and to adapt to the result. The tactical layer uses an abstract level of difficulty representing the player's skills. The player's experience evolves during play, for a moment they might be performing extremely well and the following moment they are less than average. To represent how this behavior functions during quick game play, the tactical layer exploits two main concepts. First, it uses a level of intensity; this level represents the level of difficulty that matches the player's skills best for a particular moment. According to the fast paced nature of games, the second main concept refers to the use of a desired level of difficulty. This second concept within the tactical layer is used to map the player's experience. Since their experiences and skills are not necessarily constant, the player's desired level does not evolve in a linear manner.

On the first hand, to evaluate the player's experience, the tactical layer uses an arbitrary level of difficulty; it's an abstract representation of the intensity of the game. The abstract representation is translated into a variable called *levelIntensity*, and it evolves to understand and to represent the player's experience. The game mechanics must adapt to every value *levelIntensity* could encompass. This layer works best on small game mechanics, not on the whole game difficulty. When the *levelIntensity* increases, the game mechanic that it is attributed to should be harder, when it decreases it should be easier.

According to the value of *levelIntensity* in Table 1, the difficulty is different according to the player's *levelIntensity* value. These difficulty levels are just an example but they speak for themselves, depending on the variable value the difficulty changes for the player. Some games can use anywhere from three to twenty levels according to the game design. In the tactical layer, it is crucial to implement every level of intensity. There are two possibilities: first, the difficulty is procedural, which can be done by implementing the difficulty using variables. The level of intensity changes the variable of the implemented difficulty.

Table 1. LevelIntensity chart

levelIntensity	**Difficulty**
1	low
2	medium
3	normal
4	high
5	intense

On the other hand, sometimes it is impossible to make the difficulty procedural. The second possibility would entail every level of intensity being implemented manually, case by case. There is a limit between every level of difficulty that is distinct. It is important to play test all along with both techniques.

How do we determine the value of *levelIntensity*? The value of the variable is adjusted according to the player's skills and performance in game. In order to keep track of the player's performance, the designer has to analyze the game mechanics. For example, in a calculation game, the player answers math questions. The designer should adjust *LevelIntensity* based on that action.

By introducing a new concept, *nowExperience*, it is possible to evaluate the outcome of the player's actions. When the player completes an action the value of *nowExperience* increases. When it hits a certain value, 10 for instance; the value of *levelIntensity* is increased by one, thus changing the difficulty level of that action. On the other hand, when the player fails at that same action, *nowExperience* decreases. When it hits a certain value for instance, -10, the value of *levelIntensity* decreases by one. Therefore, *nowExperience* is meant to be updated quickly and *levelIntensity* depends on its evolution.

Figure 3 represents how the model works in a general manner. As the player plays the game, *nowExperience* evolves according to the success or failure of their actions. If the resulting action is positive *nowExperience* is increased. On the other hand if the player fails at that action, the variable decreases. When the value of *nowExperience* is 10 or -10, it has an increasing or decreasing influence on *levelIntensity*. Over all, this will have an impact on the difficulty of the game. *NowExperience* varies based on the player's data for every instant of the game.

For instance let us use a fast puzzle game as an example. The player's *levelIntensity* is 2, which is average for the game. They are solving all the puzzles quickly, for example 2.4 seconds per puzzle. From analysis and observations it is known that for this level of intensity a puzzle takes 3 seconds in average to be solved. Since the player is succeeding in the puzzle, *nowExperience* is increased. The following formula expresses *nowExperience* behavior:

$$nowExperience \mathrel{+}= succeed\ (+)\ or\ (-)\ averageTime\,/\,timePlayer \quad (1)$$

By dividing 3 by 2.4, it is known that *nowExperience* is increased by 1.25. That way, it is possible to follow the player's experience and adapt to it for a certain action. When *nowExperience* is over 10, *levelIntensity* is increased by one, making its value three. When the value of the *levelIntensity* changes (increases or decreases), the value of *nowExperience* is set back to zero. Moreover, it is possible to base our evaluation of *nowExperience* on other variables and/or game mechanics, for instance: the amount of gold collected, the period of time without being hit by an enemy, the amount of ammo available, etc. The important thing is to base the evaluation of *nowExperience* on the game mechanics.

This layer of our approach can be applied to several parts of the game. According to what the designers want to emphasize, they could make some parts adaptive and others not and design them in a linear fashion. For instance, in a FPS, the player could get their ammo based on the tactical layer, so if the *levelIntensity* is 2, the ammo is a little scarce, but the player already has enough ammo and they do not need it. When the player is at that level it takes 4-5 shots for them to defeat an enemy. By learning the game, after some time, the player can now vanquish enemies with only two shots, *nowExperience* would increase until it hits 10 then the *levelintensity* would become 3. Now, the ammo is scarce and the player has to search to find it on the map. At level 3, the system is expecting 3-4 shots to take down an enemy and the ammo and health packs are rare. This way, the players will experience a game that is harder and more challenging for their skill level based on flow theory. If the level of intensity increases quickly, the players could experience anxiety and failure. Failure is really important in game, as (Juul 2009) pointed out: "failure is central to the experience of depth in game, to the experience of improving skills." Therefore in order to master a game, its content and mechanics, it is important to experience failure at some point.

Figure 3. Tactical layer flow chart

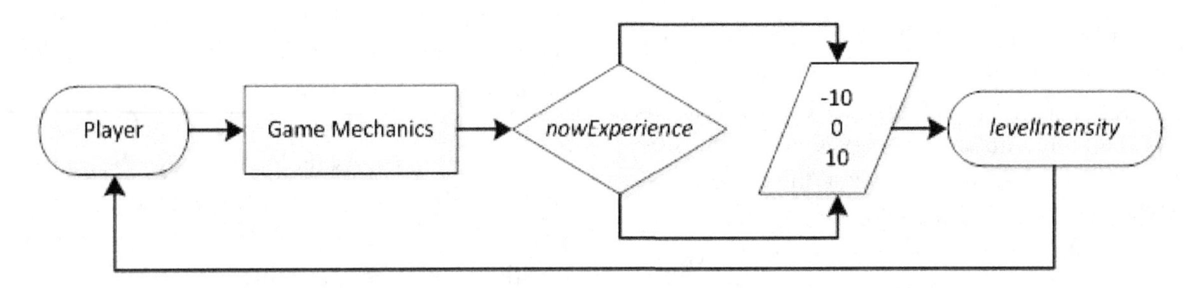

Strategic Layer

The tactical layer is really useful for analyzing small sections or actions in the game, though it is not very useful in understanding the player's whole experience. The strategic layer is not so different than the tactical layer, but it leads to different conclusions. It is especially powerful at evaluating the player's overall experience because its exploits a rating system. The main idea is to rank the player, like tennis, football and chess players are ranked. By dynamically ranking the player, it is possible to determine their skill level and how they are evolving. Their ranks could be used to control the difficulty in the game. For example, the player's rank could be used to feed the artificial intelligence difficulty level.

First, it is important to determine how the ranking system will evolve, it should be dynamic and viable. In order to complete this task, it is important to look at different approaches to rating players. Instead of designing a new approach to the rating system, it is possible to use one that already exists (Coulom, 2010): ELO, Harkness, Glicko, etc. They are widely used in the competition world of chess and football, for instance. The model used is ELO for its simplicity and its ease to implement.

Elo System (Ranking the Player)

How do we predict the result of a match based on statistics? Arpad Elo designed a way to rank players in 1939 to answer that question (Coulom, 2010). He proposed to use the normal distribution function, *FN*, to predict the result of a match. With two players ranked *R1* and *R2*, the expectation value is given by *FN(R1-R2)*. The greater the difference is between the players, the higher the expectation value for the highest ranked player. Moreover, it is used to determine the player's rank after the match. A game can have three values as a result win = 1, draw = 0,5 and loose = 0. These values are given to the player that wins or loses or both in the case of a draw. The expectation value expresses the odds of winning, for example, a player with a ranking of 1400 playing against a player ranked 1600 has an expectation value of 0.18 for *R1*, so not a really good odds of winning. Based on that, after the match the ranks of the players are incremented or decremented according to the result. If the player has a better result than expected, their rank will increase and if their result is worse than expected, their rank will decrease. The following formula expresses the rating variation:

$$RA = RA + K \cdot [SA - E(A)] \quad (2)$$

As we can see in Formula 2, *RA* is the rank of the player A, *RB* is the rank of the player B, *SA* is the score obtained for player A (1 = victory, 0,5 = draw, 0 = lose). *E(A)* is the expectation value of player A; *E(A) = Fn(CA-CB)* where *Fn* is the normal distribution function. *K* is the update coefficient. For example, *K* could be the function of the number of games during the career of player A, *K*= 30 if the players have less than 30 games, $k = 15$ if $CA < 2400$ and $K = 10$ if $CA >$ 2400. *K* is the volatility of the player. The ELO system is still used today as the rating system by the World Chess Federation. ELO is used in our project for its capacity to be easily implemented and its simplicity.

Inside the game, the algorithm used is quite simple. According to the output of the player's action (i.e. combat) the formula to increase or decrease their ranks is going to be used. We used the following assumptions: first *K* is constant equal to 30, because the rank is more volatile at 30, second, the *SA* can be assumed to be one or zero according to the case. Since the game is programmed under Action Script and there is no math library, we implemented the normal distribution according to the following:

$$E(A) = y_a + (y_b - y_a) \cdot ((A - x_a) / (x_b - x_a)) \quad (3)$$

It has been decided to manually implement the cumulative distribution function in code, where $\mu = 0$ and $\sigma^2 = 0.2$. We implemented 20 intervals in a if/else structure. For *A,* when the right interval is found, we simply compute an average (see Formula 3) to get the point defined between the two points. This is actually an approximation of the cumulative distribution. In the Formula 3, *A* is the players' rank difference, *x* and *y* are the boundaries defined by the intervals of the function. For example, for a value of $A = 84$; $x_a = 80$, $x_b = 100$, $y_a = 0.66$ and $y_b = 0.69$.

We could have used Bayesian based algorithms that are more reliable to find the expectation value of a match based on the rank (Bolstad 2007). However, a Bayesian approach is harder to implement and control than the ELO algorithm and it required more processing time and resources, therefore, it was not chosen. *TrueSkill* or *Glicki* are other ranking algorithms based on Bayesian, although they do not consider the probability distribution, but only two parameters: the proposed output of the match and its trust level. These two levels are calculated based on Bayesian inference. Even more precise algorithms could be used such as Bradley-Terry or EDO, they were not selected for our project for their complexity and for more information on such models you should consult (Dangauthier, Herbrich, *et al.*, 2007).

Using the Player' Rank

Based on the ranking of the player it is possible to control the difficulty. First, it is crucial to determine the rank level for any kind of player. By looking at tennis players' ranks; a normal player has a rank around 1400, a very good player around 1800 and a weak player around 1200. Second, these levels have to be understood by the system. As the player wants to fight AI around their difficulty level, it is important to give them what they are expecting. Therefore the AI rank is based on the player's.

$$RankAI = rank +/- 200 \times rand(0,1) \qquad (4)$$

Using Formula 4, it is possible to have a stronger, weaker or normal AI. By controlling the AI ranking level, the system controls the difficulty. The AI difficulty is based on the game mechanics, which offers a challenge to the players by introducing conflict, like in a racing game, the faster and quicker the AI is, the harder the game becomes. Another example would be in a FPS, a strong AI will have a fast response to the players' actions, better accuracy, more health, etc. The idea is to find the layers of the difficulty of interest and implement those dynamically using the determined AI rank. Imagine a RPG, the adaptive challenge uses its own rank to determine the difficulty and bonuses. To make the game challenging to the player, the system adds a certain bonus (i.e. attack).

$$Bonus = rankAI / 1000 \times 2 + rand(0, rankAI/1000 + 1) \qquad (5)$$

In Formula 5, the opponent gains a bonus according to its rank that is based on the player's rank. So imagine the player has a rank of 1400 the AI 1450, the bonus for the attack will be between 2.9 and 5.35, with a basic attack of 10 that could end up being a really interesting challenge.

The flow chart in Figure 4 leads us to a better understanding of how the strategic model works. Every time the player performs an action that matters to the ranking, the result influences the ranking system as the player's rank influences the intensity of the game mechanics. This model is used to understand in understanding the player's experience. Therefore it could be used in any kind of game. Even the strategic rating system would be useful for online gaming where players of the same skill levels could be playing against each other. In general, it is a strong and easy approach to implement that leads to dynamic difficulty in games.

Figure 4. Strategic layer

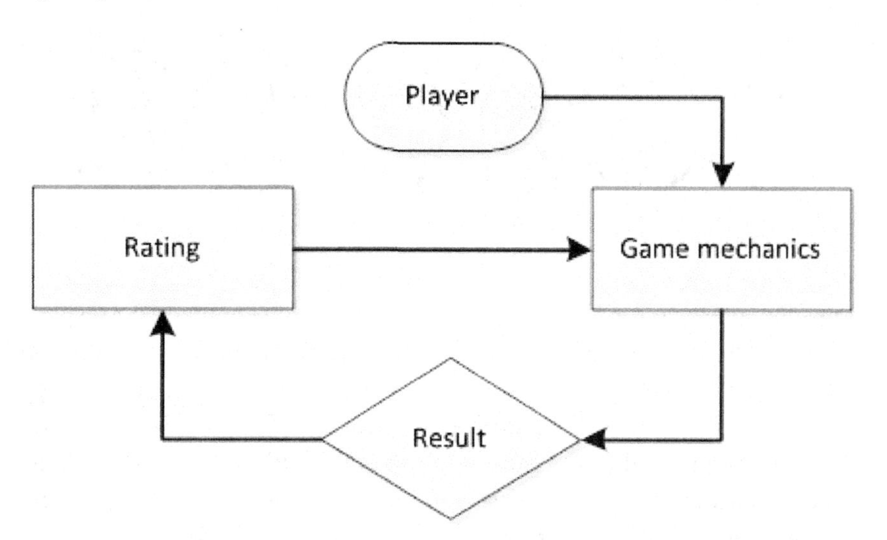

Implementation of the Model: Number to Number

The new adaptive model presented in the last section was implemented into the form of a simple calculation/fighting game named Number to Number. This game is in fact a version 2.0 of the former Number to Number Combat game, presented in (Tremblay, Bouchard, *et al.*, 2010), which implemented an earlier version of our adaptive model. It received good comments from players for its design. By choosing to keep the same basic game concept, the focus was on the players' experience based on the adaptive game mechanics and not on the design approach for enhancing their experience. It was really important to start with a design that was already interesting for the players.

The game was made using the open source Integrated Development Environment (IDE) *FlashDevelop* which is based on the MIT license. This editor allowed the developers to use the Software Development Kit (SDK) Flex released by Adobe Systems for the development and deployment of cross-platform rich internet applications based on the Adobe Flash platform. Flex was developed for programmers that found it challenging to adapt to the animation framework upon which the Flash Platform was originally designed. It allowed the developers to use websites that could utilizes streaming video, audio and a whole new set of user interactivity for online gaming. The IDE is an easy to use editor with syntax highlighting, bookmarks and tasks. It also includes a comprehensive find and replace dialog. The interface is intuitive with a very flexible panel. It is possible to add open plug-in based architecture. The *IDE* supports *ActionScript* 2 and 3, MXML, HaXe, XML and HTML. In general *FlashDevelop* is a very handy *IDE* that has been a great tool in order to complete this project.

The art assets were made using Adobe Photoshop CS5. Photoshop is a graphics editing program that uses an image layer system. A script has been developed in order to compile the layered animations into spreadsheets. As we processed all the frames of the animation in separate layers, our script program was able to take all layers and distribute them as seen in Figure 5. Inside the video game, the developers just had to load the graphics. The spritesheet contains 6 images, every image is showed one after the other during a short amount of time. This creates the illusion of an animated object.

Figure 5. Spritesheet example, minion level 3

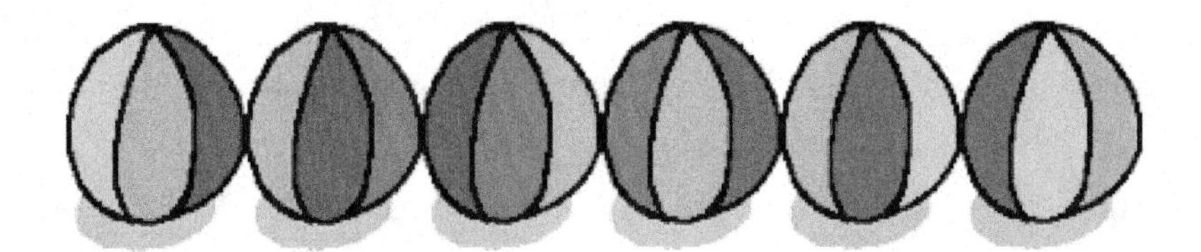

Figures 6 shows the interface for Number to Number (NtN). The game is quite simple, but some twists have been added to make sure the player does not get bored. As seen on the game play screen in Figure 6 (a), the goal of the game is to answer arithmetic questions in order to perform actions and win the combat against the opponent. Every correct answer fills up a bar on the right hand side on the screen; see Figure 6 (a).

The bar on the right hand side of Figure 6 (a) is the player's power, as you can see there is an arrow pointing to the power bar. In the game F5, F6 and F7 on keyboard correspond to the actions the player can perform; in this case F5 refers to "Punch". It means that if the player wants to perform that particular action, the level of power should be above that particular arrow. Those actions are situated over the arithmetic question, in this example, the first one is "Punch" and the two others are empty. These actions could have been "Heal", "Fireball", "Protect", etc. Therefore the player has perfect control over the actions that they want to perform. The basic idea of the game is for the player to beat the AI before it beats them.

Figure 6. (a) NTN game play (b) After combat screen in NtN

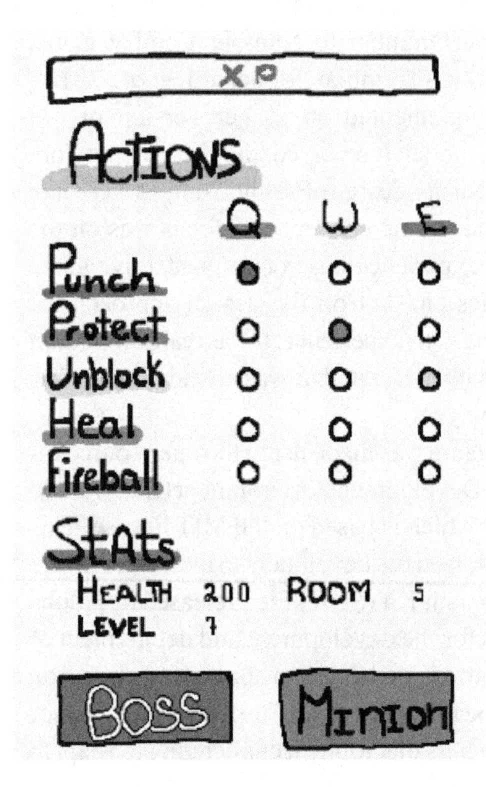

Figure 7. Adaptive flow apply to NtN

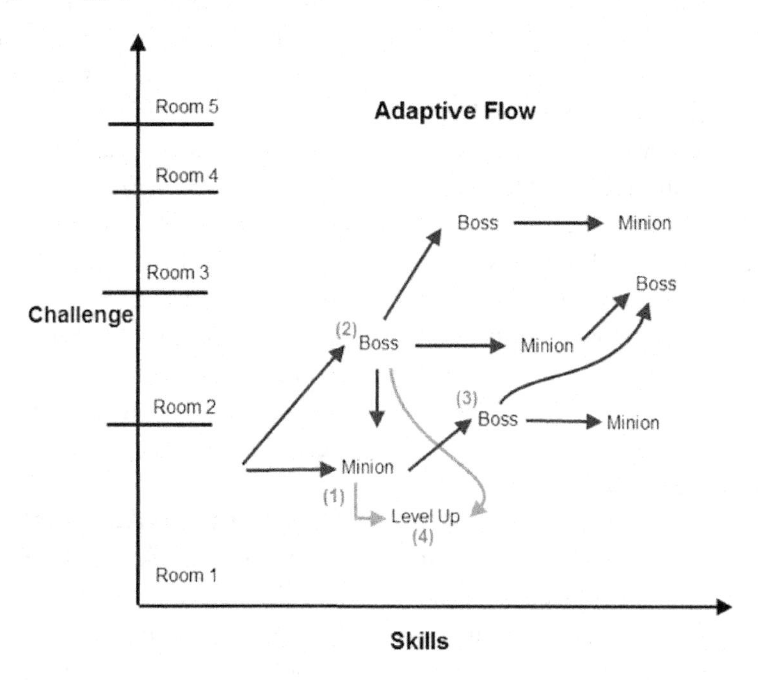

The player's health bar is on the left hand side of the screen. The player loses when their health bar is empty. They win when the AI's health bar is empty. The AI's health bar is located over the head of the monster on Figure 6 (a). After every positive combat for the player, they will gain some experience. When the player has enough experience points, they can level up. When leveling up the player will be granted new actions to perform. Figure 6 (b) represents the screen after a combat. The player can choose the actions they want to use for the next combat. The punch and fireball are used to attack the enemy. Protect and heal are used to protect themselves against the enemy. In order to move to the next room, the player must fight against the boss. The game is over when the player fights the calculator; the last boss in room 5.

Integration of Adaptive Flow

The game is separated into 5 rooms; each room has a minion and a boss to fight. The minion is always easier to vanquish than the boss. Every fight gives the player experience points and at every level the points needed to reach the next level gets higher. When the player gains a level, they have access to a wider spectrum of magic and strategies. Getting a magic power at the time helps the player to understand and master the game; they do not feel overwhelmed by the game mechanics they must learn. They start the game and have to understand the mechanics of answering math questions in order to increase their power and when they have enough power, they can attack their foe. Learning how to heal and protect themselves at the same time of learning the basic mechanics would have lead the game to fail in entertaining the player.

Leaving the choice to the player to fight against a minion or a boss creates greater room for expression in the game. The player can choose between fighting the boss directly for a greater challenge or preparing for their fight against the boss with a minion by leveling up their character. If they choose to build up their character (1) the fight in (3) is going to be less challenging then the one in (2). As you can see the player has access to a colorful expression of difficulty. All the fights in the game lead to more experience for the player and to leveling up (4).

Although the DDA system makes several little changes in the challenge of the game, the basic idea is still the same, some enemies are designed to be fought with a certain magic. If the player does not have it, it is still possible for them to succeed even though it drastically increases the challenge put forth by the enemy. In general, the possibility for the player to explore the game difficulty as they wish gives them a better control over their experience.

Core Combat Mechanics

As previously explained, it is crucial to implement the adaptive mechanics in accordance with the player's interactions. NtN is a combat game using arithmetic questions, so the main game mechanic is the arithmetic questions. The tactical layer uses 5 levels of difficulty, though the answers for every level are within the range of 0-9 on the keyboard to simplify the interaction. As a part of the review received regarding the first version of Number to Number Combat, players were expecting harder questions from the game. Table 2 represents the different intensity the arithmetic questions can attain.

To respond to this issue, harder questions were imbedded in the adaptive game mechanics. As the player starts a new combat, they have a *levelIntensity* of one. At this point, the questions are pretty simple. As the player answers questions, the *nowExperience* value follows the player's

Table 2. Tactics layer apply to NtN

levelIntensity	*Difficulty*	*Question*
1	low	1 + 3
2	medium	3 x 2
3	normal	3 + 10 - 4
4	high	(20 + 2) / 11
5	intense	3 x 10 - 23

behavior according to the time taken to solve the questions. For every question an average time to compare to the player's was designed.

Table 3 expresses the average time required to solve the level question. For instance, if the player has a *nowEperience* equal to 8.4 and a *levelIntensisty* of 3 and they answer the question $4 + 9 - 5$ in 1.8 seconds, then *nowExperience* will increase by $2.4/1.8 = 1.3$. If they succeed in answering the next question, then the level of intensity will increase. As the questions get harder, it takes longer for the player to answer them, therefore the challenge becomes harder.

Design of the Artificial Intelligence (AI) of NtN

The AI of NtN is designed based on the dynamic adaptive model proposed in this chapter. First, the difficulty should feed off the player's rank. In NtN, the difficulty is expressed by the player fighting a monster in a RPG. Therefore, the AI and difficulty is defined by: the timing rate, variety and power of the monster's attack.

Figure 8 represents the basic idea behind the AI design, based on a ranking system. The AI is

Figure 8. Working AI model

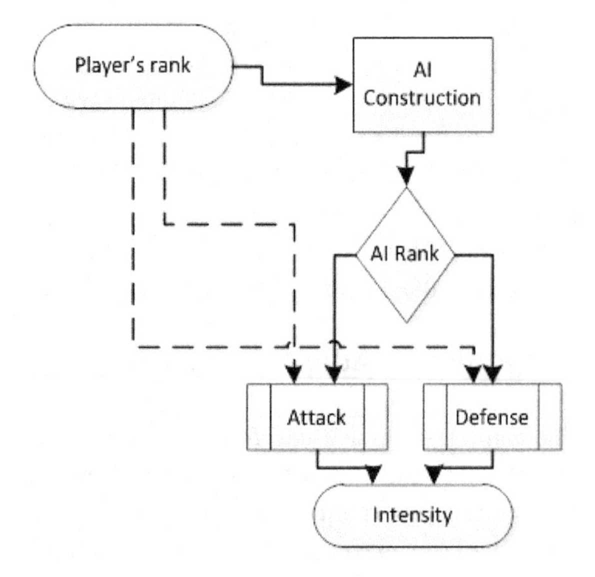

built based on its rank and the player's rank. The AI's actions are derived from the player's rank too. By using both the player's and the AI's rank, the AI can rapidly adapt to the player's level of intensity.

All opponents are designed differently. The basic monster, for example, is the red monster in Figure 6 (a). This opponent only attacks the players; it waits for a period of time, according to different variables and then hits the player. Its attack power is based on the player's data: maximum health, level, rank and number of victories. First, it is important to determine the rank of the AI. The rank is based on the following Formula 6 derived from the Formula 4. The Formula 4 randomized the difficulty to make each combat different and each AI development as well. The constants used in the formulas are from tweaking the system and play testing.

$$RankAI = playerRank - T + rand(0, 100) \times (nv + 1) \quad (6)$$

where, T (7) is a random function based on the number of the player's victories in a row.

$$T = rand(0, 30) \times nv \quad (7)$$

In Formula 6 and 7, *nv* is the player's number of victories in a row. The number of victories is used as positive feedback (Salen and Zimmerman 2003), where the new monster encountered is going to be harder every time the player wins a match. However, it is sometimes important that the player, even if they are getting stronger, play against weaker enemies, so they can build self-confidence. Then, the enemies could be weaker one third of the time. The rank is based on the following Formula 8.

$$rankAI = playerRank - T \quad (8)$$

The player will encounter weaker enemies according to the Formula 8. When the AI has ranking, it is now time to find its power and how long the time lapse between its attacks will be. The power of the attack is given by the following Formula 9.

$$Power = rand(0,1) + rand(0, rankAI/600) + min(nv/5, 5) + levelPlayer \quad (9)$$

RankAI is the monster's rank, determined by the Formula 6, *nv* is the number of victories in a row and *levelPlayer* is the rank the player is currently at. We add the *levelPlayer* to the formula because its attack is based on their level. The other part of the AI faculty is time between attacks and the resulting behavior is given by the following:

$$timeCheck = rand(0,T) + 3 - min(nv/3,3) + timeAverageMath \quad (10)$$

The variable T could be anywhere between one and three if the AI has a higher rank than the player. The value of *nv* is the number of victories the player has and *timeAverageMath* is the average time for the math question level the player is at, expressed in Table 3. To make the attack more dynamic, if the player has energy over 15 and more than three victories in a row then, as demonstrated in Formula 10, the time between attacks will decrease. *TimeCheck* is given by the following:

$$timeCheck = rand(0, 2.5) + 1 \quad (11)$$

The time between attacks is very important; it changes the player's experience by changing

Table 3. Timing for questions

levelIntensity	Time (*s*)	Question
1	1.2	1 + 3
2	1.8	3 x 2
3	2.4	3 + 10 - 4
4	4	(20 + 2) / 11
5	4.5	3 x 10 - 23

the difficulty of the game. A faster monster is a much harder monster to fight. By putting all these formulas together the developers constructed a basic AI design for the game. This design process was then used to build different AI characters with different actions. For instance, the calculator can hide from the player.

Generally speaking, the game is pretty easy to play and all those adaptive mechanics are hidden to the player so they cannot understand how it works and change the way the game evolves. NtN had pretty good reviews during the testing phase. The players really enjoyed the new twists added to the mechanics and the adaptive mechanics gave the players some real epic play sessions.

STUDY DESIGN, RESULTS AND DISCUSSION

Study Design

The primary goal with this game was to enhance the overall experience by giving the player an adaptive challenge. However, simply implementing a game using the adaptive approach does not assure that the model will fulfill its objectives. Therefore, a design study was designed to put the model to the test. The experimental protocol was quite simple. First, a group of 42 participants was recruited for the test, mainly university students. The participants were solicited to go online and play the game. At the end of their playing session, they were asked to complete an online survey of their gaming experience. This survey was carefully designed to give the needed clues about the performance of the model and the evolution of the difficulty level. Finally, the survey had, at the end, an open section where participants could write notes, comments and suggestions about the game and their gaming experience. The study was separated into two stages. The first stage was a study conducted with 32 players who tried the game presented in this paper.

The second stage of the study was conducted with 10 other students that played a version of the game that did not include DDA. Those 10 players did know they were playing a different version of the game, but were not aware of how the version they were playing differed from the original. The version without DDA was a lot easier to play than the original version, as the game was designed with one unique level of difficulty. Approximately 150 play sessions involving 42 players were conducted.

In Table 4, all the questions and the possible answers are represented. Most of the questions revolve around the player's experience according to the difficulty of the game. There were three parts to the survey: what type of players they considered themselves to be, what they thought of the game, and what they thought of the difficulty in the game. The survey took about 5 to 10 minutes to complete including the commentary section and it gave us a comprehension of the players' experience during their play session of NtN.

Result

The results are going to be presented with an emphasis on the results of the game with DDA, seeing as they are more numerous. The different kinds of results obtained from the DDA version and the non-DDA version when it is appropriate will be compared. In Figure 9 (a), most of the players considered themselves as experienced (44%), almost the same percentage considered themselves as casual (37%), and a small amount of players believed that they were inexperienced (19%). As the majority of the players were university students, and most of them from the computer science department, such results were expected. Half of the recruited testers play games for more than 10 hours a week. Hence, the majority of the testers know how to play games and are really interested in playing games. The game was really well received by the participants, 62% said that the game produced just the right amount of fun and 20% had a good time playing the game.

Table 4. Questionnaire questions

Questions	Possible answers
What kind of player are you?	Skilled; casual; new player
How many hours do you spend playing games in a week?	0-5; 5-10; 10 and more
Did you have fun playing NtN?	A lot; just enough; A little; Not at all
What is your general appreciation of the game?	Awesome; good; ok; bad
According to you, the game's difficulty was:	Too easy; just the right amount; too hard
What is your general appreciation of the game's difficulty?	Awesome; good; ok; bad
Did you realize that there was an adaptation/augmentation of the level of difficulty according to your skills?	Yes; no
Do you think adjusting the game difficulty was adequate?	Yes; no
Do you think it is important that the game difficulty followed your skill level?	Yes; no
If you were to retouch parts of the game, which ones would it be?	Art asset; AI; game balance; ergonomics; goal; actions; game title
Commentary	No limits

Those are indications that the game matched the player's expectations or at least that they enjoyed the concept of the game. As discussed, pleasure arises from the player discovering, mastering and being challenged by the game mechanics. In a general manner, it means the testers enjoyed the game, but it does not show a direct link to the game DDA. In Figure 9 (b) 81% of the players thought the game was just hard enough. This data is interesting for this study; it implies that the DDA system has the faculty to give an interesting challenge to the player. In other words, the game is designed in such a way that when the player is experiencing boredom or anxiety as seen in Figure 1 the game tries to put the player back in their flow zone.

However, out of the testers who played the game without DDA, 9 players out of 10 though the game difficulty was just the right amount. This data supports the idea that the game was well designed, but a question arises from this observation, "is enjoyment embedded in the game challenge?" The players might have been looking for childish amusement rather than challenge during the test. The game without DDA was easier to play than the original, which for a lot of players is more gratifying, as Koster (2004) points out; the player seeks for the easiest and effortless parts

Figure 9. (a) The type of player who tested the game (b) How the player perceived the game difficulty

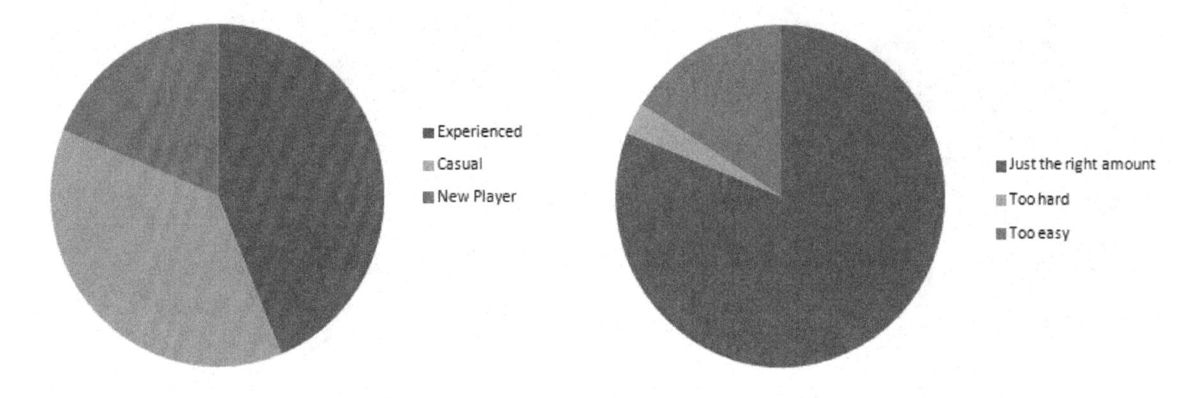

of the game. In this case, playing an easier version leads to a better general appreciation of the game, even though this appreciation is unhealthy (Koster 2004).

In addition, in Figure 10 (a), 41% of the players thought the difficulty was "awesome" and 31% thought it was good. This question allowed the tester to qualify the difficulty of the game. It supports the idea that the system is supposed to change the difficulty of the game according to the player's skill level. By giving the player an adaptive difficulty, the game gives the possibility of the player to enjoy uniqueness in their play session according to their skill level. Furthermore, out of the testers who played the version without DDA of the game, 60% thought the difficulty of the game was good and 10% said the difficulty was awesome. By comparing the two "awesome" results we can clearly see that the game using DDA has better results. The game without DDA lacks uniqueness in the play experience. Meanwhile, 81% of the players noticed an adaption of the game. This data could be explained by how the arithmetic questions would get harder and easier depending on the player's answer. Although, there is no data on which adaptation the player noticed, it could have been either the strategic or the tactical layer. Moreover, 60% of the players who played the non adaptive version of the game noticed adaptation. This data leads us to believe that the player is not aware of how to recognize adaptation and does not really understand the phenomenon of adaptation. Believing that the game has DDA could also lead the player to think the difficulty is just perfect all the time.

In addition, the majority of the participants (78%) who played the adaptation version felt the difficulty progression was adequate for the game. Furthermore, 94% of the testers said that it is important that the level of difficulty corresponds to the players' skill level. This corresponds with our observations about the need for DDA algorithms in today's games. In general, the majority of the players had fun playing the game (62%), as we can see in Figure 9 (b), as a result of the DDA, art assets and game design combined.

These results are very encouraging. This testing phase suggests that the DDA system enhances the players' experience in the game NtN mixed with the design of the game. Nevertheless, it should be pointed out that the results of a DDA system, namely the fun factor and the human experience in the game, are not trivial to evaluate and nearly impossible to quantify. It is why we have chosen an experimental approach based on a qualitative evaluation of the players' experience. Although the survey suggests that the DDA system worked as expected, the system also introduced some limitations.

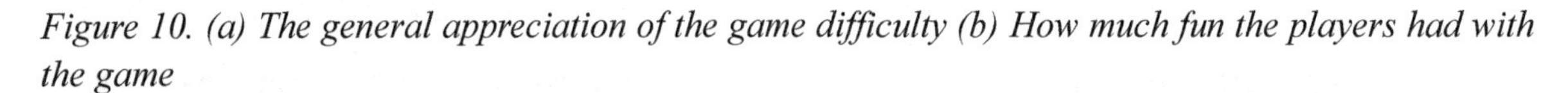

Figure 10. (a) The general appreciation of the game difficulty (b) How much fun the players had with the game

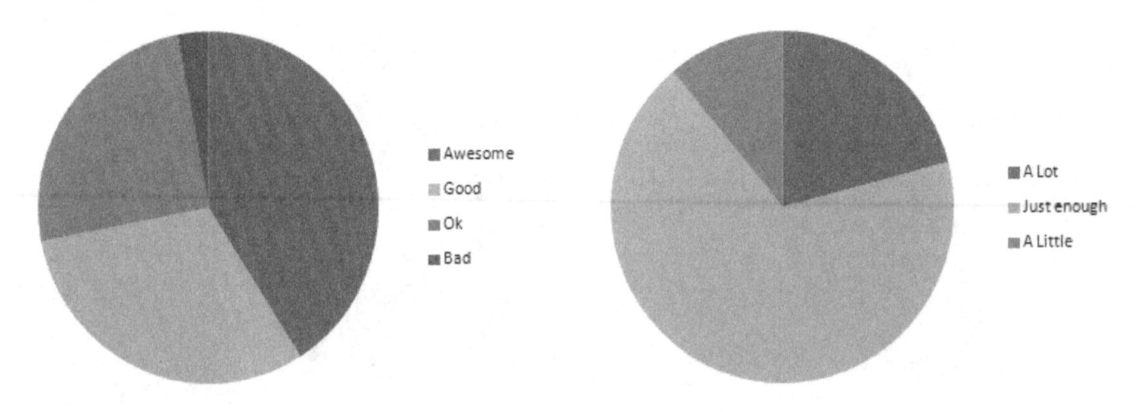

Figure 11. DDA limitations

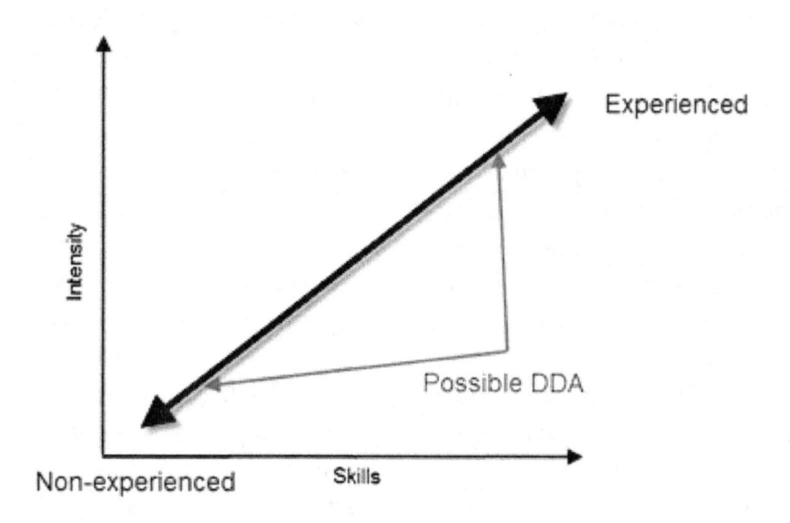

Discussion

During the survey, the last question asked the testers to tell the researchers what they would like to change in the game. These answers revealed some design limitations in the game. For instance, 60% of the players were expecting better art assets in the game, even though a lot more time was spent on the graphics than in the first prototype (Tremblay, Bouchard et al. 2010). Getting a real artist to do the art assets would be a way to solve this problem. The combat ergonomic did not please all players, 41% think it could be better. Some players pointed out that during a combat the player is only looking at one part of the screen; the arithmetic question. The players had the tendency to forget to look at the enemy and/or their own health bar. A way to solve that particular issue would be to put together the question, avatar, enemy and possible answer in the same location.

In Figure 9 (b), approximately 80% of players found the difficulty adequate to their skill level. What about the 20% that found the game too difficult or too easy? Figure 11 expresses that idea, when a very experienced player plays the game, the DDA system is not fast enough to keep pace with the player. The same phenomenon happens for non-experienced players, the game evolves too fast for them, while they might not fully understand the game.

The game should give them help, a support for their frustration of not being able to accomplish the task to be accomplished in the game. In Figure 11, the DDA system works great for the average player, which includes everyone one that falls within the two arrows. As pointed out by a tester, at some points during the game, when the combat becomes too difficult, in order to stay alive, the only possible action for the player is to heal themselves. The game should provide a support for the player so they can continue enjoying the game and fighting. In general, according to the comments and the survey, the game still stands out for its adaptive properties.

FUTURE RESEARCH DIRECTIONS

The first step for us is to get more data on the comparisons of games with DDA and without and the players' reaction to them. Based on these positive results, we plan to work on the implementation of our adaptive approach for other varieties of video games, such as puzzle, action, strategy, etc. It will be interesting to test the volatility of this approach. Moreover, the questions of learning from game mechanics could be enhanced using a DDA system as the player is experiencing flow for a longer period of time. DDA could be a clear asset in the world of serious games as players of all kinds can play them.

In a larger perspective, over time, games are going to include more and more adaptive models to reach a wider audience. It should also be noted that this work is a part of a wider project conducted by the LIARA laboratory of the University of Quebec at Chicoutimi. This project consists of implementing a larger model that includes support and DDA to serious game players such as Alzheimer patients. With this model, it will be possible to stimulate the cognitive functions of the patients, to propose adequate challenges to them, and to give assistance inside the game when they are in need. In the end, it will help slowing the degenerative process of the disease. DDA is expected to be a big part of serious gaming as well, as doctor are learning new procedures and the military is learning new tactics from video games, DDA systems will enhance their learning and gaming experience. Hopefully in the future it will be possible to see different generations being able to play the same games and experience an equal level of amusement by introducing DDA, as the game will adapt to the players and not the other way around.

CONCLUSION

As the game industry is rapidly growing and trying to reach new markets, it is essential to understand and master the process of making video games that every type of player would like to play. To achieve this, a game has to propose challenges that match the players' abilities (Csikszentmihalyi, 1990), in order to keep them in their zone of flow (Chen, 2006). Most games currently on the market do not provide any system to dynamically adjust the difficulty level to their players' abilities. Moreover, serious games that are played by the widest spectrum of player types and need to reach all of them as the instructor wants to maximize the teaching faculty of the game. Recent works on how to design and implement such systems are limited (Hunicke, 2005; Chen, 2006; Lindley & Sennersten, 2008; Goetschalckx, Missura *et al.*, 2010). In this chapter, a presentation of the concept evolving around adaptive difficulty was presented such as the player's pleasure in games, the player's expectation of challenge in games, different DDA models, etc.

In addition, a new adaptive model for dynamically adjusting, in real-time, the difficulty level of a game according to the player's performance, in order to enhance their game experience was introduced. The foundation of this model relies on a computational approach for Dynamic Difficulty Adjustment (DDA) (Hunicke and Chapman 2004) and it is an extension of a previous model proposed in (Tremblay, Bouchard et al. 2010). This model is flexible, can be adapted to multiple game genres, and offers two different layers of reasoning, short term adjustment (*tactical*) and long term adjustment (*strategic*). The *strategic* layer exploits the theory behind the well-known ranking system ELO (Coulom 2010), which allows the system to estimate the player's performance in the long term. This model has been implemented for validation in the form of a simple calculation/combat game called *Number to Number*. A study has been conducted using this prototype where 150 playing sessions have been completed by 42 players. Each player had to answer a detailed questionnaire on their playing experience. The results were very promising, showing the value of the proposed model and giving us indications for improving the model.

REFERENCES

Baker, R. S. J. D., Corbett, A. T., et al. (2006). *Adapting to When Students Game an Intelligent Tutoring System*. Proceedings from the *8th International Conference on Intelligent Tutoring Systems.* Jhongli, Taiwan

Bolstad, W. M. (2007). *Introduction to Bayesian Statistics*. Wiley-Interscience. doi:10.1002/9780470181188

Chanel, G., & Rebetez, C. Bétrancourt. M., & Pun, P. (2008). *Boredom, engagement and anxiety as indicators for adaptation to difficulty in games.* New York, NY: ACM Press

Chen, J. (2006). *Flow in games*. Unpublished Master's Thesis. School of Cinematic Arts Los Angeles, University of Southern California, California.

Coulom, R. (2008). *Whole-history rating: A bayesian rating system for players of time-varying strength. Conference on Computers and Games*, Beijing, China

Csikszentmihalyi, M. (1990). *Flow: The psychology of optimal experience*. New York, NY: Harper Perennial.

Dangauthier, P., Herbrich, R., Minka, R., & Graepel, T. (2007). Trueskill through time: Revisinting the history of chess. [Cambridge, MA: MIT Press.]. *Advances in Neural Information Processing Systems*, *20*, 337–344.

Gilleade, K. M., & Dix, A. (2004). Using frustration in the design of adaptive videogames. Proceedings from the 2004 *ACM SIGCHI International Conference on Advances in computer entertainment technology* (pp. 228-232). Singapore. Association for Computing Machinery.

Goetschalckx, R., Missura, O., Hoey, J., & Gartner, T. (2010). *Games with Dynamic Difficulty Adjustment using POMDPs*. International Conference on Machine Learning, Haifa, Israel.

Hunicke, R. (2005). The case for dynamic difficulty adjustment in *games*. Proceedings from the 2005 *ACM SIGCHI International Conference on Advances in computer entertainment technology* (pp. 429-433). Valencia, Spain: Association for Computing Machinery.

Hunicke, R., & Chapman, V. (2004). AI for Dynamic Difficulty Adjustment in Games. Proceedings from the *Challenges in Game AI Workshop, Nineteenth National Conference on Artificial Intelligence.*

Juul, J. (2009). *Fear of failing? The many meanings of difficulty in video games. The Video Game Theory Reader 2*. New York, NY: Routledge.

Koster, R. (2004). *A Theory of Fun for Game Design*. Phoenix, AZ: Paraglyph Press.

LeBlanc, M. (2000). *Formal Design Tools: Emergent Complexity, Emergent Narrative*. Game Developers Conference.

Lindley, C. A., & C. C. Sennersten (2008). *Game play schemas: From player analysis to adaptive game mechanics* (pp. 1-7). International Journal of Computer Games Technology.

Littlejohn, S. W. (1989). *Theories of Human Communication*. Belmont, CA: Wadsworth Publishing Company.

Pelland, S. (2009). *A Wii Bit of History*. Retrieved from www.Gamasutra.com

Rolling, A., & Adams, E. (2003). *Andrew Rollings and Ernest Adams on Game Design*. Berkely, CA: Pearson Education- New Riders Game series.

Salen, K., & Zimmerman, E. (2003). *Rules of play: Game design fundamental*. Cambridge, MA: MIT Press.

Saltsman, A. (2009). *Game changers: Dynamic difficulty*. Retrieved from http://www.gamasutra.com/blogs/AdamSaltsman/20090507/1340/Game_Changers_Dynamic_Difficulty.php

Schell, J. (2008). *The art of game design - A book of lenses*. Waltham, MA: Morgan Kaufman.

Sweetser, P., & Wyeth, P. (2005). GameFlow: A model for evaluating player enjoyment in games. *Computer Entertainment*, *3*(3), 3–3. doi:10.1145/1077246.1077253

Tremblay, J., Bouchard, B., & Bouzouane, A. (2010). Adaptive game mechanics for learning purposes: Making serious games playable and fun. Proceedings from the *International Conference on Computer Supported Education: session "Gaming platforms for education and reeducation"*, Valencia, Spain.

Um, S., Kim, T., & Choi, J. (2007). Dynamic difficulty controlling game system. IEEE International Conference on Consumer Electronics, 2007. ICCE 2007. Digest of Technical Papers. Las Vegas, Nevada.

Vygotsky, L. S. (1976). Play and its role in the mental development of the child. *Journal of Russian & East European Psychology*, *5*(3), 6–18. doi:10.2753/RPO1061-040505036

Wark, M. (2007). *Gamer theory*. Cambridge, MA: Harvard University Press.

ADDITIONAL READING

Arey, D., & Wells, E. (2001). *Balancing act: The art and science of dynamic difficulty adjustment*. Game Developers Conference. San Jose, CA.

Bleiweiss, A., Eshar, D., Kutliroff, G., Lerner, A., Oshrat, Y., & Yanai, Y. (2010). Enhanced interactive gaming by blending full-body tracking and gesture animation. Proceedings from *ACM SIGGRAPH ASIA 2010 Sketches* (pp.1-2). Seoul, Republic of Korea. *Association for Computing Machinery., 1-2*. doi:doi:10.1145/1899950.1899984

Bouchard, B., Bouzouane, A., & Giroux, S. (2007). A keyhole plan recognition model for alzheimer's patients: First results. *Journal of Applied Artificial Intelligence*, *22*, 623–658. doi:10.1080/08839510701492579

Brown, E., & Cairns, P. (2004). *A grounded investigation of game immersion. Proceedings from CHI '04 Extended Abstracts on Human Factors in Computing Systems* (pp. 1297–1300). Vienna, Austria: Association for Computing Machinery.

Coulom, R. (2010). Le problème des classements. *Science*, *293*, 20.

Crawford, C. (2003). *Chris Crawford on Game Design*. Berkely, CA: Pearson Education- New Riders Game series.

Dormann, C., & Biddle, R. (2008). Understanding game design for affective learning. Proceedings from the 2008 Conference on *Future Play: Research, Play, Share (pp.41-48)*. Toronto, Canada: Association for Computing Machinery.

Dunniway, T., & Novak, J. (2008). *Gameplay mechanics*. Independence, KY: Delmar Cengage Learning.

Gardner, M. (1970). Mathematical games: The fantastic combinations of john conway's new solitaire game "life". *Scientific American*, *223*, 120–123. doi:10.1038/scientificamerican1070-120

Gee, J. P. (2003). What video games have to teach us about learning and literacy. *Computers in Entertainment*, *1*(1), 20–20. doi:10.1145/950566.950595

Hunicke, R. (2005). The case for dynamic difficulty adjustment in games. Proceedings of the *2005 ACM SIGCHI International Conference on Advances in computer entertainment technology* (pp. 429-433). Valencia, Spain: Association for Computing Machinery.

Iacovides, I. (2009). Exploring the link between player involvement and learning within digital games. Proceedings from the *23rd British HCI Group Annual Conference on People and Computers: Celebrating People and Technology.* Cambridge, United Kingdom. British Computer Society: 29-34.

Jiang, C.-F., Chen, D.-K., Li, Y.-S., & Kuo, J.-L. (2006). Development of a computer-aided tool for evaluation and training in 3d spatial cognitive function. Proceedings from the *19th IEEE Symposium on Computer-Based Medical Systems* (pp. 241-244). Salt Lake City, UT: IEEE Computer Society.

Johnson, D. M., & Wiles, J. (2003). Effective affective user interface design in games. *Ergonomics*, *46*(13-14), 1332–1345. doi:10.1080/00140130310001610865

Juul, J. (2009). *Fear of failing? The many meanings of difficulty in video games. The Video Game Theory Reader 2.* New York, NY: Routledge.

Lazzaro, N., & Keeker, K. (2004). *What's my method? A game show on games. Proceedings from the CHI '04 extended abstracts on Human factors in computing systems* (pp. 1093–1094). Vienna, Austria: Association for Computing Machinery.

Lin, C., Lin, C. M., Lin, B., & Yang, M.-C. (2009). A decision support system for improving doctors' prescribing behavior. *Expert Systems with Applications*, *36*(4), 7975–7984. doi:10.1016/j.eswa.2008.10.066

Mandler, J. M. (1984). *Stories, Scripts and Scenes: Aspects of Schema Theory.* Hillsdale, NJ: Lawrence Erlbaum Associates.

Meloni, W. (2009). *Gaming's New Market Dynamics and the Importance of Middleware.* Retrieved from http://www.gamasutra.com/blogs/WandaMeloni/20090819/2789/ Gamings_New_Market_Dynamics_and_the_Importance_of_Middleware.php.

Morales, C., Martinez-Hernandez, K., Weaver, G., Pedela, R., Maicher, K., & Elkin, E. ... Nattam, N. (2006). Immersive chemistry video game. Proceedings from the *ACM SIGGRAPH 2006 Educators program* (pp. 50). Boston, Massachusetts. Association for Computing Machinery.

Nacke, L., & Lindley, C. A. (2008). Flow and immersion in first-person shooters: Measuring the player's gameplay experience. Proceedings of the *2008 Conference on Future Play: Research, Play, Share* (pp. 21-88). Toronto, Canada: Association of Computing Machinery.

Pagulayan, R. J., Keeker, K., Wixon, D., Romero, R., & Fuller, T. (2003). User-centered design in games. *The human-computer interaction handbook: Fundamentals, evolving technologies and emerging applications* (883-906). Mahwah, New Jersey: L. Erlbaum Associates Inc.

University of Pittsburgh Medical Center. (2006). Computer-based 'games' enhance mental function in patients with alzheimer's. *ScienceDaily.* Retrieved from http://www.sciencedaily.com/releases/2006/10/061023193307.htm

Siwek, S. E. (2007). *Video Games in the 21st Century: Economic contributions of the US entertainment software industry* (pp.09-10). Washington D.C., USA: Entertainment Software Association.

Srinivasan, V., Butler-Purry, K., et al. (2008). *Using video games to enhance learning in digital systems.* Proceedings of the *2008 Conference on Future Play: Research, Play, Share* (pp.196-199). Toronto, Canada: Association for Computing Machinery.

Sykes, J., & Brown, S. (2003). *Affective gaming: Measuring emotion through the gamepad. Proceedings from CHI '03 extended abstracts on Human factors in computing systems* (pp. 732–733). Ft. Lauderdale, FL: Association for Computing Machinery. doi:10.1145/765891.765957

Tychsen, A., & Canossa, A. (2008). Defining personas in games using metrics. Proceedings of the *2008 Conference on Future Play: Research, Play, Share* (pp. 73-80). Toronto, Canada: Association of Computing Machinery.

KEY TERMS AND DEFINITIONS

Design Approach: The process of designing non computational dynamic difficulty. For example, letting the player choose the enemies' bonus at the beginning of every level, i.e. *Bastion* from *Supergiant Games*.

Dynamic Difficulty Adjustment (DDA): The action of dynamically changing the difficulty of the game to suit the player's skill level (design or procedural approach).

Enjoyment, Fun and Pleasure: These terms are used to describe the euphoric feeling that a game induces in the player.

Flow: The state of mind that a participant achieves with a high degree of focus and enjoyment during any activity.

Level of Intensity: A variable that represents the player's skill level in the game.

Pleasure in Games: Feeling of euphoria created by the action of playing a game.

Procedural Approach: The action of dynamically changing the difficulty using mathematic models using data such as Level of intensity.

Chapter 6
Application and Evaluation of Artificial Intelligence Algorithms for StarCraft

Luke Deshotels
University of Louisiana at Lafayette, USA

ABSTRACT

This chapter will discuss several algorithms and techniques used in artificial intelligence that can be applied to StarCraft (Blizzard, 2009) and other similar real-time strategy (RTS) games. A significant section of the chapter is devoted to teaching the reader how to use the Brood War API (Heinermann, 2011), BWAPI. This API allows developers to create agents that can play StarCraft with or without human interaction. Each section after the tutorial introduces an algorithm or technique, explains the relevant concepts, shows an example of its implementation in StarCraft, and evaluates its performance and applications. The algorithms and techniques covered include swarm intelligence, gathering algorithms, expert systems, and hill climbing. The example implementations are kept simple to allow readers to follow along and implement them as practice.

INTRODUCTION

Overview of StarCraft

By allowing programmers to see algorithms come to life in challenging scenarios StarCraft makes an excellent environment for learning new algorithms. This game is over a decade old and is still played throughout the world. It is treated like a national sport in South Korea with competitors making hundreds of thousands of dollars each year. An application programmer interface, Brood War API that allows the programming of intelligent agents to play StarCraft has recently been released. These factors make StarCraft a prime platform for developing an RTS AI.

For readers planning to create an AI that plays competitive StarCraft the first step is learning to play the game. As with any AI, the developer should have a firm understanding of the environment his agent will act in, the tasks it must accomplish, and any obstacles or noise that could prevent

DOI: 10.4018/978-1-4666-1634-9.ch006

the task from being accomplished. In this case the environment is an alien planet, the task is world domination, and the noise is caused by monsters that want to eat the agent. Unfortunately a tutorial on playing StarCraft is beyond the scope of this chapter. However there are numerous StarCraft guides (Team Liquid, 2010) and video tutorials on the Internet.

Opportunities for AI

StarCraft and real-time strategy games in general offer new challenges for AI that chess could not. By vastly expanding possible actions, introducing time as a new dimension, and making the environment more realistic, StarCraft is a perfect environment for intelligent agents to grow, learn, and compete. At the time of this writing even the best AI could not beat professional human players at StarCraft, but this is likely to change in a few years.

A vast variety of AI algorithms and techniques can be applied to StarCraft including swarm tactics, expert systems, gathering algorithms, hill climbing, and more. By applying such algorithms intelligent agents are able to use their superior speed to produce uncanny behavior with the game's units. It is truly fascinating what sort of behaviors can emerge as an agent fights for its life.

BACKGROUND

The best way to learn StarCraft is to play the game, but a brief overview may help prepare the reader for references made in this chapter. The game takes place in a science fiction universe where space travel is trivial, but survival is not. StarCraft puts players in the position of a military leader that must decide how to manage a small colony and an army of up to 200 units in order to defeat an opponent with similar resources.

Setting

The StarCraft universe is incredibly diverse. It includes wasteland planets, ancient civilizations, thriving Earth-like areas, and more. There is a wide variety of technology available depending on the economic status and intelligence of those developing it. Backwater planets with revolver wielding marshals exist along side societies with advanced teleportation devices and monsters with psychic abilities.

Two things fuel the armies of StarCraft. Minerals are present in large deposits on most planets and are used for currency, food, energy, and building material. Vespene Gas is available from geysers and used as fuel, but it is often more difficult to gather than minerals. In the game these resources determine which units, structures, or technologies a player can afford.

There are three playable races in the game. The Terran colonies of humanity, the insect-like horde of the Zerg, and the mysterious tribes of Protoss are in constant combat. The game is well balanced so that no race has a significant advantage over another.

The Terran represent humanity in the future. Their technology is composed of advanced versions of mankind's current weapons. Tanks, stealth jets, and nuclear weapons are all still available, but invisibility, powered exoskeletons, and flying buildings are just a few of the advancements they have made. The Terran have relatively little knowledge of the other two races and spend a significant amount of time fighting among themselves.

The Zerg Swarm is based on rapidly spreading but controlled mutation. Large brain-like creatures called Cerebrates are able to control other creatures in the swarm by using psychic abilities. The Zerg travel from planet to planet consuming any resources available while searching for new specimens worth mutating to serve their purpose. For example if the Zerg infested Earth they might capture a lion and mutate it by mixing its DNA with other creatures already controlled by the Zerg. The end result might be a smarter, more aggressive,

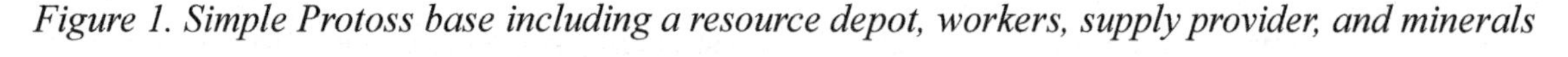

Figure 1. Simple Protoss base including a resource depot, workers, supply provider, and minerals

armor plated lion. Most Zerg units begin life as larva and morph into one of the monsters created by twisted mutations. Even Zerg structures are the result of a Zerg Drone mutating into a large immobile creature. The Zerg swarm does not use technology, but instead causes an appropriate mutation when a new situation is encountered.

Much about the Protoss is unknown. They have humanoid shapes and are similar to humans in size but do not have mouths. Their technology is more advanced than that of the Terran, but it is usually more expensive. Much of the Protoss technology is based on light with blades made of lasers being a common infantry weapon. The Protoss are also able to use invisibility, teleportation, and force fields to their advantage. Artificial intelligence is evident in a few of the robots used in Protoss armies. This race values honor, religion, and tradition with its units willing to sacrifice themselves for a greater cause. The goal of the Protoss is to eliminate the Zerg.

In short the three races are in combat with themselves and each other throughout the game's storyline which is incredibly detailed. The Zerg want to dominate the universe, the Protoss want to eliminate the Zerg, and the Terran are caught in the middle. For example, the Protoss may want to destroy a Terran planet that is starting to be invaded by the Zerg.

Gameplay

The objective for winning a typical game of StarCraft is simple. Destroy all of your opponent's structures before the opponent destroys all of yours. However the variety of units and structures that can be created and the actions they can perform complicate the game significantly.

When a normal match begins each player has four workers, a resource depot, 50 minerals gathered, enough supplies for 9 units, mineral clusters to harvest, and a Vespene Geyser for gas. The player is expected to use these resources to build a colony and create an army to destroy the enemy. How one accomplishes this can vary greatly, but most of the time players build up a strong economy first by investing in more workers. A simple base for the Protoss can be seen in Figure 1.

A player's economy is made up of three elements, minerals, gas, and supply. Supply represents the amount of units the army can support. For the Terran supply is food, water, etc. Zerg supply represents psychic control, and Protoss supply is energy produced by special structures. Players can produce additional workers to increase gathering rates, but must also increase supply to support the additional units. Maps in the game have additional mineral clusters and geysers that players can harvest by building expansions. An expansion is an additional resource depot located near the resources to minimize the distance workers must travel to deposit their goods.

Aside from determining the victor of a game structures serve valuable functions. Buildings in StarCraft hold resources, produce units, provide technology, and offer defensive functions. For example a Terran Star Port allows the production of flying units while a Protoss Photon Cannon can attack any enemy that comes near it.

The units available in StarCraft are also very diverse. Units are often divided into flying and ground types. Certain units can only fight ground units, others can only hit flying targets, and a few can attack both. Units can also be categorized as workers, fighters, and casters. Workers gather minerals and build structures, fighters directly attack the enemy, and casters can use special abilities like calling down nuclear bombs instead of standard attacks. One infamous unit is the dog-sized Zergling. Due to its low resource cost and fast movement speed it is often used to quickly blitz an opponent. The simplicity of Zerglings makes them a great example unit for several of the algorithms to be discussed.

Combat in StarCraft can become very complex when several units and skilled players are involved. Certain attacks are more effective against some targets and less effective on others, so players often try to position more resilient or less expensive targets in defensive positions while targeting their attacks at the most vulnerable targets. Placing armored units in front of snipers is an example of this strategy. Armies also benefit from coordinated attacks such as attacking the same target to kill it faster and reduce the enemies attack potential. Lastly, terrain has a significant impact on combat. Units on higher ground have advantages over those below them, and natural choke points like ramps can force units into vulnerable formations.

These options and requirements for game play produce several challenges for human players. With an unfriendly user interface and fast-paced action StarCraft can be quite difficult to learn.

SETTING UP

Installing the Brood War API

Please note that at the time of this writing the latest version of BWAPI was 3.6. The installation process may change slightly, but these instructions should remain relevant overall.

Developers will need four things before installing BWAPI.

- StarCraft Brood War 1.16.1
- Microsoft Visual C++
- Windows XP/Vista/7
- ChaosLauncher

StarCraft can be purchased and downloaded online at Blizzard.com (Blizzard, 2011). The easiest way to patch StarCraft to its latest version is to start the battle.net multi-player service provided in the game. This will link your installation to the StarCraft servers and download any missing updates.

Microsoft Visual C++ (Microsoft, 2011) is an interactive development environment that can be used to develop agents that use BWAPI. It can usually be obtained for free by students through the Microsoft Developer Network Platforms Academic Alliance, MSDNAA. The express edition of Visual C++ is also free. The 2008 version of Visual C++ seems to be more compatible with BWAPI than the 2010 version.

Unfortunately BWAPI can only be used on Windows systems. However BWAPI and StarCraft run quite well on Virtual Machines, VMs. This allows developers to keep their native operating systems and even run multiple agents on the same device.

ChaosLauncher (MasterOfChaos, 2011) is a program that allows code injection directly into StarCraft while the game is running. This program simply uses one's code to simulate an extremely fast human user, and the game should run as if it is receiving normal commands. ChaosLauncher can be downloaded at http://winner.cspsx.de/Starcraft/Tool/Chaoslauncher.zip.

Here is a step by step sequence for installing BWAPI.

1. Create a directory called chaoslauncher inside your system's starcraft directory. By default this directory should be located at C:\program files\starcraft
2. Extract the contents of Chaoslauncher.zip into the chaoslauncher directory created in step 1
3. Download the latest version of BWAPI from code.google.com/p/bwapi/
4. Extract the BWAPI directory in the downloaded zip file to the starcraft directory
5. Copy the files in the BWAPI directory's chaoslauncher directory to the starcraft directory's chaoslauncher directory
6. Copy the files in the BWAPI directory's starcraft directory to the starcraft directory
7. Copy the files in the BWAPI directory's WINDOWS directory to C:\WINDOWS

Some of the listed steps can be quite confusing because of the directories with similar names. Here is an approximate outline of how the relevant directories should look.

- C
    - WINDOWS
        - Contents of BWAPI\WINDOWS
    - program files
        - starcraft
- BWAPI
- Contents of BWAPI\starcraft
- chaoslauncher
    - Contents of Chaoslauncher.zip
    - Contents of BWAPI\chaoslauncher

Now that everything is set up the next step is running a test program. BWAPI provides example AI for testing the installation and to act as a starting point for developers.

1. Open ExampleProjects.sln in the BWAPI directory
2. Use Visual C++ to view ExampleAIModule.cpp. In the solution explorer on the left of the IDE it should be under ExampleAIModule and then under Source Files
3. Look for a drop down menu in the top center of the IDE that is set to Debug and change it to Release
4. Click Build then Build Solution
5. This should compile the example program into a dynamic-linking library, dll, called ExampleAIModule.dll which will be located in starcraft\BWAPI\Release
6. Copy the dll from starcraft\BWAPI\Release\ExampleAIModule.dll to starcraft\bwapi-data\AI\ExampleAIModule.dll
7. Start the chaoslauncher by running starcraft\chaoslauncher\Chaoslauncher.exe. Be sure to run Chaoslauncher as an administrator, otherwise strange errors will appear.
8. Click the settings tab and make sure the Installpath points to your starcraft directory
9. Click the Plugins tab and make sure BWAPI Injector (1.16.1) RELEASE has a check in its box
10. Highlight BWAPI Injector (1.16.1) RELEASE and click on the Config button
11. In the bwapi.ini window that appears replace the lines

ai = bwapi-data\AI\ExampleAIModule.dll, bwapi-data\AI\TestAIModule.dll
ai_dbg = bwapi-data\AI\ExampleAIModuled.dll, bwapi-data\AI\TestAIModuled.dll

with the following

ai = bwapi-data\AI\ExampleAIModule.dll
ai_dbg = bwapi-data\AI\ExampleAIModuled.dll

12. Close and save bwapi.ini
13. Click the Start button for Chaoslauncher
14. Create a single player custom match in StarCraft
15. When the game starts the AI should cause your workers to begin harvesting minerals and your resource depot should begin producing a worker.

Do not give up hope if BWAPI does not work on the first try. The installation process is incredibly complex since the goal is to implement significant hacks for the game. Sometimes the best thing to do is start over and step carefully through each part of the process.

Automating BWAPI Tests

Fortunately only steps 4, 6, 7, 13, and 14 of running the test program need to be repeated to run code after the initial set up. Since developers will need to test their agents often it is beneficial to automate as many of these steps as possible.

The next step that can be automated is the game creation process. By configuring the bwapi.ini file located in starcraft\bwapi-data\bwapi.ini programmers can automate menu selections such as races, game types, and maps. There are even configurations to automatically start a new game when one finishes. Here is an example of settings that would automatically create a single player game for an AI playing as Zerg against a Terran opponent on the map called Ice Floes. Assume settings not listed are left at their default values.

```
auto_menu = SINGLE_PLAYER
map = maps\BroodWar\(2)Ice Floes.scx
race = Zerg
enemy_count = 1
enemy_race = Terran
game_type = MELEE
```

Developers familiar with scripting languages should be able to easily construct a script to move the dll file to the appropriate location and start Chaoslauncher. This script can also be configured to load preset BWAPI automenu configurations for testing different situations.

Lastly, typing the command "/speed 0" without the quotes during gameplay will cause a game to run as fast as the host machine's CPU can process it. Choosing a higher value than 0 can be done if a slower speed is desired. For example "/speed 50" will set the game to a speed easier to follow in a live demonstration.

Additional Tools

There are additional libraries available that save developers from reinventing the wheel when it comes to several tasks necessary in most StarCraft matches. Two worth mentioning are the Broodwar Terrain Analyzer (Heinermann, 2010), BWTA, and the Broodwar Stand Alone Library (Heinermann, 2011), BWSAL.

BWTA is a very helpful tool that simulates a player studying a map before a match. This library is able to analyze any StarCraft map and provide detailed information about resources, choke points, terrain types, and even base locations. Aside from learning the map without exploring it the BWTA assists agents in calculating minimum ground distance and finding shortest paths over ground. This is a highly recommended tool that seems very stable and does not interfere with anything else an agent may need to do. The only cost is taking time to analyze maps played for the first time. This process could take a few minutes and may look like the game has frozen. Once a map is analyzed the data is stored locally and the process should not need to be repeated.

BWSAL is more controversial than BWTA. By simplifying common tasks throughout every match such as placing buildings, managing workers, and spreading expansions BWSAL's goal is to make the API as simple as possible so that developers can focus on the unique strategies their bots will employ. While the Terrain Analyzer simply provides data about the map, the Stand Alone Library requires control of the agent's units to be helpful. Developers will need to decide whether any loss in flexibility is worth the benefits presented by BWSAL.

THE BASICS

Overview of Programming With BWAPI

By using extensions BWAPI agents can be programmed with Python or Java, but by default it is programmed with C++. The Python extension seems to cost significant computational overhead, but the Java extension is very practical and was used by the 2010 world champion StarCraft AI quite successfully. For simplicity all examples provided in this chapter will be written in pseudo code, but feel free to translate them into your favorite language supported by BWAPI.

The simplest way to make an agent is to start with the example AI used for testing the installation. It provides a good foundation for how to select units and issue commands. This section will step through several fundamental techniques necessary for building a well-rounded agent that can produce and use a variety of units. The following examples will refer to specific StarCraft content, so an understanding of the game is highly recommended. All of the following examples will utilize the Zerg race, but each technique should be easily adaptable to the other races.

The Zerg are an insect-like alien race that is able to mutate very quickly. The base structure is very similar to an ant hill. A large central structure called the hatchery acts like a queen. Workers gather resources and feed them to the hatchery which produces larva for the army. These larva can be mutated into soldiers, workers, or controllers. The need for controllers is proportional to the amount of units in the Zerg army. Additional structures can be made to gain access to more powerful mutations at the cost of resources and time. Managing resources and choosing which units to produce are a few of the basic techniques your AI will need to perform in most games.

There are two very important functions in the ExampleAIModule that will be used for most of the coding necessary to make an agent. The function onStart performs a series of commands when a match starts and is a great place for determining the value of game dependent variables such as starting locations, the race of the enemy, and which map is being used. The most important function in the starter code is onFrame. This function will run several times each second and is used to dynamically issue commands throughout each game. Do not underestimate the power of an AI when adding more commands to this function. Computers can easily issue tens of thousands of actions per minute, APM, while even the best players can only manage a few hundred.

Produce Workers

The first thing most players do is produce additional workers in order to increase the resource gathering rate. Luckily the ExampleAIModule already provides an example of creating a worker in the onStart function. Developers will want to place code similar to the pseudo code (Code 1) in

Code 1. Producing workers

```
for each unit in self.units
{
  if unit.type is resource_depot
  {
    if unit.race is not Zerg
    {
      unit.train(worker);
    }
    else
    {
      for each larva in self.larva
      {
        larva.morph(Zerg_Drone);
        break;
      }
    }
  }
}
```

the onFrame function in order to produce workers throughout the game.

Cycling through units with a for loop is very useful with BWAPI. It creates an iterator, unit, that traverses the list of all the agents units. A similar loop can be constructed to iterate through all enemy units. Note that in BWAPI buildings, soldiers, workers, and controllers are all units. The function getType can be used to check what type of unit the iterator is currently referencing. Since the hatchery is the Zerg resource depot we want to provide commands when the iterator finds it. If the current race were not Zerg the command would simply be to have the ResourceDepot train a worker. However Zerg rely on larva, so the getLarva function is called to provide a list of available larva. The first larva is then given the command to morph into a Drone, the Zerg worker unit.

Assign Workers to Harvest Minerals

Now that a worker has been produced it needs to start contributing to the hive by harvesting minerals. The starter code (Code 2) also provides an example of this in the onStart function.

This code is similar to the previous example in that it iterates through all units and issues commands to the appropriate types of units. The target units to receive commands are workers that are not doing anything. This prevents the code from commanding workers already gathering minerals to do it again on every frame. Such loops can cause units to freeze and become useless. The code uses each idle worker's current location to find the closest mineral and assigns the worker to begin harvesting it.

Increase Supply Limit

Each race must manage a supply limit. Without control the Zerg army cannot handle more units. Therefore if programmers implement the previous two examples they will not be able to have more than nine workers at a time without producing a controller. Code 3 provides a simple example that will create controllers when more supply is required.

The Overlord provides control for the Zerg. Readers should notice the code to create an Overlord is very similar to the code necessary for creating a worker. The main differences are the condition that checks to see if the supply limit is about to be reached and the morph command that causes a larva to mutate into an Overlord.

Build an Extractor

StarCraft players must manage time, supply, minerals, and Vespene Gas throughout the game. While it is possible to play the game without gathering gas, it would severely limit a player's options and is not advised. In order to gather gas a special structure must be built on a Vespene Geyser. This structure is comparable to a drilling rig above a source of oil. The most difficult part of building structures with BWAPI is choosing where to place them. The simplest means is to use the StarCraft campaign editor

Code 2. Harvesting minerals

```
ClosestMineral = NULL;
for each unit in self.units
{
  if unit.type is worker && unit is not busy
  {
    for each mineral in game.minerals
    {
      if closestMineral is NULL || distance from unit to
      mineral < distance from unit to closestMineral
      {
        closestMineral=mineral;
      }
    }
    if closestMineral != NULL
    {
      unit.harvest(closestMineral);
    }
  }
}
```

Code 3. Creating a controller

```
for each unit in self.units
{
  if unit.type is resource_depot
  {
    if unit.race is Zerg
    {
      for each larva in self.larva
      {
        if self.supplyTotal - self.supplyUsed <= 4
        {
          larva.morph(Zerg_Overlord);
          break;
        }
      }
    }
  }
}
```

to place a layout of the buildings you plan to create during a game and record their tile positions. This process will need to be done for each starting position and your agent's current location will need to be determined before choosing the correct set of building placements. It should be noted that in StarCraft maps the X coordinate increases from left to right, and the Y coordinate increases from top to bottom. Therefore a very low Y value is closer to the top of the map. (Code 4)

This example references the player's starting location and the size of the map to determine if the player is on the top half or the bottom half of a map with two starting positions. Structures in BWAPI use the TilePosition data type to de-

Code 4. Building an extractor

```
top=true;
if self.startLocation.y > game.mapHeight / 2
{
  top=false;
}
if top is true
{
  extractorPosition = (top_X_target, top_Y_target);
}
else
{
  extractorPosition = (bottom_X_target, bottom_Y_target);
}
for each unit in self.units
{
  if unit.type is worker
  {
    unit.build(extractorPosition, Zerg_Extractor);
    break;
  }
}
```

termine their location. Tile positions represent cells laid out in a grid across the map. Players may notice them when trying to manually place a building during a match. It should be noted that the Tile Position covered by a building's top left corner determines that building's location. The other data type that represents a location in BWAPI is Position. Position represents a finer degree of accuracy and is used for detecting where a player clicked on the battleground. TilePosition coordinates can be converted into Position coordinates by multiplying the x and y values by 32. For example, TilePosition(10, 20) represents the same location as Position(320, 640).

Harvesting Vespene Gas

Players will only want a few workers harvesting gas at each base. Three harvesters is the magic number since there is a limited amount of access to the Extractor. The example (Code 5) uses a few useful tricks to assign three workers to harvest gas. In fact, if a gas harvester should be blown up while performing its duty another will be pulled from gathering minerals to take its place.

Several new techniques are introduced in this example. The integer pacer counts how many frames have passed since the beginning of the game. It can be useful for creating delays by taking its mod in a conditional. The variable gassers is an integer that counts how many workers are gathering gas. In this example there must be only one extractor and the goal is to keep 3 workers gathering gas at all times.

Iterating through all units is the easiest way to count how many of each unit are in a player's army. Note that the condition to assign the variable extractor checks whether the Zerg Extractor is being constructed. Structures will appear as units before they are done being built. This must be handled carefully as it cannot serve the functions it is expected to until it is no longer being constructed. A pointer to a unit can be saved by assigning the iterator's value to a Unit* data type. The conditional checking for the value of frames-

Code 5. Harvesting Vespine gas

```
pacer ++;
gassers = 0;
extractor = NULL;

for each unit in self.units
{
  if unit.type is worker && unit is gathering gas
  {
    gassers ++;
  }
  if unit.type is Zerg_Extractor && unit is not being built
  {
    extractor = unit;
  }
}

for each unit in self.units
{
  if pacer % 10 == 0
  {
    if unit.type is worker && unit is mining minerals
    && gassers < 3 && extractor != NULL
    {
      worker.harvest(extractor);
      break;
    }
  }
}
```

Passed mod 10 and the break statement are both used to prevent BWAPI from assigning too many or even all workers to harvesting gas. Imagine if the break statement were removed. The iterator would tell every worker to begin harvesting gas because until the for loop finished the value of gassers would not become greater than or equal to three.

It is recommended that readers practice these basic techniques until they become proficient at producing a respectable army and even resisting attacks from StarCraft's built in AI.

SWARM INTELLIGENCE

Overview of Swarms

It is quite impressive that simple creatures such as ants and bees can exhibit incredibly well co-ordinated attacks when they become threatened. Enough individuals following just a few simple rules can produce intelligent behavior. Even fish traveling in schools have been found to follow simple rules. A fish in a school knows not to get too far from any other fish and not to get too close to anything. The fish only has to process a little information, but any observers see a dizzying array of incredibly synchronized swimmers exhibiting what seems to be a hive mind. The cumulative intelligence of the school with no centralized source can be referred to as swarm intelligence.

These techniques are often successfully applied in AI to produce emergent behavior. Simple agents following just a few rules are able to mimic life-like behavior of simple animals.

Swarm intelligence is a simple, but extremely powerful and aggressive means of exploring the map and assaulting the enemy. Most players experienced with the game will agree that Zerglings are ideal for swarming. They are small, fearless, and very fast. These traits make them similar to ants, and when using swarm intelligence they act just like them.

Efficient Searching

StarCraft uses a feature called the fog of war to prevent players from seeing enemy actions unless that player has a unit present to see what is happening. This allows enemies to hide structures and units in hopes that they won't be attacked. Therefore, it is critical to explore the map efficiently throughout the game and gather as much information about the enemy as possible. Code 6 uses Zerglings and swarm behavior to traverse most of the map very quickly.

This deceptively simple code causes behavior similar to that of ants when someone steps on their hill. Every Zergling will run to a random point on the map attacking any enemy it sees. When it reaches the designated point it will be assigned a new random point to find. One Zergling following this algorithm will just look confused, but fifty of them can explore a large area in very little time. An example of several Zerglings exploring a maze is presented in Figure 2.

Coordinated Attacks

One of the most powerful behaviors ants exhibit is coordinated attacks. When an ant is crushed it releases an alarm pheromone that drives other ants into a frenzy and attracts them to its location. This technique can be implemented in BWAPI with the onDestroy function. Code 7 will alert all Zerglings if one of their own is killed and where the act occurred. This causes one of the most impressive displays of swarm behavior StarCraft has to offer as dozens or even hundreds of Zerglings smell blood and seek revenge.

This code runs every time a unit is destroyed. If the killed unit was one of the agent's Zerglings then this algorithm commands all of the agent's remaining Zerglings to rush to the location where the unit died and attack any enemy found on the way there. This usually forms very effective chain reactions. For example, when a single Zergling is killed by guards on the outskirts of an enemy base, the swarm is alerted and begins pouring into the area. As they move closer in order to kill the guards Zerglings will be killed inside the base and more will rush to those locations. This cycle should continue until either the enemy or all of the Zerglings are dead. Put more simply, if a Protoss warrior were to kill one Zergling, dozens more would assault his position. Figure 3 illustrates this result.

While this technique is powerful it is still too simple to beat more advanced AI or human players. The most obvious weakness is that Zerglings alone cannot fight flying or invisible units. Therefore, an enemy with superior technology will usually have a significant advantage. Secondly it is not always best to rush in blindly in StarCraft. There are some situations where caution could be beneficial. A few units such as Siege Tanks have devastating attacks that can be used at very far distances. If a Zergling is killed at long range by a Siege Tank then his brethren will run blindly into a trap as they gather upon a graveyard while being bombarded from afar. Swarm intelligence can be improved by applying it to a more versatile group of units and trying new rules for deciding when, where, and how to attack. Thresholds can be introduced as well. With this code even if the agent has only two Zerglings they will act as a tiny swarm. Perhaps it would be best to wait until there are twenty before starting to explore the map.

GATHERING ALGORITHMS

Reasons to Gather

As emphasized in the previous section there is power in numbers. However it does no good to send in units single file and have them killed

Code 6. Simple Zergling behavior

```
for each unit in self.units
{
  if unit.type is Zerg_Zergling
  {
    if unit is not busy
    {
      target_X = randomNumber % game.mapWidth * 32;
      target_Y = randomNumber % game.mapHeight * 32
      unit.attack(target_X, target_Y));
    }
  }
}
```

Figure 2. Zerglings following the rules for scouting are able to rapidly explore a maze

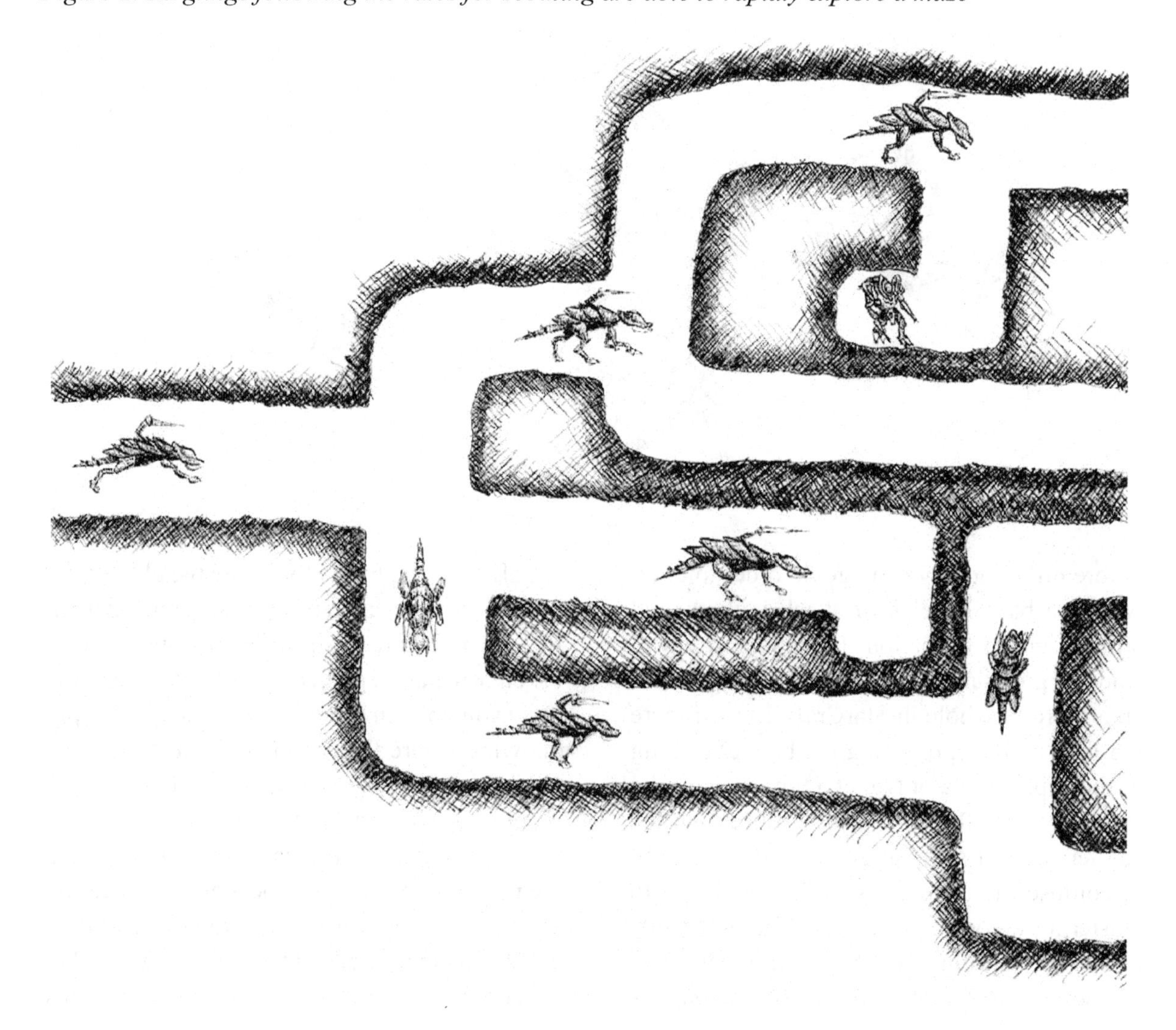

Code 7. Zergling swarm behavior

```
void onUnitDestroy(destroyedUnit)
{
  if destroyedUnit.type is Zerg_Zergling && unit.player is self
  {
    for each unit in self.units
    {
      if unit.type is Zerg_Zergling
      {
        unit.attack(destroyedUnit.position);
      }
    }
  }
}
```

Figure 3. When a Zergling is destroyed it sends out an alert calling all other Zerglings to the location

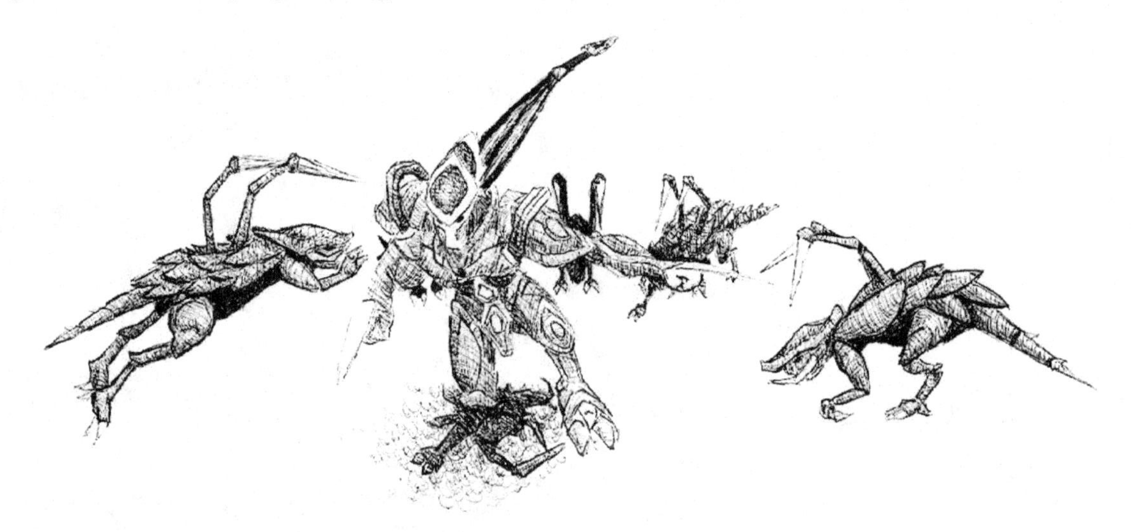

before inflicting any damage. In order to overwhelm a base it helps to attack with several units in group formation. Imagine a group of soldiers guarding a bunker, the equivalent of an armored foxhole in StarCraft. Even if there are fifty units approaching the bunker as long as they appear one or two at a time they can be picked off. If all fifty assault at once then the soldiers do not stand a chance. It can be easy to confuse the algorithms in this section with clustering algorithms. It should be noted that clustering algorithms focus on categorizing units into which clusters they belong to, and gathering algorithms physically create the clusters by taking scattered units and moving them together. In StarCraft time is a precious resource, so an efficient means of gathering units into an army is quite valuable. In some cases it is useful simply knowing where the center of an army is. This section provides examples of and evaluates several gathering algorithms.

The simplest means of gathering units is to have them all move to the same location. This location should be chosen with respect to the location of all units and the intended target the units need to assault as a group. Two means of choosing this

Figure 4. Due to poor formation this group of Hydralisks will die before dealing any significant damage

rendezvous point will be demonstrated and evaluated with a group of Hydralisks.

The Hydralisk is a versatile Zerg unit with the ability to launch poisonous spines at enemies. They are significantly stronger than Zerglings and can attack flying units. However with a higher cost and slower movement it is not acceptable to send Hydralisks to attack in small numbers. The Hydralisks in Figure 4 are attacking a Terran Bunker and are steadily being defeated as they approach in single file.

It is also necessary to assign a detector to the group of Hydralisks in order to locate invisible units. The Zerg units responsible for providing control, the Overlord, can also sense invisible enemies within a wide range. It is ideal for the Overlord to stay with the Hydralisk army where it can detect enemies while being protected. By using one or more of these algorithms every few hundred frames in conjunction with an intended target to kill, the group will stay together and be more likely to attack the target as a unit.

Foundation Code for Gathering Examples

Code 8 serves as a foundation that takes the rendezvous point referred to as HunterKiller and has the units use it to gather. The other two code examples in this section will provide different means of determining the position value for the HunterKiller variable. The name HunterKiller is a reference to a stronger than average Hydralisk in the game, and the variable HuntersMark indicates the next target position the army should be assaulting. HuntersMark could be where a unit died, the location of a

Code 8. Foundation for gathering examples

```
if pacer % 10 == 0 && hcount > 0
{
  for each unit in self.units
  {
    if unit.type is Zerg_Overlord
    {
      unit.move(HunterKiller);
      break;
    }
  }

  if pacer % 150 == 0 && pacer / 50% 2 == 0 && hcount < 40
  {
    for each unit in self.units
    {
      if unit.type is Zerg_Hydralisk && unit is not attacking
      {
        unit.attack(HunterKiller);
      }
    }
  }

  if pacer % 150 == 0 && pacer / 50% 2 == 1
  && HuntersMark.x != -1 && hcount > 10
  {
    for each unit in self.units
    {
      if unit.type is Zerg_Hydralisk && unit is not attacking
      {
        unit.attack(HuntersMark);
      }
    }
  }
}
```

threatening unit, or any other location where the army is needed. The hcount variable represents how many Hydralisks a player currently controls and the pacer is simply a counter for how many frames have passed since the beginning of the game.

Closest to Target

An aggressive means of gathering that is similar to techniques humans use involves grouping around the unit closest to the intended target. Human players will often move armies in short increments that simulate this type of behavior. All units move toward the specified area, bunch up, and then move again to the next spot. By doing this instead of trying to move all the way to the target in one motion the army better maintains its formation. Code 9 uses the Hydralisk closest to the target as the HunterKiller for all units to gather around.

This method performed quite well in matches. The army moved quickly while maintaining its power in numbers. The Overlord was able to spot invisible units, but since the Hunter-

Code 9. Gathering: closest to target

```
mindist = 999999;
for each unit in self.units
{
  if unit.type is Zerg_Hydralisk
  {
    distance = distance from unit to HuntersMark;
    if mindist > distance
    {
      HunterKiller = unit.position;
      mindist = distance;
    }
  }
}
```

Killer was always on the front lines of battle, so was the Overlord. This caused the army's detector to die and need to be replaced often. The other drawback of this algorithm becomes apparent if the army is defeated. Imagine a scenario in which the Hydralisks meet fierce resistance and their numbers begin to dwindle. Even if there is only one Hydralisk left fighting this algorithm will send all newly created Hydralisks to assist and form a chain reaction in which the troops are sent to the enemy a few at a time. Overall this algorithm was effective and won quickly if the enemy did not have significant defenses. However, if the army's first attack was resisted the algorithm had little chance of recovering. It is recommended for early rushing attacks when mobility is crucial and the enemy is unlikely to survive the first assault. An example of Hydralisks attacking as a team is illustrated in Figure 5.

Pack Leader

A slightly more complex but often more reliable solution involves finding the Hydralisk with the least distance from all other Hydralisks. This causes an inch-worm effect as the army alternates between approaching the target and gathering around the "pack leader" which is usually near the front lines with the majority that had been approaching the target. The pack leader is in no way stronger or significant in any way except that at the relevant point in time it was closest to every other Hydralisk. The Overlord is also commanded to stay as close to the pack leader as possible. (Code 10)

This algorithm is more reliable. While it causes the army to double back frequently in order to regroup it is able to keep the Overlord in a safe position as it detects for most of the army. It even causes its own retreating behavior when the army's numbers are dwindled. Imagine a scenario in which there are 4 recently spawned Hydralisks near the agent's base and only three fighting the enemy. The HunterKiller should be chosen as one of the units near the base causing the front line and the Overlord to retreat and regroup with the next wave. This algorithm's weakness is its loss of mobility. If the army must suddenly return to base to fend off a surprise attack it is not likely to make it in time and may even send newly created units away from the base in order to gather with the army.

Developers should feel free to test their own gathering algorithms as well as create hybrids and modifications of the ones presented. A simple solution to reduced mobility when responding to surprise attacks is to turn off any gathering behavior if the target is on the agent's side of the map. If one needs to preserve the aggressive behavior of gathering with the unit closest to the

Figure 5. The group of Hydralisks was able to attack together and defeat the Terran Bunker

Code 10. Gathering: pack leader

```
mindist = 999999;
for each unit in self.units
{
  if unit.type is Zerg_Hydralisk
  {
    distance = 0;
    for each otherUnit in self.units
    {
      if otherUnit.type is Zerg_Hydralisk
      {
        distance += distance from unit to otherUnit;
      }
    }
    if mindist > distance
    {
      HunterKiller = unit.position;
      mindist = distance;
    }
  }
}
```

target while protecting supporting units such as a detector then there is nothing wrong with using one gathering technique for the supporting units and another for the soldiers.

EXPERT SYSTEMS

Overview of Expert Systems

There are several situations in life that require consulting an expert. If one is feeling ill or one's car will not start an expert can often determine the cause by making inferences based on the information available. These inferences can be as simple as a one step process (high temperature implies fever) or incredibly complex (symptoms, genetic background, age, weight, lifestyle, previous ailments, and more could all be factors in diagnosing an illness). Expert Systems are artificially intelligent agents that simulate the same process in order to solve problems and answer questions that would normally require experts.

The inferences used by expert systems are generally complex sets of if and then statements referred to as rules. These rules are usually generated by consulting experts until a sufficient amount of rules have been constructed and the machine is able to make decisions similar to those the consulted experts would have made. These systems have been quite successful and can be fairly threatening to some professionals. Suddenly physicians and others that were held in high regard for wielding a vast store of knowledge are in danger of being replaced by machines that can quickly reference databases and inference engines in order to make the same decisions.

Expert systems make a great addition to any StarCraft agent's arsenal of algorithms. These expert systems could consist of massive inference engines that identify and counter popular strategies used in the game or a few simple rules that help units make decisions. The following examples implement a few functions that serve as very basic expert systems.

When to Use Special Attacks

Several units in StarCraft can use special abilities with unique effects such as infecting enemies with diseases, entangling enemies in webs, or even using mind control to make a unit betray his allies. These abilities require a limited resource called energy, so they should be used sparingly and when they will be most effective.

One such technique can instantly destroy any ground unit with organic parts by rapidly mutating it into two savage Zerg abominations called Broodlings. The technique is call Spawn Broodling and is used by the Zerg Queen, a flying unit that uses abilities instead of attacks. This ability has one of the highest energy costs of all abilities and can strike only one unit each time it is used, so targets must be chosen carefully. Therefore a simple expert system was created for choosing targets by consulting a few casual gamers and an online StarCraft guide (Team Liquid, 2010) and implementing a few rules. Code 11 examples demonstrate a means of commanding queens to use the ability and using an expert system to determine which units are worthy targets.

While any respectable expert system would fill dozens of pages with inferences, this simple implementation should serve as a concrete example for developers hoping to implement more complicated expert systems for their agents.

With a limited state space it was simplest to create a white list of units that are vulnerable to Spawn Broodlings and are expensive to produce. Although the machine is simply following a set of rules the end result seems to be intelligent decision making.

When to Burrow

With very little health and no ranged attack Zerglings on solo missions aren't expected to survive

Code 11. Commanding a Zerg Queen to use a special attack

```
for each unit in self.units
{
  if unit.type is Zerg_Queen
  {
    for each enemy in enemy.units
    {
      if shouldBrood(enemy) is true
      {
        unit.cast(Spawn_Broodlings, enemy);
        break;
      }
    }
  }
}

bool shouldBrood(enemy)
{
  if enemy.getType is Terran_Siege_Tank_Siege_Mode
  || Terran_Siege_Tank_Tank_Mode || Terran_Goliath
  || Protoss_High_Templar || Protoss_Dark_Templar
  || Protoss_Dragoon || Zerg_Defiler || Zerg_Lurker
  || Zerg_Ultralisk
  {
    return true;
  }
  else
  {
    return false;
  }
}
```

encounters with the enemy. However due to a very primitive, but effective technique they can be quite resilient. They hide by burrowing. Several Zerg ground units have the ability to quickly dig shallow holes and hide in them while still sensing their surroundings. This is useful for ambushing, spying, and avoiding combat.

Zerglings make excellent scouts, but it can be frustrating to replace them if they keep dying. It is also vital to use Zerglings for harassing workers and buildings anytime these are undefended. Therefore Code 12 helps Zerglings decide when to burrow and unburrow. Figure 6 shows a burrowed Zergling spying on a Terran Marine who is unaware of its presence.

This example commands all Zerglings to burrow when the expert systems decides that they should. The appropriate time to burrow is determined by a simple set of rules acting as an expert system in the shouldBurrow function.

The rules suggest burrowing if there is an enemy nearby that is alive, could attack the Zergling, is not a worker, and is not near the agent's main base. These are very helpful by allowing Zerglings to harass the enemy and even fight to the death when defending the hive. This is an example of behavior human players have trouble reproducing because it would require monitoring every scout at all times, but it is very effective at preserving units and frustrating the enemy.

Figure 6. This Zergling will remain burrowed and spying until there are no threats within its sight range

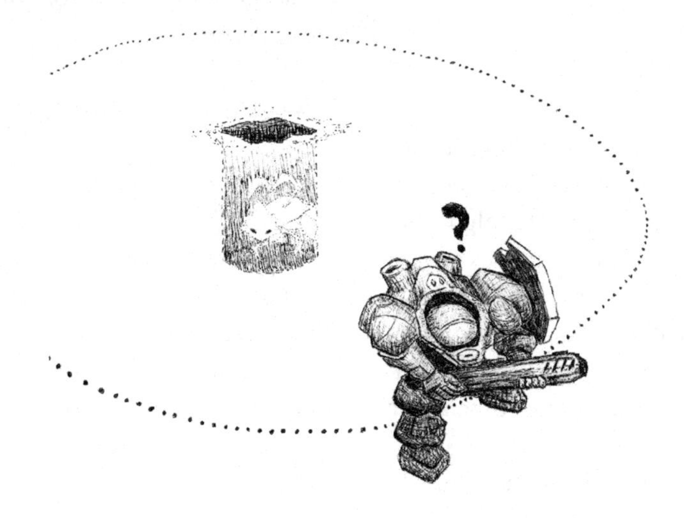

Code 12. Zergling burrow functionality

```
for each unit in self.units
{
  if unit.type is Zerg_Zergling
  {
    if shouldBurrow(unit) is true && unit is not burrowed
    {
      unit.burrow();
    }
    if shouldBurrow(unit) is false && unit is burrowed
    {
      unit.unburrow();
    }
  }
}

bool shouldBurrow(unit)
{
  for each enemy in enemy.units
  {
    if enemy is alive
    && enemy is near unit
    && enemy can attack unit
    && enemy.type is not worker
    && enemy is not near self.baseLocation
    {
      return true;
    }
    else
    {
      return false;
    }
  }
}
```

HILL CLIMBING

Overview

Hill climbing is a local search algorithm used for finding local minima or maxima and is valued for its simplicity and intuitiveness. The algorithm takes a starting point and iteratively moves toward the "best" location one unit away. It then repeats this process until all locations one unit away from the current one are "worse" than the current location. The final result is a local minima or maxima depending on the goal of the search.

As an example consider an ant that wants to reach the top of a two dimensional hill. The ant is currently located near the bottom of the left side of the hill. If it uses the hill climbing algorithm it would check to see if going left or right would put it in a higher position. Since moving right would go higher, it would take one step in that direction and repeat the process. Eventually the ant will reach the top of the hill and stop because stepping left or right will only put it at a lower elevation.

This algorithm is not without weaknesses, but it is still widely used despite them. Even when moving one step at a time it is possible for an agent to overstep its true goal. The ant is unlikely to stop on the exact top of the hill. Once it is a single ant step from any point that would set it at a lower elevation the position is judged as good enough. This can be addressed by calibrating the step size or by running multiple tests and taking their average results. Overall it must be noted that hill climbing will only produce a near optimal result. The other weakness is apparent when there are multiple hills to be climbed and one of them is higher than the rest. By only looking one step ahead an agent will know if it has reached the top of the hill it is currently on while knowing nothing about any other hills in the area. This problem can be addressed with the use of multiple runs at random starting points, but in its simplest implementation hill climbing should not be used for finding global minima or maxima.

Avoiding Threatening Units

There is a rock, paper, scissors feel to most of the StarCraft units. Players often debate how much this matters if the units are managed properly, but some units simply have advantages over others. One example is Terran Missile Turrets against Zerg Queens. None of the Queen's abilities are effective against missile turrets while surface to air missiles are quite adept at killing any flying unit.

Code 13 example uses a basic hill climbing algorithm to keep queens away from turrets. If a queen comes near the firing range of a turret it quickly evaluates its surroundings and moves in the direction that will put it farthest away from the turret. Imagine the turret being at the bottom of a pit the Queen has slipped into. The hill climbing is a little reversed, but still useful in reaching high points on the edge of the pit, which are far enough from the turret to be safe. A queen using this algorithm to recognize and evade a Terran Missile Turret is shown in Figure 7.

This example simplifies the state space by using only eight directions to "step" in. The true interface for StarCraft offers a full 360 degree range of movement, but using more options quickly becomes unfeasible during a match.

A significant advantage of using the iterative method of hill climbing is its convenient handling of dynamic conditions. Imagine the Queen is being chased by a threatening unit. During each frame it will move to the position farthest from its threat and display a fleeing behavior until it has reached a safe distance or something else eliminates the threat. This algorithm contains a clause that cancels evasive behavior if the Queen has been assigned to cast Spawn Broodlings. Once its mission is complete the evasive behavior will begin and

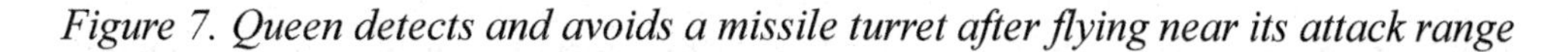
Figure 7. Queen detects and avoids a missile turret after flying near its attack range

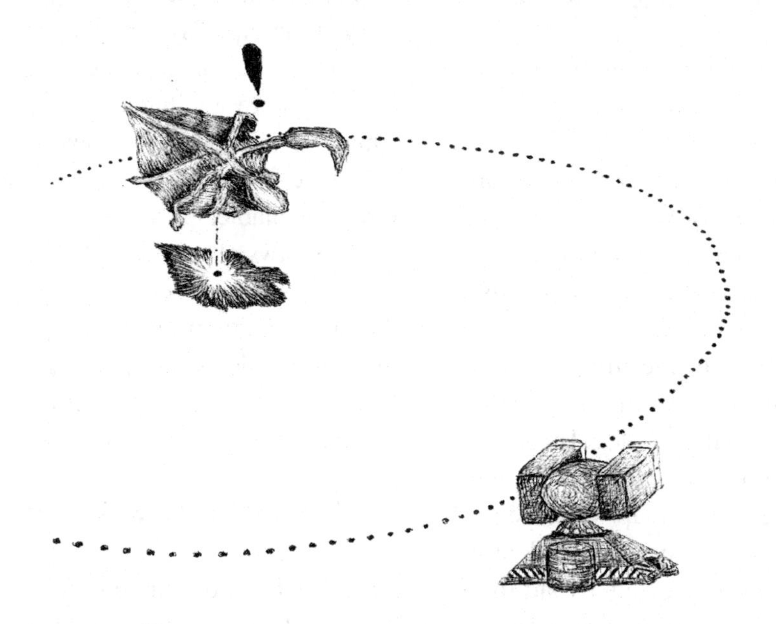

Code 13. Keeping Queens away from turrets with hill climbing

```
for each unit in self.units
{
  if unit.type is Zerg_Queen
  {
    totalDist = 0;
    for each enemy in enemy.units
    {
      if enemy can attack unit
      && unit is near enemy
      && unit is not trying to cast SpawnBroodlings
      {
        temp = unit.position;
        for each direction in (N,S,W,E,NE,NW,SE,SW)
        {
          target = one tile in direction from unit.position;
          if distance from target to enemy
          > distance from temp to enemy
          {
            temp = target;
          }
        }
        unit.move(temp);
      }
    }
  }
}
```

the Queen will run away from any threats it had previously ignored.

One weakness of this implementation manifests itself when multiple threats are present. Imagine a Queen being flanked on the east and west by two threats. This algorithm will process a command to evade the first threat then overwrite this order with a command to evade the second target. While trying to distance itself from the second threat the Queen would only move closer to the first. Another flaw occurs if the queen is near the edge of the map. It may determine that a non-legal position is the safest place to be. The queen will then keep pushing the edge of the map while being attacked by the threat that cornered it. These disadvantages can be addressed by adding exceptions to this example or by implementing a more sophisticated technique called potential fields.

FUTURE RESEARCH

Expanding Provided Implementations

Several of the algorithms have been simplified for use as introductory experiments for programmers to use as practice. It is highly encouraged that developers improve these algorithms to be more robust, more efficient, and more effective overall.

The implementation of Swarm Intelligence can easily be improved by implementing more rules or adding exceptions to the current ones. For example if a Zergling is killed by an invisible unit a detector should be sent to support the approaching swarm.

Gathering algorithms can be optimized by finding the best parameters for where and how often to gather. With the use of automated tests or machine learning, optimal parameters could be found to suit a combination of units on a given set of maps.

Perhaps the most obviously expandable algorithm is the expert system. By simply implementing more rules programmers can usually make their AI more intelligent. These rules could be simple or incredibly complex depending on the inference engine used and the computational power available.

The flaws of the example hill climbing algorithm have already been discussed. By expanding its functionality programmers should be able to make it handle avoiding multiple threats simultaneously as well as preventing it from being "cornered" on the edge of a map.

Implementing New Algorithms

Not all of the algorithms presented are the optimal solutions for the problems they were used to solve. There are more complex and sometimes simply more appropriate techniques available for the challenges provided by StarCraft. For situations where algorithms are not available there are opportunities to create new ones.

For example the gathering algorithms could be greatly improved or even made unnecessary with the implementation of marching and formation algorithms. If the units were able to maintain their group structure while moving then there would be less need to regroup. However due to obstacles present in most StarCraft maps such as choke points, canyons, and enemy units, maintaining formation is often difficult for ground units. This seems to be an open problem in the field of StarCraft AI, but hopefully readers of this chapter will be up to the challenge.

Man vs. Machine

As of the writing of this chapter professional StarCraft players are still able to consistently defeat even the best StarCraft AI. Despite the near omnipotent speed and access to game data that StarCraft AI possess they are still no match for the adaptability and creativity used by human competitors. Kasparov's defeat by an AI in chess was a major breakthrough for artificial intelligence, and defeating a world champion with an AI in StarCraft should be an even bigger accomplishment for the field.

CONCLUSION

In summary, it has been demonstrated that StarCraft makes an excellent environment for practicing, improving, and even creating algorithms that can be applied to other video games. With this chapter developers will be able to utilize the BroodwarAPI and work their way up from simple algorithms to more complex and more powerful agents. The competitive nature of the game should motivate developers to continue developing stronger agents for years to come.

The first technique covered was swarm intelligence. It was demonstrated that with just a few simple rules emergent behavior could be produced. This makes a great algorithm for beginners. By combining quick results with unpredictable behavior implementing this algorithm makes for an awesome introductory project.

Gathering algorithms allow an agent's army to maintain formations and take advantage of strength in numbers. By adjusting where and how often an army gathers a programmer can control how aggressive an army is as well as assign supporting units to position themselves where they can be most useful.

The effectiveness of expert systems was demonstrated by using them to help units make decisions throughout the game. Expert systems are a practical and intuitive means of helping an AI replicate the same decisions a human might make. Applications of sophisticated systems were discussed as having high potential.

Practical applications for the hill climbing algorithm were explored for evasion of threats. While simpler solutions for avoiding threats are possible, implementing this algorithm in the StarCraft environment makes it very easy to observe and understand how the algorithm works.

The installation guide, basic tutorial, and algorithm implementations provided by this chapter should be more than enough to allow anyone with prior knowledge of programming to construct an AI capable of defeating Blizzard's built-in AI and novice human players. Artificial intelligence and StarCraft enthusiasts will find building a custom AI to be both dangerously addictive and incredibly rewarding.

Bibliographical Remarks

The titles StarCraft and StarCraft Brood War and all referenced units and structures of StarCraft are the property of Blizzard Entertainment. The figures were drawn by Leia Kagawa and are adaptations of images owned by Blizzard Entertainment. While the algorithms implemented in this chapter were applied to StarCraft by the author and his team of researchers, they in no way claim credit for the original theory of such algorithms. For example, we made our own shouldBurrow function but did not invent expert systems.

REFERENCES

Blizzard Entertainment. (2009). StarCraft Brood War. Version 1.16.1.

Blizzard Entertainment. (2011). Blizzard Entertainment. Retrieved from http://us.blizzard.com/en-us/.

Heinermann, A. (2010). BWTA. Retrieved from http://code.google.com/p/bwta/wiki/BWTA.

Heinermann, A. (2011). BWAPIManual. Retrieved from http://code.google.com/p/bwapi/wiki/BWAPIManual?tm=6.

Heinermann, A. (2011). BWSAL Overview. Retrieved from http://code.google.com/p/bwsal/.

MasterOfChaos. (2011). Chaoslauncher. Version 0.5.3. Retrieved from http://wiki.teamliquid.net/starcraft/Chaoslauncher.

Microsoft. (2011). Microsoft Visual Studio. Retrieved from http://www.microsoft.com/visualstudio/en-us.

Team Liquid. (2010). Liquipedia: The StarCraft Encyclopedia. Retrieved from http://wiki.teamliquid.net/starcraft/Main_Page.

ADDITIONAL READING

Blizzard Entertainment. (2009). StarCraft Brood War. Version 1.16.1.

Blizzard Entertainment. (2011). StarCraft II Wings of Liberty. Version 1.3.6.

Blizzard Entertainment. (2011). Blizzard Entertainment. Retrieved from http://us.blizzard.com/en-us/

Buro, M. (2003a). Real-time strategy games: A new AI research challenge. Proceedings of the Eighteenth International Joint Conference on Artificial Intelligence (pp. 1534-1535). Acapulco, Mexico: Morgan Kaufmann.

Buro, M. (2003b). RTS games as a test-bed for real-time AI research. In Proceedings of the 7th Joint Conference on Information Science(pp. 481–484). K. Chen et al.

Buro, M. (2004). Call for AI Research in RTS Games. Retrieved from http://skatgame.net/mburo/ps/RTS-AAAI04.pdf

Buro, M. (2010). ORTS. Retrieved from http://skatgame.net/mburo/orts/orts.html

Buro, M., & Furtak, T. M. (2004). RTS games and real-time AI research. In *Proceedings of the BRIMS Conference* (pp. 34–41). Arlington, VA: USA.

Dorigo, M., & Stützle, T. (2004). *Ant colony optimization*. Cambridge, MA: MIT Press.

Farkas, B., & Blizzard Entertainment. (1998). *Starcraft: Prima's official strategy guide.* Roseville, CA: Prima Games.

Hagelback, J., & Johansson, S. J. (2008). Using multiagent potential fields in real-time strategy games. In Padgham, L., and Parkes, D. (Eds.). Proceedings from *the Seventh International Conference on Autonomous Agents and Multi-agent Systems*. Estoril, Portugal.

Heinermann, A. (2010). BWTA. Retrieved from http://code.google.com/p/bwta/wiki/BWTA

Heinermann, A. (2011a). BWAPI Manual. Retrieved from http://code.google.com/p/bwapi/wiki/BWAPIManual?tm=6

Heinermann, A. (2011b). BWSAL Overview. Retrieved from http://code.google.com/p/bwsal/

Huang, H. (2011). Skynet meets the swarm: How the berkeley overmind won the 2010 starcraft AI competition. Retrieved from http://arstechnica.com/gaming/news/2011/01/skynet-meets-the-swarm-how-the-berkeley-overmind-won-the-2010-starcraft-ai-competition.ars/3

Jackson, P. (1986). *Introduction to expert systems*. Reading, MA: Addison-Wesley.

Laursen, R., & Nielsen, D. (2005). Investigating small scale combat situations in real-time-strategy computer games. Master's thesis. Department of computer science. University of Aarhus. Denmark.

McCoy, J., & Mateas, M. (2008). An Integrated Agent for Playing Real-Time Strategy Games. Proceedings from the *Twenty-Third AAAI Conference on Artificial Intelligence.* Chicago, IL: AAAI.

Micić, A., Arnarsson, D., & Jónsson, V. (2011). Developing Game AI for the Real Time Strategy Game StarCraft. Retrieved from http://skemman.is/en/stream/get/1946/9143/22925/1/Final_Report.pdf

Ponsen, M., & Spronck, P. (2004). *Improving adaptive game AI with evolutionary learning. Computer Games: Artificial Intelligence, Design and Education* (pp. 389–396). Reading, UK: University of Wolverhampton.

Ponsen, M. J. V., Muñoz-Avila, H., Spronck, P., & Aha, D. W. (2005). Automatically acquiring domain knowledge for adaptive game AI using evolutionary learning. To appear in *Proceedings of the Seventeenth Conference on Innovative Applications of Artificial Intelligence*. Pittsburgh, PA: AAAI Press.

Pottinger, D. C. (2000). Terrain analysis in realtime strategy games. In *Proceedings of Computer Game Developers Conference.* San Jose, CA: USA.

Rørmark, R. (2009). Thanatos - A learning RTS Game AI. Master Thesis. University of Oslo, Norway.

Russell, S. J., & Norvig, P. (2003). *Artificial intelligence: A modern approach* (2nd ed.). Upper Saddle River, NJ: Prentice Hall.

Team Liquid. (2010). Liquipedia: The StarCraft Encyclopedia. Retrieved from http://wiki.teamliquid.net/starcraft/Main_Page

Walther, A. (2006). AI for real-time strategy games. Master Thesis. IT-University of Copenhagen, Denmark.

KEY TERMS AND DEFINITIONS

Brood War API (BWAPI): An application programming interface that can be used to create intelligent agents capable of playing StarCraft without human interaction.

Brood War Stand Alone Library (BWSAL): A library that automates several tedious processes necessary for most StarCraft matches. At the cost of reduced control it can make building an agent with BWAPI much simpler.

Brood War Terrain Analyzer (BWTA): A library used for gathering data about maps to be used in conjunction with BWAPI.

Expert System: A combination of if and then statements constructed by referencing human expertise and for the purpose of making decisions the same way a human expert would.

Hill Climbing: A local search algorithm used for finding local minima or maxima. The algorithm takes a starting point and iteratively moves toward the "best" location one step at a time until it reaches a local minima or maxima, and no better options are left.

StarCraft: A world famous real-time strategy game created by Blizzard Entertainment. The game is set in a futuristic universe and features three races that fight for world domination. The human Terran, the insect-like Zerg, and the mysterious, humanoid Protoss.

Swarm Intelligence: The phenomenon that occurs as a result of several units exhibiting intelligent behavior by following a few simple rules. Examples in nature include swarms of bees, and schools of fish.

Chapter 7
Evolving Bots' AI in Unreal™

Antonio M. Mora-García
University of Granada, Spain

Juan Julián Merelo-Guervós
University of Granada, Spain

ABSTRACT

A bot is an autonomous enemy which tries to beat the human player and/or some other bots in a game. This chapter describes the design, implementation and results of a system to evolve bots inside the PC game Unreal™. The default artificial intelligence (AI) of this bot has been improved using two different evolutionary methods: genetic algorithms (GAs) and genetic programming (GP). The first one has been applied for tuning the parameters of the hard-coded values inside the bot AI code. The second method has been used to change the default set of rules (or states) that defines its behaviour. Moreover, the first approach has been considered at two levels: individual and team, performing different studies at the latter level, looking for the best cooperation scheme. Both techniques yield very good results, evolving bots (and teams) which are capable of defeating the default ones. The best results are obtained for the GA approach, since it just performs a refinement considering the default behaviour rules, while the GP method has to redefine the whole set of rules, so it is harder to get good results. This chapter presents one possibility of AI programming: building a better model from a standard one.

INTRODUCTION

First Person Shooters (FPS) are action games where the player can only see the hands and the current weapon of his character, and has to fight against enemies by shooting at them. These games appeared at the end of the eighties in PCs as one of the new pseudo-3D games, evolving concepts previously seen in others such as Maze Wars (1974). After the first, and famous, Wolfenstein™ and DOOM™ games, FPS games began to be played by millions of video gamers in individual player mode until the appearance of games which included multiplayer modes. Unreal™ [Unreal:Site, Unreal:Wikipedia], released for PCs by Epic Games in 1998, had great success

DOI: 10.4018/978-1-4666-1634-9.ch007

since it incorporated the best multiplayer mode to date. In that mode, up to eight players (on the same PC or connected through a network) fought among themselves, trying to defeat as many of the others players as possible, and getting the so-called *frag* for each defeat. The players move in a limited scenario or arena, where weapons and other useful items appear. The players can be human or automatic and autonomous ones, known as *bots*.

Each player has a life level which is decreased every time he receives a weapon impact; this decrement depends on the weapon power, the distance, and damaged area on the character. In addition, there are some items that can be used to increase this level or to protect the player. Also, many FPS games let the programmers modify part of their source code or engine, to build new maps, weapons or characters, and even change the enemies' artificial intelligence (AI) schemes, to get new autonomous bots.

In the latest FPS games, a change in the confrontation has been introduced: the *team battle*. There are many team modes, such as death match, conquer the hill, capture the flag or hunt and escape. The common aim in all of them is the cooperation of the individuals in each team to obtain a global gain.

That is, the main objective is to get good team behaviour, rather than a good individual conduct. But in principle, it is difficult to predict how an improvement in the individual AI of a bot can profit the whole team.

Following these ideas, in this chapter there have been implemented (and presented) two general approaches to evolve bots inside Unreal™. The first one was applied for tuning up a set of parameters, corresponding to some hard-coded values inside the bot AI code. The second method was implemented to change and improve the default set of rules (or states) that defines its behaviour.

Moreover, additional work has been performed on the implementation of *two approaches of team-based evolutionary methods*, devoted to optimizing the behaviour of the whole team, in order to get the maximum number of frags against other teams.

In all cases we have implemented bots with a Genetic AI, or Genetic Bots (*G-Bots*). Evolutionary Algorithms, such as Genetic Algorithms [Goldberg, 1989] and Genetic Programming [Koza, 1992] have been used to improve the Unreal™ AI, given their well-known optimization capability. This way, each G-Bot improves its AI by playing a game and getting a better global behaviour in time, that is, defeating as much enemies as possible (getting frags) and being defeated as little as possible.

In the team-based case, every bot on the same team shares a common chromosome. As stated, the objective is to get bots whose behaviour would be good for the team profit.

The methodology described could be useful in AI designing tasks, starting from a common standard hard-coded or rule-based AI implementation, which would be significantly improved by means of evolutionary techniques.

UNREAL™ GAME FEATURES

As previously commented, Unreal™ is a very famous FPS for PCs published in 1998. It presented a very good single player mode, but the multiplayer possibilities gave it great success. Currently, there are many games which include multiplayer modes against humans or bots, but there are some features which made Unreal™ the best framework for developing our work:

- It includes bots with a high level AI, which was the best for a long time, since it introduced some novel techniques (such as predefined scripts, navigation points or states and events)
- It includes a proprietary engine programming language, called *UnrealScript* [UnrealScript], which combines C and Java syntax, with some useful features,

such as a garbage collector. It is object-oriented and handles implicit references, which means that every object is represented by a pointer.

- This language includes the declaration and handling of *states*, which are the most powerful feature of the Unreal bots' AI. They model the status and behaviour of the bots at the same time and are defined as classes. During the game and depending on the bot location and status, the current state of the bot is changed, and the functions defined in/for it are performed.
- In addition to the game, a programming environment, named *UnrealEditor* is included, which simplifies the management of the hierarchy of classes, as well as the creation of new classes which inherit from existing ones.

Thus, it is possible to change an existing class by creating another one which inherits from it and modifying the code for the desired functions or methods; this new class is known as a *mod* (since it is a modification in one of the components of the original game). What we have created in this work are some mods of the existing class for bots.

On the other hand, UnrealScript has some limitations that make it less powerful and flexible than would be desirable:

- It can only handle one dimensional arrays.
- The number of elements in each array is limited to approximately 60 elements.
- It is very difficult to debug a running program since it is only possible to write short messages to the game screen or in a log file, which is usually written once the game execution has ended.
- The number of iterations in a loop is limited as well, so if more than around 100000 iterations are performed, the program is finished. This is a constraint included to avoid infinite loops which can delay or freeze the game.

All these limitations have to be taken into account when programming a native bot in the first version of the Unreal Engine [UnrealEngine:Wikipedia], which we have used in this chapter. This engine is implemented in Unreal™ and Unreal Tournament™ (UT) games, which are the two environments where the algorithms have been implemented and tested. The first game has been used for the individual bot improvement, and the latter for the evolution of team tactics, since it is one of the games in this series focused on the battlematch modes, where teams are the most important factor. It preserves the excellent bot's AI (even better in some aspects), and UnrealScript can be also used, with less flaws than in the first version.

In addition, the UT version of the language has some improvements such as the possibility to write (but not read) in specific log files or to accelerate the run, but this option can change the results so it should be applied carefully.

In this game, the most traditional combat mode is "Death Match", in which the player must eliminate as many enemies as possible before the match ends, avoiding being defeated by other players. Depending on the combat mode, the player would belong to a team or play alone.

Each player interacts with others, trying to kill enemy players and help allies. These players can be either humans or bots, which have a limited number of health points taken from them as they get shot. If the health counter goes to zero, the player is defeated, and a frag is added to the last player who shot him. After being killed, the player is then respawned (usually in the same place or quite near) in the game, until his maximum number of lives is reached.

In order to aid players, some elements appear periodically in the arena: *weapons* to defeat the enemies (with limited ammunition, and an associated power), and *items*, that provide the player

Figure 1. A flow diagram of Bot's Roaming state. The states are represented by stars and the sub-states by circles.

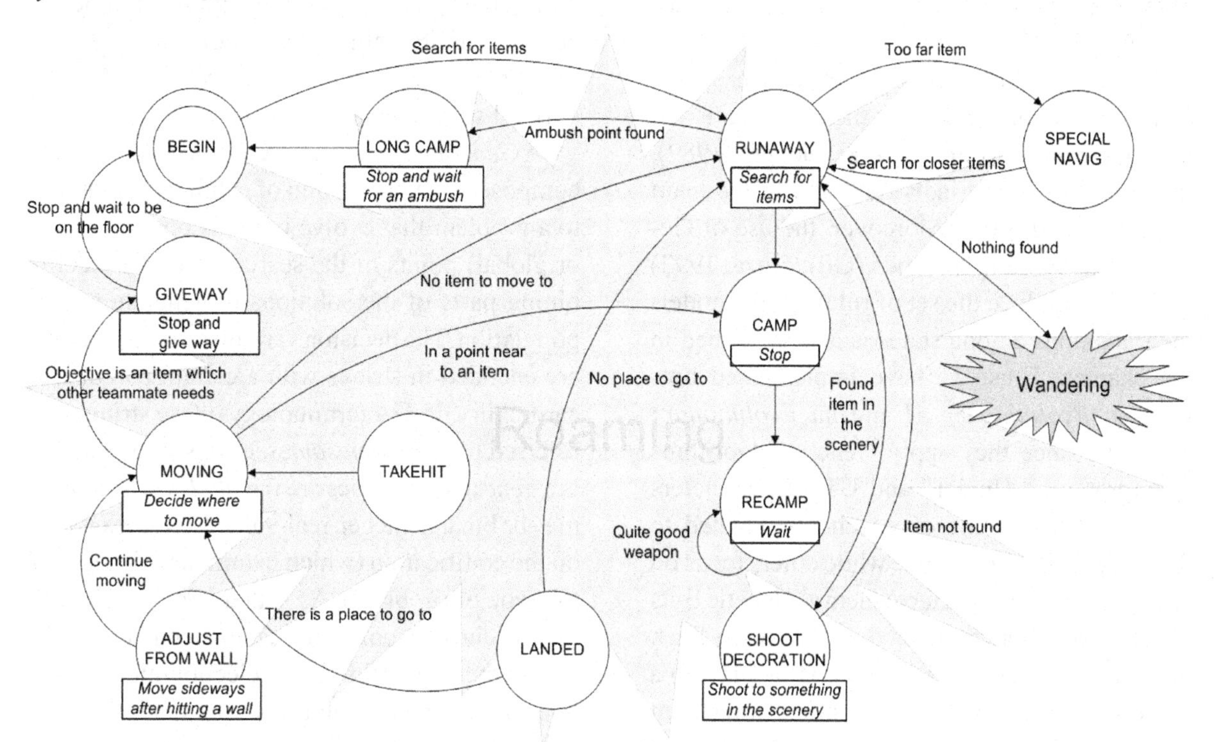

with some advantages (such as extra health, high jump, invisibility or ammunition). In team battles, the objectives can be very diverse, but the typical one is to get the maximum number of frags by adding the quantities obtained by each member of the team. A match ends when the termination conditions (typically a limit of winning frags or time) are reached.

The powerful bot AI implementation which Unreal™ and UT present is based on a Finite State Machine [Booth, 1967], representing many states, some of which have several substates. Each state models the behaviour of the bot when it has a specific status, location on the map, or relationship with the rest of players (enemies or team-mates). The substates represent the different "steps" that the bot's AI can follow inside a global state, determining for example, if the flow continues in the state or changes to another one. They are implemented as functions in the bot's AI code.

Figure 1 shows an example of the flow diagram of one of the states. Specifically, it is the *Roaming State*, one of the main states which the bots follow when they are searching for items or moving around the arena, deciding the next state to pass to.

During a game, a bot changes its current state based on factors present in its surroundings and on its own status and location. That is, the transition from one state to another depends on a number of parameters which determine the final behaviour of the bot, most of these parameters are thresholds used to trigger bot state changes (for instance, the distance to an enemy or the bot's health level). These variables are usually related to the individual behaviour, but some of them are also devoted to modeling the behaviour of the bot as part of a team.

The thresholds are typically compared with hard-coded values inside the source code of the bot's AI. Therefore, state changes (and the power of the bot's AI) ultimately depend on some set of constant values.

The topic addressed in this chapter has been mainly the improvement of these constants using a Genetic Algorithm (GA)(Goldberg, 1989), focused both in the individual and in the team behaviour and profit. Moreover, the use of Genetic Programming methods (GP) (Koza, 1992) to improve/reduce the set of rules which models the transitions among states is also described in the chapter. Thus, we have implemented bots with an *Evolutionary AI* and an *Evolutionary Team AI*, since they apply Genetic Algorithms to improve the Unreal™ and UT AI parameters and thresholds, some of which are devoted to the individual performance while others focus on team behaviour). We refer to them as Genetic Bots (*G-Bots*), and Genetic Team Bots (*GT-Bots*). Each of them improves its AI or team AI by playing a game, evolving its values and getting better team behaviour in time: defeating bots (in other teams if this is the case), being defeated as little as possible, getting items, weapons and ammunition to its team-mates, and so on.

EVOLUTIONARY ALGORITHMS

Evolutionary Algorithms (EAs) are a class of direct, probabilistic search and optimization algorithms gleaned from the model of darwinistic evolution (Bäck, 1996). This class of algorithms is usually divided into three sub-classes: Genetic Algorithms, Evolution Strategies and Genetic Programming, but the main features are common to all of them: a population of possible solutions (individuals) of the target problem, a selection method that favours better solutions and a set of operators that act upon the selected solutions. After an initial population is created (usually randomly), the selection and operators are successively applied to the individuals in order to create new populations that replace the older one. This process guarantees that the average quality of the individuals tends to increase with the number of generations. Eventually, depending on the type of problem and on the efficiency of the EA, an optimal solution may be found.

A *Genetic Algorithm* (GA) (Goldberg, 1989) is composed of a population of candidate solutions to a problem that evolve towards optimal (local or global) points of the search space by recombining parts of the solutions to generate a new population. The decision variables of the problem are encoded in strings with a certain length and cardinality. In GA terminology, these strings are referred to as *chromosomes*, each string position is a *gene* and its values are the *alleles*. The alleles may be binary, integer, real-valued, etc, depending on the codification (which in turn may depend on the type of problem). As stated, the "best" parts of the chromosomes are guaranteed to spread across the population by a selection mechanism that favours better solutions. The quality of the solutions is evaluated by computing the fitness values of the chromosomes; this fitness function is usually the only information given to the GA about the problem.

A standard GA's procedure is as follows. First, a population of chromosomes is randomly generated. All the chromosomes in the population are then evaluated according to the fitness function. A pool of parents is selected by a method that guarantees that fitter individuals have a better chance of being in the pool. Some of the possible selection methods are *tournament selection* (Goldberg *et al.*, 1989), *fitness proportionate selection*, also *known* as *roulette-wheel selection* (Goldberg, 1989a) and *stochastic universal sampling* (Bäck, 1996). After selection, a new population is generated by recombining the genes in the parents' population. This is usually done with a *crossover operator* (1-point crossover, or uniform crossover, amongst many proposals that can be found in the Evolutionary Computation literature) that recombines the genes of two parents and generates two offspring according to a crossover probability p_c

that is typically set to values between *0.6* and *1.0* (if the parents are not recombined, they are copied to the offspring population). After the offspring population is complete, the new chromosomes are mutated before being evaluated by the fitness function. *Mutation* operates at gene level and its probability p_m is usually very low (for instance, p_m is usually set to $1/l$ in many binary GAs, where l is the chromosome length).

After the evaluation of the offspring population the algorithm starts the replacement of the old population. There are several techniques for replacement, that is, for combining the offspring population with the old population in order to create the new population. Generational replacement, for instance, replaces the old population by the offspring. A steady-state strategy will replace a fraction (typically, two individuals) of the old population by the best individuals in the offspring population. Sometimes, an *e*-elitism strategy is used, i.e., the best *e* chromosomes from the old population are copied without mutation to the new population. The remaining individuals are selected according to any method.

This process goes on until a stop criterion is met. Then, the best individual in the population is retrieved as a possible solution to the problem. Algorithm 1 shows the pseudo-code of a standard GA.

Genetic Programming (GP) (Koza, 1992) is also based on the idea that in nature structures undergo adaptation. Thus, GP is a structural optimisation technique where the individuals are represented as hierarchical structures, such as a tree. The size and shape of the solutions are not defined a priori as in other methods from the field of evolutionary computation; instead, they evolve along the generations. So, the main difference with regard to GAs is the individual representation and the genetic operators to apply, which are mainly focused on the management (and improvement) of this kind of structure.

The flow of a GP algorithm is the same as any other evolutionary technique (see Algorithm 1): a population is created at random, each individual in the population is evaluated using a fitness function, the individuals that performed better in the evaluation process have a higher probability of being selected as parents for the new population than the rest and a new population is created once the individuals are subjected to the genetic operators of crossover and mutation with a certain probability. The loop is run until a predefined termination criterion is met.

STATE OF THE ART

When they arose in the late 1980s (with Wolfenstein™ in 1987), FPS games were devoted to single player modes. Later, most of them offered

Algorithm 1. Pseudo-code of the Standard Genetic Algorithm (SGA)

```
Standard Genetic Algorithm (SGA)
initialize Population(P)
evaluate Population(P)
while (not termination condition) do
      select P' individuals from P
      recombine individuals P' to generate offspring population O
      mutate individuals in O
      evaluate population O
      replace all (or some) individuals in P by those in O
end while
```

multiplayer possibilities but always against other human players, such as DOOM™ in 1988. The first known game to include autonomous characters, or bots, was Quake™ in 1992. It presented not only the option of playing against machine-controlled bots, but also the possibility of modifying them, in appearance or a few other aspects, or creating new ones. In order to do this, the programmers could use a programming language named QuakeC (QuakeC), widely used in those years, but which presented some limitations, such as strong structure and array sizes, hard typing constraints and open access to few game aspects. The bots created using QuakeC showed a simple AI based on fuzzy logic, and it was not possible to implement more complex techniques such as evolutionary algorithms. Unreal™ appeared some years later, being the first game to include an easy to use programming environment and a more powerful language. Of the bots developed using Unreal™, most were based on predefined hard-coded scripts, and only a few applied metaheuristics or complex AI techniques. Currently, there are many games that offer similar possibilities, but almost all of them are devoted to the creation of new maps or characters, being mainly focused on graphical modifications such as changing the topology or appearance, respectively.

Studies involving computer games and the improvement of some components of the characters' AI appeared some years ago (Laird, 2001), but the use of metaheuristics to study and improve the behaviour of FPS bots has risen in the last few years. We started our research in this field in 2001, publishing our results in national conferences (Montoya *et al.*, 2002) (in Spanish); in this chapter we applied evolutionary techniques to improve the parameters in the bots AI core and to change the way bots are controlled. Some other evolutionary approaches have been published, such as (Priesterjahn *et al.*, 2006), where evolution and co-evolution techniques have been applied, (Small and Bates-Congdon, 2009) in which an evolutionary rule-based system has been applied, or the multi-objective approach presented in (Esparcia-Alcázar, 2010) which evolves different specialized bots. The two latter have also been developed under the Unreal™ framework (Unreal Tournament™ 2004, UT2004).

Some years ago, several papers appeared in which the authors used self-organizing maps and multilayer perceptions(Thurau *et al.*, 2003), or machine learning (Zanetti & Rhalibi, 2004) to achieve human-like behaviour and strategies in Quake™ 2 and 3, respectively. Recent studies related to computational intelligence are based on bots controlled by neural networks (NNs). For instance, the authors in (Cho *et al.*, 2006, Soni & Hingston, 2008) train NNs by reinforcement learning, and those in (Schrum &Miikkulainen, 2009) evolve NNs searching for multimodal behaviour in bots.

Regarding the evolution of cooperative bots, or team behaviour improvements, there are some studies such as the one in (Bakkes *et al.*, 2004), which presents TEAM, an evolutionary approach to improve the behaviour in "Capture the Flag" mode in Quake™ 3. The key idea is that bots don't evolve by themselves, but as a team with a centralized agent control mechanism. In (Priesterjahn *et al.*, 2005) the authors performed a study considering team improvement in pathfinding, considering a communication mechanism using the scenery (simulating a stigmergic communication, like ants do (Deneubourg, 1983)).

Three years ago, in the paper (Doherty & O'Riordan, 2008) the authors explore the effects of communication on the improvement of team behaviour. They showed that using communication in difficult environments increases the effectiveness of the team.

With respect to the possibilities of programming bots in Unreal, a project called Pogamut [Pogamut 2] has recently appeared which defines an interface to program bots using Java. It is based on GameBots (GB) 2004 which is a mod for the game UT2004, written in UnrealScript and which provides a network text protocol for connecting to UT2004 and controlling bots. With GameBots, the user can control bots with text commands and simultaneously receive information about the game environment in text format.

The work presented in this chapter is devoted to the study of two ideas. The first is individual improvement based on the evolution and refinement of a standard AI, considering both parameters/thresholds and the rules defining the states and substates. The second objective is to improve the team behaviour as a whole by evolving the set of parameters which determines the behaviour of every team of bots. For the research presented here, the authors chose the Unreal Tournament™ (UT) game. It contains a more desirable feature set than Unreal™ and allowed us to extend our previous research within the same context.

GENETIC BOTS

Methodology Description

As previously stated in the Introduction, the objective is to improve the behaviour of an Unreal standard bot by changing its AI algorithm. Specifically we have tested two approaches. The first one consists of modifying the default AI by improving the values considered in the conditions assessed to transition from one state to another. The second one is related to the improvement of the finite state machine which the bot's AI follows.

This way, the Genetic Algorithm-based Bot (*GA-Bot*), tries to optimise the values of a set of parameters which represent each of the hard-coded constants that are in the bot's AI code. These parameters determine the final behaviour of the bot, since most of them are thresholds used to govern when bot state changes occur (for instance, the distance to an enemy or the bot's health level).

First, it was necessary to determine the parameters to optimise. After a deep analysis of the bot's AI code, 82 parameters were identified. This was too many parameters, since UnrealScript has a maximum length 60 floats for an array. In addition, it is difficult to evolve such a large array inside a game because the evaluation function depends on the results of the game, and it would require too many individuals and generations. Since the number had to be reduced, some parameters were redefined as function of others, and some of the less relevant parameters, which are used in minor states that are less relevant for the success of the bot, were not considered in the GA individual. As a result, 43 parameters were included in the array that represents an individual. Thus, each chromosome in the population is composed of 43 genes represented by normalised floating point values, which makes it a real-coded GA. This way, each parameter moves in a different range, depending on its meaning, magnitude and significance in the game, but all of them are normalised to the [0,1] range. The limits of the range have been estimated, forming a wide interval when the parameters are just modifiers, and a short one if they are considered as the main factor in the bot's decision taking. In this way, extremely bad behaviour is avoided if an important conduct parameter is completely changed. Thus, every chromosome/individual models one approach to the bot's AI, and the GA evolves 'the behaviour of the bot by evolving the trigger values to change between states.

The evaluation of one individual is performed by setting the corresponding values in the chromosome as the parameters for a bot's AI and placing the bot inside a scenario to fight against other bots and/or human players. This bot fights until a number of global frags is reached since it is not possible to set a time play for a bot in Unreal. Once the frags threshold has been reached and the current bot is defeated the next bot's AI will correspond to the next chromosome in the GA.

Two-point crossover and *simple gene mutation,* which change the value of a random gene (or none) by adding or subtracting a random quantity in [0,1], have been applied as genetic operators (see (Michalewicz, 2009) for a survey). The GA follows the *generational + elitism* scheme, considering as the *selection probability (SP)* for one individual a value calculated using its rank in the population depending on the fitness value, instead of calculating it directly considering the fitness function. This way, superindividuals, those with a very high fitness value, are prevented from dominating the next population and causing premature convergence. This method is known as *linear order selection*. Once the SP has been calculated, a probability roulette wheel is used to choose the parents in each crossover operation. Elitism has been implemented by replacing a random individual in the next population with the current global best. The worst is not replaced in order to preserve diversity in the population.

The *fitness function* was defined following expert players' knowledge. It includes all the terms which should be considered in order to have success in a game. These terms are:

- *frags*, the number of enemies defeated by the bot
- *W*, the number of weapons the bot has picked up
- *P*, the associated power to these weapons
- *I*, the number of items the bot has collected
- *d*, the number of times the bot has been defeated
- *t*, game time the bot has been playing

So, the fitness function equation for the chromosome *i* is:

$$F_i = \frac{frags_i + \left[\frac{P_i}{d_i} + \left(\frac{W_i \cdot 10}{d_i}\right)^{-1}\right] + \frac{I_i}{10} - \frac{d_i}{10}}{t_i} \quad (1)$$

where the constant values are used to decrease the relative importance of each term. Therefore, *frags* is the most important term in the formula. The next factor, inside square brackets, is related to the weapons and is composed of two terms. The first term measures the importance of the associated power of the weapons the bot has picked up on average, since a player loses all the weapons once it is defeated. The second term weighs the average number of weapons collected by the bot. In addition, this term is inverted since it should take low values when the bot has collected many weapons. The objective of this whole factor is to assign a higher weight to a bot which has picked up fewer but more powerful weapons, since searching for them is a risky task and takes some extra time.

The other two factors are devoted to weighing the collected items and the number of times the bot has been defeated, with a negative weight. All the terms are divided by the time the bot has been playing to normalise the fitness of all the chromosomes since they play for different lengths of time. Since it is the first approach, we have decided to consider an aggregative function instead of a multi-objective one. It is easier to implement and test and requires less iterations. The multi-objective approach will be addressed in future works.

The Genetic Programming-based Bot (*GP-Bot*) works in a different way, since it is based on a genetic programming algorithm (Koza, 1992) which can be used to evolve a set of rules. The idea is to redefine the initial set of rules, or transitions between states, which determine the behaviour of the bot. The first approach tried to consider all possible inputs and outputs and looked at the whole set of states that a default bot can manage. However, due to the quantity and complexity, this would mean a huge set of rules to evolve, and would be unmanageable by an algorithm defined inside UnrealScript, due to its array size constraints and limited resources. Instead, we decided to only consider the two most important states:

- *Attacking*, in which the bot decides between some possible actions as: search, escape, move in a strategic way, attack and method (from distance, nearby, close, from a flank).
- *Roaming*, in which the bot searches for weapons and items.

The flow diagrams of both states are respectively shown in Figure 1 and Figure 2.

Then, only the functions applied in the decision process (since there are some others just used to show an animation in the game, for instance, such as HitWall) were studied, in order to get the inputs and the outputs. These values are used to determine when to transition from the current state to the next one. Each of the functions is divided into a set of sub-functions or *inputs*, and, depending on the *outputs,* the corresponding transitions to the next state/substate. For example, there is a function to check if the bot has found an item (ItemFound), which returns a 'TRUE' or a 'FALSE' value depending on what the bot has found. Its return value can be considered as an input. One possible output could be GotoState('Roaming', 'Camp').

All the possible inputs and outputs for these states are used to define rules in the form:

```
IF <INPUT> [AND <INPUT>] THEN <OUTPUT>
```

After several experimental runs, it was determined that rules having three (or more) inputs were never triggered. Therefore, a two input maximum was imposed. These rules are modelled as trees. The parent nodes represent IF and AND expressions and are connected through RL nodes. Leaf nodes are the input and output values. An example tree can be seen in Figure 3, where a bot's AI is modelled with four rules:

```
IF X1 THEN Y7, IF X3 AND X5 THEN Y2,
IF X8 THEN Y1, and IF X9 THEN Y9.
```

So, every GP-Bot would have an AI structure based in the main set of states, but instead of the two previously mentioned, they consider one tree of rules, which should be obtained by the evolution of the individuals in the GP algorithm. Thus, every individual in the algorithm is represented by a tree of rules and also by one chromosome like those considered in the GA-Bot. In this way, the evolution is performed over two different AI aspects: the rules, and the parameters, which means it is a *GPGA-Bot.*

During the evolution some genetic operators are used on the parameters (crossover and mutation) which are those presented in the GA-Bot. In addition, two specific operators are considered to perform the evolution of the rule trees. The *tree-crossover operator* has been implemented by choosing two different nodes in the parents and interchanging the sub-tree below each of these nodes. However, there are some restrictions to preserve the tree coherence. For example, an IF node cannot be a leaf node. or two AND nodes cannot be parent and child. The *tree-mutation operator* just chooses a random node and substitutes the whole sub-tree below it by a randomly generated, but coherent, one. The *selection mechanism* is the same the linear order selection used in the GA-Bot.

Since both the tree of rules and the parameter configuration have value in determining the *fitness* value, the function has been updated to consider two additional factors:

- *S*, the number of shots the bot has fired. This parameter is included to reward bots which pick up weapons and use them. Some bots do not use the weapons they pick up, or do not do it correctly.
- *rR*, the repetition of rules. Tries to avoid the excessive repetition of rules in the behaviour of a bot. This value is increased for every 5 repetitions of a rule.

The complete fitness function equation for the chromosome *i* is:

$$F_i = \frac{frags_i + \left[\frac{P_i}{d_i} + \left(\frac{W_i \cdot 10}{d_i}\right)^{-1}\right] + \frac{I_i}{10} - \frac{d_i}{10} + \frac{S_i}{50} - \frac{rR_i}{500}}{t_i} \quad (2)$$

Figure 2. A flow diagram of Bot's Attacking state. The states are represented by stars.

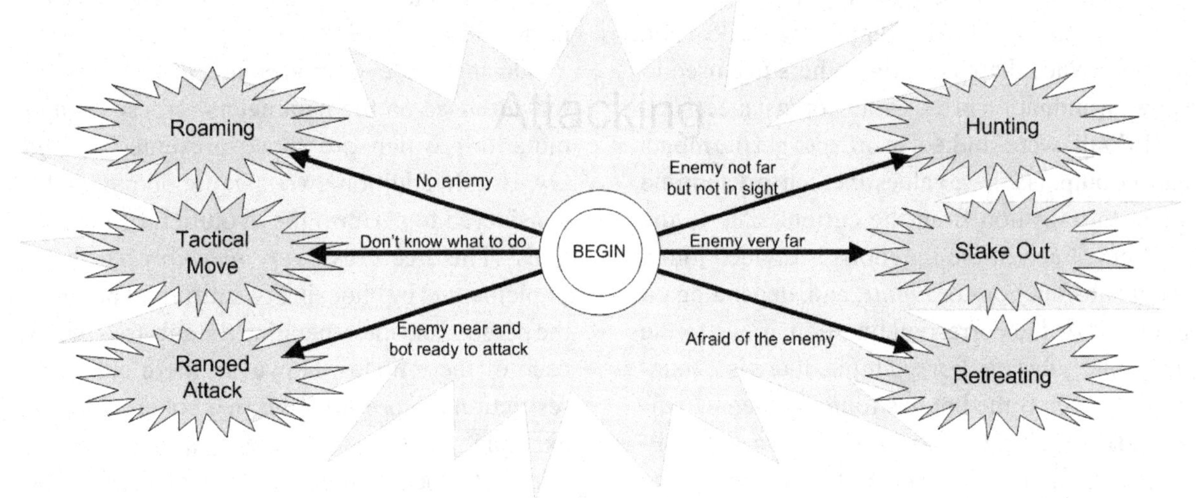

Figure 3. An example Tree of rules for a Bot's AI. The rules are defined by IF nodes, Xs are inputs and Ys are outputs.

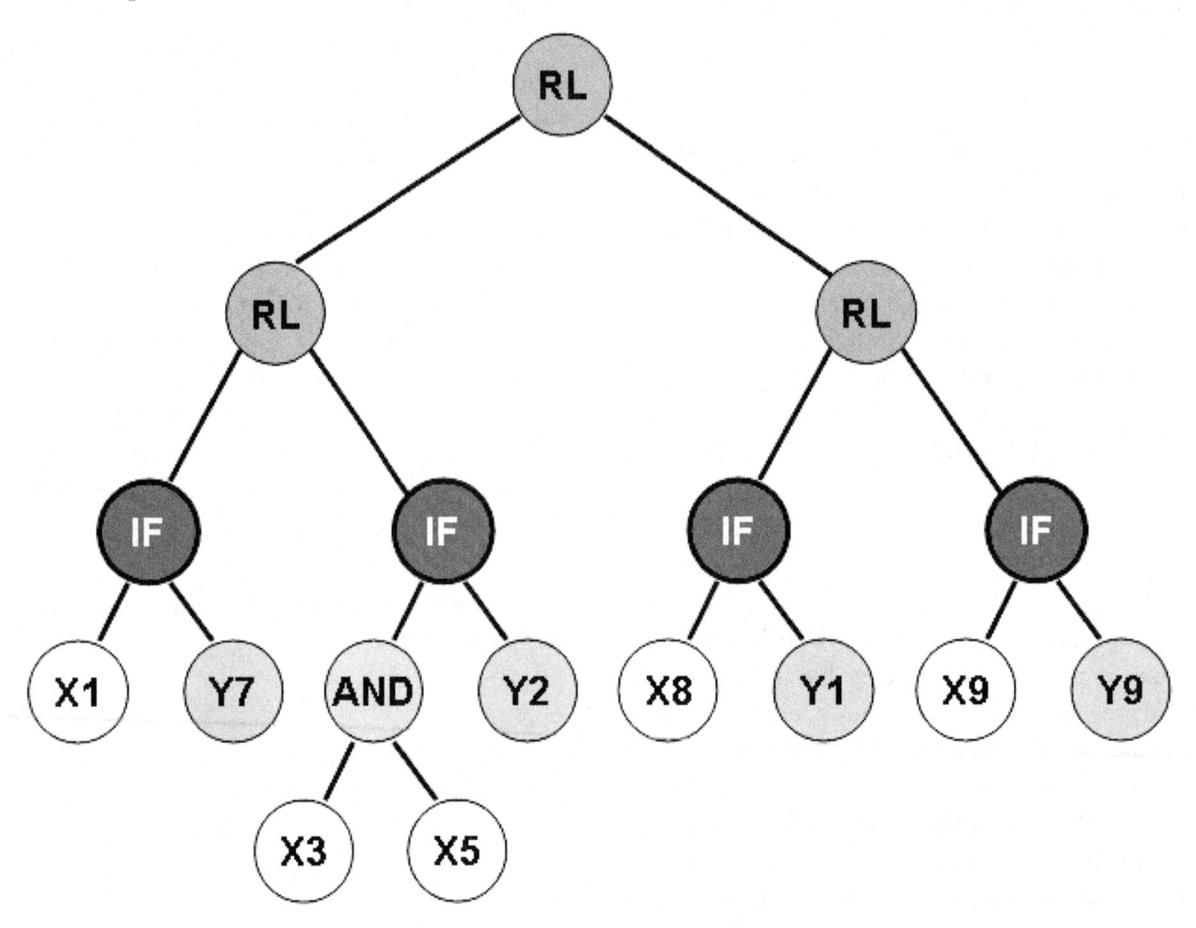

where all the factors are the same as in Equation 1 except the last two which have been included to evaluate the behaviour more accurately. Although neither of these additional factors has a high relevance, their inclusion is required to assign a more precise value to every set of rules.

Experiments and Results

We have performed some experiments to test the algorithms. Each of them consists of launching a game match for eight players, where seven are Unreal standard bots, but could be human players, and a GA-Bot or GPGA-Bot. Each run takes around one hour per generation on average, because every individual using our algorithms is playing until a number of defeats is reached and the match is played in real-time. The average run time also depends heavily on the map where the bots are fighting. Larger maps require more time to reach the required number of defeats and change to the next individual. We have considered the parameters shown in Table 1, which have been defined starting from the 'standard' EA values, and tuned through systematic experimentation.

Table 1. Parameters of GA and GP-GA algorithms used in G-Bots

Number of individuals	30
Mutation Probability	0.01
Crossing Probability	0.6
Number of defeats per chromosome	40

In this chapter, the experiments have been designed to test the good behaviour of the algorithms (and also of the bots), since they cannot be compared with results yielded by other algorithms.

Four maps have been considered, and five bots of each type (GA-Bot and GPGA-Bot) have been tested in each one of these maps. A classical run takes on average around 20 hours for 15 generations. The evolution of the algorithms with regard to the fitness has been as expected, with some fluctuations in the average due to the classical diversification of the evolutionary algorithms. These results can be seen in the Figure 4.

In this figure, a clear evolution in the average fitness is shown for all the cases. This evolution

Figure 4. Example results for two GA-Bots and two GPGA-Bots in two different maps

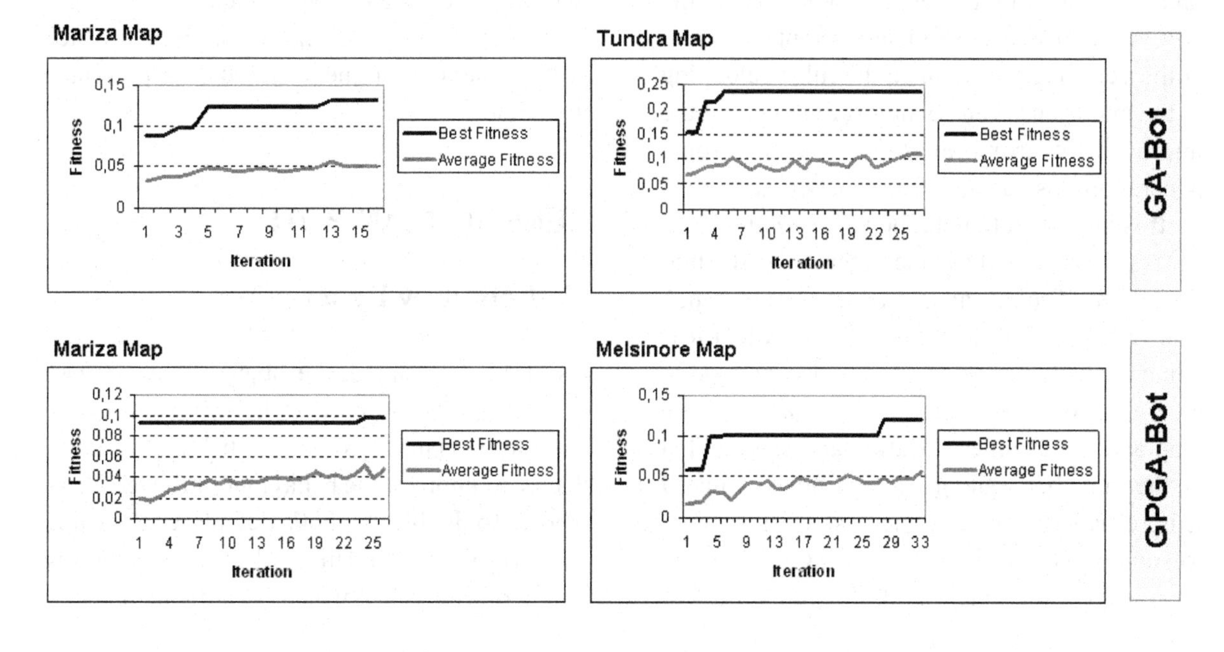

Figure 5. Screenshots of the final score (after some generations) for one GA-Bot (GeneticBOT, on the left) and one GPGA-Bot (PGeneticBOT, on the right).

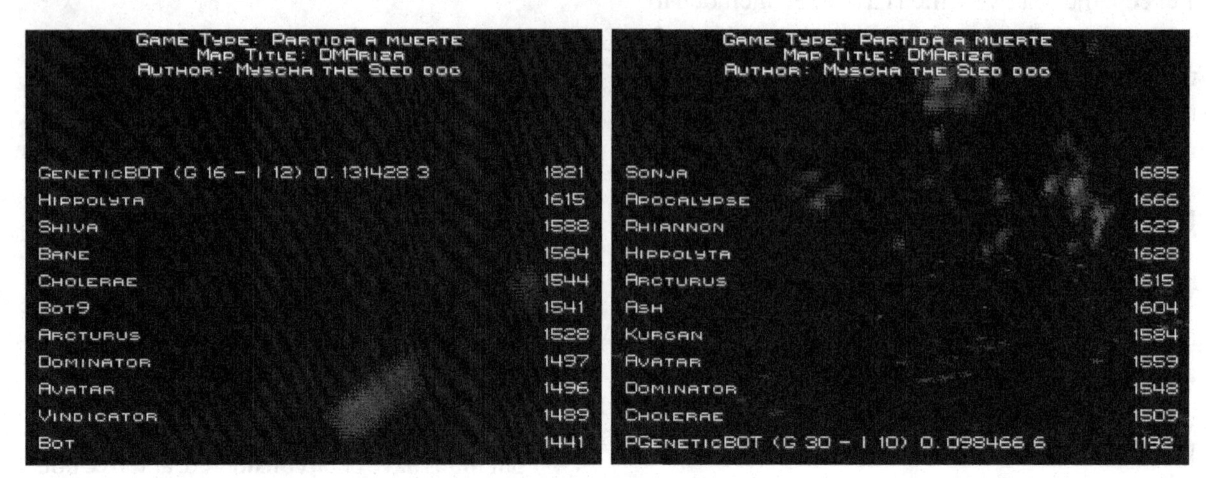

is more marked in the GPGA-Bots, since there are stronger changes to the AI in this algorithm because it evolves the bot's state transition rules. The improvements in the behaviour are much more obvious, and follow a clear progression. In the GA-Bots this change is less marked since the behaviour is quite similar, but a bit optimised in each generation. These bots show a very coherent behaviour from the first generation, because their behaviour rules remain the same, and just the decision parameters are changed evolved within predetermined limits. On the other hand, the GPGA-Bots usually exhibit incoherent behaviour at the very beginning, since the rules belonging to the main states can be almost random, but the improvements are more easily obvious after some generations, as has been previously noted.

It is important to note the *noisy nature* of the fitness function, which increments the fluctuations in the fitness calculation more than usual. The reason is that the value of a bot when it is being evaluated could vary from a very good to a very bad one, depending on its situation in the map and the enemies' location and status. That is, the same bot could get a high fitness value if it is respawned close to the best weapon, near to some important items, or seeing some weak enemies. On the other hand, it could get a low fitness value if it appears close to some enemies with good weapons.

With regard to the game score, the GA-Bots always beat their rivals, getting a high number of frags in some generations. The GPGA-Bots are at a disadvantage since the bad behaviour presented in the first generations leads to low frags and makes them an easy target for their enemies. Two screenshots are shown in Figure 5, which shows the classification for both types of bots after some generations. In both cases we have implemented the final AI configuration yielded by each of the algorithms into two definitive bots both of which perform better than the standard ones presented in Unreal.

GENETIC TEAM BOTS

Methodology Description

As previously stated, the objective in this part of the work is to improve the behaviour of the standard bots in UT when they belong to a team, that is, to improve the behaviour of the team as a whole. To do this, the bot's default AI algorithm should be improved. This can be accomplished by optimizing the values which govern state changes.

Following the GA-Bot approach presented in the previous section, the idea is to use a Genetic Algorithm that evolves toward optimal values for a set of parameters which represent some of the hard-coded constants included in the bot's AI code. Since most of them are probabilities, weights or thresholds considered in the behaviour functions (substates) of the bot, these parameters determine the final behaviour of the bot by determining when the bot state changes, such as the distance to an enemy or the bot's health level.

The first step is the selection of the set of parameters, which is quite different to the one considered in the previous approaches, since this set should be focused not only on the improvement of the bot's AI, but also on the benefit of the team to which the bot belongs.

Again, due to UnrealScript's restriction on array size, a set of 40 parameters has been chosen instead of the ideal 60-80 parameter set. After a deep analysis of the Unreal Engine 1 code [UnrealEngine:Wikipedia], some parameters were redefined as a function of others, and less relevant ones were not taken into account. The set of relevant parameters is shown in Table 2. They model the behaviour of the bots and can be classified as contributing to team benefit, both team and individual benefit (those devoted to move to or to attack the enemies in offensive states), or individual benefit only.

As with the previous GA-Bot, the Genetic Algorithm Team-based Bot (*GT-Bot*) defines each individual in the population as an array with a set of values, corresponding to each of the parameters shown in Table 2.

Thus, each chromosome in the population is composed of 40 genes represented by normalized floating point values (it is again a real-coded GA). Each parameter moves in a different range, depending on its meaning, magnitude and significance in the game. The limits of the range have been estimated, forming a wide interval if the parameter is just a modifier and a narrow one if it is an important value in the state transitions.

The *two-point crossover* and the *simple gene mutation* have been applied as genetic operators again. Moreover, the GA follows the same *generational + elitism* scheme, and applies the *linear order selection* method to calculate the *selection probability (SP)* for every individual (see (Michalewicz, 2009) for details). As in G-Bots, the elitism has been implemented by replacing a random individual in the next population with the current best. The worst is not replaced in order to preserve the diversity of the population.

In the previous approaches, the evaluation of one individual was performed by assigning the corresponding set of values in the chromosome as the parameters for a bot's AI, and placing the bot inside a scenario to fight against other bots and/or human players until a number of frags was reached.

In the present team study, two different approaches related to the chromosome and bot's AI relationship have been implemented, since the goal is the improvement of the whole team's behaviour:

- ***one chromosome per bot*** *(cr-Bot)*, where every bot in the team is assigned a different chromosome in the algorithm. When a bot finishes playing, the next available chromosome in the population is assigned to the bot. This simulates a kind of coevolution.
- ***one chromosome per team*** *(cr-Team)*, where there is just one chromosome which is shared by all the bots in the team, meaning all of them have the same behaviour.

In addition, in UT it is possible to define a time limit for playing each bot, so they compete for a similar number of seconds. It is not strictly the same because once the time limit has been reached, the bot can play until it is defeated again, which means they can stay for a different additional time, but they compete for a similar amount of time.

The *fitness function* has again been defined in terms of the main factors that determine a bot's score or a team's score. These factors have been

Table 2. Parameters to consider in the Bot's AI, grouped depending on their influence on the team or individual behaviour

PARAMETER	DESCRIPTION	TYPE
WeaponAIhelp	Power of the current weapon to decide if can help a teammate	Team
BestWWeightHelp	Maximum distance to help a teammate	Team
RoamFollowDist2	Maximum distance to follow a teammate	Team
MaxShareLocation	Maximum distance to share an item (let it go) with a teammate	Team
DiffHealthShare	Health points difference with a teammate to decide let him a health recovery pack	Team
WeaponAI2	Used to decide if the current weapon is enough powerful or the bot needs another one	Team/Indiv.
HealthPickDest	Considered to decide if the bot has enough health to attack	Team/Indiv.
NumHuntPathPickDest	Considered in the state Hunting to decide if attack an enemy	Team/Indiv.
MaxDistAttack	In the state Attacking, maximum distance to attack an enemy	Team/Indiv.
distItemRetreat1	In the state Retreating, minimum distance to pick up an item	Team/Indiv.
probAttRetreat1	Probability of attack in the state Retreating	Team/Indiv.
BestWeightAttack	Weights the best moment to attack	Team/Indiv.
pondSkillCharg1	Probability to move from the state Charging to RangedAttack	Team/Indiv.
probRangedAttackPro	Probability to move from the state TacticalMove to RangedAttack	Team/Indiv.
SkillAdvancedTactics	Weight of the bot's skill, considered to decide if the bot can apply advanced tactics	Team/Indiv.
LocationKeepingAttack	Distance to enemy, considered to decide if attack him and move to state Hunting, to Tactical Move or nothing	Team/Indiv.
WeightTacMov1	Weights the value of an item to decide if go to pick it in the state TacticalMove	Team/Indiv.
probTMTacMov1	Considered to decide if in the state TacticalMove, the bot perform a 'StrafeMove' or a 'DirectMove'	Team/Indiv.
pondSkillTacMovPickDest	Considered to decide what TacticalMove do, according to the time passed since it took the last item	Team/Indiv.
MaxDistTacMov1	Maximum distance to enemy, considered to decide the movement in the state TacticalMove	Team/Indiv.
pondSkillTacMovPickRegDest	Weights the skills to consider in the decision of the tactical move	Team/Indiv.
probSkillTacMovPickRegDest	Probability to consider in the decision of the tactical move	Team/Indiv.
distEnemyTacMov	Distance to enemy to consider in the decision of the tactical move	Team/Indiv.
pondAgressionTacMov1	Weights the type of aggression to an enemy	Team/Indiv.
pondCombatStyleTacMov1	Weights the combat style to fight with an enemy	Team/Indiv.
pondLocRoWaChTMH	Distance to have fear from an enemy or concrete area	Individual
probPickWandering	Probability to move from state Roaming to Wandering	Individual
probChargTacMov1	Probability to move from state Tactical Move to Charging	Individual
LastTimeInvPickDestAttack	Used in the state Hunting to note the last time it takes an item	Individual
pondCollisionRadiusRoam1	Weights the collision radius	Individual
pondWHuntAttack1	Weights the distance considered to move from the state Attacking to Hunting	Individual
probWarnRetreat1	In the state Retreating, it is considered to dodge shots	Individual

continued on following page

Table 2. Continued

PARAMETER	DESCRIPTION	TYPE
pondLocationFallB1	Weights the changing of enemy when the bot is in the state Falling	Individual
VSizeAttackStakeout	Distance to enemy to change the state Hunting to Stakeout	Individual
lastSeenTimePickDestAttack	Time stamp marking the last time the bot saw an enemy (to decide if the bot tries to attack him)	Individual
probAdvancedTacticsCharg	Probability of perform an advanced tactic in the state Charging	Individual
distFindNewStakeOut	Maximum distance to a point to be watched over by the bot in the state StakeOut	Individual
pondFindNewStakeout	Weights the time stamp of the last time an enemy was seen to decide a change in the watching over point in the state StakeOut	Individual
probChangTacMov1	Probability to move from the state TacticalMove to Charging	Individual

identified through experimental observation, and are the same as in the GA-Bot case (*frags*, W, P, I, d, and t). The fitness function formula for a chromosome (bot) i is defined by Equation 1. The fitness function associated to a team is obtained by adding the fitness of every bot in that team, using the following the equation:

$$F_T = \sum_{i \in 1..N_T} F_i \quad (3)$$

where N_T is the number of bots in the team T.

Experiments and Results

Several experiments have been performed to test the GT-Bots. Each experiment consists of launching a game match in the *Death Match* mode for eight to sixteen players, grouped into a different number of teams. Each team has either two players or four players. The players are all bots but could also be human players. The bots in one (or more) of the teams are GT-Bots, so their AI is modelled using a chromosome. The chromosome can be the same or different for all the bots in the team, depending on the method being tested. For all the runs of the GA, the parameters used are the ones shown in Table 3. The parameter values start with the standard GA ones, and are tuned through systematic experimentation.

Each run takes around 20 hours, since every individual in the algorithm is playing for 90 seconds, plus the remaining time of the last life. The match is played in 'pseudo' real-time, since it is possible to increase the run speed in a game, but it is increased in only one point in order to avoid false results. The time limit is an advantage with regard to the number of defeats limit considered in the previous work. Since the run time also depended heavily on the map where the bots were fighting, a larger map took longer to reach this number and change to the next individual.

In this chapter, several experiments have been performed. To summarize:

- *cr-Bot based team of two*: one team of two GT-Bots, with a different chromosome per bot, versus three standard teams of two bots.
- *cr-Team based team of two*: one team of two GT-Bots, with the same chromosome

Table 3. Parameters of GA algorithm used in GT-Bots

Number of individuals	27
Mutation Probability	0.01
Crossing Probability	0.6
Number of generations	27
Time limit per chromosome (seconds)	90

for all the bots, versus three standard teams of two bots.

- *cr-Bot based team versus cr-Team based team of two*: One team of each type fighting in the same battlefield as two teams of two standard bots.
- *cr-Team based team of four*: one team of four GT-Bots, with the same chromosome for all the bots, versus three standard teams of four bots.

Each experiment consisted of three runs and considered four maps (battlefields) in order to test the value of the results. Figure 6 shows the team fitness function evolution during the running of the GT-Bots, considering the *cr-Bot* approach. Both plots (Best and Mean) in every case correspond to the average value of the fitness yielded in the three runs. There is a two GT-Bots team, fighting against three teams of two standard bots, in each one of the maps. In this figure, clear evolutions in the best and mean team fitness are shown in all cases. The best fitness function follows a clear progression, while the mean fitness function fluctuates due to both the noisy nature of the evaluation function and the stochastic component present in the genetic algorithms. This led to a worse team fitness value in any individual and generation, but produced better results for the team in later runs.

This evolution is beneficial and indicates that the algorithm behaves as expected in all the runs and all the maps. A very smooth progression can be noticed in the best fitness function due to the bot's cooperation to evolve and improve. Since each bot uses a different chromosome in a generation, and all of them are devoted to improve the team performance, because of the fitness function, there is an implicit *co-evolution*, which improves the fitness in every generation.

The team fitness function evolution for the *cr-Team* approach is presented in the Figure 7. Again, both plots (Best and Mean) in every case correspond to the average value of the correspondent fitness in the three runs. The experiment performed in the DECK16 map includes a four GT-Bots team and three standard bot teams with four bots in each one. In the rest of the maps, there is a two GT-Bots team, fighting against three teams of two standard bots.

Looking at this figure, it can be noticed again the progression in both team fitness functions, but it is less smooth this time. The reason is that all the bots share the same chromosome at a time, so the evolution is slower than in the previous case because there is no co-evolution. Even so, the changes in the fitness functions are as expected, showing again some fluctuations in the mean fitness, due to the classical diversification in the evolutionary algorithms. At the end of every experiment, the best individual is considered as the set of parameters which models the final behaviour for the bot which has been evolving in the team. It continues playing until the game is stopped by the user since there are no stop criteria in the match.

In Tables 4 and 5, the final scores, representing the number of frags, of one run for each of the performed experiments are presented, to show the team's value. The number of frags achieved by each bot in the teams is shown in addition to the global score and the mean per team. In Table 4, the final scores yielded by the *cr-Bot* based teams in three maps, are displayed.

As can be seen in the table, the GT-Bots always beat their rivals, both at a bot level comparison, and at a team level comparison.

In Table 5, the scores for the *cr-Team* based teams are presented, showing again a clear supremacy of the GT-Bots. Even in the case of having four bots per team, where the best bot in terms of score belongs to a standard team, the best global score again belongs to the evolutionary team.

Finally, Table 6 shows a comparison between the final scores of both approaches (*cr-Bot* and *cr-Team*).

Looking at this table, it can be noticed that the *cr-Team* based team yields much better results than the *cr-Bot* based team. It is just a slight

Figure 6. Team fitness evolution with cr-Bot approach in three different maps. Both fitness plots are the average (of best and mean) of three runs. There are four teams of two bots (one including GT-Bots).

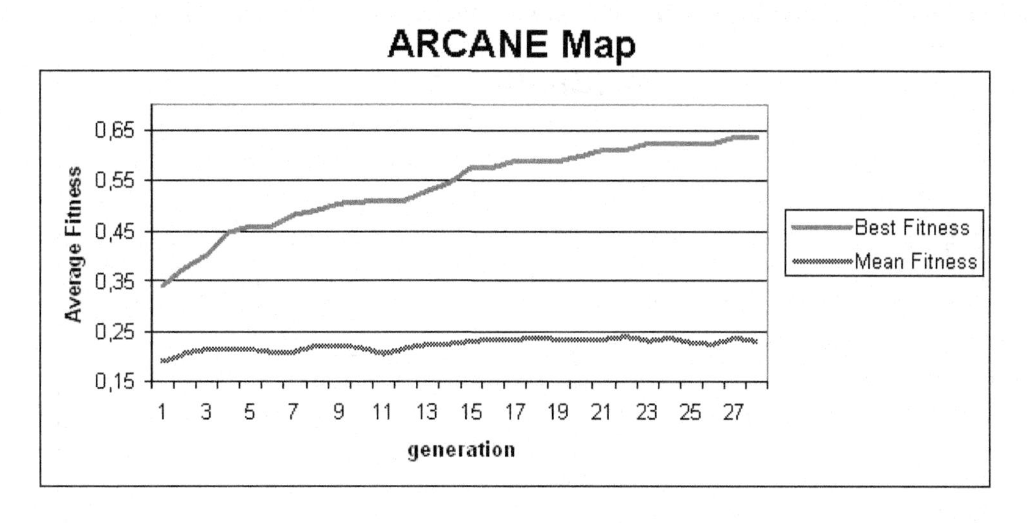

CONVEYOR Map

Average Fitness

0,64
0,54
0,44
0,34
0,24
0,14

1 3 5 7 9 11 13 15 17 19 21 23 25 27

generation

Best Fitness
Mean Fitness

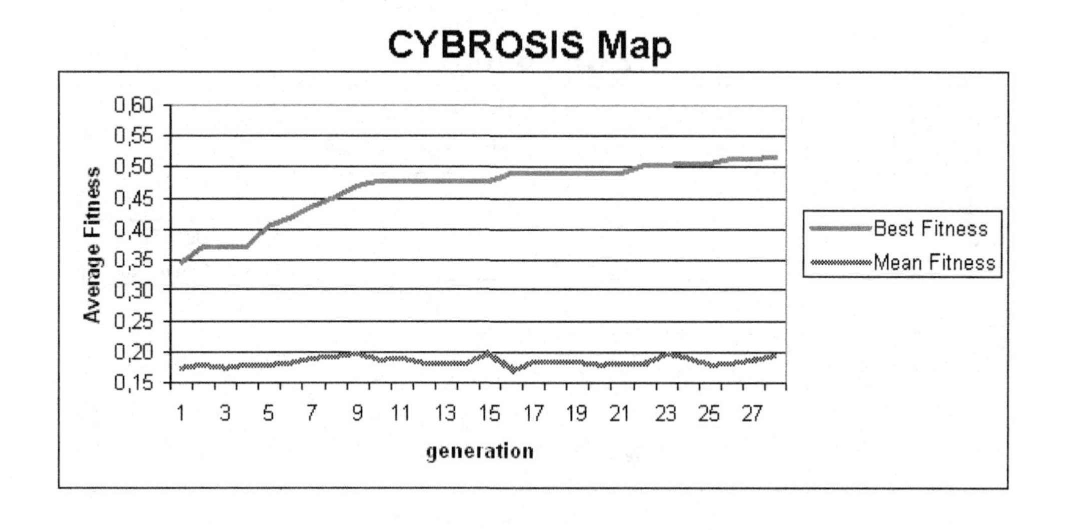

Figure 7. Team fitness evolution with cr-Team approach in three different maps. Both fitness plots are the average (of best and mean) of three runs. There are four teams of two bots in the first two maps and four teams of four bots in the last one. In every case, one of the teams includes GT-Bots.

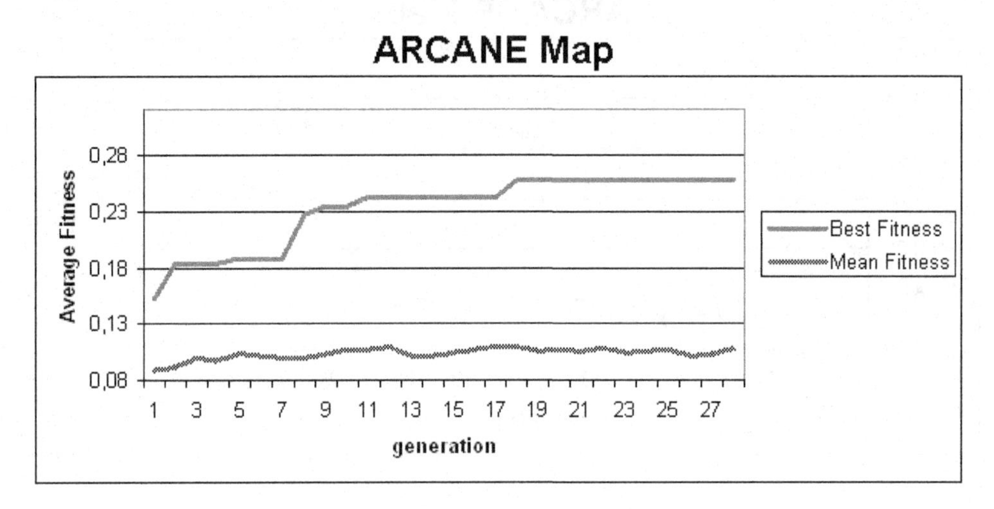

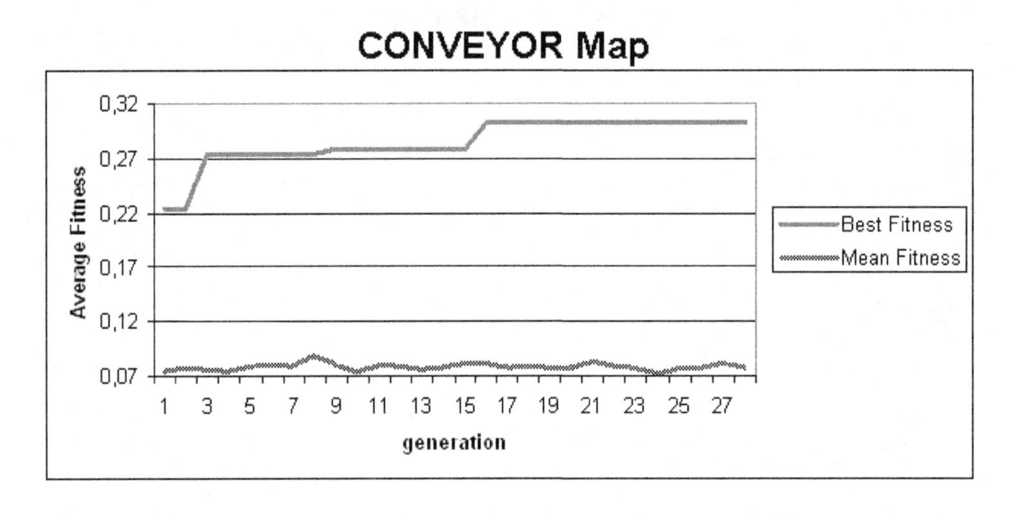

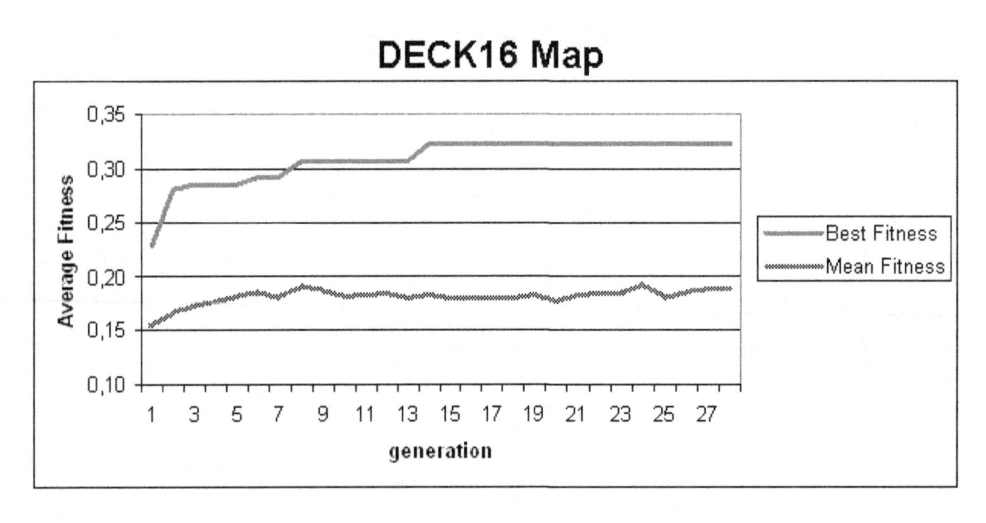

Table 4. Final scores (number of frags) for three standard and the cr-Bot based teams in three different maps

ARCANE Map				
	Team1	***Team2***	***Team3***	***Team GT-Bots***
Bot1	1615	1725	1787	1901
Bot2	1573	1682	1755	1858
total	3188	3407	3542	**3759**
mean	1594.00	1703.50	1771.00	**1879.50**
CONVEYOR Map				
	Team1	***Team2***	***Team3***	***Team GT-Bots***
Bot1	1074	1236	1162	1308
Bot2	965	1196	1052	1242
total	2039	2432	2214	**2550**
mean	1019.50	1216.00	1107.00	**1275.00**
CYBROSIS Map				
	Team1	***Team2***	***Team3***	***Team GT-Bots***
Bot1	1402	1388	1432	1575
Bot2	993	1198	1267	1555
total	2395	2586	2699	**3130**
mean	1197.50	1293.00	1349.50	**1565.00**

Table 5. Final scores (number of frags) for three standard and the cr-Team based teams in three different maps

ARCANE Map				
	Team1	***Team2***	***Team3***	***Team GT-Bots***
Bot1	2831	2783	2931	3147
Bot2	2298	2526	2904	3064
total	5129	5309	5835	**6211**
mean	2564.50	2654.50	2917.50	**3105.50**
CYBROSIS Map				
	Team1	***Team2***	***Team3***	***Team GT-Bots***
Bot1	1448	1616	1557	1723
Bot2	1267	1356	1412	1650
total	2715	2972	2969	**3373**
mean	1357.50	1486.00	1484.50	**1686.50**
DECK16 Map				
	Team1	***Team2***	***Team3***	***Team GT-Bots***
Bot1	1898	2054	1962	2036
Bot2	1813	1857	1866	2013
Bot3	1776	1804	1854	2009
Bot4	1680	1769	1834	1927
total	7167	7484	7516	**7985**
mean	1791.75	1871.00	1879.00	**1996.25**

Table 6. Comparison of the final scores (number of frags) between the cr-Bot and the cr-Team based teams in one map

CYBROSIS Map				
	Team1	***Team2***	***Team cr-Bot***	***Team cr-Team***
Bot1	4530	4573	5251	5612
Bot2	3997	4243	5102	5513
total	8527	8816	10353	**11125**
mean	4263.50	4408.00	5176.50	**5562.50**

comparison, but the high differences in the results show that *cr-Team* is a better approach, as was expected, since it evolves a whole team, while the *cr-Bot* evolves individual bots which belong to a team. These results have been checked long after the algorithms runs have finished, since *cr-Bot* evolves faster, as shown in previous figures, and gets better scores in less time. So a comparison in the first steps of the algorithms, or even just when they have finished their process, would be different.

CONCLUSION AND FUTURE WORK

In this chapter, two different evolutionary algorithms have been described to improve the AI of the default bots in a PC game named Unreal™. The approaches employed a genetic algorithm to optimize the decision parameters in the bots and a genetic programming method to optimize the set of rules which the bots consider in their AI. In addition, another genetic algorithm has been implemented to improve the team AI of the default bots in the game Unreal Tournament™. It has been also applied again to evolve a set of parameters in the bots that were mainly focused on those parameters related to team behaviour and performance. Two different approaches have been studied: the first one, named *cr-Bot,* considers a different set of parameters for each bot in the team while the second one, called *cr-Team,* works with the same set of values for all the bots in the team.

The results obtained indicate that in the first case both algorithms work as expected, reaching a clear improvement, and yielding final AI configurations which get the best scores in matches against the standard bots. In the second case, the approaches yield final team AI configurations which get the best scores in the matches against the teams of standard bots.

The *cr-Bot* method application implies a co-evolution in the search for the best team fitness function, while the *cr-Team* one has proven to be a better option to improve team AI, since it evolves the team as a whole. These results indicate that the use of evolutionary methods could be very beneficial when player AI is being designed and implemented, possibly starting from a standard hard-coded set of rules which could be evolved and improved using this method.

This is our first approach to the problems of individual and team bot improvement, so there are many future lines of work starting from this point. The first one is the implementation of different methods to evolve the individual and team AI bots, in order to compare the results with those yielded by the approaches presented in this chapter. In addition, we have to perform some studies to find the best parameter setting for the genetic algorithm, in order to improve its performance. Another task to address is the implementation of these algorithms inside a newer engine, such as Unreal Tournament 2004/2008, or using different tools, such as Pogamut, in order to avoid the constraints which obstruct a better problem definition and solution, such as limited arrays and number of iterations in loops. In addition, it would be possible to study some other team level problems, such as communication and coordination of bots. The third line of improvement is related to the fitness function which is currently an aggregative function, so it could be separated into different functions, transforming the problem into a multi-objective one which is closer to the real problem to address for getting good bot and team AI.

We also want to note that this is a rather 'noisy' problem, where each of the individuals has a different value for the fitness function at every time, since it depends on many factors which continuously change in time and can be different between two evaluations for the same bot i.e., the position of the bots during game play, their weapons, the situation of the new weapons, or the position of a bot when it appears in the map). This can complicate the evolution in the algorithms. Perhaps a dynamic approach could yield better results.

The last idea is related to the performance study of a 'pure' co-evolutionary approach, rather than a co-evolution based on teams with different individuals. The consideration of a team level co-evolution, including, for instance, heterogeneous teams would be also fruitful.

ACKNOWLEDGMENT

This work has been developed with the collaboration of R. A. Montoya and M.A. Moreno.

The authors would like to thank Dr. Jim Etheredge for his final review and editing of the manuscript.

REFERENCES

Bäck, T. (1996). *Evolutionary algorithms in theory and practice*. New York, NY: Oxford University Press.

Bakkes, S., Spronck, P., & Postma, E. O. (2004). Team: The team-oriented evolutionary adaptability mechanism. In Rauterberg, M. (Ed.), *ICEC, ser* (*Vol. 3166*, pp. 273–282). Lecture Notes in Computer Science New York, NY: Springer.

Booth, T. L. (1967). *Sequential machines and automata theory*. New York, NY: John Wiley & Sons, Inc.

Cho, B. H., Jung, S. H., Seong, Y. R., & Oh, H. R. (2006). Exploiting intelligence in fighting action games using neural networks. *IEICE – Transactions on Information Systems. E (Norwalk, Conn.)*, *89-D*(3), 1249–1256.

Deneubourg, J. L., Pasteels, J. M., & Verhaeghe, J. C. (1983). Probabilistic behaviour in ants: A strategy of errors? *Journal of Theoretical Biology*, *105*, 259–271. doi:10.1016/S0022-5193(83)80007-1

Doherty, D., & O'Riordan, C. (2008). Effects of communication on the evolution of squad behaviours. In Darken, C., & Mateas, M. (Eds.), *AIIDE*. Paso Alto, CA: The AAAI Press.

Esparcia-Alcázar, A. I., Martínez-García, A. I., Mora, A. M., Merelo, J. J., & García-Sánchez, P. (2010). Controlling bots in a first person shooter game using genetic algorithms. In *IEEE Congress on Evolutionary Computation* (pp.1-8). Washington, DC: IEEE Press.

Goldberg, D. E. (1989). *Genetic Algorithms in search, optimization and machine learning*. Reading, MA: Addison-Wesley.

Goldberg, D. E., Korb, B., & Deb, K. (1989). Messy genetic algorithms: Motivation, analysis, and first results. *Complex Systems*, *3*(5), 493–530.

Hesprich, D. (1998). QuakeC Reference Manual. Retrieved from http://pages.cs.wisc.edu/~jeremyp/quake/quakec/quakec.pdf

Koza, J. R. (1992). *Genetic Programming: On the programming of computers by means of natural selection*. Cambridge, MA: MIT Press.

Laird, J. E. (2001). Using a computer game to develop advanced AI. *Computer*, *34*(7), 70–75. doi:10.1109/2.933506

Michalewicz, Z. (1996). *Genetic algorithms + data structures = Evolution programs* (3rd ed.). Berlin, Germany: Springer Verlag.

Montoya, R., Mora, A., & Merelo, J. J. (2002). Evolución nativa de personajes de juegos de ordenador. In E. Alba, F. Fernández, J. A. Gómez, F. Herrera, J. I. Hidalgo, J.-J. Merelo-Guervós, and J. M. Sánchez, (Eds.) *Actas del Primer Congreso Español de Algoritmos Evolutivos*, (pp.212-219) AEB′ 02, Universidad de Extremadura.

Pogamut 2. (n.d.). Retrieved from http://artemis.ms.mff.cuni.cz/pogamut/tiki-index.php

Priesterjahn, S., Goebels, A., & Weimer, A. (2005). Stigmergetic communication for cooperative agent routing in virtual environments. *In International Conference on Artificial Intelligence and the Simulation of Behaviour* (AISB'05).

Priesterjahn, S., Kramer, O., Weimer, A., & Goebels, A. (2006). Evolution of human-competitive agents in modern computer games. In *IEEE Congress on Computational Intelligence 2006, CEC'06* (pp. 777-784). Vancouver, Canada.

Schrum, J., & Miikkulainen, R. (2009). Evolving multi-modal behavior in NPCs. In *IEEE Symposium on Computational Intelligence and Games, CIG 2009,* (pp. 325-332). Milan, Italy.

Small, R., & Bates-Congdon, C. (2009). Agent Smith: Towards an evolutionary rule-based agent for interactive dynamic games. In *IEEE Congress Evolutionary Computation, CEC '09,* (pp. 660-666). Trondheim, Norway.

Soni, B., & Hingston, P. (2008). Bots trained to play like a human are more fun. In *IEEE International Joint Conference on Neural Networks, IJCNN 2008, (IEEE World Congress on Computational Intelligence)* (pp. 363-369). Hong Kong, China.

Thurau, C., Bauckhage, C., & Sagerer, G. (2003). Combining self organizing maps and multilayer perceptrons to learn bot-behaviour for a commercial game. In Mehdi, Q. H., Gough, N. E., & Natkine, S. (Eds.), *GAME-ON* (pp. 119–123). EUROSIS.

Unreal Engine. (2011). Wikipedia: The free encyclopedia. Retrieved from http://en.wikipedia.org/wiki/Unreal_Engine

Unreal (n.d.). Retrieved from http://www.unreal.com

Unreal Script. (n.d.). Unreal Engine Site. Retrieved from http://www.unrealengine.com/features/unrealscript/

Unreal(2009). Wikipedia: The free encyclopedia (2009). Retrieved from http://en.wikipedia.org/wiki/Unreal

Zanetti, S., & Rhalibi, A. E. (2004). Machine learning techniques for FPS in Q3. In *Proceedings of the 2004 ACM SIGCHI International Conference on Advances in computer entertainment technology, ACE '04* (pp. 239-244). New York, NY: ACM.

ADDITIONAL READING

Cole, N., Louis, S., & Miles, C. (2004). Using a genetic algorithm to tune first-person shooter bots. In *Congress on Evolutionary Computation, CEC'2004, 1,* 139-145.

Eshelman, L. J., & Schaffer, J. D. (1993). Real-coded genetic algorithms and interval-schemata. In D. L. Whitley (Ed.) *Foundations of Genetic Algorithms, 2*, 187-202. San Mateo, CA: Morgan Kaufmann.

First Person Shooter. (2011). Wikipedia: The free encyclopedia. Retrieved from http://en.wikipedia.org/wiki/First-person_shooter

Game Bot. (2011). Wikipedia: The Free Encyclopedia. Retrieved from http://en.wikipedia.org/wiki/Computer_game_bot

Goldberg, D. E. (1989). *Genetic Algorithms in search, optimization and machine learning*. Reading, MA: Addison- Wesley.

Hong, J.-H., & Cho, S.-B. (2005). Evolving reactive NPCs for the real-time simulation game. In *Proceedings of the 2005 IEEE Symposium on Computational Intelligence and Games (CIG05)*. Essex, UK: IEEE Press.

Jacobs, S., Ferrein, A., & Lakemeyer, G. (2005). Controlling unreal tournament 2004 bots with the logic-based action language GOLOG. In *Proceedings from Artificial Intelligence for Interactive Digital Entertainment Conference 2005.* Marina del Ray, CA.

Karpov, I., D'Silva, T., Varrichio, C., Stanley, K., & Miikkulainen, R. (2006). Integration and evaluation of exploration-based learning in games. In *2006 IEEE Symposium onComputational Intelligence and Games* (pp. 39-44). Reno, NV: IEEE Press.

Koza, J. R. (1992). *Genetic Programming: On the programming of computers by means of natural selection*. Cambridge, MA: MIT Press.

McPartland, M., & Gallagher, M. (2008). Creating a multi-purpose first person shooter bot with reinforcement learning. In *IEEE Symposium on Computational Intelligence and Games, CIG '08,* (pp. 143-150). Perth, Australia.

van Hoorn, N., Togelius, J., & Schmidhuber, J. (2009). Hierarchical controller learning in a first-person shooter. In *Proceedings of the 2009 IEEE Symposium on Computational Intelligence and Games*. Milano, Italy.

KEY TERMS AND DEFINITIONS

AI Evolution: Improvement of the game AI (at any level) by means of evolutionary algorithms.

Bot: Autonomous player in a FPS.

Co-Evolution: Cooperation between individuals in an evolutionary algorithm to get a common profit.

Evolutionary Algorithms (EAs): Extended metaheuristic based in the natural evolution for performing optimization tasks. The most common are Genetic Algorithms and Genetic Programming.

First Person Shooter (FPS): A game where the player can only see the hands of the character which he/she controls. Usually it is based in fight against another players or enemies in arenas, using weapons to kill them.

Game AI: Set of parameters, rules, algorithms, and so on, which guide the behaviour of one autonomous player in a game.

Genetic Algorithm-Based Bot (GA-Bot): Bot which evolves a set of parameters by means of a Genetic Algorithm.

Genetic Algorithms (GAs): Evolutionary technique used in optimisation problems, which evolves a population of solutions modelled as sets/arrays of values.

Genetic Bots (G-Bots): Bots which evolve a set of parameters that models its behaviour as an independent individual using evolutionary algorithms.

Genetic Programming (GP): Evolutionary method used in the so-called program optimisation problems, in which a population of hierarchical structures (usually trees) is evolved.

Genetic Programming with Genetic Algorithm-Based Bot (GPGA-Bot): Bot which evolves both, a set of rules and a set of parameters using a GP algorithm and a GA at each level respectively.

Genetic Programming-Based Bot (GP-Bot): Bot which evolves a set of rules (modelled as a tree) by means of a Genetic Programming Algorithm.

Genetic Team Bots (GT-Bots): Bots which evolve a set of parameters that models its behaviour inside a team using evolutionary algorithms.

Team AI: AI which is devoted to yield the best behaviour of a team of autonomous players.

Section 2
Game Physics

Chapter 8
Practical Introduction to Rigid Body Linear Complementary Problem (LCP) Constraint Solvers

Ben Kenwright
Newcastle University, UK

Graham Morgan
Newcastle University, UK

ABSTRACT

This chapter introduces Linear Complementary Problem (LCP) Solvers as a method for implementing real-time physics for games. This chapter explains principles and algorithms with practical examples and reasoning. When first investigating and writing a solver, one can easily become overwhelmed by the number of different methods and lack of implementation details, so this chapter will demonstrate the various methods from a practical point of view rather than a theoretical one; using code samples and real test cases to help understanding.

INTRODUCTION

With the ever-increasing visual realism in today's computer-generated scenes, it should come as no shock that people also expect the scene to move and react no less realistically. With the computational power available today, the ability to run physically accurate real-time simulations is required to hold a players attention.

Simulating scenes, using physics-based methods, is important because it enables us to produce environments that respond to unpredictable actions and simulate situations that are indistinguishable from real life, e.g. buildings collapsing. However, it is very difficult to simulate reliably, large number of objects and complex articulated structures as shown in Figure 1.

DOI: 10.4018/978-1-4666-1634-9.ch008

Figure 1. Simulation screenshots demonstrate stable stacking (left), articulated joints for characters (middle) and chains of objects (right)

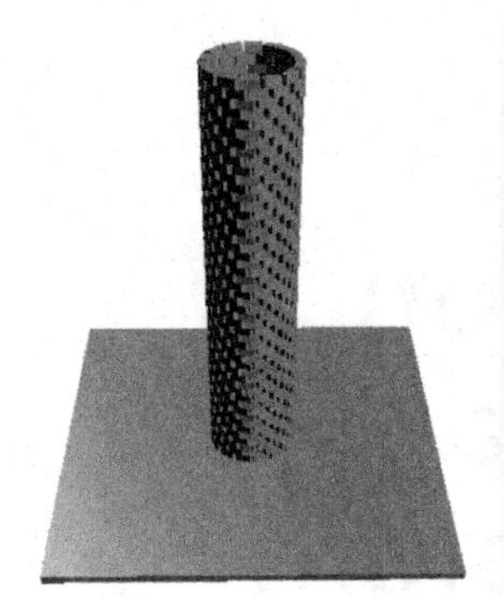

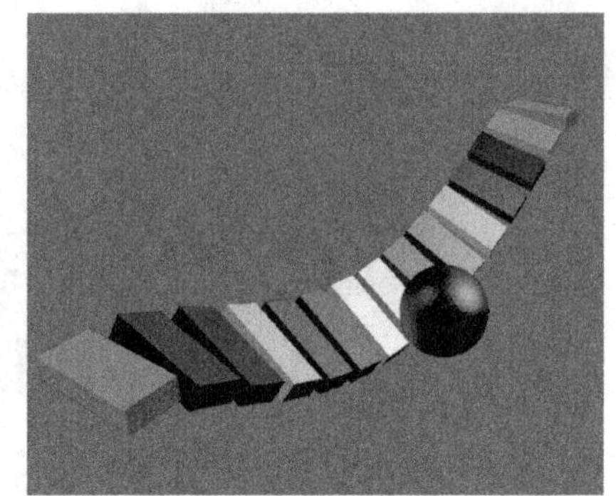

Writing a flexible, scalable rigid body simulator is a challenging task because you need strong background knowledge in programming and Newtonian mechanics. While there are several approaches (i.e. penalty methods, impulse methods), solvers offer numerous advantages, such as requiring less user tuning and the ability to handle highly coupled configurations (e.g. large stacks or chains).

The reader, after being introduced to how solvers operate and how they are constructed, is introduced to dependent techniques; such as sparse matrices. And while we try and explain everything from the bottom-up, it is still required that the reader is at least familiar with basic Newton's laws and calculus techniques. After completing this chapter, the reader should have a basic understanding of what a solver is, how to implement one, and how to use it.

BACKGROUND

Rigid Body Dynamics

Rigid body dynamics is a well understood and documented field, and as such, will not be covered here. For background information, we recommend reading (Baraff, 1999; Eberly, 2004; Hecker, 1998), which gives details on unconstrained dynamics and concepts such as body mass, acceleration, velocity and the equations of motion, which we use throughout this chapter.

While we introduce the reader to writing a practical LCP solver, there are also commercial and open source engines, which can be taken advantage of, and we would recommend them for the purposes of background knowledge. Some well-known LCP simulation engines are:

- Open Dynamics Engine (ODE), (Smith, 2004)
- PhysX (NVIDIA, 2011)
- Newton Game Dynamics (Jerez & Suero, 2003)
- Havok (Havok, 1998)
- Crisis Physics Engine (Vondrak, 2006).

Block Matrix Methods

The equations that make up our dynamic system and constraints can be large, cumbersome, error-prone and difficult to manage, and to help alleviate this problem, we represent them in matrix form. This gives us a more manageable high-level view of the system and its components, which is more intuitive to work with.

In our simulator, a large majority of the computation cost is in calculating an inverse matrix for our solution. As with real numbers, when you have 'ax=b' you can solve for 'x' by dividing both sides by 'a' to get 'x=b/a', which is acceptable, as long as 'a' is not zero. Similarly, when working with matrices, we formulate the equation 'AX=B', and divide both sides by 'A' to get 'X=B/A'. But instead of dividing both sides by 'A', we calculate the inverse of 'A' (i.e. A^{-1}) and multiply both sides, to achieve the same result.

This chapter uses numerical methods to calculate the inverse of the matrix, which is central to achieving a usable solver. However, on occasions, we are unable to calculate the inverse, i.e. we get a *singular* matrix, similar to a divide by zero with real numbers. When this occurs, usually some ill-conditioned configuration has developed or perhaps some numerical problem has occurred, which we detect as a singular matrix (i.e. determinant is zero) and we determine why it has arisen so that we can fix it and prevent it happening again.

Lagrange Multiplier Formulation

We have the unconstrained dynamic equations of motion from classical mechanics, which describes how the rigid bodies move, and a set of constraint conditions – which describe how they cannot move. We then combine these two equations and solve the unknowns by using a powerful technique of multi-variable calculus, known as 'Lagrange multipliers'.

We create the equations for our system using Newton's second law 'f = ma', in combination with constraint equations which we form through differentiation and substitution to establish a combined problem, which is solved using Lagrange's multiplier methods.

Equations of Motion

Each unconstrained body has six *Degrees of Freedom* (DOF), three for translation and three rotation. For a group of rigid bodies (m), the total DOFs is 6m. Since constraints restrict the relative motion, the total number of DOFs of a group of rigid bodies, with constraints, is less than 6m. The rigid body dynamics in collaboration with the constraint configurations form the *Equations of Motion* (EOM), that describe how the group of rigid bodies will move as time changes.

We can categorise the EOM into two types: *Maximal* and *Reduced* coordinates.

Maximal coordinates use Cartesian space, so each body has 6m state variables, and requires 6m-n constraint equations (where n represents the number of constraints). These constraints *explicitly* remove extraneous DOF through their formulation. For further reading on explicit constraint methods see (Shabana, 1994), and further details on maximal coordinate methods are available in (Baraff, 1999; Eberly, 2004; Hecker, 1998).

Reduced coordinates use an *implicit* incorporation of the constraints to formulate the equations of motion. The system uses n state variables to represent the various DOF, so for example, if we have a single object which can only rotate around the 'y' axis (no translation or x-z rotation), then the system state only needs a single state variable to represent the system (i.e. the object's angle). It has a major drawback, whereby for each unique configuration we need to derive by hand a set of equations for that particular arrangement.

Both *Maximal* and *Reduced* coordinate methods are able to run in 'O(n)' time. Maximal coordinates are more popular because of their modularity and straightforwardness to understand and implement. Although maximal coordinates operate in Cartesian space, we still sometimes need to use awkward conversions, to convert between constraint and Cartesian space. Maximal coordinates can drift due to numerical errors and integration inaccuracies, in addition they need a minimum of 6m state variables to represent the system, so optimised methods need to be used to reduce memory and bandwidth impacts (sparse matrices). Due to the modular flexible nature of maximal coordinates, we use this method in this chapter, so we can formulate constraints once, and use them again and again for various configurations.

Simulation Approaches

Three main types of constraint methods exist: *Penalty methods*, *Impulse methods* and *Global methods*.

Penalty methods (springs) – are the easiest technique for formulating constraints, whereby the violated constraint error is fed back into the system as restoring forces to correct the error. They have the ability to be simple, fast and intuitive, and can be combined with other methods to add controllability. Their downside is that the constraints rely on error feedback forces, and suffer from stability issues, which require small time-steps (offline) or computationally expensive integration techniques to remain stable. The reader can refer to (Kaˇ, Nordenstam, & Bullock, 2003) for a practical example of penalty methods being used to create constraints.

Impulse methods (velocity impulses) – use instantaneous force changes, known as velocity impulses to implement constraints. These velocity impulses are repeatedly applied to each constraint, where you solve one constraint after the other, re-evolving the system to satisfy all constraint conditions, until the system converges, or you reach a maximum update limit. Further reading can be found in (Guendelman, Bridson, & Fedkiw, 2003) which presents a realistic impulse-based simulator with stacking.

Global methods (analytical methods) – which we use and apply in this chapter, involve computing the exact magnitude of force that will satisfy the constraints at every step of the simulation. It is accurate and requires minor parameter tuning by the user and can maintain stability for relatively large integration steps. It works, fundamentally, by constructing a linear system of the form.

SOLVER (LCP)

The Linear Complementarily Problem (LCP) is a special kind of problem that aims to find a solution to a set of equations, subject to constraint limits. The type of LCP we focus on in this chapter is the box-constrained LCP, which aims to find a solution to the form ‘y = Ax+b’, subject-to limits on ‘y’:

$$\begin{aligned} &\textit{where} && y = Ax + b \\ &\textit{subject to} && y \geq 0 \;\rightarrow\; x = x_{lower} \\ & && y \leq 0 \;\rightarrow\; x = x_{upper} \\ & && y = 0 \;\rightarrow\; x_{lower} < x < x_{upper} \end{aligned} \tag{1}$$

We can broadly classify LCP solvers into two method types, *iterative* methods and *pivoting* methods.

Pivoting methods use recursion; where the solution to the problem depends on solutions to smaller instances of the same problem, which can be solved in a finite number of steps. While pivoting methods can be fast, it is our experience that for a large number of constraints, they can produce erroneous results for perfectly valid systems due to floating point errors. Further reading and examples of pivoting methods are Lemke’s algorithm used in (Eberly, 2004), and Dantzig’s algorithm, used in (Baraff, 1994).

Iterative methods alternatively do not terminate but converge on a solution finitely, where convergence depends on a number of factors such as the initial starting value. In addition, because iterative methods move closer to the solution with every update, if they were interrupted early, the current result can be good enough for the simulation to continue. Furthermore, we can take advantage of the starting guess and coherency between frame updates to accelerate convergence, by feeding the previous frame’s solution to the next. This property makes them ideal for real-time applications, where we can break out at varied times, trading accuracy for speed. Finally, iterative methods in practice are able to find acceptable results in ill-conditioned or singular configurations; due to contacts or overly constrained systems that allow the simulation to continue and recover.

Further reading about LCP methods for solving can be found in (Cottle, Pang, & Stone, 1992).

To restate, we use iterative methods because:

- The stopping criteria can be adjusted to trade accuracy for performance.
- In practice, they are more stable and reliable.
- They are simple to implement compared to other direct methods.
- There is a greater potential for optimisation (exploiting matrix sparsity for speedups).

Two popular iterative methods for solving *systems of linear equations* are '*Gauss-Seidel*' and the '*Jacobi method*', which can be modified to handle equality constraints for our complementary problems, such as, contacts. For this chapter, we use a modified '*Gauss-Seidel*' algorithm, called the '*Projected Gauss-Seidel*', which offers a simple and intuitive implementation with good convergence rates. An in-depth explanation on how systems of linear and complementary equations are solved can be obtained by reading (Hager, 1988), (Cottle *et al.*, 1992) and (Erleben, 2004), also for a more detailed explanation of the Gauss-Seidel and its differences to projected Gauss-Seidel, see (Catto & Park, 2005).

The *Gauss-Seidel* equation is shown below, followed by its implementation in code.

$$x_i^{(k+i)} = \frac{1}{a_{ii}}\left(b_i - \sum_{j=1}^{i-1} x_j^{(k+1)} - \sum_{j=i+1}^{n} x_j^k\right) \quad (2)$$

where the subscript i and j indicates the column and row of the matrix elements from our linear equation Ax=b. It works by starting from some initial value (e.g. 0), then iteratively updating the answer repeatedly using the result from the previous step to converge on a solution. In the equation above, we have x^k as our current result, and x^{k+1} as the next.

The convergence on an acceptable answer depends upon the size and complexity of the configuration, where it can take anywhere from two or three iterations to hundreds, depending on the topology and initial starting value. For example, highly coupled configurations such as stacks of objects or long chains take longer to converge than less densely coupled ones.

We add an extra step to our vital *Gauss-Seidel* method to incorporate boundary conditions and enforce the complementary constraints with an additional projection step:

$$x_j = \max(\min(x_j, x_j^{upper}), x_j^{lower}) \quad (3)$$

Stop Condition

The sample code uses a fixed iteration count, but a check can be added to determine if the solution is within a certain tolerance and provide an early breakout. We can calculate this value using the equation below:

$$\frac{\| b - Ax^{(k)} \|}{\| b \|} < \varepsilon \quad (4)$$

Acceleration or Velocity Level

We can broadly classify LCP solvers into two main types, *acceleration-based* (Baraff, 1989, 1994; Lötstedt, 1984) or *velocity-based* (Anitescu & Potra, 1996; Stewart & Trinkle, 1996), where the solver is classified according to the level it operates on to solve its constraints. In the next few sections, we review both the acceleration and velocity level solvers, outlining their similarities and differences, but towards the end of this chapter, we will focus on *velocity-based* solvers due to their added simplicity and practicality.

We give a brief comparison of both methods and review their advantages and disadvantages. To begin with, we demonstrate the two main sets of equations and the steps for formulating constraints at the various levels to illustrate their differences.

Velocity

$$JM^{-1}J^{T}\lambda + J\dot{q} + JM^{-1}F_{ext} \geq 0$$

using

$$\dot{q}_{n+1} = \dot{q}_{n} + M^{-1}(F_{ext} + F_{c})" t$$

$$F_{c} = J^{T}\lambda$$

Acceleration

$$JM^{-1}J^{T}\lambda + \dot{J}\dot{q} + JM^{-1}F_{ext} \geq 0$$

using

$$\ddot{q}_{n+1} = M^{-1}\left(F_{ext} + F_{c}\right)$$

$$F_{c} = J^{T}\lambda$$

The constraint formulation steps are as follows:

Velocity

1. Create positional constraint C.
2. Differentiate C with respect to time to obtain the velocity constraint $\dot{C}$.
3. Isolate and extract the Jacobian from $\dot{C}$.

Acceleration

1. Create positional constraint C.
2. Differentiate C with respect to time to obtain the velocity constraint $\dot{C}$.
3. Isolate and extract the Jacobian from $\dot{C}$.
4. Solve for the derivative of the Jacobian ($\dot{J}$) w.r.t. time.

From the constraint formulation steps above, the reader may notice that the initial steps are very similar, but the acceleration level requires additional work to calculate the Jacobian derivative. The velocity approach is basically a subset of the acceleration level, where it moves the acceleration problem into the LCP integration step in order to obtain a discrete problem, having velocities as unknowns rather than accelerations.

The velocity-based method has the slight advantage of only having to compute the first differential for the constraint equation due to the velocity method not needing the Jacobian derivative to calculate the unknowns. Earlier solvers were formulated at the acceleration level, but had problems where friction constraints could produce unstable systems. Later, the problem was modified so it could be solved at the velocity level where this problem was solved.

If principles of how to resolve constraints at the velocity level are understood, then the knowledge is there to implement it for the acceleration level (and vice versa). For the remainder of this chapter, we will focus on deriving and implementing methods for the velocity level, and where appropriate, mention any extraneous details that might be relevant to the acceleration level.

CONSTRAINTS

Overview

The formulation of the equations we use for our simulator can be broken down into four easy steps as shown in (Hecker, 1998) and reproduced in Figure 2. We introduce the steps briefly here, showing how the solver equations fit together, to attain a physics-based simulator. Even though they may not be completely understood initially, we reintroduce them again with further details and applied examples as we advance through the chapter.

Equality and Non-Equality Constraints

There are two types of constraints, *Equality* "==" and *Inequality* ">,>=,<,<=". Examples of *equality* constraints are rigid-ropes and ball-joints, while *inequality* constraints would be contacts and collisions.

Equality constraints have a fixed solution and *cannot* be varied along the constraint direction. They're formed by setting the constraint condition 'c' to zero, using the equality sign.

Inequality constraints can have numerous solutions and are formed by using a greater than or equal operator in constraint equations.

In the steps mentioned earlier for solving the unknown constraint values, we used an equal sign for $A\lambda = b$. For inequality conditions, we modify our linear solver to handle linear complementary problems. The modification adds the condition '$\geq$', and only accepts values greater or equal to zero, essentially clamping λ so it is always positive:

Equality

$$A\lambda + b = 0$$

$$c = 0$$
$$\dot{c} = 0$$

Inequality

$$A\lambda + b \geq 0$$
$$\lambda \geq 0$$
$$\left(A\lambda + b\right) x = 0$$

$$c \geq 0$$
$$\dot{c} \geq 0$$

Due to inequality constraints having multiple results, we can check if we need them before we add them to our solver, i.e.:

```
if c≤0:
     add ċ≥0 to constraint solver
else
     ignore
```

Jacobian

Forward kinematics enables us to define functions that convert between Cartesian space (i.e. positions and orientations) to constraint space, and the Jacobian represents how this Cartesian-constraint space relationship changes with respect to time, (e.g. Cartesian-constraint space *velocity* relationship).

The Jacobian gives the instantaneous transformation between the constraint velocities and the rigid body velocities. Where, for objects in 3D, each rigid body has six DOF, which represents the three linear and three angular velocities, for (m) rigid bodies we are able to form a (6m) column vector containing all the rigid body velocities, in addition to our (n) constraints, to form a (6m x n) Jacobian matrix, describing how all the objects in our system are *allowed* to move.

Jacobian is a matrix function of the form:

$$\dot{c} = J\,\dot{q} \tag{5}$$

where $\dot{q}$ (linear and angular) Cartesian space velocities and $\dot{c}$ the constraint velocities.

We can also express $\dot{q}$ as its separate angular and linear velocity components:

$$\dot{q} = [\,v,\ \omega\,] \tag{6}$$

Figure 2. Basic four-step breakdown of acceleration and velocity based methods

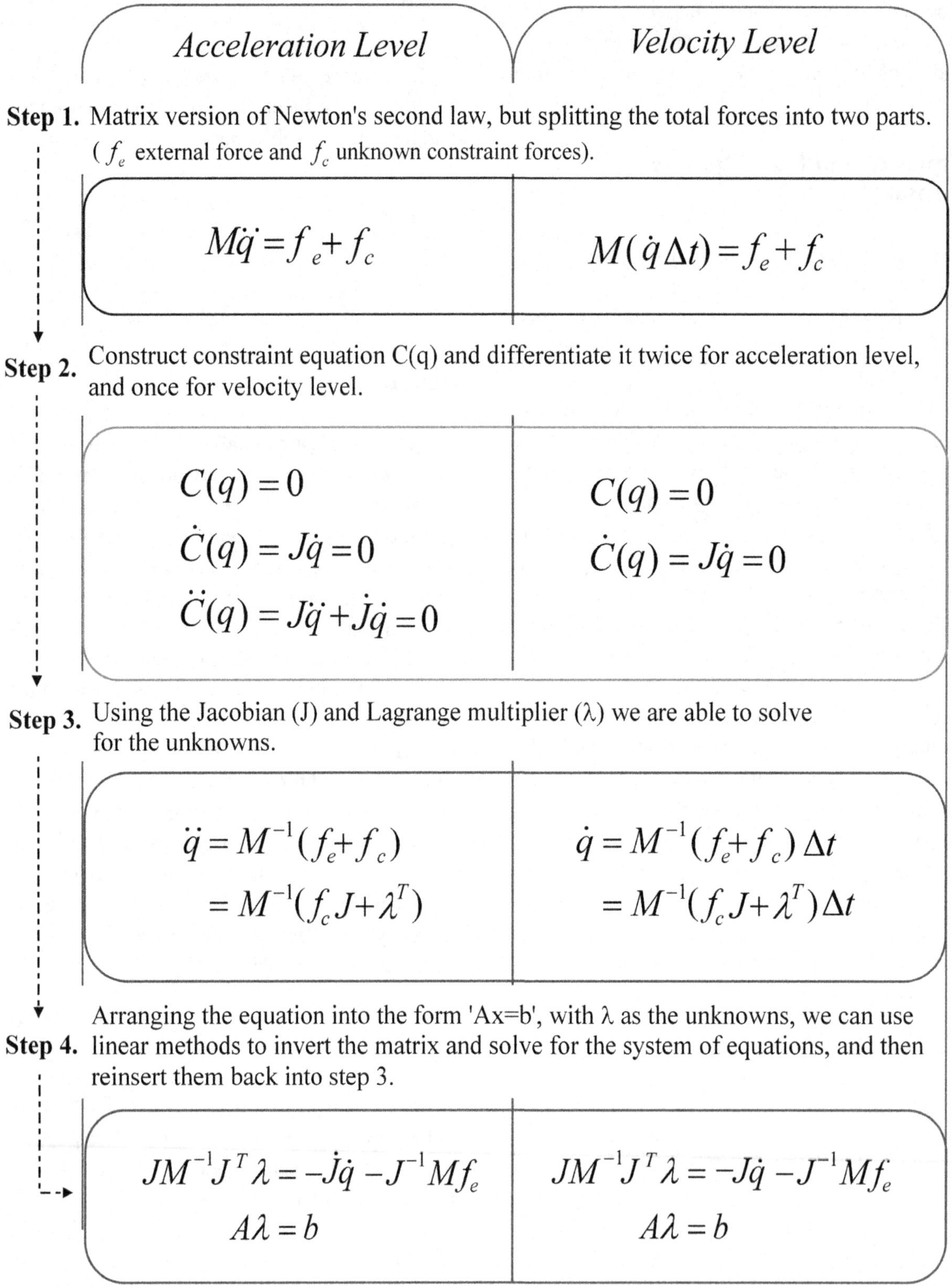

Acceleration Level	*Velocity Level*

Step 1. Matrix version of Newton's second law, but splitting the total forces into two parts. (f_e external force and f_c unknown constraint forces).

$M\ddot{q}=f_e+f_c$	$M(\dot{q}\Delta t)=f_e+f_c$

Step 2. Construct constraint equation C(q) and differentiate it twice for acceleration level, and once for velocity level.

$C(q)=0$ $\dot{C}(q)=J\dot{q}=0$ $\ddot{C}(q)=J\ddot{q}+\dot{J}\dot{q}=0$	$C(q)=0$ $\dot{C}(q)=J\dot{q}=0$

Step 3. Using the Jacobian (J) and Lagrange multiplier (λ) we are able to solve for the unknowns.

$\ddot{q}=M^{-1}(f_e+f_c)$ $=M^{-1}(f_c J+\lambda^T)$	$\dot{q}=M^{-1}(f_e+f_c)\Delta t$ $=M^{-1}(f_c J+\lambda^T)\Delta t$

Step 4. Arranging the equation into the form 'Ax=b', with λ as the unknowns, we can use linear methods to invert the matrix and solve for the system of equations, and then reinsert them back into step 3.

$JM^{-1}J^T\lambda=-\dot{J}\dot{q}-J^{-1}Mf_e$ $A\lambda=b$	$JM^{-1}J^T\lambda=-J\dot{q}-J^{-1}Mf_e$ $A\lambda=b$

for each body, $\dot{q}$ is a 6x1 Cartesian velocity vector, (3x1) linear vector and (3x1) rotational vector stacked together.

We can also write:

$$J = \left[\, J_v,\ J_\omega \right] \tag{7}$$

i.e.

$$\begin{aligned} v &= J_v \dot{q} \\ \omega &= J_\omega \dot{q} \end{aligned} \tag{8}$$

As we said earlier, a single unlinked object moving in 3D, can move in six possible ways, three linear and three angular. If you wanted to constrain the object from moving in a certain direction, the 'z' direction, for example, then you would set the 'z' velocity to zero, effectively limiting translation movement to the x-z plane. This is a simplified example of how a Jacobian works, and describes how constraints operate to remove DOF to achieve a desired motion trajectory. While the Jacobian is at the heart of our analytical method, it allows us to relate the classical equations of motion (f=ma) with Lagrange's multiplier to solve systems of constraints. Later in the chapter we give numerous examples on how to describe and derive a mixture of common constraints and their Jacobian.

Because the Jacobian is central to our constraint formulation, we will spend a bit more time explaining its mechanics in detail with examples to give a rock-solid understanding. The Jacobian, sometimes called the *constraint Jacobian*, is a matrix which allows us to specify which motions are *not allowed*. The number of rows of the matrix determines the order of the constraint or the number of degrees of freedom removed from the system. For most constraints we calculate J on a frame by frame basis as it depends on the body's positions and orientations.

$$J = \frac{Change\,Input}{Change\,Output} = \frac{\partial In}{\partial Out} \tag{9}$$

We can combine Jacobian matrices to formulate more complex ones, by which we mean, we can construct simple constraints and combine them to build more difficult ones.

To repeat the most important equation of this section, the *Jacobian velocity constraint* is shown again below in Equation (10), where $\dot{q}$ is the system body velocities, and $\dot{c}$ is the derivative of the positional constraints c with respect to time, and produces constraints by *removing* degrees of freedom from the system.

$$J\dot{q} = \dot{c} \tag{10}$$

To derive the Jacobian, we need to obtain $\dot{c}$, which we attain by constructing a positional constraint 'c' and differentiating it. The positional constraint is constructed by representing how the body is allowed to move with a kinematic equation such that when it is satisfied, it evaluates to zero:

$$c = 0 \tag{11}$$

If we differentiate our positional constraint equation (c) with respect to time, it will describe the constraint velocity properties. Also as 'c' is equal to zero, the derivative should be zero:

$$\dot{c} = 0 \tag{12}$$

Since we have said that c and $\dot{c}$ will be zero and have explained that the Jacobian has six elements for each body which constitute how the six velocity components change with respect to time, we can conclude that any non-zero values in J will affect the corresponding body velocities. Remember that these are relative to the body's point of reference (its position and rotation at that instance in time and can need re-calculating as the object moves).

Using this knowledge, the Jacobian can be calculated in three straightforward steps; firstly,

Figure 3. Lagrange multiplier modified to fit a system of linear equation to solve our constraints

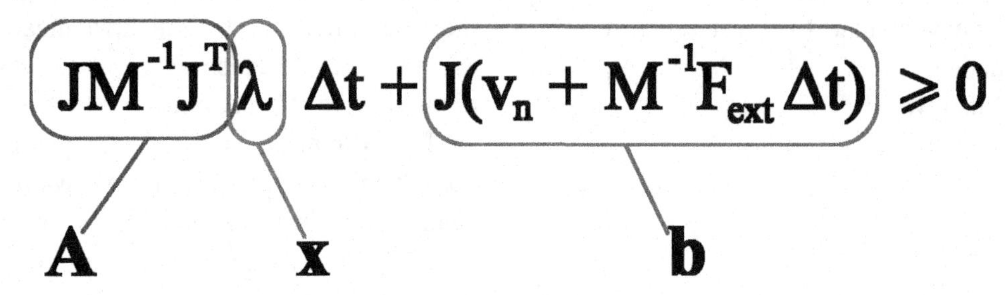

by constructing an equation for the positional constraints, secondly by differentiating it with respect to time to get the velocity properties, and finally by isolating and extracting the Jacobian.

In some cases, you can compose the Jacobian constraint matrix by visually looking at the system, but for more complicated constraint schemes, you will need to follow the steps above, whereby you'll find that the ability to construct the Jacobian gets easier with practice. We give examples of simple Jacobian constraint matrices in the next section.

Note: When a constraint only affects a single object, and does not affect or rely upon any other objects, they are referred to as 'unary constraints', whilst a constraint involving another point-mass or rigid body is termed a 'binary constraint'. In addition, the Jacobian for binary constraints is usually the same but reversed.

Constraint Equations

We now focus on explaining how we resolve the constraints at the velocity level and the formulation of our system of equations that we use to represent our procedure. Even though we focus on the velocity level, we can apply the same principles and practice to the acceleration level without too much effort.

During the simulation update, we calculated a set of constraint forces, which we combined with the external forces, to keep the constraints valid. The main equation, which we construct and use to determine these cancelling constraint forces, is shown in Figure 3.

The equation can be initially overwhelming, but once the reader understands what it does and how it works, it is very satisfying and rewarding. The LCP equations are usually presented in the form:

$$\begin{aligned} b &= Ax + q \\ b, x &\geq 0, \quad b^T x = 0 \end{aligned} \tag{13}$$

This may seem straightforward, but notice the complementary conditions ('b' is 0 when 'x' is not and vice versa) and the non-negativity conditions ('b' and 'x' are >= 0), which is what makes the equations so useful and powerful.

But where does this equation come from, and what does it mean? We will start at the beginning and introduce the problem and how this solves it for us.

It might not be obvious, but if a snapshot of a rigid body simulator taken at any moment in time, the system can be represented, using a set of equations that illustrates how it behaves. If the objects are connected (e.g. through contacts or joints), then their individual EOM will be connected, and so we build up a single large equation (i.e. using matrices). This large equation follows a set of rules, which represent how the system of objects can move, allowing us to predict how the system will change with time.

Our formulation of a solver relies heavily on the principles of linear algebra. To restate, a basic system of linear equations in matrix form, would be:

$$Ax = b \tag{14}$$

where we know 'A' and 'b', and we are trying to find 'x'. 'A' is a symmetric, positive definite sparse matrix. The linear system can have three possible outcomes:

- A single solution which converges
- No solution
- Infinite solutions.

We use the linear equation as the starting foundation from which we build our solver, expanding the basic linear equation above so that it can be used for equality and inequality constraints, i.e.:

$$Ax \geq b and x \geq 0 \tag{15}$$

(From our 'Ax = b' linear equation above, if 'Ax' is not equal to 'b', then we have to reformulate it into something more complex, i.e. 'Ax-b = s', and if we make some assumptions, such as mentioned above, 'Ax>=b' and 'x>0', then we have introduced some constraints.)

The large equation used earlier actually comes from basic algebraic principles. Since we are focusing on the velocity level, we start with the simple velocity integration equation and convert it through substitution and common sense into the equation above.

At each time-step, we can use the Euler integration method to predict how the velocity will change, based on the applied acceleration and time-step. We use this updated velocity to predict how our position will change during our time-step, e.g.

$$\begin{aligned} v_{n+1} &= v_n + a\Delta t \\ x_{n+1} &= v_{n+1} + a\Delta t \end{aligned} \tag{16}$$

According to Newton's second law (i.e. f=ma), we are able to modify our velocity update to:

$$v_{n+1} = v_n + M^{-1}F\Delta t \tag{17}$$

With the velocity integration scheme, we extrapolate it to solve for the applied acceleration changes, which are proportional to the applied forces.

It is important to comprehend how we go from basic velocity integration to the constraint equation. Whereas a majority of the texts skim over the subject, we will give a step-by-step justification in Figure 4.

The two main steps above, are 2 and 4, which we describe further.

In Step 2, we integrate our Jacobian into the velocity equation by multiplying it across both sides. Recalling that the Jacobian defines how our objects will move, adding the Jacobian vector enables us to define the movement of our objects.

Step 4, is one of the biggest steps, as it brings together the equations for classical mechanics and the Lagrange multiplier through the use of the Jacobian. We arrive at our final equation by splitting the constraint force into its magnitude and direction. (This is a very noteworthy step, and should be remembered for later when we reuse it to reinsert our solved values).

$JM^{-1}J^T$ is referred to as the effect mass matrix, and we give it the symbol 'K'. For basic cases we can pre-compute the effective mass matrix and simplify it to show additional properties in the way the constraint behaves.

Note: When a solution for lambda (λ) is obtained, it can be used to give additional information about the system, e.g. the stress or strain between constraints. Additionally, during implementation, the $JM^{-1}J^T$ matrix should have no zeros in the diagonal.

Figure 4. Four steps for deriving the constraint equation from the velocity integration step

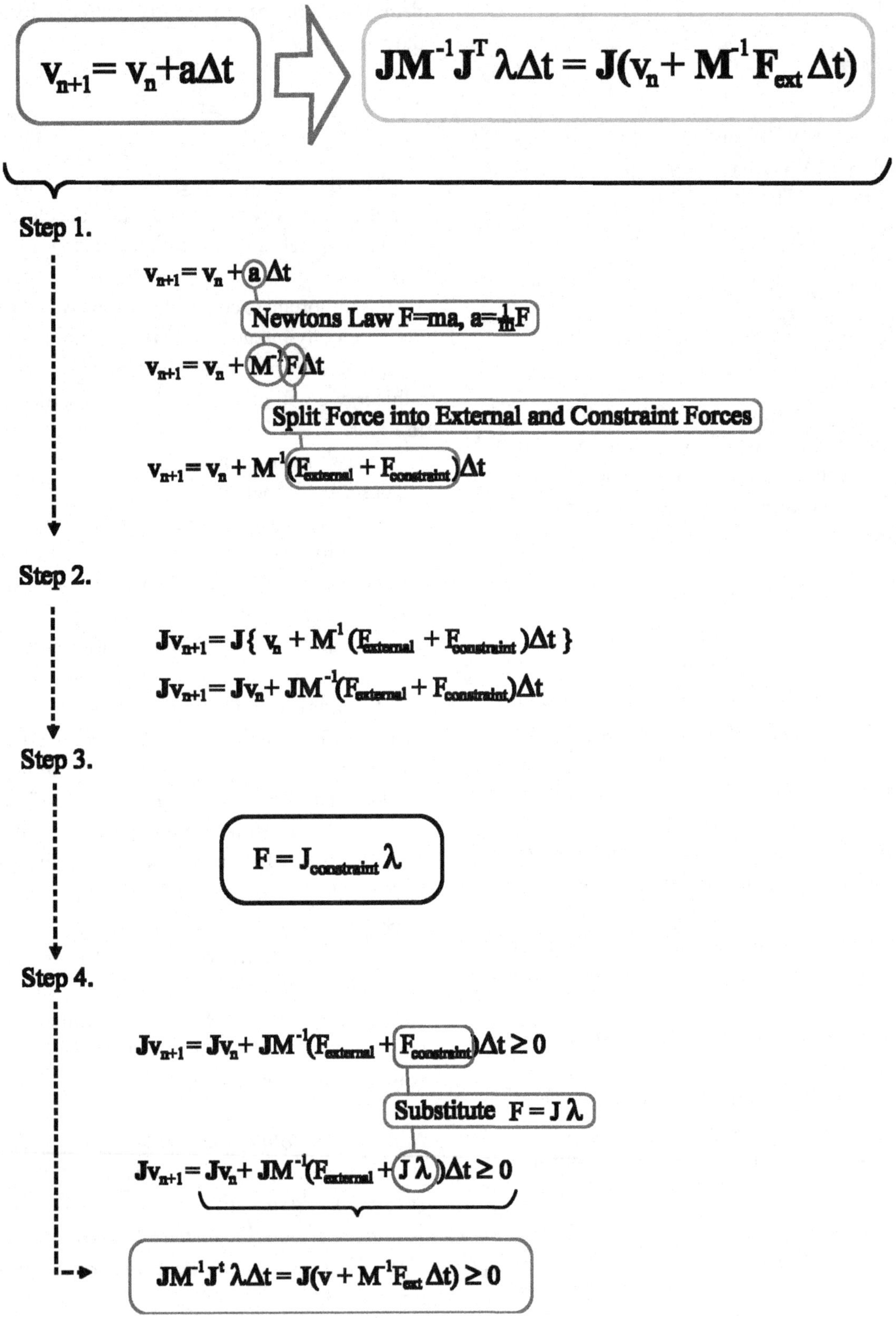

$$v_{n+1} = v_n + a\Delta t \Rightarrow JM^{-1}J^T\lambda\Delta t = J(v_n + M^{-1}F_{ext}\Delta t)$$

Step 1.

$$v_{n+1} = v_n + a\Delta t$$

Newtons Law F=ma, $a=\frac{1}{m}F$

$$v_{n+1} = v_n + M^{-1}F\Delta t$$

Split Force into External and Constraint Forces

$$v_{n+1} = v_n + M^{-1}(F_{external} + F_{constraint})\Delta t$$

Step 2.

$$Jv_{n+1} = J\{ v_n + M^{-1}(F_{external} + F_{constraint})\Delta t \}$$

$$Jv_{n+1} = Jv_n + JM^{-1}(F_{external} + F_{constraint})\Delta t$$

Step 3.

$$F = J_{constraint}\lambda$$

Step 4.

$$Jv_{n+1} = Jv_n + JM^{-1}(F_{external} + F_{constraint})\Delta t \geq 0$$

Substitute $F = J\lambda$

$$Jv_{n+1} = Jv_n + JM^{-1}(F_{external} + J\lambda)\Delta t \geq 0$$

$$JM^{-1}J^t\lambda\Delta t = J(v + M^{-1}F_{ext}\Delta t) \geq 0$$

Solving the equations gives us lambda, which when multiplied by our Jacobian, gives us the constraint forces. We add the computed constraint forces to the applied external forces before integrating. This ensures that the constraints will constantly stay valid, even when large external forces are applied.

The rigid body matrix representations for our simulation are shown in Figure 5; where we collect together similar attributes, such as position and orientation, linear and angular velocities, into groups of matrices.

We use quaternions to represent our rigid body's orientation, and hence our incremental update to angular velocity, using:

$$\dot{q}_{n+1} = \frac{1}{2}\omega_n \dot{q}_n \tag{18}$$

So to achieve this same result with a matrix multiplication, we need to use a special matrix to represent our quaternion orientation. We refer to this matrix as the '*Q*' matrix, which is inside the 'S' matrix and is of the form:

$$Q = \frac{1}{2}\begin{bmatrix} -x & -y & -z \\ w & z & -y \\ -z & w & x \\ y & -x & w \end{bmatrix}$$

where 'w, x, y, z' are scalar and vector components of the quaternion. To phrase it another way, the Q matrix is essentially a sub-matrix representing the rotation using quaternions. (We highlight the 'Q' matrix above inside the 'S' matrix formulation).

Traditionally, using the Euler integration method, we would have integrated each component (linear and angular) for each rigid body using Equations (19) and (20).

$$\begin{aligned} v_{n+1} &= v_n + \frac{1}{m}F_{ext}\Delta t \\ x_{n+1} &= x_n + v_{n+1}\Delta t \end{aligned} \tag{19}$$

$$\begin{aligned} \omega_{n+1} &= \omega_n + \omega_n I_{InvWorld}\ \Delta t \\ q_{n+1} &= q_n + \frac{1}{2}\omega_{n+1} q_n\ \Delta t \end{aligned} \tag{20}$$

We move our rigid body into matrices and perform the integration update using Equations (21) and (22) below.

$$u_{n+1} = u_n + \Delta t\, M^{-1} F_{ext} \tag{21}$$

$$s_{n+1} = s_n + \Delta t\, S\, u_{n+1} \tag{22}$$

This is the basic integration without any intervention from our constraint forces, which we add into the modified block matrix integration:

$$u_{n+1} = u_n + \Delta t M^{-1} F_{ext} - (\Delta t\, M^{-1} J^T x) \tag{23}$$

$$s_{n+1} = s_n + \Delta t\, S\, u_{n+1} \tag{24}$$

where $(\Delta t M^{-1} J^T x)$ uses the cancelling force magnitude in 'x' to feedback into the integrated update and keep the constraints valid. The code snippets (Codes 1 and 2) show the method implemented in code.

Block matrix methods make it possible to group together system state variables, such as forces, velocities and positions, into intuitive manageable pieces and reduces the amount of code disarray. In the sample above, we use a matrix to represent the position and rotation of each body; alternatively, you can exclude this fragment and merely solve for the constraint forces, then apply them as you would to the simulator. The reasons for this may be because the constraint solver is part of a larger system where objects are unable to be merged into matrix formation, or because it is easier to manage data some other way, due to optimisation.

Whilst we use methods to combine angular and linear components into single larger matrices, it can sometimes be more efficient to separate them into in-

Figure 5. Matrix configuration and contents (linear and angular components)

6 D.O.F (3 Linear, 3 Angular) *3 D.O.F (3 Linear)*

$$M = \begin{bmatrix} m & 0 & 0 & 0 & 0 & 0 \\ 0 & m & 0 & 0 & 0 & 0 \\ 0 & 0 & m & 0 & 0 & 0 \\ 0 & 0 & 0 & I_{xx} & I_{xy} & I_{xz} \\ 0 & 0 & 0 & I_{yx} & I_{yy} & I_{yz} \\ 0 & 0 & 0 & I_{zx} & I_{zy} & I_{zz} \end{bmatrix} \qquad M = \begin{bmatrix} m & 0 & 0 \\ 0 & m & 0 \\ 0 & 0 & m \end{bmatrix}$$

Rotation Inertia Matrix

Mass Matrix (Angular and linear components).

6 D.O.F (3 Linear, 3 Angular) *3 D.O.F (3 Linear)*

$$F = \begin{bmatrix} F_x & 0 & 0 & 0 & 0 & 0 \\ 0 & F_y & 0 & 0 & 0 & 0 \\ 0 & 0 & F_z & 0 & 0 & 0 \\ 0 & 0 & 0 & \tau_x & 0 & 0 \\ 0 & 0 & 0 & 0 & \tau_y & 0 \\ 0 & 0 & 0 & 0 & 0 & \tau_z \end{bmatrix} \qquad F = \begin{bmatrix} F_x & 0 & 0 \\ 0 & F_y & 0 \\ 0 & 0 & F_z \end{bmatrix}$$

Force Matrix (Linear Force and Angular Force, aka Torque).

6 D.O.F (3 Linear, 3 Angular) *3 D.O.F (3 Linear)*

$$V = \begin{bmatrix} v_x & 0 & 0 & 0 & 0 & 0 \\ 0 & v_y & 0 & 0 & 0 & 0 \\ 0 & 0 & v_z & 0 & 0 & 0 \\ 0 & 0 & 0 & \omega_x & 0 & 0 \\ 0 & 0 & 0 & 0 & \omega_y & 0 \\ 0 & 0 & 0 & 0 & 0 & \omega_z \end{bmatrix} \qquad V = \begin{bmatrix} v_x & 0 & 0 \\ 0 & v_y & 0 \\ 0 & 0 & v_z \end{bmatrix}$$

Velocity Matrix (Linear and Angular Velocitys)

$$S = \begin{bmatrix} 1 & 0 & 0 & 0 & 0 & 0 \\ 0 & 1 & 0 & 0 & 0 & 0 \\ 0 & 0 & 1 & 0 & 0 & 0 \\ 0 & 0 & 0 & q_x & q_y & q_z \\ 0 & 0 & 0 & q_w & q_z & q_y \\ 0 & 0 & 0 & q_z & q_w & q_x \\ 0 & 0 & 0 & q_y & q_x & q_w \end{bmatrix} \qquad q_{n+1} = \frac{1}{2}\omega q_n$$

Q Matrix

S Matrix (combine quaternion matrix and linear components).

6 D.O.F (3 Linear, 3 Angular) *3 D.O.F (3 Linear)*

$$X = \begin{bmatrix} p_x & 0 & 0 & 0 & 0 & 0 \\ 0 & p_y & 0 & 0 & 0 & 0 \\ 0 & 0 & p_z & 0 & 0 & 0 \\ 0 & 0 & 0 & q_x & 0 & 0 \\ 0 & 0 & 0 & 0 & q_y & 0 \\ 0 & 0 & 0 & 0 & 0 & q_z \end{bmatrix} \qquad X = \begin{bmatrix} p_x & 0 & 0 \\ 0 & p_y & 0 \\ 0 & 0 & p_z \end{bmatrix}$$

$$X_{n+1} = X + \frac{1}{2}\Delta t S V$$

Exclude S if no Rotation

Position Matrix (Angular component is a quaternion).

Code 1. Basic Euler integration step using matrices

```
// Basic Integration without constraints or collisions
u_next = u + dt*MInverse*Fext;
s_next = s + dt*S*u_next;
```

dividual matrices so as to disable and enable rotational components for debugging, or alternatively develop a point mass simulator which contains no rotational components. For example, the mass matrix is a combination of the linear and angular mass components (centre of mass and inertia tensors), which could be broken up into two separate matrix functions.

Initially, matrix methods in combination with analytical solutions can seem like more work compared to the penalty and impulse methods, but the results give us a reliable, flexible simulator, which requires little or no user tweaking.

Our solver uses classical mechanics in combination with Lagrangian multiplier techniques, so we are able to formulate systems of equations to calculate the necessary forces to apply at each frame and enable our constraints to remain legal. For example, a ball resting on the floor has a downward force applied to it, known as gravity. We cannot allow this downward force to update the velocity and let the ball move translate downwards. Hence, we add a contact constraint to the system to prevent this penetration violation. This constraint would produce a cancelling upward force, keeping the ball resting on the ground. The complementary part of the solution is when the ball is falling with velocity, and we need to add an impact force, causing the ball to bounce upwards and not stick to the ground (it can move up but not down).

Algorithm Steps (Code 3)

1. Apply forces and torques (i.e. from springs, gravity, etc.).
2. Build matrices for rigid body masses/velocities/positions.
3. Build Jacobian constraint matrices representing various constraints.
4. Evaluate your system by solving for lambda.
5. Build constraint forces using lambda and Jacobian matrices.
6. Calculate and apply constraint forces.
7. Integrate and update as usual.

Note: Making objects immovable by setting their mass to zero or making their mass near infinite (e.g. 10000000) can crash the simulator. This is because the solver needs to invert the matrix, so large or zero values can result in a non-invertible matrix. Large values may not crash the simulation, but it can require a very large number of iterations to find its inverse solution.

'1D' Numerical Example

The best way to get familiar with solvers is practice, which we now do using simplified straightforward test cases that we analyse with common sense

Code 2. Modified integration step with our added constraint forces

```
// Basic Integration without constraints or collisions
u_next = u + dt*MInverse*Fext - MInverse*Jt*x;
s_next = s + dt*S*u_next;
```

Code 3. Steps for updating code

```
void Simulation_c::Update(float dt)
{
    // Add gravity, forces from springs etc.
    ComputeForces(dt);
    // Calculate constraint forces and apply them to our system
    ComputeJointConstraints(dt);
    // Step forwards in time - integration done inside ComputeJointCon-
straints(..)
    // using block matrices, but you can disable it and integrate using
    // an alternative integrator
    //Integrate(dt);
}
```

and understanding. Common mistakes to keep an eye out for initially, are errors due to wrong number sign (+ instead of a -) or invalid divide by zero troubles.

We start with simple 1D examples, where we do not need to worry about rotation, just the translation, and can focus on the force magnitudes and directions. Whereby, our matrix formulations will be greatly simplified (e.g. the mass matrix in 1D is cut down to a single variable for each object), so we can evaluate the equations by hand. It is also recommended that this simplified approach is used to work through additional problems which will demonstrate first-rate principles of how constraint solvers achieve reliable simulations, such as jitter free stacking.

The example shown in Figure 6 is constructed by taking a single object and letting it rest on an immovable surface. Then when we apply downward force due to gravity, the top object will continue to stay fixed. Since it is only a straightforward example, we use a very simple uncomplicated Jacobian for our constraint, which prevents any movement in the single-axis direction. Of course, to make it a little more interesting, the top object also has an internal velocity, so the calculated resultant force, when integrated into the velocity, will cancel it out. Hence, our constraint solver keeps our resting object on the surface as the simple constraint demands.

If the question; why do we get [1,1] for the resulting force and not [0,0], is asked, this is as we mentioned earlier. Our object has an initial velocity of [-1,0], so a positive force is necessary to cancel out this downward velocity and keep our constraint valid. If we have had zero velocities, the downward force would have cancelled out, leaving us with [0,0].

It should be well-known by now that the solver needs to run each frame before the integration step, which updates the velocities and positions. This pre-step evaluation calculates the correcting constraint forces which, when applied to the input forces, (e.g. from gravity, wind, springs) keeps the constraints legal.

1D Example: Multiple Bodies

Continuing to keep the examples as simple as possible, we can build on the previous example to illustrate stacking. The configuration shown in Figure 7 consists of three objects stacked upon one another; each having equal mass.

We set all the objects to have zero velocities initially, except the second to top one, which we set to having an internal downward velocity. As

Figure 6. Numerical example

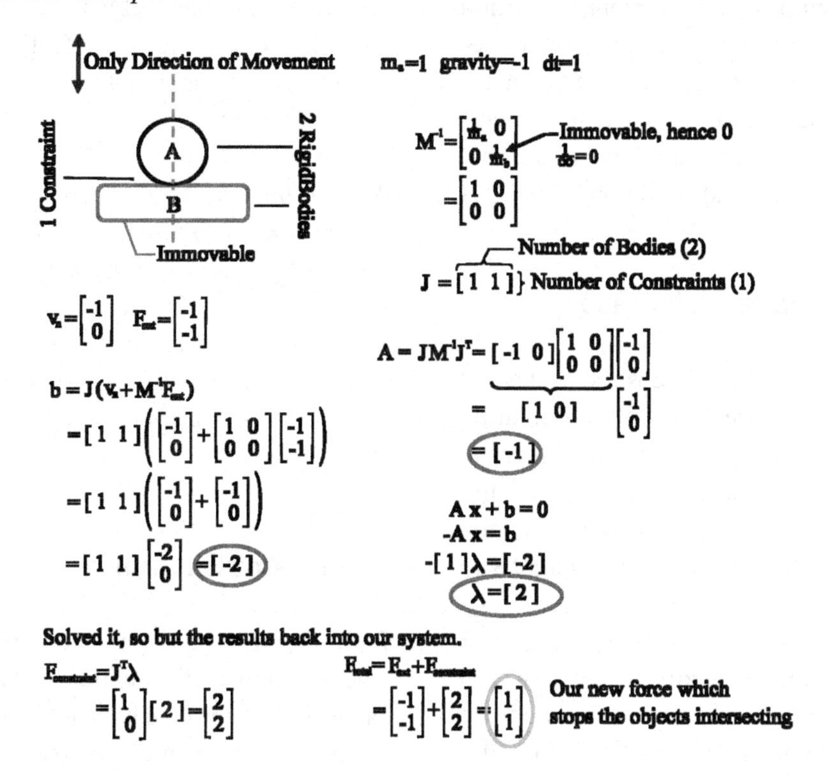

Figure 7. Numerical example

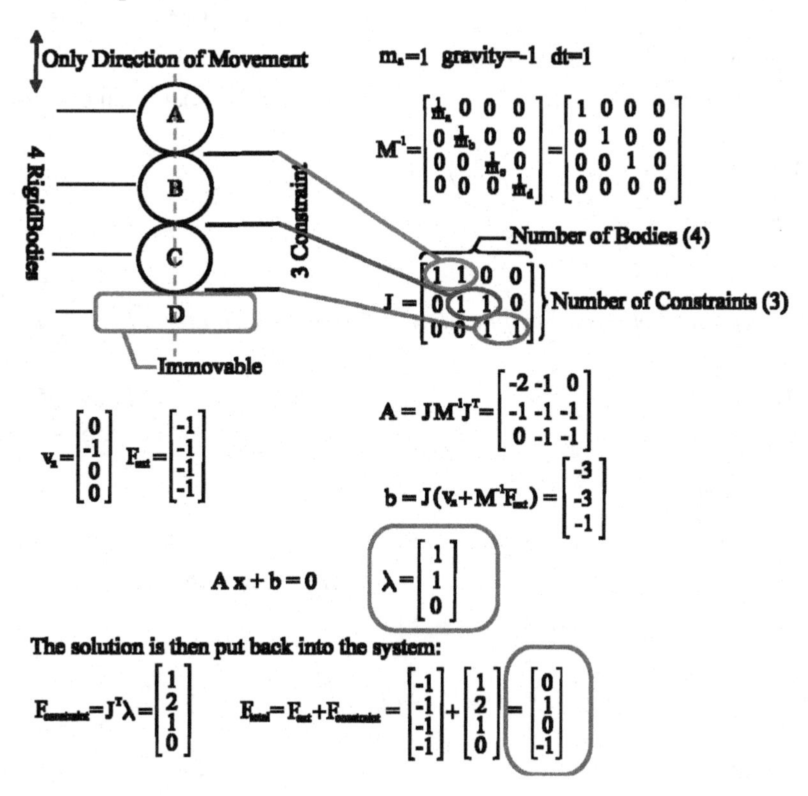

above we use a simplified constraint formulation to calculate the constraint force to prevent movement of the objects. The solved example calculates the modified force to stop the objects moving downwards, including an additional force to cancel the second to top body's velocity.

Drifting (Baumgarte Stabilization)

The constraint calculation keeps the system of constraints valid by cancelling out any violating forces/velocities. However, due to numerical inaccuracies in calculations, or bad starting values, the constraints can become invalid, and so we need to modify our calculation slightly to correct this.

To solve for this, we modify our solver equation by adding a feedback term, which gradually corrects drifting errors. We achieve this correction by using the position constraint 'c', to feed back into the velocity constraint, causing the simulation constraints to remain valid at all times.

Modifying our original velocity constraint from Equation (10) to give feedback, we get:

$$\dot{c} = J\dot{q} + \beta c = 0 \tag{25}$$

where β is the bias factor $(0 < \beta < 1)$.

This corrective term is effectively the same as adding a spring, due to its remedial nature being proportional to the positional error. (Code 4)

Note: If the simulation begins with invalid constraint configurations, the Baumgarte feedback term will cause the system to try and correct itself. Additionally, we can use this effect to modify our contact constraints, so that if they penetrate, the constraint will apply a correcting penalty force to push the shapes out of penetration.

Constraint Examples

Formulating the constraint condition 'C', we can calculate how to differentiate and extrapolate the Jacobian to determine how our system moves and to restrict its movement to create our constraint.

Example: Fixed With No Translational or Rotational Movement

Formulate the positional constraint:

$$c = \left[q_x - p_x,\ q_y - p_y,\ q_z - p_z,\ q_{\theta x} - p_{\theta x},\ q_{\theta y} - p_{\theta y},\ q_{\theta z} - p_{\theta z}\right] = 0 \tag{26}$$

Code 4. Drifting correction: achieved by appending a drift correction term to the base solver equation

```
float beta = 0.1f;                          // Drifting correction factor
DMatrix Jt = Transpose(J);                  // Formulation of the solver equa-
tions
DMatrix A  = J*MInverse*Jt;
DMatrix b  = J*(u + dt*MInverse*Fext) +     //
             beta*c;                         // Drifting (Baumgarte) appended on
end

DMatrix x(A.GetNumRows(), b.GetNumCols());
// Solve for x
LCPSolver(A, b, &x);                        // Solver, equality constraint case
```

where $q_x, q_y, q_z, q_{\theta x}, q_{\theta y}, q_{\theta z}$ are the position and orientation of the body, and $p_x, p_y, p_z, p_{\theta x}, p_{\theta y}, p_{\theta z}$ are the global translation and angular *constants*. Noting that '0' represents a zero vector [0,0,0,0,0,0], if we then differentiate this constraint with respect to time, we obtain the velocity constraint:

$$\dot{c} = \left[\dot{q}_x, \dot{q}_y, \dot{q}_z, \dot{q}_{\theta x}, \dot{q}_{\theta y}, \dot{q}_{\theta z}\right] = 0 \tag{27}$$

Since our constraint position and orientation are constant, they simply cancel out during differentiating, hence enabling us to compare and isolate the Jacobian:

$$J\dot{q} = \dot{c} = \begin{bmatrix} 1 & 0 & 0 & 0 & 0 & 0 \\ 0 & 1 & 0 & 0 & 0 & 0 \\ 0 & 0 & 1 & 0 & 0 & 0 \\ 0 & 0 & 0 & 1 & 0 & 0 \\ 0 & 0 & 0 & 0 & 1 & 0 \\ 0 & 0 & 0 & 0 & 0 & 1 \end{bmatrix} \dot{q} = 0 \tag{28}$$

We make the constraint from six rows and takes away six degrees of freedom, hence the rigid body cannot move in any direction or orientation.

Example: No Movement along the X-Axis

From the previous example, we can then build upon the fact that if we use only a single row, we can eliminate a single degree of freedom.

$$\begin{bmatrix} 1 & 0 & 0 & 0 & 0 & 0 \end{bmatrix} \dot{q} = 0 \tag{29}$$

Through visual analysis, we can make out that the velocity in the x-axis is cancelled out by having a single 1 in the first column, hence no velocity in this direction results in no movement. Since the other velocity components are not affected, we can deduce that they will be unaffected. This constraint formulation only affects the changing velocity to prevent movement in a single direction, and not the actual position at each moment.

Example: No Rotation Around Up Vector (Y-Direction)

Whereas we removed translation in the x-axis previously, we extend this further by removing a single axis of rotation. It should be visually clear that the first three columns are the positional constraints, and the final three columns make up the rotational movement.

$$\begin{bmatrix} 0 & 0 & 0 & 0 & 1 & 0 \end{bmatrix} \dot{q} = 0 \tag{30}$$

As shown above, adding a one to the fifth column prevents any angular velocity in the y-axis and hence any rotational movement similarly.

Example: Ball and Socket Constraint

When dealing with pairs of bodies, we need to use two Jacobian matrices to represent the constraint condition. Below, we show the formulation of a ball-joint, by constructing its positional constraint then differentiating it, so we can extract the Jacobian for each object. Notice that the rows are a combination of translational and rotational information; making the constrained movement more complex.

$$c = \left[(p_1 + r_1) - (p_0 + r_0)\right] = 0 \tag{31}$$

where p represents the object centre, and r the relative offset from it.

$$\dot{c} = \left[(v_1 + \omega_1 \times r_1) - (v_0 + \omega_0 \times r_0)\right] = 0 \tag{32}$$

$$J_o = [I, r_1^*], \quad J_1 = -[I, r_0^*] \tag{33}$$

$$\begin{bmatrix} 1 & 0 & 0 & 0 & r_{0z} & -r_{0y} \\ 0 & 1 & 0 & -r_{0z} & 0 & r_{0x} \\ 0 & 0 & 1 & r_{0y} & -r_{0x} & 0 \end{bmatrix} \dot{q}_0 - \begin{bmatrix} 1 & 0 & 0 & 0 & r_{1z} & -r_{1y} \\ 0 & 1 & 0 & -r_{1z} & 0 & r_{1x} \\ 0 & 0 & 1 & r_{1y} & -r_{1x} & 0 \end{bmatrix} \dot{q}_1 = 0 \quad (34)$$

where r_1^*, r_0^* represents a 3x3 skew matrix which is equivalent to the cross product.

Example:Fixed Point (Nail)

If we took our object and just stuck a nail through any point, it would stay locked at that point while still able to rotate. We effectively use the same method as above (ball and socket), except, we only have a single body. The Jacobian formulation is shown below, and shows how the object cannot move in the 'x', 'y' or 'z' direction but can still rotate relative to a specific point.

$$\begin{bmatrix} 1 & 0 & 0 & 0 & r_z & -r_y \\ 0 & 1 & 0 & -r_z & 0 & r_x \\ 0 & 0 & 1 & r_y & -r_x & 0 \end{bmatrix} \dot{q} = 0 \quad (35)$$

where r_x, r_y, r_z represent the offset from the object's centre to the point we are rotating around.

Example: Distance Constraint (No Rotation)

A simple and useful constraint, especially when working with point-masses is a distance constraint whereby the distance between any two points remains fixed.

Setting up the conditions under which a constraint would be valid, we can say that the distance must always be equal to some length (l):

$$|| p0 - p1 || = l \quad (36)$$

where ||p0-p1|| is the length between the two points, 'p0' and 'p1'.

Constructing the position constraint 'c', we have:

$$c = || p0 - p1 || - l = 0 \quad (37)$$

Knowing 'c', we can differentiate it to get $\dot{c}$, and using the definition for the Jacobian velocity constraint, extrapolate the Jacobian component:

$$J\dot{q} = \dot{c} \quad and \quad \dot{q} = \begin{bmatrix} \mathrm{v0} \\ \mathrm{v1} \end{bmatrix} \quad (38)$$

$$J = \begin{bmatrix} -\hat{n}^T & \hat{n}^T \end{bmatrix} \quad (39)$$

e.g.

$$\begin{bmatrix} \hat{n}_x & \hat{n}_y & \hat{n}_z \end{bmatrix} \dot{q} = 0 \quad (40)$$

where 'n' is the unit vector between the two points:

$$\hat{n} = \frac{(p1 - p0)}{|| p1 - p0 ||} \quad (41)$$

Example: Contact Constraint

We formulate a contact constraint which uses the contact normal 'n' to prevent objects intersecting. The added difference between this constraint and the previous constraint examples is the added boundary condition. Whereas we used the equality condition for previous examples, we now apply an inequality condition for the solution λ to be greater than or equal to zero, warranting objects to move away from each other. For the calculations below we take for granted that the normal vector is pointing from body0 to body1.

$$c = \left[(p_1 + r_1) - (p_0 + r_0)\right] \cdot \hat{n}_1 \geq 0 \tag{42}$$

$$\dot{c} = \left[(v_1 + \omega_1 \times r_1) - (v_0 + \omega_0 \times r_0)\right] \cdot \hat{n}_1 + \left[(p_1 + r_1) - (p_0 + r_0)\right] \cdot (\omega \times \hat{n}_1) \geq 0 \tag{43}$$

$$J_o = [\hat{n}_1, r_0 \times \hat{n}_1], \qquad J_1 = -[\hat{n}_1, r_1 \times \hat{n}_1] \tag{44}$$

$$\begin{bmatrix} n_{1x} & 0 & 0 & 0 & a_{0z} & -a_{0y} \\ 0 & n_{1y} & 0 & -a_{0z} & 0 & a_{0x} \\ 0 & 0 & n_{1z} & a_{0y} & -a_{0x} & 0 \end{bmatrix} \dot{q}_0 - \begin{bmatrix} n_{1x} & 0 & 0 & 0 & a_{1z} & -a_{1y} \\ 0 & n_{1y} & 0 & -a_{1z} & 0 & a_{1x} \\ 0 & 0 & n_{1z} & a_{1y} & -a_{1x} & 0 \end{bmatrix} \dot{q}_1 = 0 \tag{45}$$

where 'r' is the relative offset from the object centre to the contact point and $a_0 = r_0 \times \hat{n}_1$, $a_1 = r_1 \times \hat{n}_1$.

When implementing this constraint, it is crucial to remember $\lambda \geq 0$ for the boundary conditions. We can additionally add friction by modifying our constraint to add a cancelling force along the tangential direction of movement.

IMPLEMENTATION

Writing the Code

Code 5 is the code for the rigid body class, which encapsulates the properties of each object. We separate out the angular and linear components, so that we can easily modify it to work with full 3D or point-mass simulations.

Note: Disable Rotation – If the rigid body simulator contains bugs, it is best to go back to basics, if angular problems are suspected, remove the rotation, then various degrees of freedom and add them back when it is known that they are working.

Iterative LCP Solver

From the Code 6, you can see that the LCP solver consists of nested loops, within which we modify our starting approximation iteratively and converge gradually towards a solution. Where the sample code uses a preset maximum number of iterations (maxIterations) for straightforwardness, we can alternatively, add an early breakout condition to accelerate the simulation, which we introduce later.

Note: Fewer Iterations – If you notice in the example code for the 'Gauss-Seidel' iterative solver, it clears the solution matrix to zero at the start. This makes the assumption that our initial guess is zero. A faster method is not to clear the matrix, instead just using the values that are still in it from the previous frame. This will give us a good starting guess, enabling us to find a correct result with fewer iterations due to very small changes happening between frames.

Constraints that 'Snap-Together'

One important aspect when writing a solver is to write the code in such a way that it can easily be adapted and the code expanded to handle new constraints and configurations. We do this by constructing a common virtual base class, which all constraints inherit from and implement, so that we are able to 'plug-in' any constraint into our simulator as and when we need it.

Code 5. Basic Euler integrator

```
class RigidBody_c // DirectX Implementation
{
    //<---------LINEAR----------------><-------------ANGULAR-------------->
    D3DXVECTOR3 m_position;           D3DXQUATERNION  m_orientation;
    D3DXVECTOR3 m_linearVelocity;     D3DXVECTOR3     m_angularVelocity;
    D3DXVECTOR3 m_force;              D3DXVECTOR3     m_torque;
    float       m_invMass;            D3DXMATRIX      m_invInertia;
    void AddForce(const D3DXVECTOR3& worldPosForce,
                  const D3DXVECTOR3& directionMagnitude)
    {
        DBG_VALID(worldPosForce);
        DBG_VALID(directionMagnitude);
        //<---------LINEAR----------------->
        m_force         += directionMagnitude;

        //<-----------ANGULAR-------------->
        D3DXVECTOR3 distance = worldPosForce - m_position;
        D3DXVECTOR3 torque   = Cross(distance, directionMagnitude);
        AddTorque(torque);
        DBG_VALID(m_force);
        DBG_VALID(m_torque);
    }

    void AddTorque(D3DXVECTOR3 worldAxisAndMagnitudeTorque)
    {
        DBG_VALID(worldAxisAndMagnitudeTorque);
        m_torque += worldAxisAndMagnitudeTorque;
    }
    D3DXMATRIX CreateWorldII()
    {
        D3DXMATRIX orientationMatrix = CreateMatrixFromQuaternion(m_orienta-
tion);
        D3DXMATRIX inverseOrientationMatrix  = Transpose(orientationMatrix);
        D3DXMATRIX inverseWorldInertiaMatrix = inverseOrientationMatrix * m_
invInertia *
                                                                        ori-
entationMatrix;
        return inverseWorldInertiaMatrix;
    }
};
```

Code 6. Iterative solver: Showing how simple it can be to invert a matrix

```
void GaussSeidelLCP(DMatrix& a, DMatrix& b, DMatrix* x, const DMatrix* lo,
const DMatrix* hi)
{
    int maxIterations = 10; // Test Max value
    x->SetToZero();  // Clear our matrix to start with (slow, only for debug)
    const int n = x->GetNumRows();

    float sum = 0.0f;
    while(maxIterations--)
    {
        for(int i = 0; i < n; i++)
        {
            sum = b.Get(i);
            for(int j = 0; j < n; j++)
            {
                if(i != j)
                {
                          sum = sum - (a.Get(i,j) * x->Get(j));
                }
            }
            // If a.Get(i,i) is zero - you have a bad matrix!
            DBG_ASSERT(a.Get(i,i)!=0.0f);
            x->Set(i) = sum/a.Get(i,i);
          // Only do condition to check if we have them
        }
       // If we have boundary conditions, e.g. >= or <=, then we modify our
basic Ax=b,
       // solver to apply constraint conditions
       // Optional - only if bounds
       if (lo || hi)
       for (int i=0; i<n; i++)
       {
         if (lo)
         {
            DBG_ASSERT(lo->GetNumCols()==1); // Sanity Checks
            DBG_ASSERT(lo->GetNumRows()==n);
            if (x->Get(i) < lo->Get(i)) x->Set(i) = lo->Get(i);
         }
         if (hi)
         {
            DBG_ASSERT(hi->GetNumCols()==1); // Sanity Checks
            DBG_ASSERT(hi->GetNumRows()==n);
```

continued on following page

Code 6. Continued

```
                if (x->Get(i) > hi->Get(i)) x->Set(i) = hi->Get(i);
            }
        }
    }
    // We've solved x!
}
```

```
class Constraint_c
{
public:
    virtual ~Constraint_c(){}
    virtual DMatrix   GetPenal-
ty()                        {   DBG_
HALT;    return DMatrix(0,0);};
    virtual DMatrix
GetJacobian(const RigidBody_c* rb) {
DBG_HALT;    return DMatrix(0,0);};
    virtual int        GetDimension()
const                  {    DBG_HALT;
return 0;              };
};
```

If equality and inequality constraints are to be implemented, a base method should be added to differentiate them (i.e. as shown below). These are so the inequality conditions can incorporate additional boundary checks when solving the system of equations.

```
     virtual bool  IsEquality ()
const                        {    return
false;                        };
```

'Heart' of the Simulator

Code 7 shows the function 'ComputeJointConstraints', which is responsible for asking every constraint about its dimensions, then constructing large sparse matrices, which correspond to the system configuration and calculate the corresponding constraint forces:

Of course, the solver is the one that is going to consume most of the time, as it has to determine the actual solution to the matrix. The code of our iterative solver, however, is really quite simple. As mentioned earlier, if one is dealing with real-time applications, an iterative solver should always be used.

Why Do Solvers Break?

When implementing a solver, there are a few things to keep an eye out for:

- Overly constrained systems.
- Numerical error (drifting, floating point accuracy).
- Divide by zero errors.
- Impossible constraints.
- Invalid coordinates (e.g. placing rigid bodies at identical locations).
- Bad code (e.g. memory leaks, memory corruption, and incorrect implementation of algorithm).

Note: Diagonal Zeros – If there are zeros on the diagonal, something is wrong. It can cause the LCP solver to break, giving solutions that go to infinity.

Bad Constraints

There are times when the simulation constraints cannot be solved, usually because of human error

Code 7. Section of code that actually does the magic, by building the large sparse matrices, passing them to the solver which returns the corrected the forces and torques which we re-inject back into the simulator

```
void Simulation:: ComputeJointConstraints ()
{
    // Magic Formula
    //
    // J * M^-1 * J^t  * lamba = -1.0 * J * (1/dt*V + M^-1 * Fext)
    //
    // A x = b
    //
    // where
    //
    // A = J * M^-1 * J^t
    // x = lambda
    // b = -J * (1/dt*V + M^-1 * Fext)
    //
    const int numBodies         = m_rigidBodies.Size();
    const int numConstraints    = m_constraints.Size();
    if (numBodies==0 || numConstraints==0) return;
    //-------------------------------------------------------------------------
--
    // 1st - build our matrices - very bad to build them each frame, but
    //-------------------------------------------------------------------------
--
    // simpler to explain and implement this way
    DMatrix s(numBodies*7, 1);          // pos & qrot
    DMatrix u(numBodies*6, 1);          // vel & rotvel
    DMatrix s_next(numBodies*7, 1);     // pos & qrot after timestep
    DMatrix u_next(numBodies*6, 1);     // vel & rotvel after timestep
    DMatrix S(numBodies*7, numBodies*6);
    DMatrix MInverse(numBodies*6, numBodies*6);
    DMatrix Fext(numBodies*6, 1);
    for (int i=0; i<numBodies; i++)
    {
        const RigidBody_c* rb = m_rigidBodies[i];
        s.Set(i*7+0) = rb->m_position.x;
        s.Set(i*7+1) = rb->m_position.y;
        s.Set(i*7+2) = rb->m_position.z;
        s.Set(i*7+3) = rb->m_orientation.w;
        s.Set(i*7+4) = rb->m_orientation.x;
        s.Set(i*7+5) = rb->m_orientation.y;
        s.Set(i*7+6) = rb->m_orientation.z;
        u.Set(i*6+0) = rb->m_linearVelocity.x;
```

continued on following page

Code 7. Continued

```
        u.Set(i*6+1) = rb->m_linearVelocity.y;
        u.Set(i*6+2) = rb->m_linearVelocity.z;
        u.Set(i*6+3) = rb->m_angularVelocity.x;
        u.Set(i*6+4) = rb->m_angularVelocity.y;
        u.Set(i*6+5) = rb->m_angularVelocity.z;
        const D3DXQUATERNION& q = rb->m_orientation;
        DMatrix Q(4,3);
        Q.Set(0,0)=-q.x;        Q.Set(0,1)=-q.y;        Q.Set(0,2)=-q.z;
        Q.Set(1,0)= q.w;        Q.Set(1,1)= q.z;        Q.Set(1,2)=-q.y;
        Q.Set(2,0)=-q.z;        Q.Set(2,1)= q.w;        Q.Set(2,2)= q.x;
        Q.Set(3,0)= q.y;        Q.Set(3,1)=-q.x;        Q.Set(3,2)= q.w;
        Q = 0.5f * Q;
        DMatrix Idenity(3,3);
        Idenity.SetToZero();
        Idenity.Set(0,0) = Idenity.Set(1,1) = Idenity.Set(2,2) = 1.0f;
        S.SetSubMatrix(i*7+0,   i*6+0,   Idenity);
        S.SetSubMatrix(i*7+3,   i*6+3,   Q);
        DMatrix M(3,3);
        M.Set(0,0) = M.Set(1,1) = M.Set(2,2) = rb->m_invMass;
        const D3DXMATRIX& dxm = rb->CreateWorldII();
        DMatrix I(3,3);
        I.Set(0,0)=dxm._11;  I.Set(1,0)=dxm._12;  I.Set(2,0)=dxm._13;
        I.Set(0,1)=dxm._21;  I.Set(1,1)=dxm._22;  I.Set(2,1)=dxm._23;
        I.Set(0,2)=dxm._31;  I.Set(1,2)=dxm._32;  I.Set(2,2)=dxm._33;
        MInverse.SetSubMatrix(i*6,   i*6,   M);
        MInverse.SetSubMatrix(i*6+3, i*6+3, I);
        DMatrix F(3,1);
        F.Set(0,0) = rb->m_force.x;
        F.Set(1,0) = rb->m_force.y;
        F.Set(2,0) = rb->m_force.z;
        D3DXVECTOR3 rF = rb->m_torque;
        DMatrix T(3,1);
        T.Set(0,0) = rF.x;
        T.Set(1,0) = rF.y;
        T.Set(2,0) = rF.z;
        Fext.SetSubMatrix(i*6,   0,  F);
        Fext.SetSubMatrix(i*6+3, 0,  T);
    }
    //-----------------------------------------------------------------------
--
    // 2nd - apply constraints
    //-----------------------------------------------------------------------
```

continued on following page

Code 7. Continued

```
--
    // Determine the size of our jacobian matrix
    int numRows = 0;
    for (int i=0; i<numConstraints; i++)
    {
        const Constraint_c* constraint = m_constraints[i];
        DBG_ASSERT(constraint);
        numRows += constraint->GetDimension();
    }
    // Allocate it, and fill it
    DMatrix J(numRows, 6*numBodies);
    DMatrix e(numRows, 1);              // Error Correction
    int constraintRow = 0;
    for (int c=0; c<numConstraints; c++)
    {
        Constraint_c* constraint = m_constraints[c];
        DBG_ASSERT(constraint);
        for (int r=0; r<numBodies; r++)
        {
            const RigidBody_c* rigidBody = m_rigidBodies[r];
            DBG_ASSERT(rigidBody);
            DMatrix JMat = constraint->GetJacobian(rigidBody);
            if (JMat.GetNumCols()==0 && JMat.GetNumRows()==0)
                continue;
            DBG_ASSERT(JMat.GetNumCols()!=0);
            DBG_ASSERT(JMat.GetNumRows()!=0);
            J.SetSubMatrix(constraintRow, r*6, JMat);
            DMatrix errMat = constraint->GetPenalty();
            e.AddSubMatrix(constraintRow, 0, errMat);
        }
        constraintRow += constraint->GetDimension();
    }
    float beta = 0.1f; // Error correction term
    DMatrix Jt = Transpose(J);
    DMatrix A  = J*MInverse*Jt;
    DMatrix b  = J*(u + dt*MInverse*Fext) + beta*e;
    DMatrix x(A.GetNumRows(), b.GetNumCols());
    DMatrix* lo = NULL; // Don't set any min/max boundaries for this demo/sam-
ple
    DMatrix* hi = NULL;
    // Solve for x
    LCPSolver(A, b, &x, lo, hi);
```

continued on following page

Code 7. Continued

```
    //dprintf(A.Print());
    u_next = u - MInverse*Jt*x + dt*MInverse*Fext;
    s_next = s + dt*S*u_next;
    // Basic integration without - euler integration standalone
    // u_next = u + dt*MInverse*Fext;
    // s_next = s + dt*S*u_next;
    //----------------------------------------------------------------------
--
    // 3rd - re-inject solved values back into the simulator
    //----------------------------------------------------------------------
--
    for (int i=0; i<numBodies; i++)
    {
        RigidBody_c* rb = m_rigidBodies[i];
        rb->m_position.x            = s_next.Get(i*7+0);
        rb->m_position.y            = s_next.Get(i*7+1);
        rb->m_position.z            = s_next.Get(i*7+2);
        rb->m_orientation.w         = s_next.Get(i*7+3);
        rb->m_orientation.x         = s_next.Get(i*7+4);
        rb->m_orientation.y         = s_next.Get(i*7+5);
        rb->m_orientation.z         = s_next.Get(i*7+6);
        rb->m_linearVelocity.x      = u_next.Get(i*6+0);
        rb->m_linearVelocity.y      = u_next.Get(i*6+1);
        rb->m_linearVelocity.z      = u_next.Get(i*6+2);
        rb->m_angularVelocity.x     = u_next.Get(i*6+3);
        rb->m_angularVelocity.y     = u_next.Get(i*6+4);
        rb->m_angularVelocity.z     = u_next.Get(i*6+5);
        rb->m_force                 = D3DXVECTOR3(0,0,0);
        rb->m_torque                = D3DXVECTOR3(0,0,0);
        // Just incase we get drifting in our quaternion orientation
        D3DXQuaternionNormalize(&rb->m_orientation, &rb->m_orientation);
    }
}
```

or some unforeseen circumstances. This commonly occurs when constraints fight or violate each other, so that no one single solution exists to keep all the constraints valid. An example is illustrated below, whereby two objects are placed out of reach of one another using nail constraint, plus an additional joint constraint connecting them. These constraints contradict each other because the objects cannot be kept out of reach and also connected. For instance, in the example shown in Figure 8, illustrates an impossible constraint situation.

These ill-conditioned constraint configurations can introduce erratic behavior into the simulation because there is no real solution. However our it-

Figure 8. Impossible constraint situations

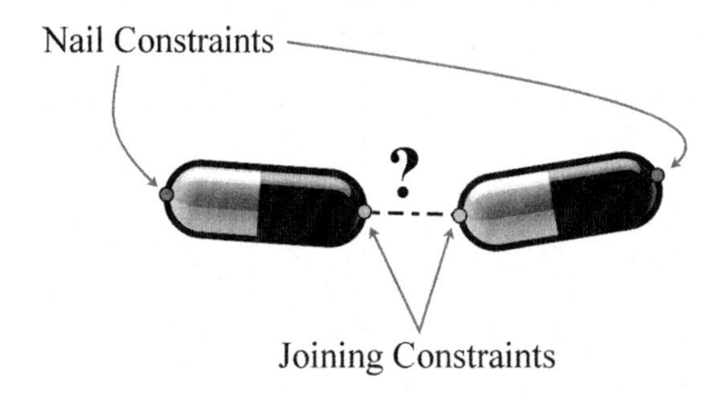

erative solver, more often than not, aims to find a best-fit approximation, which will allow us to move forward in the hope that the configuration fixes itself.

Point-Mass Constraints (2D, No Rotation)

Using a simplified constraint solver, rotations can be cut out, just working with position constraints, enabling complex structures to be constructed in 2D or 3D just using point-mass objects. Simple constraints (e.g. nail and distance) let the solver be tested before moving onto rigid bodies and adding in rotations. Figure 9 and Figure 10 illustrate examples of point-mass constraints for constructing articulated structures.

The code, written for simplicity, has no optimisations, is inefficient and slow, but has been written to introduce a working solver that can be played with to gain understanding. All the particles have a mass of 1, but they can easily be set to different values – maybe draw larger spheres for larger mass? We used only two constraint types, fixed (nail) and distance (rods), but other constraints and springs can be mixed in. The simulation runs in 3D, so all the points and constraints actually solve for 'x', 'y' and 'z', but we clamp the 'z' to zero and render in orthogonal to give the appearance of 2D; so elaborate 2D or 3D structures can be constructed.

We set a fixed number of iterations for the solver demo, so the constraints will stretch slightly for large forces. We thought this was acceptable, but the maximum iterations can be increased and the error can be monitored.

The mass-point solver in 2D/3D environments can be used to create dynamic structures, from which orientation and rotation information can be extracted, by using positions of point masses to build up a reference orientation.

Point Mass Demo (3D, No Rotation)

Expanding the simple point-mass demo to 3D, with a sphere dropping onto a mesh surface, we added constraints at each frame between the rigid body spheres to keep their distance to the sum of their radius, as shown in Figure 11.

Stable Stacking Demo (3D, With Rotation)

A good test for any physics simulator is the ability to deal with large stacks of objects. Creating a stable stacking configuration is in part down to how reliable the contact information is, and how much it changes between frames. To create reliable contact information, reliable contact manifolds need to be built, which are in contact across multiple frames. When we stack objects that are not moving, the collision information between frames should remain the same, and not jitter around due to floating point errors and recalculation of collision points. (Figure 12)

SPARSE MATRICES

Applying conventional *dense matrix* techniques to solve systems of constraints presented in this chapter is an exceedingly ill-advised thing to do. As mentioned earlier, the matrices used in our solver are sparse, and we need to use this sparsity to our advantage. You can see a simple demonstration of the sparsity of a matrix illustrated in Figure 13, where we plot the non-zero matrix values to the right of the simulation screenshot in the figure.

Figure 9. Simulation screenshots for Point-Mass solver. (Spheres represent connecting joint constraints, and the lines represent point-mass distance constraints)

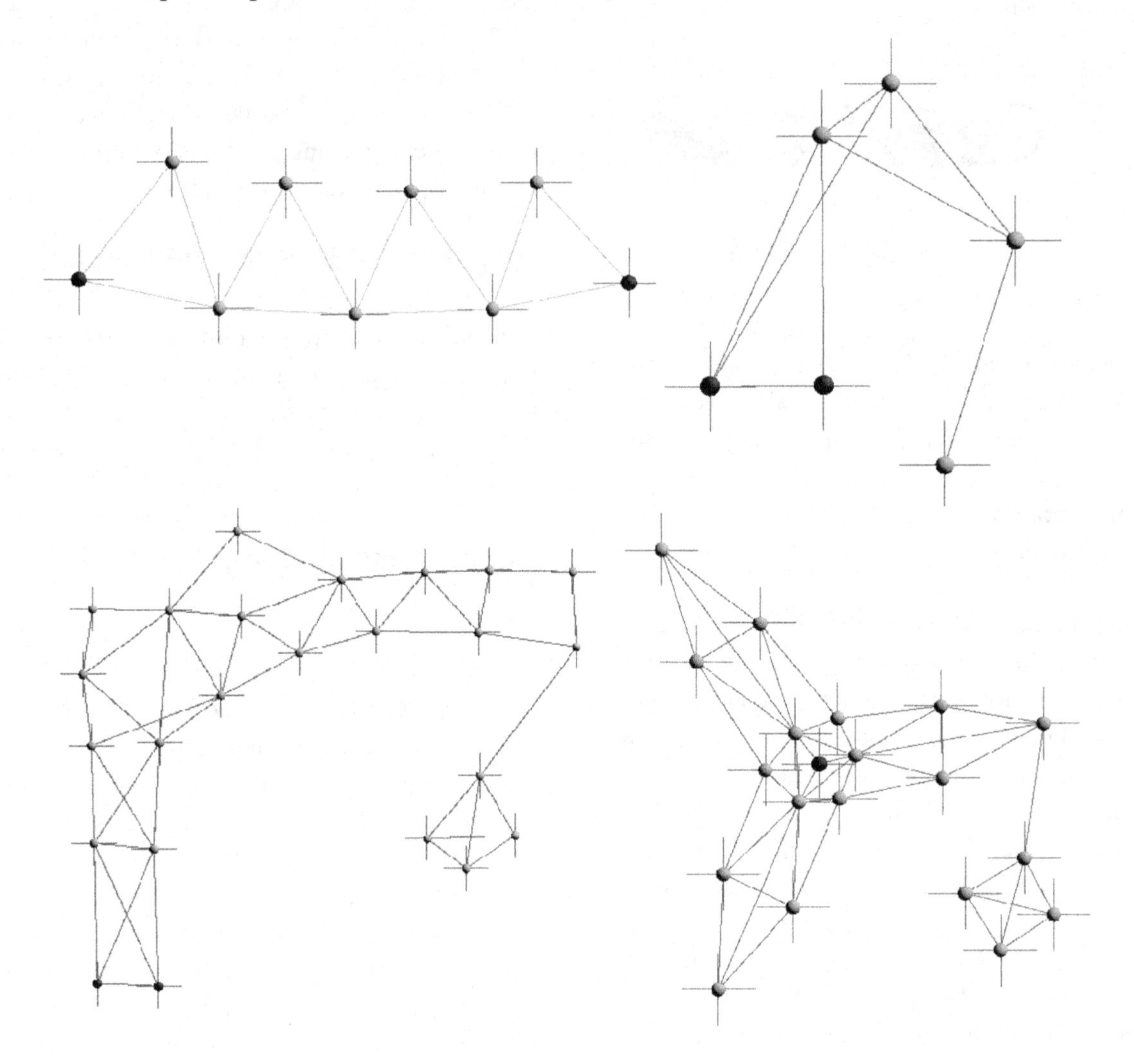

Figure 10. Use point-mass configurations to extract orientation (rotation) information for graphical models or rigid bodies

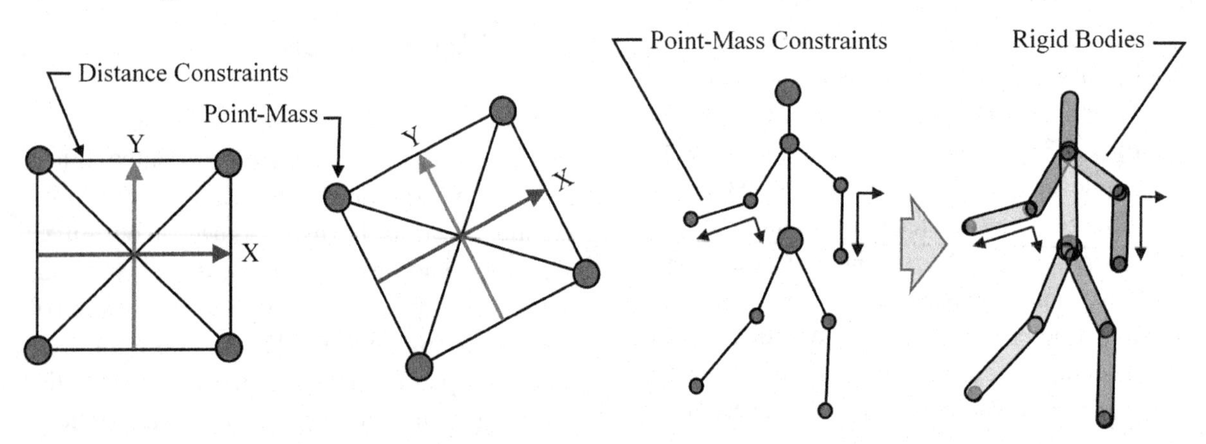

Figure 11. (Left) Demo screenshot: Large sphere falling onto the distance constrained net. (Right) Simplified diagram showing how we handled collisions between spheres and mass-points

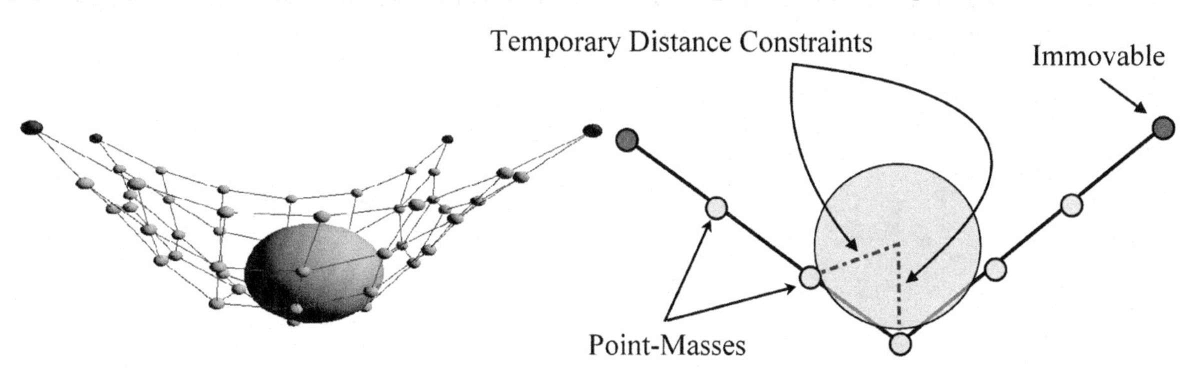

Taking advantage of operator overloading, especially the multiplication operator '*', allows us to create more robust code with the ability to track down problems early on. Asserts and halts enable runtime problems to be detected, both by other people using the code, or just silly typing mistakes. These good coding techniques will help prevent memory corruption and random unpredictable crashes in unrelated code. Reliable code will 'stop' and tell us *why*. If possible, code that ignores a situation without telling us and attempts to cope with the problem and continue, should be avoided.

When implementing the matrix code, a wrapper class should be used so that various implementations for comparison and profiling can be swapped

Figure 12. Stacking demo, 700 rigid body cubes in a circular tower

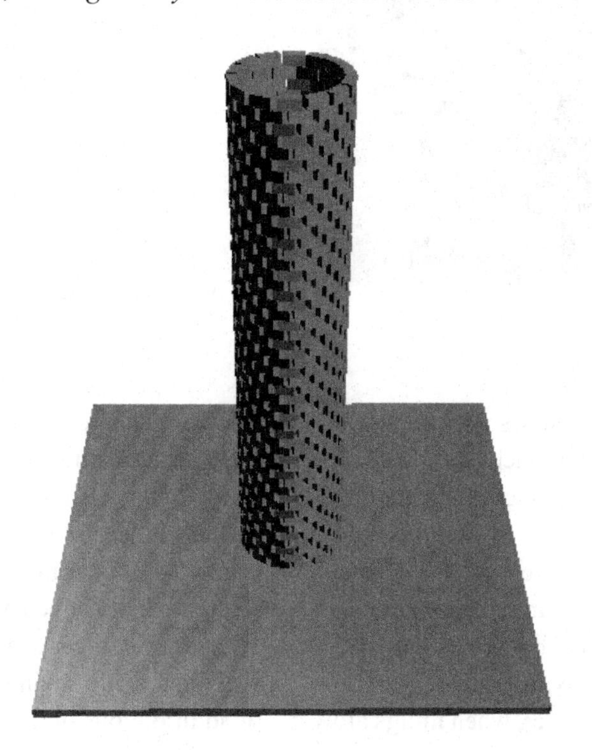

Figure 13. Sparsity pattern of the $JM^{-1}J^{T}$ matrix for constraint simulations. Top, mass-point distance constraints constructing a net (matrix is less than 1% filled). Bottom, circular stack of rigid bodies. The graphs to the right illustrate the matrix density

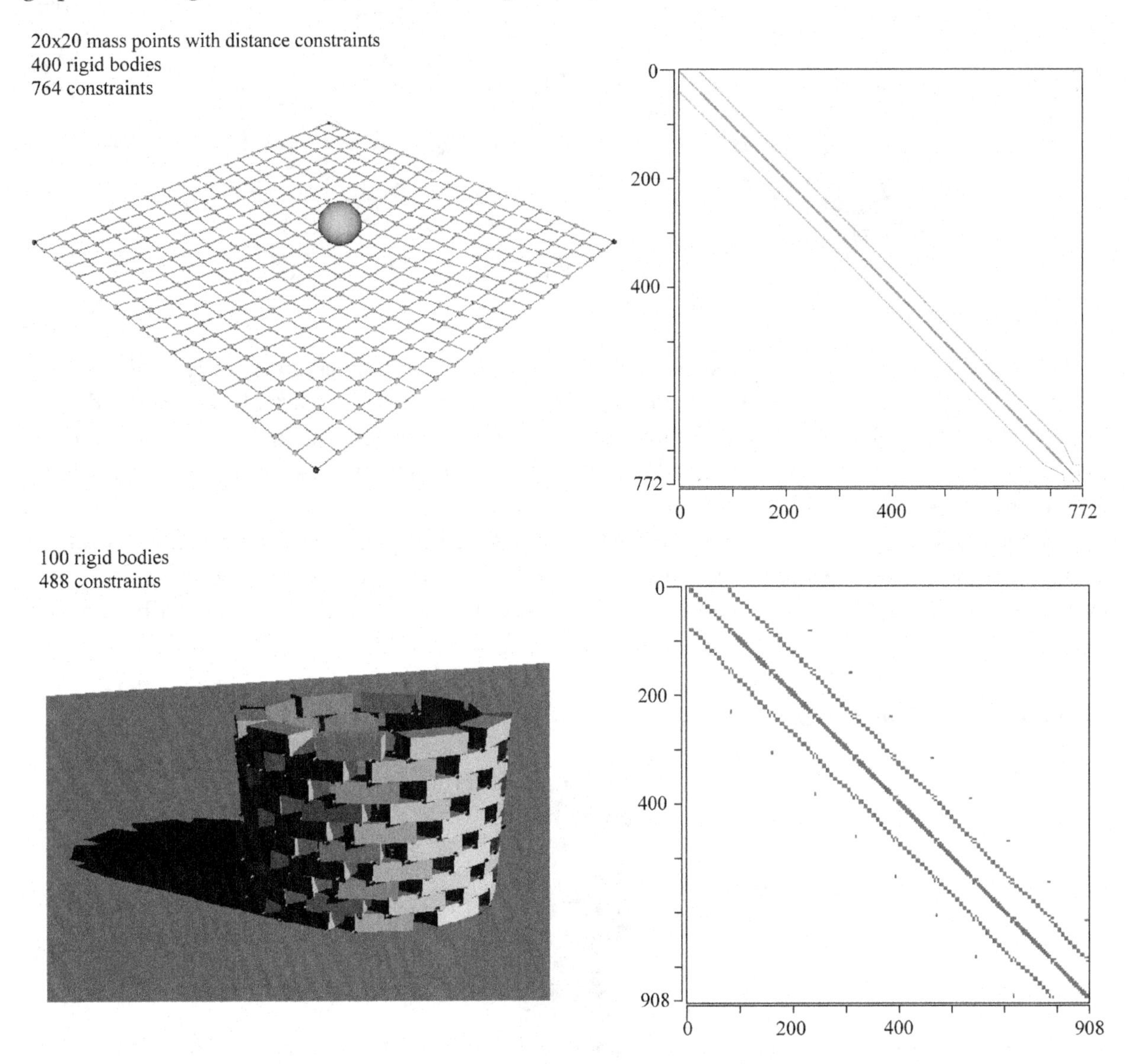

out. There are numerous matrix solutions freely available (Dongarra, 2009), each having their own advantages and disadvantages, and trading between various features such as memory and access times. We introduce a few methods here, giving the source for an STL mapping method.

Initially, the aim should be to get the LCP constraint solver working, without worrying too much about optimisation. Later, when things are working as expected, everything can be sped up. The first implementation will use a dense matrix method, and will initially be the biggest slowdown, because 90% of the time it will be managing the matrix data, where massive chunks of memory will be created, destroyed and copied in each frame.

Sparse matrices enable us to gain optimisation speedups while saving memory and bandwidth. In addition, our linear iterative systems are also able to exploit sparsity, by skipping unnecessary calculations, to improve performance, where even trivial operations like filling a $n \times m$ matrix with '0s' would take 'O(nm)' time, for a dense matrix can be avoided.

Note: A matrix wrapper class will give safety checks, an array of out of bound checks, and enable various internal optimisations (e.g. memory speedups) to be added without modifying the working solver code. If multiple platforms are worked on, a lot of platform specific code inside the matrix class can be embedded without complicating the solver. The matrix code can be expanded to use templates, so it can work with more data types, other than just floats, e.g. doubles, or any other objects, so it can be expanded to other situations, such as image processing or fluid dynamics.

One of the biggest slowdowns that will be encountered in solvers, other than actually solving the system of equations by inverting the matrix, is the dynamic nature of the matrices. Dynamic constraints, such as collisions and contacts can appear and disappear from frame to frame, which means resizing and updating the large matrix. Modifying the matrix by each frame, means introducing additional slowdowns, as the matrix class is allocating and reallocating large chunks of memory each frame, and if using the assignment operator, it is also being copied. For static constraints, there is the added benefit of coherency between frames, which we can exploit for speedups. (Code 8)

In a poorly implemented solver with little consideration for matrix management, the biggest slowdown will be the matrices, where a large number of rigid bodies and constraints can produce very large matrices which are difficult to manage and manipulate and trigger large memory allocations and copies.

We solve our slow matrix implementation by taking advantage of our custom matrix wrapper class, where we are able to optimise things under the hood, using methods such as custom caching and memory pools. Internally, the matrix class should have a MatrixManager, which we shared across matrix instances, and stores common sized matrices, so instead of creating them, the manager would have them loaded and ready for use.

In those cases when the matrix has not actually been destroyed and created, but exists as a member variable, or a single instance, you can have it keep a large chunk of memory internally, so when the matrix is resized to make it smaller or bigger, it will keep enough memory internally for a larger matrix. So if the matrix through experimentation is no larger than 100, it could be set to allocate

Code 8. Showing a slice of code, where working with matrices is as easy as working with base variables like ints and floats

```
// Build ...
DMatrix MInverse(numBodies, numBodies);
DMatrix MFe(numBodies, 6);
DMatrix MVel(numBodies,6);
DMatrix JMatrix(constraints, numBodies);
DMatrix JTransposeMatrix = Transpose(JMatrix);
// .. fill with data
DMatrix A =  JMatrix * MInverse * JTransposeMatrix;
DMatrix b = JMatrix * (MVel + MInverse * MFe);
// .. solve etc
```

Code 9. 'Raw' starting point: The essentials only

```
class DMatrix
{
public:
    float*  m_data;       // rows x cols array of data
    int     m_numRows;
    int     m_numCols;
}; // End DMatrix Class
```

space internally to 100x100, but to the outside world it would appear to be whatever size it was resized to, e.g. 2x3, 30x60 etc., but it will have allocated 100x100 internally and avoid memory resize slowdowns.

Overloading the assignment operator is another good way to achieve speedups, where we can copy large chunks at a time internally instead of byte by byte, and if we are using sparse data representations for our matrix data, the amount of data that we will copy is a noticeably less, as we are only copying the non-zero data instead of the massive chunk of memory.

Dense Matrix (Brute Force)

So to begin with, we start with a basic dense matrix class, which we will call the DMatrix (i.e. Dynamic Matrix). Three essential variables are needed to begin with: the pointer to the data, the number of columns and the number of rows, as shown in Code 9.

A few helper functions need to be added, so that all the data operations inside the class are managed. For example, when the DMatrix class is created, we need to be able to specify a default size and maybe the starting data, to initialize the array. We add some helper functions in Code 10, including a constructor, and add numerous assert checks.

If you look at the code above, you will notice that a large majority of the code is for checks. This is important, as it is far too easy for a mistaken value to propagate between functions and variables, to a point where it is impossible to track down the cause, leaving the system in an unrecoverable state. To reiterate, it is best to assert and stop as soon as a problem appears, so that the situation can be analysed and fixed.

There are various 'sanity' checks in the code, which are not vital, but give us an early warning that something might be incorrect. One such example of an early warning assert, is if the matrix size is greater than 500x500, we know something might have gone wrong, before attempting to allocate memory for some unexpected size matrix (i.e. 10000x10000, which can really be harmful). An additional check would be that zero size arrays are not created, and hence assert if the code attempts to, so we can track it down and repair it. These checks give us a 'robust' matrix wrapper class that defines a set of common access functions from which we can swap in a better sparse matrix implementation at a later time.

Notice as well that the data and size variables have been made private. This is deliberate, so that all our internal workings for the matrix class remain hidden from the outside world, and use access functions to set or get the information within. Even simple functions such as 'Get', 'Set' and 'GetNumRows', can be used to give us extra checks, and enable us to change the code internally for any reason (i.e. data management optimisations). So to reiterate, this enables us to modify and improve our implementation without actually changing our simulator code, only the

Code 10. Constructor and sanity checks: Starting pieces

```
class DMatrix
{
public:
    DMatrix(const int numRows, const int numCols, const float* data=NULL)
    {
        Init(numRows, numCols, data);
    }
    ~DMatrix()
    {
        delete[] m_data;
    }
    inline int GetNumRows() const   { return m_numRows; }
    inline int GetNumCols() const   { return m_numCols; }
    inline float Get(int row, int col=0) const
    {
      DBG_ASSERT(row>=0 && row<m_numRows);
      DBG_ASSERT(col>=0 && col<m_numCols);
      const int indx = GetDataIndex(row,col);
      return m_data[indx];
    }
    inline float& Set(int row, int col=0) const
    {
      DBG_ASSERT(row>=0 && row<m_numRows);
      DBG_ASSERT(col>=0 && col<m_numCols);
      const int indx = GetDataIndex(row,col);
      return m_data[indx];
    }
    inline int GetDataIndex(int row, int col) const
    {
      const int indx = col + m_numCols*row;
      DBG_ASSERT(indx>=0 && indx<m_numRows*m_numCols);
      return indx;
    }
private:
    void Init(const int numRows, const int numCols, const float* data=NULL)
    {
        DBG_ASSERT(numRows>=0 && numRows<500); // Sanity checks - temp, do we
really have size
        DBG_ASSERT(numCols>=0 && numCols<500); // 0 or >500 array or did some-
thing go wrong?
        if (numRows==0) { DBG_ASSERT(numRows==0 && numCols==0); }
        if (numCols==0) { DBG_ASSERT(numRows==0 && numCols==0); }
```

continued on following page

Code 10. Continued

```
        m_data      = NULL;
        m_numRows   = numRows;
        m_numCols   = numCols;
        if (numRows>0 && numCols>0)
        {
            m_data = new float[numRows * numCols];
            DBG_ASSERT(m_data);
            if (data)
            {
                memcpy(m_data, data, sizeof(float)*numRows*numCols);
            }
            else
            {
                memset(m_data, 0, sizeof(float)*numRows*numCols);
            }
        }
    }
    float*  m_data;
    int     m_numRows;
    int     m_numCols;
}; // End DMatrix Class
```

matrix code needs be changed. The dense matrix class gives a solid test bed from which results can be compared and checks made to ensure optimisations and improvements work correctly.

STL Map Matrix

Here, we present a practical, exploitable sparse matrix implementation that uses the Standard Template Library (STL). Code 11 gives a skeleton implementation which can built upon, and uses a technique based on Compressed Row Storage (CRS) (i.e. similar to the method MatLab uses). The STL map matrix saves space by using more pointers to represent the data. (Table 1)

Linear Linked-List Sparse Matrix

Linked-list matrices are an alternative way of building a sparse matrix class, and while probably not one of the fastest, they do offer large memory savings. One way the linked-list matrix would take advantage of sparsity, is by storing only non-zero elements in a list, and access them by it. It offers large memory savings but access times can be slow, since we need to find the index in the list, which can mean iterating over the list to find it.

This is a good solution for extremely sparse matrices, but as the matrix becomes less sparse, its memory overheads and access times grow exponentially. For example, the memory overhead of an empty linked-list class is can be as little as '12 bytes', no matter how many rows or columns, and whereas a dense matrix class of size 100x100 usually uses around 40,000 bytes, for a linked-list version, if every element contained a non-zero value, would use approximately 280,000 bytes. This is due to the linked-list overheads (i.e. a ratio of 28:1).

Table 1. Comparing standard brute force matrix memory usage with that of an STL map one.

Matrix using Brute Force	Matrix using STL Map
(0%, 1%, 10% Random Fill) Size: 1 x 1, Bytes Used: 4 Size: 10 x 10, Bytes Used: 400 Size: 100 x 100, Bytes Used: 40000 Size: 1000 x 1000, Bytes Used: 4000000 Size: 10000 x 10000, Bytes Used: 400000000	(0% Fill) Size: 1 x 1, Bytes Used: 36 Size: 10 x 10, Bytes Used: 36 Size: 100 x 100, Bytes Used: 36 Size: 1000 x 1000, Bytes Used: 36 Size: 10000 x 10000, Bytes Used: 36 (1% Random Fill) Size: 1 x 1, Bytes Used: 36 Size: 10 x 10, Bytes Used: 120 Size: 100 x 100, Bytes Used: 6456 Size: 500 x 500, Bytes Used: 89520 Size: 1000 x 1000, Bytes Used: 299028 Size: 10000 x 10000, Bytes Used: 24481620 (10% Random Fill) Size: 1 x 1, Bytes Used: 36 Size: 10 x 10, Bytes Used: 756 Size: 100 x 100, Bytes Used: 28932 Size: 500 x 500, Bytes Used: 600924 Size: 1000 x 1000, Bytes Used: 2343684 Size: 10000 x 10000, Bytes Used: 229075404

So what if 50% is used? Approximately 70,000 bytes would be used then; so as you would expect, the sparser the matrix, the better, not just with linked-list matrices but any sparse matrix implementation.

In conclusion, linked-lists offer a flexible data-storage solution that can be combined with other methods to achieve hybrid configurations, which offer both speed and memory savings. For example, speeding up element access times, by using nested linked-lists, or fast search algorithms, are just some ideas to explore.

Binary Tree (Search Speedup)

The binary tree version is basically a way of speeding up the linked-list version by storing the data in a binary tree formation. Whereby, when we search for a particular index, we search in a binary tree methodology, hence drastically reducing our search time.

Hashing

In order to achieve an alternative way of building a sparse matrix class to achieve exceptionally good access times, 'hashing' can be used, whereby indices are mapped to specific memory addresses using a hash lookup table and results are achieved in almost near instant access times. A typical implementing allocates a large number of elements, which are mapped to a hash lookup table, so each time an element is accessed, the hash algorithm generates the index offset within the large array. If there is no element at that address (i.e. if it does not exist), then it is added to the hash lookup list. Furthermore, because our matrix is sparse, the large chunk of elements we set aside should only be a fraction of the size of what a dense matrix allocates.

Note: Debugging – When working with solvers and matrices it is worth investing some time in learning about the various matrix types, e.g. adding code to check for singular matrices, and how 'bad' matrices can be tested for. One such example is a matrix where one of its rows is all zeros, and understanding what this means.

Code 11. STL map matrix

```
//  Sparse Matrix Implementation using STL Maps
#include <cstdlib>
#include <map>
#include <vector>
#define STL_CONST const
class DMatrixSTL
{
public:
    typedef std::map<size_t, std::map<size_t, float> >     mat_t;
    typedef mat_t::const_iterator                           row_iter;
    typedef std::map<size_t, float>                         col_t;
    typedef col_t::const_iterator                           col_iter;
    DMatrixSTL(size_t numRows, size_t numCols, const float* data=NULL)
    {
        m_numRows=numRows;
        m_numCols=numCols;
        DBG_ASSERT(numRows>=0 && numRows<1501);
        DBG_ASSERT(numCols>=0 && numCols<1501);
        if (numRows==0) { DBG_ASSERT(numRows==0 && numCols==0); }
        if (numCols==0) { DBG_ASSERT(numRows==0 && numCols==0); }
        if (data)
        {
            m_mat.empty();
            for (int i=0; i<(int)numRows; ++i)
            {
                for (int k=0; k<(int)numCols; ++k)
                {
                    const float val = data[ i + k*numRows ];
                    Set(i,k) = val;
                }
            }
        }
    }
    inline float Get(size_t row, size_t col=0) const
    {
        if(row<0 || col<0 || row>=m_numRows || col>=m_numCols) { DBG_HALT; }
        #if 0
        // Destroys sparsity and prevents the use of const
        // If the element doesn't exist, it inserts one.
        return m_mat[row][col];
        #else
        mat_t::const_iterator it;
```

continued on following page

Code 11. Continued

```
        it = m_mat.find(row);
        float val = 0.0f;
        if (it != m_mat.end())
        {
            col_t::const_iterator itc = (*it).second.find(col);
            if (itc != (*it).second.end())
            {
                val = (*itc).second;
            }
        }

        return val;
        #endif
    }
    float& Set(size_t row, size_t col=0)
    {
        if(row<0 || col<0 || row>=m_numRows || col>=m_numCols) {  DBG_HALT  };
        return m_mat[row][col];
    }

    void SetToZero()
    {
        m_mat.empty();
    }
    inline int GetNumRows() const   { return m_numRows; }
    inline int GetNumCols() const   { return m_numCols; }
protected:
    DMatrixSTL(){}
public:
    mat_t   m_mat;
    size_t  m_numRows;
    size_t  m_numCols;
};
// Declarations for essential helper functions and operators overloading. e.g:
DMatrixSTL operator*(DMatrixSTL &a, DMatrixSTL &b);
DMatrixSTL Transpose(STL_CONST DMatrixSTL& m);
DMatrixSTL operator+ (STL_CONST  DMatrixSTL &a, STL_CONST DMatrixSTL &b);
inline DMatrixSTL operator- (STL_CONST DMatrixSTL  &a, STL_CONST DMatrixSTL
&b);
inline DMatrixSTL operator*(float a, STL_CONST DMatrixSTL &b);
```

Thinking Out of the Box

It is worth experimenting and profiling to aim for the lowest possible memory usage and fastest achievable access times. Simple things like caching and coherent memory accesses are good things to keep an eye out for. If you access an element at 'x', you might cache it so it and its neighbours can be accessed immediately if the next access is itself or its neighbour.

Memory allocations are another factor, which can cause a major slowdown, so adding in a memory pool to the array creation/destruction helps with memory delays and fragmentation. Furthermore, using a memory manager in conjunction with the sparse matrix would enable partitioning of the memory and keeping track of usage; improving bandwidth and giving further control.

Cross platform computability is always a side thought, for example, working with matrices on the GPU, in which case the focus needs to be on parallelizing the data and modifying it to take advantage of the high number of cores. Further reading on exploiting LCP solvers on the GPU is available in (Nguyen, 2007).

STABILITY AND RELIABILITY

One of the biggest problems with writing physics simulators is reliable stable contact/collision information. One such example is an object resting on the ground. If the object moves within a minimum threshold, the contact information from previous frames should be reused so that it remains constant, and the objects converge and settle down. Some collision detection algorithms only return a single contact point between frames. Hence, this contact point needs to be stored and a contact manifold over numerous frames needs to be built up, which will keep the object from jittering and give a stable rigid body stacking.

When writing a simulator, asserts and sanity checks should be used whenever possible, especially checks for NaNs, which can arise, more often than not, due to the large number of mathematical operations. This will allow the simulation to be halted at the point where it went wrong, so an investigation of what went wrong at that moment in time can be made, instead of looking at debug prints or, even worse, randomly re-running the code and hoping for it to happen again and guessing what could have gone wrong.

CONCLUSION

In the scope of this chapter, we introduced LCP solvers as a practical solution for rigid body constraint simulations, with an emphasis on clarity and simplicity. It introduced the basic algorithms and their implementation so that the reader, having grasped the principles, can go on to more esoteric constructs.

We discussed and compared the two main solver types, acceleration and velocity, upon which we outlined their differences and finally went on to focus on the velocity method in the remainder of the chapter, due to its effectiveness and simplicity.

The chapter gave special emphasis towards the implementation design, whereby the reader could go away and construct a modular flexible simulator with the ability to extend it over time and add original joint types effortlessly; where, the crucial steps necessary for constructing new constraints types using numerous examples and their Jacobian formulation were presented in a simplified clear-cut approach.

FUTURE RESEARCH DIRECTIONS

This chapter has laid the practical foundations for using solvers to represent a robust constraint simulator. Further work entails numerous enhancements to increase performance, either by investigating and implementing more sophisticated algorithms (e.g.

Successive Over Relaxation Method), or taking advantage of today's highly parallel processors (i.e. CPU or even better the GPU cores).

While we only introduced a few basic constraints, it is worthwhile experimenting and creating a wider range of unique constraints, both equality and inequality. When formulating constraint types, alternative derivatives of the equations should be researched and how they perform (e.g. length squared instead of length for distance constraints) should be compared.

This chapter has only been a springboard into the world of solvers and it is highly recommend that the reader look at the latest literature and explore the subject.

Possible side projects to continue with might include:

- Profiling and optimising code.
- Displaying kinetic/potential energy of objects.
- Creating complex constraint structures (bridges, ragdolls, catapult, ropes).
- Adding further distinctive constraint types.
- Displaying external and internal constraint forces.
- Using varying masses for point-mass distant constraint simulations.
- Analysing and optimising sparse matrix operations, and explore bandwidth speedups.

REFERENCES

Anitescu, M., & Potra, F. A. (1996). Formulating dynamic multi-rigid-body contact problems with friction as solvable linear complementarity problems. *Computer*, *93*, 1–21.

Baraff, D. (1989). Analytical methods for dynamic simulation of non-penetrating rigid bodies. *Computer*, *23*(3), 223–232.

Baraff, D. (1994). Fast contact force computation for non-penetrating rigid bodies. *Proceedings of the 21st annual conference on Computer graphics and interactive techniques - SIGGRAPH '94*, 23-34. New York, NY: ACM Press. doi: 10.1145/192161.192168

Baraff, D. (1999). Physically based modeling course notes. *ACM SIGGRAPH*, (2-3).

Catto, E., & Park, M. (2005). Iterative Dynamics with Temporal Coherence, 1-24. Retrieved from http://erwincoumans.com/ftp/pub/test/physics/papers/IterativeDynamics.pdf

Cottle, R. W., Pang, J. S., & Stone, R. E. (1992). *The linear complimentarity problem*. New York, NY: Academic Press.

Dongarra, J. (2009). Free Matrix Library List. Retrieved from http://www.netlib.org/utk/people/JackDongarra/la-sw.html

Eberly, D. H. (2004). *Game Physics*. San Francisco, CA: Morgan Kaufmann.

Erleben, K. (2004). Stable, robust, and versatile multibody dynamics animation. Unpublished doctoral dissertation. University of Copenhagen, Denmark. Retrieved from http://www2.imm.dtu.dk/visiondag/VD05/graphical/slides/kenny.pdf

Guendelman, E., Bridson, R., & Fedkiw, R. (2003). Nonconvex rigid bodies with stacking. *ACM Transactions on Graphics*, *22*(3), 871. doi:10.1145/882262.882358

Hager, W. W. (1988). *Applied Numerical Linear Algebra*. Upper Saddle River, NJ: Prentice Hall.

Havok. (1998). Havok Inc. Retrieved from http://www.havok.com

Hecker, C. (1998). *Rigid body dynamics*. Retrieved from http://chrishecker.com/Rigid_Body_Dynamics

Jerez, J., & Suero, A. (2003). *Newton*. Retrieved from http://www.newtondynamics.com

Kacic-Alesic, Z., Nordenstam, M., & Bullock, D. (2003). A practical dynamics system. In *Proceedings of the 2003 ACM SIGGRAPH Eurographics Symposium on Computer animation* (pp. 7-16). Aire-la-Ville, Switzerland: Eurographics Association.

Lötstedt, P. (1984). Numerical simulation of time-dependent contact friction problems in rigid body mechanics. *Society for Industrial and Applied Mathematics: Journals on Scientific Computing, 5*(2), 24.

Nguyen, H. (2007). *GPU gems 3*. Reading, MA: Addison-Wesley.

NVIDIA. (2011). *PhysX.* Retrieved from http://www.nvidia.com/object/physx_new.html

Shabana, A. (1994). *Computational dynamics*. Chichester, UK: John Wiley & Sons, Inc.

Smith, R. (2004). *Open dynamics engine*. Retrieved from http://www.ode.org

Stewart, D., & Trinkle, J. C. (1996). An implicit time-stepping scheme for rigid body dynamics with coulomb friction. *International Journal for Numerical Methods in Engineering, 39*, 2673–2691. doi:10.1002/(SICI)1097-0207(19960815)39:15<2673::AID-NME972>3.0.CO;2-I

Vondrak, M. (2006). *Crisis physics library.* Retrieved from http://crisis.sourceforge.net/.

ADDITIONAL READING

Bourg, D. (2002). *Physics for Game Developers*. O'Reilly & Associates, Inc.

Cottle, R. W., & Dantzig, G. (1968). Complementarity Pivot Theory of Mathematical Programming. *Linear Algebra and Its Applications, 1*, 103–125. doi:10.1016/0024-3795(68)90052-9

Dantzig, G. B. (1963). *Linear Programming and Extensions*. Princeton University Press.

Eberly, D. (2003). *Game Physics (Interactive 3D Technology Series)*. Morgan Kaufman.

Erleben, K. (2005). *Physics-Based Animation*. Charles River Media.

Hecker, C. (2000). How to Simulate a Ponytail, In: http://chrishecker.com/How_to_Simulate_a_Ponytail/

Kipfer, P. (2007). GPU Gems 3, Chapter 33. LCP Algorithms for Collision Detection using CUDA

Kokkevis, E. (2004). Practical Physics for Articulated Characters.

Murty, K. G. (1988). *Linear complementarity, linear and nonlinear programming*. Berlin, Germany: Heldermann Verlag.

Smith, R. (2004). Open Dynamics Engine. Retrieved from: http://www.ode.org/

Vondrak, M. (2006). Crisis Physics Engine. Retrieved from http://crisis.sourceforge.net/

KEY TERMS AND DEFINITIONS

Acceleration: The rate of change of velocity.

Centre of Mass (COM): Also known as the centre of gravity, and represents a centroid point where the mass of an object is balanced in all directions.

Constraint: Is a limitation or restriction on a Degree of Freedom (DOF) of the system.

Degrees of Freedom (DOF): The number of independent ways an object may move, and consequently the number of measurements necessary to document the kinematics of the object.

Equations of Motion (EOM): Is a set of equations, which describe how the objects of the system will move as time changes.

Inertia: The tendency of an object in motion to remain in motion, and of an object at rest to remain at rest.

Linear System: A linear system is a collection of linear equations.

Mass: A measure of inertia, indicating the resistance of an object to a change in its motion; including a change in velocity. A kilogram is a unit of mass.

Moment of Inertia: The resistance of a body to change its state when rotating.

Newton's Laws of Motion: The principles, formulated by Sir Isaac Newton (1642-1727), which state how objects move.

Velocity: The rate of change of position.

Weight: A measure of the gravitational force on an object; the product of mass multiplied by the acceleration due to gravity (equal to 9.8 m per second per second).

Chapter 9
Rocket Jump Mechanics for Side Scrolling Platform Games

Golam Ashraf
National University of Singapore, Singapore

Kenny Lim
National University of Singapore, Singapore

Ho Jie Hui
National University of Singapore, Singapore

Esther Luar
National University of Singapore, Singapore

Luo Lan
National University of Singapore, Singapore

ABSTRACT

This chapter uses the Mechanics, Dynamics and Aesthetics (MDA) framework as a practical guide to incorporate the Rocket Jump mechanic in a side-scrolling platform game. The authors systematically approach the design problem by encoding physically based rules for Rocket propulsion (Mechanics), and then creating appropriate obstacles to construct progressively difficult levels (Dynamics). A comparison is then made to the Rocket Jump with the Conventional Jump with a subjective questionnaire on game difficulty (Aesthetics) using the NASA TLX template. Participants report that they find the rocket jump mechanic more mentally demanding, and thus better immersed in the game. Game-play data is also evaluated to analyze performance differences between the two mechanics modes and find that there is negligible difference. This proves that the Rocket Jump implementation is balanced and quite compatible with conventional platform games. This article aims to provide useful insights and information for the construction of challenging levels with Rocket Jump.

INTRODUCTION

The revival of the platformer's popularity in the recent years could be attributed to new breakthroughs in the area of game design. We have seen new and interesting game mechanics in *Braid* (2008) and *Super Meat Boy* (2010) that had breathed in new life into the genre. We can see that the platformer genre is reinterpreting game mechanics and design that was popularized in other game genres. For example, Braid's idea of time manipulation was heavily influenced by other time manipulation games such as *Prince of Persia: Sands of Time* and *Blinx* (Totilo, 2007). Similarly, other mechanics like parachute jumping, or gliding have been used to inject more fun

DOI: 10.4018/978-1-4666-1634-9.ch009

into platformers. Encouraged by these examples, we are looking to incorporate an existing game mechanic from another genre into the platformer.

Rocket Jump is a form of emergent gameplay that had gained popularity in First Person Shooters (FPS) like *Quake* (1996) *Team Fortress 2* (2007). Rocket Jump was originally born from a game glitch that allowed the player wielding the rocket to be propelled across distance. The popularity of Rocket Jump was quickly noticed by many game developers and was subsequently incorporated as a core mechanic for many FPS games.

We are interested in applying the Rocket Jump as a novel movement mechanic for side scrolling platform games. In this game experiment, we compare the Rocket Jump mechanic with the conventional movement mechanics found in most platformers. Also, to aid in the design of the levels, we apply physics equations to the design of our game levels for the tweaking of level difficulty. The details of the Rocket Jump mechanic is discussed in conjunction with links to level design that afford balanced play and progressive difficulty, in a 2D side scrolling platform game. Throughout this process, we will be using the Mechanics, Dynamics and Aesthetics (MDA) framework (Hunicke, LeBlanc, & Zubek, 2004) as a design guide.

We first explore the underlying physical constraints of rocket propulsion (Mechanics). Following that, we create game levels that afford progressive difficulty (Dynamics). We feel that it is important to objectively visualize the challenge posed to the players even before user trials are done. Typically, level obstacle design is done manually by iteratively tweaking important variables and testing the resulting game-play. This approach has obvious limitations in addressing puzzles with many variables and different player skills but as a reference for a technical discussion of a mechanics implementation, we will not consider the design aspects of the mechanics. We describe a simple parabolic motion estimation framework that allows designers to calculate permissible ranges for successful jumps, and thus allowing them to create scalable difficulty without any guesswork. We suggest some values for key rocket mechanics and obstacle parameters for the reader's reference and predict the corresponding game-play challenge. Besides evaluating the difficulty of the obstacles (Dynamics) by looking at the performance of the players, we also assess the difficulty of the Rocket Jump versus the Conventional Jump, to assess its pros and cons as a new mechanic replacing a well-accepted one.

This chapter guides the reader on how to design platformer levels to suit the Rocket Jump mechanic. The focus is not really on game design, but more on a practical framework for assessing difficulty and comparing the player experience.

BACKGROUND

Why Platformer?

Once best sellers in the gaming industry, platform games (or platformers) have now been reduced to only 2% of the market in 2002 (Boutros, 2006). While one may attribute the massive decline to the emergence of other genres of games, one could not deny that the platformer had been too caught up in the technological advancements that brought about more glitter and glitz into games, preferring to dazzle players with improved graphics or better audio quality. From the classic 2D single screen and side-scrolling platformer, the genre has branched into 2.5D, where the game combines 2D gameplay with 3D visual effects. Unable to compete with the novelty of emerging genres and without much advancement in terms of game design to further the potential of platform games, the sales of platformers languished and had never quite recovered. It is evident from the popularity of recent platformers such as *Braid* (2008) and *Super Meat Boy* (2010) that the genre is far from its end. The truth behind the success of the platformer, like any other games, lies in its game design and the ability to innovate within the constraints of the genre. We hope that the revival of the platformer genre would inspire renewed interest in novel puzzle and mechanics design. In this article, we adapt a 3D FPS mechanic to side scrolling platformer, basing the design of the game on mechanics equations.

Mechanics, Dynamics and Aesthetics

The relationship between the mechanics and overall game design is described in Hunicke, LeBlanc and Zubek's Mechanics, Dynamics and Aesthetics (MDA) framework (2004). The MDA framework formalizes components of games design into Mechanics, Dynamics and Aesthetics as a direct relation to a game's Rules, System and Fun respectively. Mechanics describes the particular components of the game, at game level of data representation and algorithms. Dynamics describes the run-time behavior of the mechanics acting on player inputs and agent outputs over time. Aesthetics describes the desirable emotional responses evoked in players, when they interact with the game system. The framework supports an iterative approach to design and tuning that allows designers to reason explicitly about particular design goals and to anticipate the impact of changes in mechanics to the resulting design/implementation. The MDA framework's segmentation of the design process into a three-tiered model brings to front the correlation between mechanics and emotive responses. Unfortunately, the primary usage of the framework has mostly been confined to the academia. It has been commented that the MDA framework does not provide any structure that allows it practical usage in game design and games testing (Cook, 2010). However, we demonstrate that the framework is indeed useful in analysis, implementation and refinement of game designs with novel mechanics and puzzles.

Game mechanics encompass a combination of rules and parameters that define the response, interaction and outcome of events triggered by players and elements within the game. In Nintendo's Super Mario Bros. (1994), the player controls the character Mario to navigate through the game world in search of the princess. Mario's mechanics can be decomposed into his movements (height of jump, speed of run), damage capabilities (generally one-hit kills for most items) and his interactions with other items such as coins and fire flowers. Let us now describe the Rocket Jumping mechanics a bit more in detail.

What is Rocket Jump?

Rocket jumping, a movement mechanic popularized in First Person Shooters, is the technique of harnessing the explosive force of a rocket launcher fired close to the character's body in order to cross large distances. Conventional Jump is usually constrained by muscle power, running speed and gravity. Some variants allow longer, higher or successive jumps, with prolonged or repeated key-press events. On the other hand, the horizontal and vertical distance of a Rocket Jump is dependent on the angle of the Rocket and the force of propulsion. Thus, unlike the Conventional Jump mechanic found in most platformers, Rocket Jumping requires greater control. This may introduce new challenges in game design.

The Rocket Jump mechanic could include other variables such as the decrease of health due to the explosion and the damage inflicted when player lands. As our aim is to design obstacles and challenges that cater to the unique movement set of the Rocket Jump, we consider only the motion variables pertaining to character displacement and remove or simplify all other mechanics that would hinder our assessment. Readers who are considering incorporating this mechanic into their games are encouraged to add extra variables that would increase the fun factor (e.g. power-ups, damage to self, etc.).

Figure 1. Resolving the initial velocity vectors

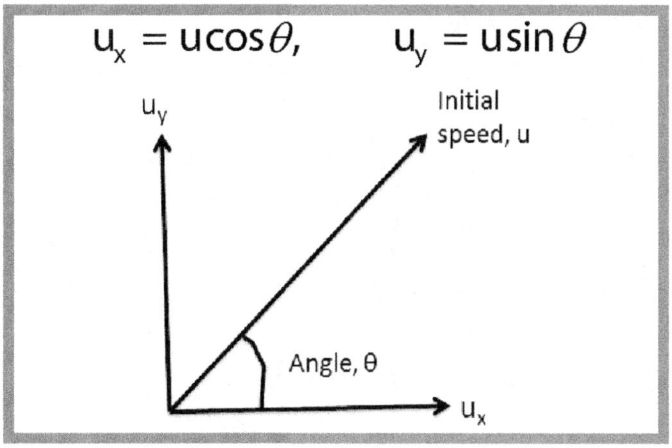

$$u_x = u\cos\theta, \qquad u_y = u\sin\theta$$

The Game

Lerpz the Rocketeer (Beat It! Games, 2011) is a game project developed by a team of four undergraduate students that aims to explore the implementation of Rocket Jump into the 2D side scrolling platformer as compared to the conventional jump mechanics. As a result, the game level designs were purposely kept simple with few variations in the game obstacles to have a fair comparison between the usages of the two mechanics.

The player plays Lerpz, an alien robot who is trying to reach his space ship that is positioned at the end of the game. Lerpz travels through the game world by two methods - the conventional jump and Rocket Jump. The conventional jump mechanic is implemented as a linear mapping of the x-y translation of the character to the duration of a key-press on the keyboard. Rocket Jump is implemented with a set of parabolic equations. The controls for this mechanic are the mouse and the left-right buttons of the keyboard.

ROCKET LAUNCHER MECHANICS

In the game, the rocket launcher is attached to the back of the Lerpz. The height and distance travelled by Lerpz is proportional to the initial speed of the launched rocket and the launch angle. We had fixed the following parameter values to achieve a believable distance and height within the levels we have designed.

A constant initial propulsion velocity = $30ms^{-1}$

Thrust reload time = 0.5s (which means that there is a possibility of successive jumps)

As shown in Figure 1, we can resolve the initial speed of Rocket Jump into u_x and u_y, the horizontal and vertical initial velocity of the jump respectively.

The horizontal initial velocity, u_x, remains constant during the jump, as air resistance is negligible. As a result, the horizontal distance (Range R) travelled for a jump is simply:

$$R = u_x t = (u\cos\theta)t$$

Due to the effects of gravity, the vertical velocity of the jump changes (from u_y at take-off, it becomes 0 at the peak, and then gains magnitude as the character freefalls under gravity). The projectile motion under uniform acceleration (or deceleration due to gravity) can be described as:

$$v_x = u_y + gt$$

Figure 2. Deriving Jump angles from Rocket Launcher angles

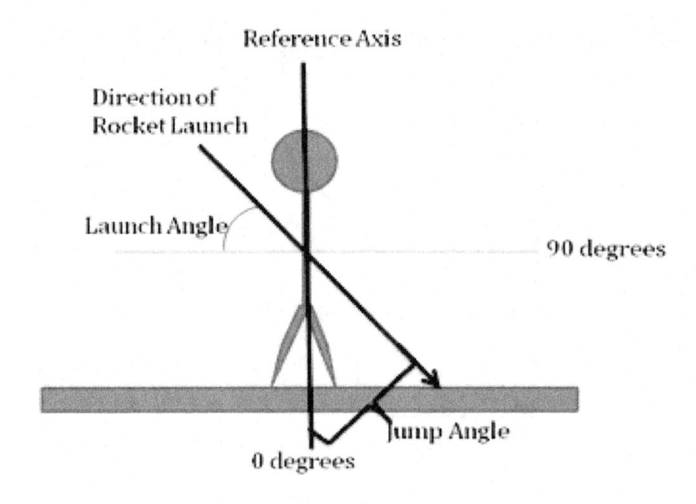

$$H = u_y t - \frac{1}{2} g$$

$$v_y^2 = u_y^2 + 2gH$$

where:

H = height of jump
v = final velocity
g = gravity
t = time taken

The jump angle is derived from the player controlled rocket launch angle. Running vertically down the character is the reference axis for rocket angle calculations. Figure 2 illustrates how the rocket launch angle and jump angles are calculated.

Table 1 presents three scenarios after substituting key angles into the projectile motion equations presented above. Collision distance refers to the point of impact of the rocket propulsion fluid on the takeoff platform. This parameter is relevant for narrow platforms, or when the character is close to the edge. Figure 3 explains the relationship between rocket launch angles and collision distance.

Besides the projectile motion of the Rocket Launcher, we also need to consider attributes of the player-controlled avatar. Let us enumerate some key parameter values chosen for the avatar in our game:

- Walk speed = 10 ms^{-1}. Thus, when the character is walking and launching a rocket at the same time, his total launch velocity

Table 1. Values obtained after solving motion equations for the Rocket Jump mechanic

Launch Angle (Degrees)	Max. Horizontal Distance (Meters)	Max. Vertical Distance (Meters)	Time Taken to Complete a jump (Seconds)	Collision Distance (Meters)
0	0	7.5 (maximum height attainable)	0.5	0.2
45	15 (maximum horizontal distance attainable)	3.75	0.35	0.9
16	7.9	0.57	0.13 (shortest time taken)	2.9 (Maximum Collision distance)

Figure 3. Calculating key distances and angles from the projectile equation

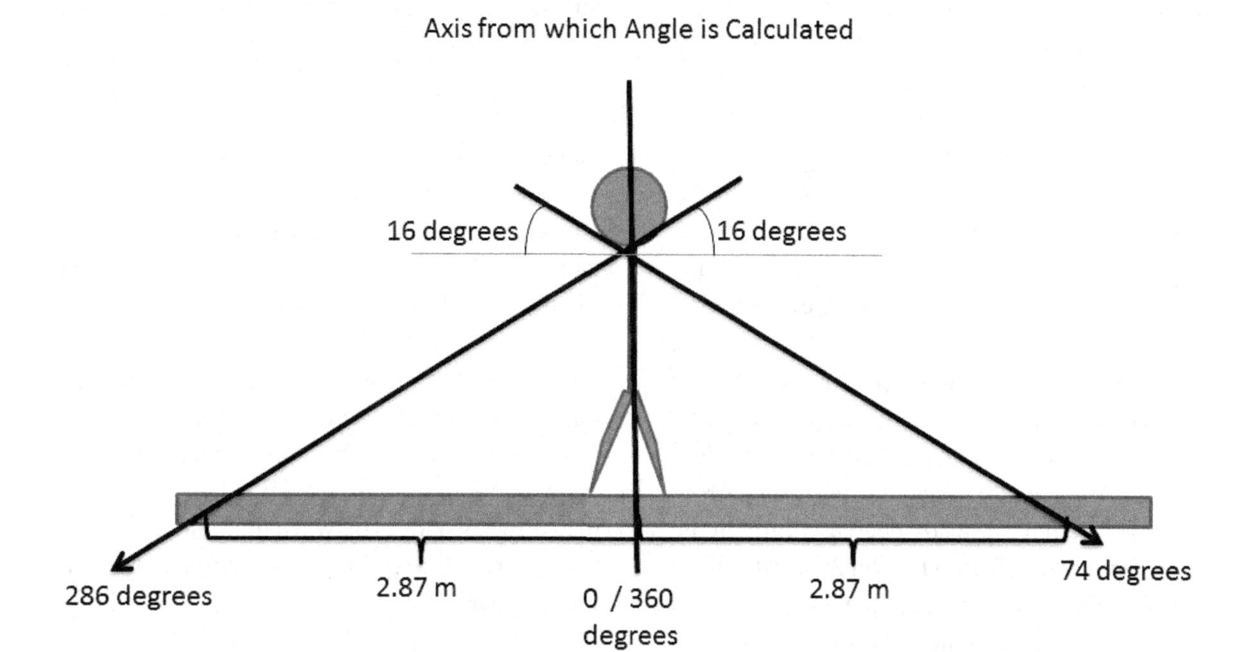

would be $(30 + 10/\cos\theta)$ ms^{-1}, θ being the launch angle.

- Max fall speed = 70 ms^{-1} (this is to model terminal velocity)
- Since the character is able to jump because of the propulsive force of the rocket hitting the ground, we need to consider the maximum collision distance the rocket needs to hit the ground for the propulsive forces to affect him. This maximum collision distance hence limits the angles possible for a jump and further restricts the distance the character is able to travel.

OBSTACLE DESIGN

We now describe the design of obstacles using the variable values in the previous sections as a reference. An orthographic side view is used for the game scene. After trying out different gravitation acceleration constants, we chose $60m/s^2$ as it gave us the desired "zing" for the character motion. We implement four types of obstacles: Static, Angled platforms, Timed obstacles, and Rhythm obstacles.

As previously mentioned, the distance achievable by the character is dependent on the permissible range of rocket launch angles, which may be sometimes constrained by the launch platform size. Furthermore, there may be obstacles in the path of the jump trajectory that the player might need to avoid. There are a number of ways graded difficulty can be introduced into the obstacle design: 1) range of angles available to complete a successful jump can be constrained (reduce takeoff/landing platform sizes, and add obstacles in the path of a range of jump trajectories); 2) number of times player is required to change the launch angle, over a sequence of obstacles; 3) response time could be constrained for moving platforms. We now describe the details of these obstacles, and evaluate the variables and conditions that contribute to the game balance.

Lerpz the Rocketeer has six levels in all. The first three levels consist of static platforms that were meant to be a warm-up for players on the usage of Rocket Jump. The first is of increasing horizontal distances, the second of vertical heights and distances and the third is of a combination of heights, distances and angles. In the next three levels, we implemented timed platforms, distances that were capped by low ceilings, angled platforms and a combination of angled platforms and low ceilings. We had expected the performance of the players to have a steady decrease through the levels, with levels 4 and 5 having the least performance due to obstacles where players had to deal with low ceilings on timed platforms (change in angle per second), jumping from small moving platforms to another similar platform (change in angle per second) and jumping from an angled platform to a moving platform. We intended the angled platforms and timed platforms to be the most difficult levels because players would have problems estimating the angles and keeping up with the changing angles. This design indeed provided the intended progression in challenge, and helped make our game fun, apart from being just an academic study of a new mechanics.

Static Obstacles

Static obstacles are one of the easiest obstacles found in platformers. These obstacles require players to jump over distances to cross the obstacle. Gap widths and heights affect the launch angles chosen by the players. The crux in designing for horizontal obstacles lies in the range of angles permitted to make the jump. Given that initial rocket velocity and gravitational constant are known, and the additional character velocity is either 0m/s or 10m/s, we could easily work out a range of angles that enable successful jumps. As mentioned earlier, constraining the range of launch angles leading to a successful jump could increase the challenge. A wider range of angles means that players would have a greater chance of firing successfully. Designers can limit the permissible range of angles in a number of ways: 1) limit the height of the jump by putting a ceiling above the obstacle; 2) increase the height of the target platform; 3) increase the distance between platforms; 4) reduce the size of the launch and/or the target platform.

From Figure 4, it can be seen that there is a range of angles available for the player to make a successful jump. The approximate range of permissible angles can be obtained by substituting the known distances between the launch platform and the two extremities of the target platform, into the parabolic motion equations listed in the Rocket Launcher mechanics section. We can use the following formula for the horizontal distance R:

$$R = \frac{u^2 \sin 2\theta}{g}$$

If we put in the known variables, namely permissible horizontal landing distances R_{min} and R_{max}, initial speed and gravity, we get:

$$\theta = \frac{1}{2} \sin^{-1} \frac{Rg}{u^2}$$

Note that there are two possible launch angles for landing at the same horizontal distance, as shown in Figure 4a. The second angle = $90° - \theta$. Thus we have two sets of angles whereby a jump that is executed with an angle within this range will be successful. In Figure 4b, the launch angles included between the middle two arrows would result in an overshoot, whereas angles below the right-most arrow, and above the left-most arrow, would result in an undershoot.

The design of static obstacles also applies to vertical obstacles where the target platform is

Figure 4. Range of launch angles to execute successful jumps

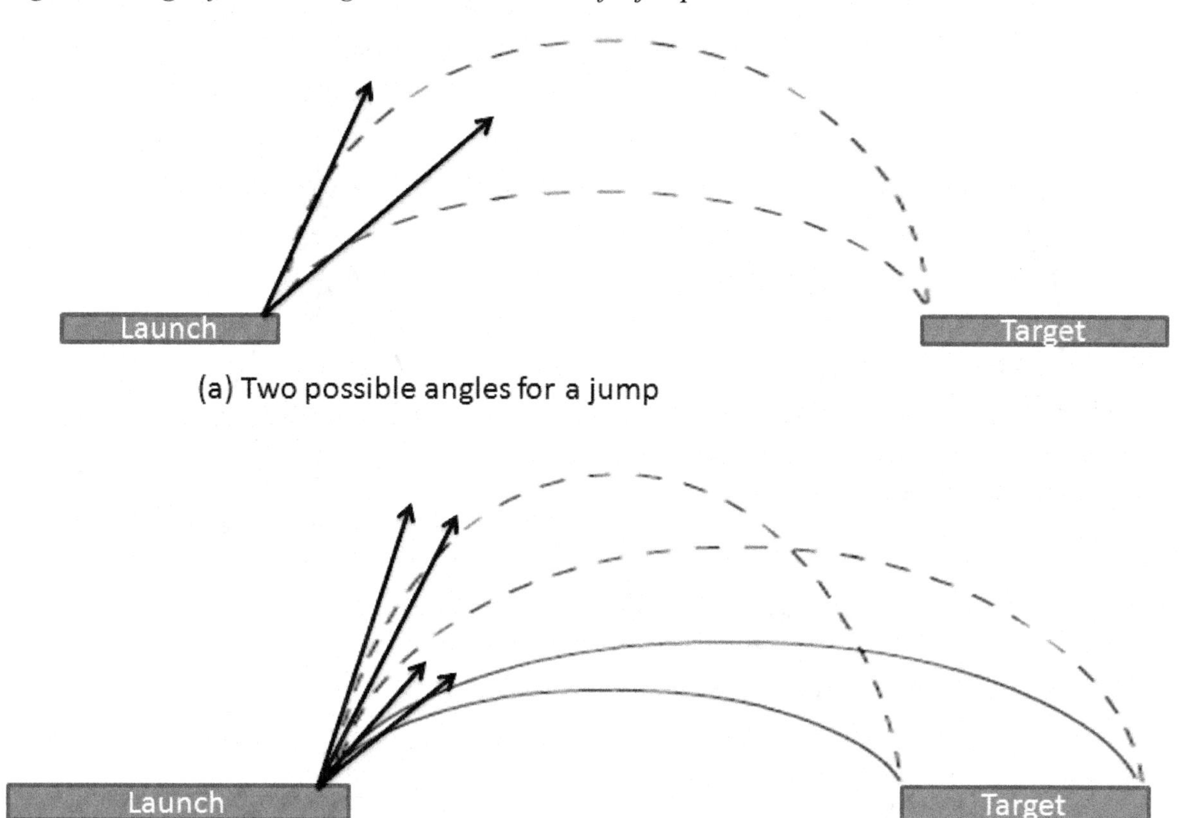

placed at a distance and height away from the launch platform. Such obstacles increase the difficulty of static obstacles as players are now forced to use larger launch angles to clear the raised height and the distance of the target to land successfully. Given the target platform's horizontal gap, x, and height, y, we solve the following quadratic equation defining parabolic motion, to find the minimum and the maximum angles to clear the x-y distance offsets:

$$\frac{gx^2}{2u^2}\left(1+\tan^2\theta\right)-x\tan\theta+y=0$$

We notice that game balance scales better to tweaks in distances between platforms, rather than changes in platform sizes. A wider launch platform means that players are able to employ a much wider range of angles compared to a smaller launching platform, where players are only able to use larger angles (see Figure 5). When coupled with larger gaps between platforms, this affords a very narrow range of permissible launch angles, and thus non-linearly scales up the difficulty.

Example of Difficulty Analysis using Range of Permissible Angles

From the above explanation of the jumps, we can relate the difficulty of the obstacle with the range of permissible angle the obstacle allows for the player to make a successful jump.

Figure 5. Effects of gap and platform size variation on permissible launch angles

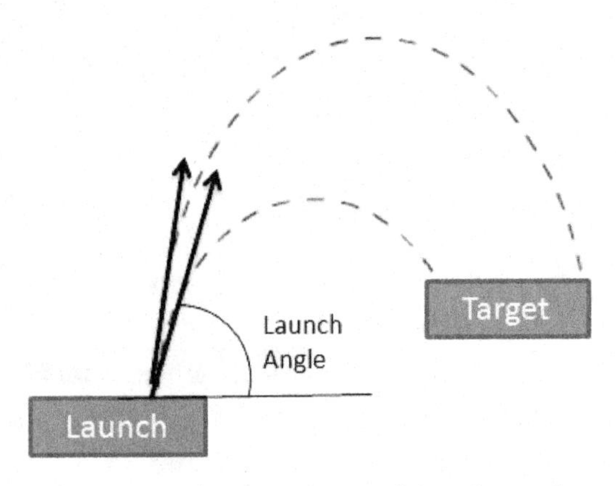

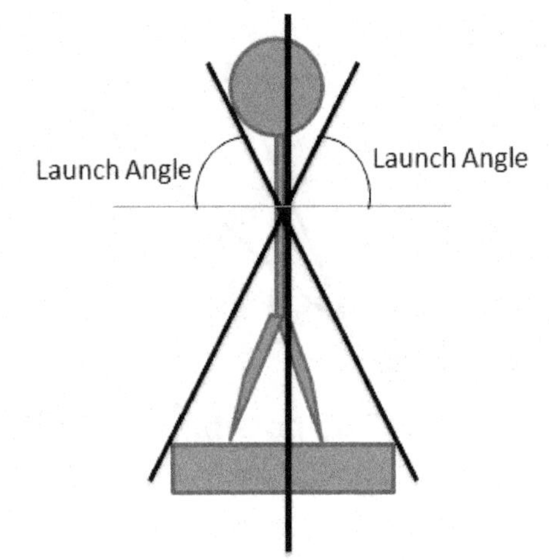

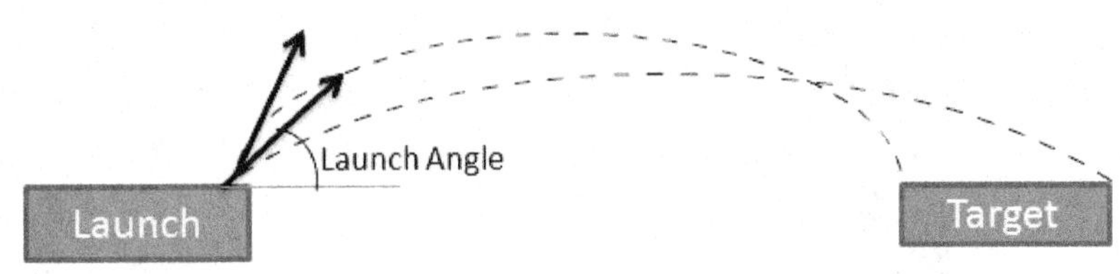

For a target platform that is of width 3 meters and at a distance of 5 meters away,

Angles for a successful landing to the nearest edge of the target platform:

Horizontal distance, R = 5m;

Initial velocity when Lerpz is at rest, u = 30 ms^{-1}

Gravity = 70 ms^{-1}

We shall now find the angle required to make the jump, θ.

Applying the equation, $R = \frac{u^2 \sin 2\theta}{g}$

$$\sin 2\theta = \frac{5(70)}{30^2} = 0.389$$

Hence the smaller angle is:

$$\theta = \frac{\sin^{-1} 0.389}{2} = 11.44°$$

And the corresponding bigger angle is:

$$90° - 11.44° = 78.56°$$

Angles for a successful landing to the farthest edge of the target platform:

Horizontal distance, R, is now 8m away from the edge of the launching platform

Hence:

Applying the equation, $R = \frac{u^2 \sin 2\theta}{g}$

$$\sin 2\theta = \frac{8(70)}{30^2} = 0.622$$

Hence the smaller angle is:

$$\theta = \frac{\sin^{-1} 0.622}{2} = 19.24°$$

And the corresponding bigger angle is:

$$90° - 19.24° = 70.76°$$

The range of permissible angle to make such a jump is hence:

From 11.44° to 19.24° and from 70.76° to 78.56°

We can quantify difficulty as the percentage of angles allowed by the player. It can be calculated from the following:

Biggest angle – smallest angle = 78.56°-11.44° = 67.12°

Range of angles where an overshoot will occur from 70.76°-19.24°=51.52°

Range of permissible angles: 67.12°-51.52°=15.6°

Difficulty of obstacle:

$$\frac{Range\,of\,angles}{360°} \times 100\%$$
$$= \frac{15.6°}{360°} \times 100\% = 4.33\%$$

Angled Platforms

In Conventional Jump, vertical displacement usually consists of a default height and an additional height that is dependent on how long the player holds down the jump button. The horizontal displacement is usually dependent on how long the player holds down the direction keys (e.g. "*WASD*" or arrow keys). With such simplifications, we normally do not see much of the projectile physics in the character motion. At best, some sliding forces are added to characters while they are still on an inclined platform. On the other hand, Rocket Jump is very much dependent on the angle of the launch from the platform. The jump trajectory is different if the player were to jump from a horizontal launch platform than if the player jumps off an inclined one at the same launch angle, as shown in Figure 6.

The projectile motion vectors change when the launching platform is not horizontal. There is now a component of gravity acting upon the character along both the platform-tangent and platform-normal components, as shown in Figure 7. Horizontal motion is no longer uniform as it is under the influence of the gravity component. However, the theory behind the motion is similar to that if the projection was done on a horizontal plane but with an extra component of angle of incline. We include a section on the equations for projectile motion for upward and downward inclined platforms.

To prevent confusion, we distinguish between the launch angle θ, and the angle of incline, α. For projection up an incline, the angle of projection in relation to the incline is $\theta - \alpha$. For downward inclines, we only need to change the polarity of "α" to "$-\alpha$". Let us describe the upward incline scenario here.

The initial velocity (u_x, u_y) and gravitational acceleration constants (g_x, g_y) are:

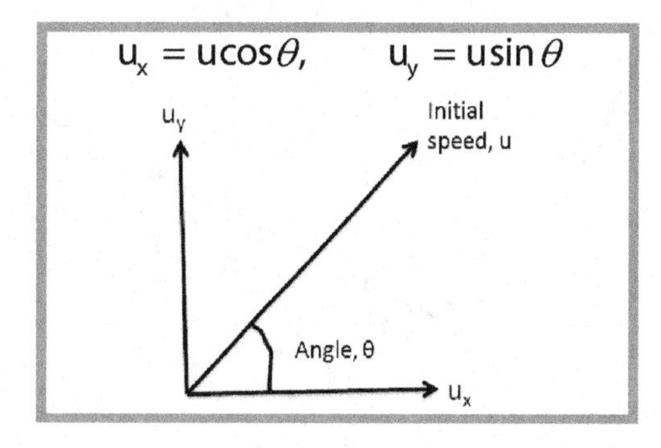

After substituting these into the motion equations, we can calculate the airborne time (t), range (R) and maximum height (H) of the rocket-launched character:

$$t = \frac{2u\sin(\theta - \alpha)}{g\cos\alpha}$$

$$R = \frac{u^2}{g\cos^2\alpha(\sin(2\theta - \alpha) - \sin\alpha)}$$

Figure 6. Differences in normal jumps and rocket propelled jumps on inclined platforms

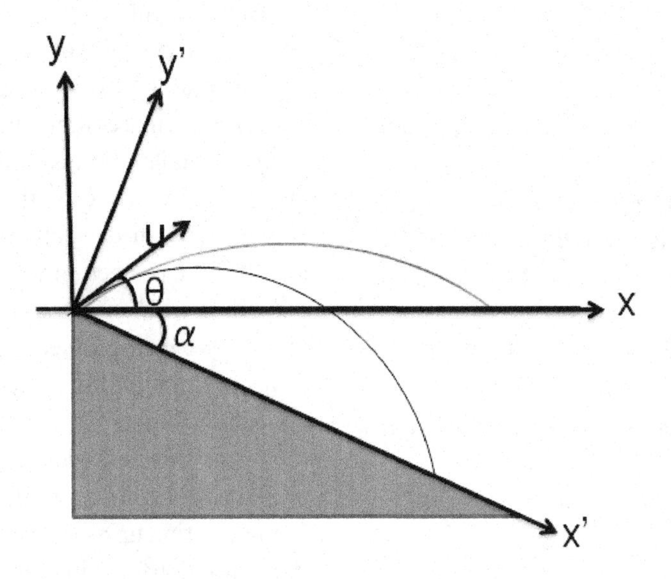

Figure 7. Incorporating gravity for inclined planes

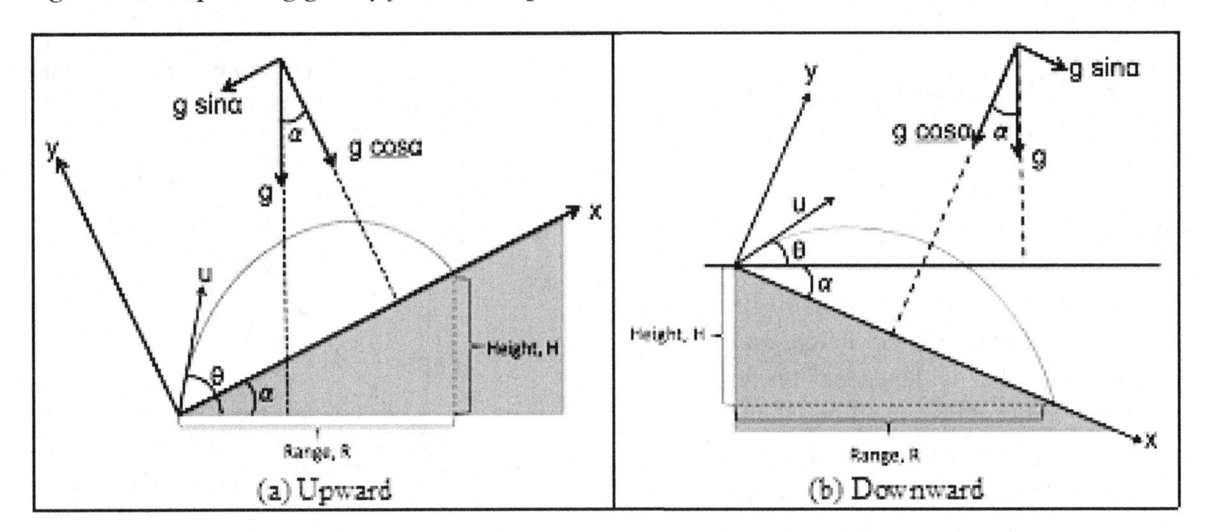

$H = R \sin \alpha$

Maximum range, R_{max} occurs at $\max(\sin(2\theta - \alpha))$ implying that $\theta = (\frac{\pi}{4} + \frac{\alpha}{2})$. Thus R_{max} can be simplified as:

$$R_{\max} = \frac{u^2}{g \cos^2 \alpha (1 - \sin \alpha)}$$

Depending on the angle of incline and the distance from the platform, we can create an easier or more difficult obstacle. Recall, that an easy obstacle is one with wide range of permissible launch angles. This could be achieved by placing the target platform's leading edge well within the maximum distance range available for the shooting point (see top row, Figure 8). Also, inclined target platforms afford longer landing

Figure 8. Effect of changing platform inclination on game difficulty

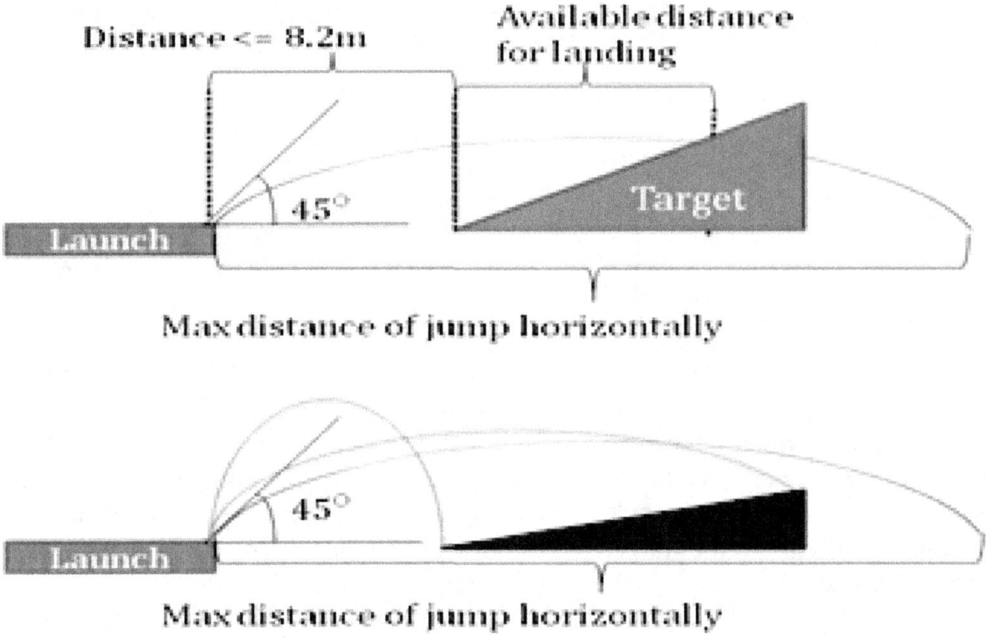

lengths, while flatter targets are more prone to overshooting (bottom row, Figure 8). On the other hand, players find it trickier to estimate a successful launch angle on more inclined slopes.

Timed Obstacles

Timed obstacles change position over time. The changing distance from launch to target platforms means that the angle of jump needs to be adjusted regularly. At least, it makes most players pause and think before executing the jump. The difficulty increases when both launch and target platforms are moving as the range of valid angles are changing at a faster rate than if one platform is static, as shown in Figure 9.

Rhythm

The presence of several moving platforms lined over a stretch of distance where players jump onto them in succession is often seen in many platformers. In sequences of obstacles where players are required to make several jumps in succession, the presence of a regular rhythm makes the obstacle easier to traverse. Similarly, if the rhythm is irregular or if there are breaks, players tend to find it more difficult. Compton and Mateas (2006) noted that:

"... level design in platform games relies heavily on rhythm. Rhythmic actions help the players reach a "flow" state in games, a state of heightened concentration (Csikszentmihalyi, 1990). When a player is "in the flow" or "in the rhythm" of a game, making jumps requires not only distance calculation, but also timing. Rhythmic placement of obstacles creates a rhythmic sequence of player movements, making each individual jump easier to time."

When players are sufficiently adept in estimating distances and angles with Rocket Jump that the above obstacles we have discussed are relatively easy for them to overcome, rhythm – or the disruption of rhythm, can be used to add another layer of challenge to the game.

Figure 9. Effect of moving platforms

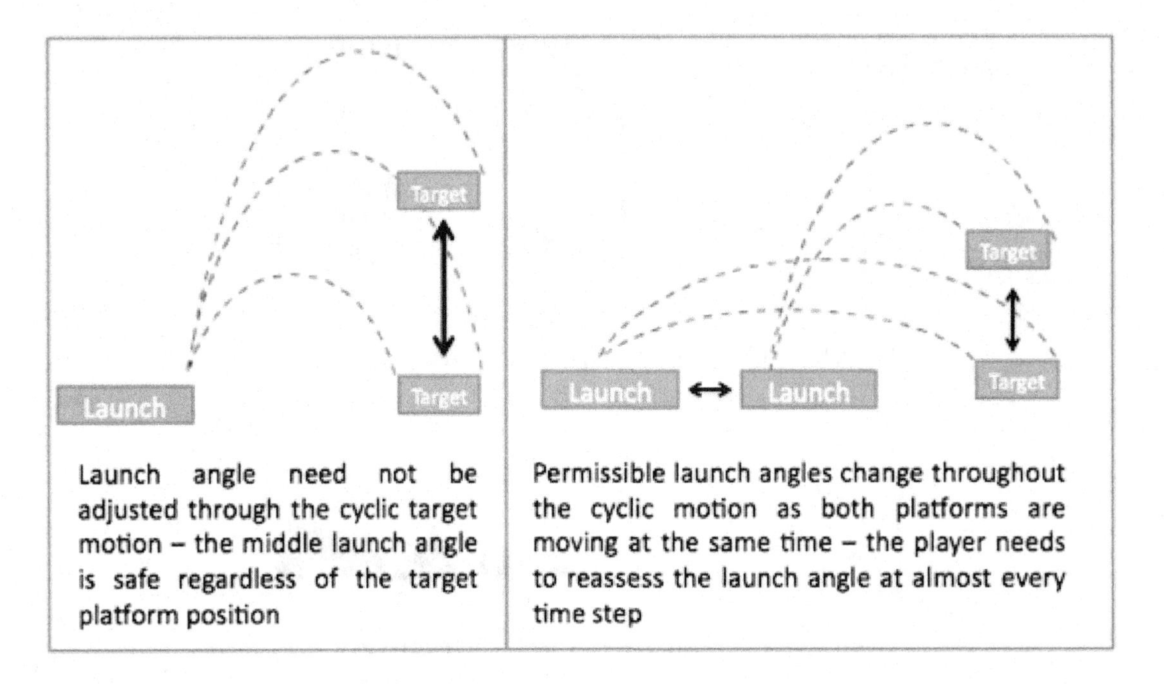

From our analysis of the Rocket Jump, we can calculate the time taken for each jump. As time taken for a jump is intrinsically linked to the distance travelled, and in relation, the angle of the jump, we can use the motion equations described in the Rocket Jump Mechanics section to determine the placement and motion of target platforms. Rhythm then can be implemented as a logical pattern or repetition of launch angles or jump timing. For moving platforms, rhythm could be established by jumping from platform to platform with the same speed. By oscillating the platforms at the same frequency but at different phases, we could achieve a basic obstacle that could be easily navigated if player can tune into the frequency. If we were to change this frequency within the same sequence of moving platforms, the players would be thrown off rhythm and thus may find it more difficult to overcome the obstacle. Care needs to be taken while assigning the platform motion speed, so that it is not too high to disorient/frustrate players, and not too low to bore them.

Procedural Implementation of Level Difficulty

Compton and Mataes (2006) proposed a method for the procedural generation of levels. Their work focuses on the representation of the level through Components, Patterns, Cells and Cell Structures. This article focuses on the orientation, placement and difficulty associated with obstacles; namely, platforms and gaps; in close alignment to motion equations. From the above discussion, there are multiple considerations for the design of obstacles: 1) the range of permissible jump; 2) the angle of target and launch platform; 3) the rate of change of angle that the player needs to make and 4) an established rhythm for of the obstacles.

PERCEIVED CHALLENGE AND PLAYER PERFORMANCE

As mentioned at the start of this chapter, one of the aims of this paper is to compare the conventional movement mechanics against our implementation of the Rocket Jump. For a meaningful comparison, we have chosen to turn to the MDA Framework again. Hunicke. LeBlanc and Zubek described Aesthetics as the "desirable emotional responses evoked in players, when they interact with the game system". One such desirable emotional response is Challenge. Several prominent game designers have noted Challenge to have a key importance in the player's enjoyment of games (Byrne, 2005; Nicollet 2004) as it promotes player's engagement with the game world and creates a motivating force for the player to remain playing. We evaluate challenge with a user survey, and later player performance with game-play data analysis for different obstacles. As such, the comparison we did was based on the level of Challenge that players felt between the different mechanics.

Comparison using NASA TLX Survey

We first did a workload test to find out if the Rocket Jump is more challenging than the Conventional Jump. For this, we have implemented the Conventional Jump on to the same set of game levels that we have designed for the Rocket Jump and have invited 16 testers, 8 Experts and 8 Novices, to play the game and evaluate the mechanics based on a subjective workload index. All participants are required to finish all levels before answering the questionnaire.

For this, we employed the NASA Task Load Index (NASA TLX) as a subjective method for comparing different aspects of workload, or mechanic-specific aspects which contributes to Challenge. The NASA TLX is an assessment index that measures work load for a particular activity on six different subscales – mental demand, physical demand, temporal demand, performance, effort and frustration (Cao, Chintamani, Pandya and Ellis, 2009). Players rated the mechanics on a seven-point Likert scale based on their interaction with the mechanics.

From the gathered results, we ran a Between-Subjects Multivariate ANNOVA statistics test using SPSS and found that the only difference in workload between Rocket Jump and conventional Jump was the mental demand on players. Mental demand was the only component that had a significant difference (p-value = 0.047, which is lesser than 0.05) and that both Experts and Novice players had consistently rated the Rocket Jump as being more mentally demanding than the Conventional Jump.

Table 2. Results of the NASA TLX questionnaire

NASA TLX components	Significance between Conventional and Rocket Jumping (p-value)	Average rating for Conventional Jumping		Average rating for Rocket Jumping	
		Experts	Novices	Experts	Novices
Mental Demand	0.047	4.125	4.375	4.875	5.625
Physical Demand	0.753	4.5	4.875	3.625	5.375
Temporal Demand	0.584	4.375	5.25	4.125	4.875
Performance*	0.308	5.25	4	4.75	3.375
Effort	0.682	4.375	5.25	4.375	5.75
Frustration	0.590	4.25	4.375	4.625	4.625

Note*: For the rating of Performance, the lower the score, the worse the performance. For others, the higher the score, the more challenging the mechanic is in terms of the component.

Game-Play Data Analysis

After the first round of focus testing and workload testing, we did a more general second round of data collection on the final polished version of our game. We gathered game-play data from 50 invited participants who played through the levels, in an online version. We implemented the game in Unity and collected game data via a backend PHP – mySQL system. For this phase of the experiment, we did not separate our players into Experts and Novices, as we preferred to capture data from interested players only, most of whom had played our game earlier during the alpha testing stages. Participation was entirely voluntary and they could leave at any particular time if they chose. We allowed this, as we only wanted to capture data closely resembling actual player behavior. Readers can play this game by surfing to the NUS Game Portal link (*Lerpz the Rocketeer, 2011*).

Figure 10 shows the 25 obstacles set up over 6 different levels in the game. As we can see from Table 3, the percentage difference of failed jumps between conventional jumping and our rocket jump mechanic is negligible. Failure rates for rocket jumps are just 1.31% – 4.94% higher than conventional jumps. This shows that though the players report more mental challenge for our rocket-jumping mechanic, they perform nearly as well as the conventional jump mechanic. Thus we can safely conclude that our puzzles and rocket mechanics are well designed, and provide adequate tension between risk and reward. Also, we can see that there is lower variance in the failure rates for rocket jumps, indicating a flatter difficulty progression curve. This may be due to the lack of familiarity of the mechanic that resulted in errors even in the easier levels. We hope to collect more data from returning players in another study to confirm this hypothesis.

Lastly, let us take a closer look at the player performance with respect to obstacle types. We find that angled and timed jumps posed some of the tougher challenges for rocket jumping. Highest failure rates (above 9.5%) are noted at on jumps involving an angled platform in level3 (302→303 and 303→304), and timed platforms in level4 (403→404 and 404→405). Interestingly, players using conventional jumps also experience highest failure rates for timed platforms, but not as high for angled platforms. Though the differences in failure rates are not very high, they support our claims that more accurate slope projectile physics does increase the difficulty for players (see Figures 6 and 7).

SUMMARY

We have implemented a Rocket Jump mechanic for a 2D side scrolling platform game, and compared its design with conventional jumping. We have described ways of procedurally creating puzzles with graded difficulty. The main idea here is to compute a range of permissible rocket launch angles that result in successful jumps, and simply tweak the design so that the permissive range gets more constrained. Furthermore, timed jumps following the concept of rhythm in game design, also increases the challenge noticeably, as the player has limited time to judge the launch angle. We have proof from the NASA TLX survey results that the amount of mental processing the player needs to undergo to determine the rocket launch angle is statistically significant. This ability is affected by the perception of launch angles due to inclined platforms and moving platforms.

CONCLUSION

More balancing work needs to be done to counter the effects of: 1) player perception; 2) more experience with a new mechanic; 3) rhythm. However, our current findings do back up our puzzle generation heuristic. We have presented a pragmatic approach towards procedural game design with new mechanics; i.e. using core mechanics equa-

Figure 10. Level design with progressive challenge. Obstacles tagged for player jump analysis

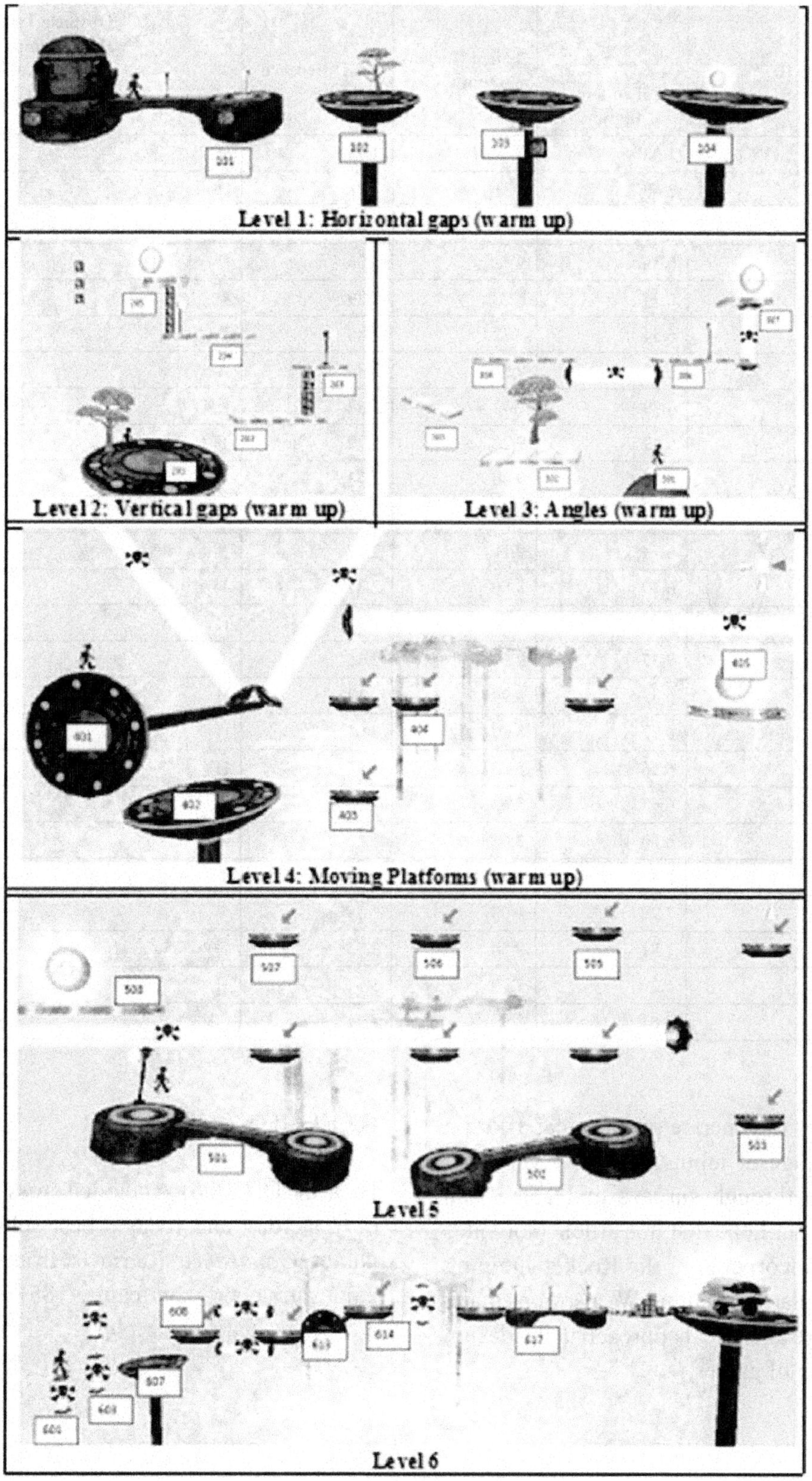

Level 1: Horizontal gaps (warm up)

Level 2: Vertical gaps (warm up)

Level 3: Angles (warm up)

Level 4: Moving Platforms (warm up)

Level 5

Level 6

Table 3. Jump failure percentages at different obstacles sorted by performance difference

Obstacle ID	Failed Usual Jump%	Failed Rocket Jump%	% Difference	Obstacle Type H: Horizontal V: Vertical A: Angled T: Timed
J1(101to102)	2.08	7.02	4.94	H
J1(102to103)	2.60	7.31	4.71	H
J4(401to402)	5.05	8.91	3.86	V
J3(304to306)	6.36	9.42	3.06	H
J5(506to508)	6.34	9.22	2.89	T
J5(505to506)	6.34	9.22	2.88	T
J3(303to304)	6.72	9.56	2.84	A
J4(402to403)	6.59	9.42	2.82	HVT
J5(503to505)	6.49	9.28	2.79	T
J5(502to503)	6.68	9.40	2.73	AT
J3(302to303)	7.04	9.66	2.62	HVA
J5(501to502)	6.74	9.11	2.37	A
J2(204to205)	6.49	8.86	2.37	V
J2(203to204)	6.75	8.96	2.21	HV
J4(403to404)	7.37	9.53	2.15	T
J6(603to607)	6.83	8.94	2.12	H
J6(607to608)	6.77	8.79	2.01	H
J2(201to202)	6.56	8.53	1.96	HV
J6(608to613)	6.67	8.63	1.96	T
J6(613to614)	6.80	8.62	1.81	T
J6(614to617)	6.82	8.62	1.81	T
J2(202to203)	7.29	9.08	1.79	V
J4(404to405)	7.81	9.56	1.75	HT
J1(103to104)	6.29	7.79	1.51	H
J3(301to302)	7.67	8.99	1.31	HV

tions and rules to generate puzzles that afford a permissive range of inputs, and later verifying the challenge through surveys and game-data analysis. It is our hope that this article motivates new designs incorporating the Rocket Jumping mechanic for platform games. We also hope that it inspires a more analytical approach in the design and evaluation of games.

REFERENCES

Boutros, D. (2006). A detailed cross-examination of yesterday and today's best selling platform games. *Gamasutra*. Retrieved from http:///www.gamasutra.com/view/feature/1851/a_detailed_crossexamination_of_.php

Cao, A., Chintamani, K. K., Pandya, A. K., & Ellis, R. D. (2009). NASA TLX: Software for assessing subjective mental workload. *Behavior Research Methods*, *41*(1), 113–117. doi:10.3758/BRM.41.1.113

Compton, K., & Mateas, M. (2006). Procedural level design for platform games. Paper presented at the *Second Artificial Intelligence and Interactive Digital Entertainment International Conference (AIIDE)*. Marina del Rey, CA.

Hunicke, R., LeBlanc, M., & Zubek, R. (2004). MDA: A formal approach to game design and game research. Paper presented at *AAAI Workshop Challenges in Game Artificial Intelligence*. San Jose, CA.

Totilo, S. (2007). A higher standard – Game designer jonathan blow challenges super mario's gold coins, "unethical" mmo design and everything else you may hold dear about games. *MTV Networks*. Retrieved from http://multiplayerblog.mtv.com/2007/08/08/a-higher-standard-game-designer-jonathan-blow-challenges-super-marios-gold-coins-unethical-mmo-design-and-everything-else-you-may-hold-dear-about-video-games/

ADDITIONAL READING

Beat It! Games (2011). Lerpz the rocketeer. *SoC Games Portal*. Retrieved from http://games.comp.nus.edu.sg/lerpz/

Blow, J. (2008). *Braid. Windows. id Software (1996)*. Quake.

Nintendo (1985). Super Mario Bros.

Nitsche, M., et al. (2006, October). *Designing Procedural Game Spaces: A Case Study*. Paper presented at Proceedings of FuturePlay, London, UK and Ontario, Canada.

Sicart, M. (2008). Defining Game Mechanics. *The International Journal of Computer Games Research*, *8*(2).

Smith, G., Cha, M., & Whitehead, J. (2008, August). *A Framework for Analysis of 2D Platformer Levels*. Paper presented at Sandbox Symposium, Los Angeles, CA.

Team Meat. (2010). *Super Meat Boy*. Windows.

Valve (2007). Team Fortress 2.

KEY TERMS AND DEFINITIONS

2D Side-Scrolling Platformer: A genre of video games in the platform games category where the game world is presented with an orthographic, two dimensioned view. The camera is fixed to move only in the left-right directions.

Aesthetics: Describes the desirable emotional responses evoked when players interact with the game.

Conventional Jump: Describes the most common type of jump executed on key-presses. The distance jumped depends on the speed of the characters and duration of key-press events.

Dynamics: Describes the run-time behavior of the mechanics acting on player inputs and game object interactions.

Mechanics: Describes interactions, attributes, rates of change, and behavior of game entities.

Rocket Jump: Involves pointing a rocket launcher or other similar explosive weapons at the ground or at a wall then firing and jumping at the same time. The rocket's explosion propels the player to greater heights and distances than otherwise possible.

Section 3
Collision Detection in Games

Chapter 10
Collision Detection in Video Games

Benjamin Rodrigue
University of Louisiana at Lafayette, United States of America

ABSTRACT

This chapter will describe several methods of detecting collision events within a 3D environment. It will also discuss some of the bounding volumes, and their intersection tests that can be used to contain the graphical representation of objects in a video game. The first part of the chapter will cover the use of Axially Aligned Bounding Boxes (AABBs) and Radial Collision Volumes. The use of hierarchies with bounding volumes will be discussed. The next section of the chapter will focus on Object Oriented Bounding Boxes (OOBs). The third section is concerned with the Gilbert-Johnson-Keerthi distance algorithm (GJK). The last three sections will focus on ways of optimizing the collision detection process by culling unnecessary intersection tests through the use of type lists, sorted lists and spatial partitioning.

INTRODUCTION

Collision detection is a difficult aspect of a game engine's architecture. Knowing when two or more objects are colliding is crucial to ensure that a proper response occurs within the game's space and mechanics. The response can be a loss of health of a player character or the toppling of a lamp off of a table. The response that helps immerse the player in the game's world first starts with the detection of the collision. This chapter will cover some of the methods in use by the game industry for detecting a collision event within a game. This chapter will not discuss how to respond to a collision event as that will vary depending on the game at hand just as the choice in collision detection will vary depending on the needs of the game.

A collision detection system should be designed early in a game's development. It should not be

DOI: 10.4018/978-1-4666-1634-9.ch010

put off to be added at a later time; this can result in inefficient changes to the engine architecture, and worst case scenario, become an execution bottleneck. There is no generic algorithm that works best for all collision detection scenarios because games have different demands for detection. A game with relatively few collidable objects will not need the same detection system as a game with thousands of objects.

All bounding volumes described in this chapter are pre-computed before execution time. Predictive collision checks are assumed in the sense that an object's next position should be pre-calculated one loop before it is applied. The collision detection system will then be doing checks one frame ahead of where the object will be in order to stop an object from entering into a collision. The alternative method of moving the object out of the collision can result in more complicated and inefficient situations. Moving an object outside of a collision could result in it moving into a new collision with another object.

All pseudo code examples within this chapter are based off of C++.

BACKGROUND

This chapter is organized into three parts. The first section is a math primer on vectors and matrices. The math primer is not meant to be comprehensive, but it will present a couple of the rules for matrices and vectors. The second topic will focus on collision detection volumes that improve detection performance over pixel by pixel overlapping tests. The collision volumes are Axially Aligned Bounding Boxes (AABB), Radial Volumes, Binary Volume Hierarchies (BVH), Object Oriented Boxes (OOB) and the Gilbert-Johnson-Keerthi distance algorithm (GJK). These bounding volumes are arranged in order of increasing complexity and precision.

The last topic will describe ways to reduce the number of necessary collision tests. This part is split up into three sections. These sections describe the ways of reducing the number of detection tests based on object type and object location. The first part of this section describes how to use dynamic and static object lists. The next section is concerned with the sort and sweep algorithm. The last section describes spatial partitioning using Quadtrees and Octrees.

Math Primer

Vectors

This chapter will make use of two mathematical structures: vectors and matrices. A vector is an ordering of two or more real numbers in a set. Throughout this chapter, two and three dimensional vectors will be used to explain the different ways to check for collisions. In the pseudo code examples throughout this chapter, Vector3 will indicate a three dimensional vector while Vector2 will indicate a two dimensional vector. This data structure is useful in collision detection because it can be used to represent a position, direction, and velocity among other things in two or three dimensions. The vector, **v** is a positional vector in two dimensional space. The normalized form of **v,** represents the direction from the origin through **v** and is represented with two vertical bars, ||**v**||. The normal of **v** is calculated using the magnitude of **v**, |**v**|.

```
v = (2, 2)
//position vector
```

The magnitude is calculated by taking the square root of the addition of each element squared. An example of calculating the magnitude:

```
|v| = SquareRoot(v[0]^2 + v[1]^2)
//magnitude
|v| = 2.8284
```

The way to get the normal of a vector is to divide each component of a vector by the vector's magnitude, |**v**|.

```
||v|| = (v[0] / |v|, v[1] / |v|)
||v|| = (.7071, .7071)
//normalized vector
```

The normal of a vector has a unique property in that its magnitude is one. Vectors can be added and subtracted from each other so long as they have the same number of elements. Multiplication between two vectors is not defined in the same manner that addition and subtraction is defined and therefore cannot be done, but a vector can be multiplied by a scalar. A scalar is another name for any real number. When a scalar is multiplied to a vector, each element in the vector is multiplied by the scalar.

Dot Product

The first important mathematical operation on vectors that make them useful for collision detection systems is the dot product. This operation is commutative, can only be done on vectors with the same number of elements greater than 2, and results in a scalar value. The scalar result represents the magnitude of the two vectors multiplied together and multiplied by the cosine of the angle between them. An example equation of a dot product is:

```
a. b = |a| * |b| * cos(theta).
```

The alternate method of calculating the dot product involves multiplying each element of the two vectors together and adding them together. For example:

```
a[] = {2, 3, 1}
      b[] = {1, 1, 0}
      c   = a. b
           = a[0] * b[0] + a[1] * b[1]
+ a[2] * b[2]
           = 2 * 1 + 3 * 1
+ 1 * 0
           = 5
```

Cross Product

The next useful operation between vectors is the cross product. The cross product is similar to the dot product in that it requires that both vectors have the same number of elements, but it is not similar in that the vectors being crossed need three or more elements and is not commutative; **a** x **b** != **b** x **a**. The result of a cross product is a vector that is perpendicular to both of the original vectors. The order in which the two vectors are being crossed effects the direction of the resulting vector. For example, **a** = [1, 0, 0] and **b** = [0, 1, 0] are unit vectors for the x-axis and y-axis respectively. The cross product, **a** x **b**, results in **c** = [0, 0, 1] whereas **b** x **a** results in **d** = [0, 0, -1]. The operation that is taking place involves the multiplication and subtraction between the elements of **a** and **b**:

```
c[] = { a[1]*b[2] - a[2]*b[1],
        a[2]*b[0] - a[0]*b[2],
        a[0]*b[1] - a[1]*b[0] }
```

Example:

```
a[] = {1, 0, 0}
b[] = {0, 1, 0}

c[] = { a[1]*b[2] - a[2]*b[1],
        a[2]*b[0] - a[0]*b[2],
        a[0]*b[1] - a[1]*b[0] }
    = {   0 * 0   -   0 * 1,
          0 * 1   -   1 * 0,
          1 * 1   -   0 * 0   }
    = {     0     -     0,
            0     -     0,
            1     -     0     }
c[] = { 0, 0, 1 }
```

Matrices

The second mathematical structure is a matrix. A matrix is not similar to a vector in that its columns and rows are both greater than one. It is, like a vector, an ordered set. Matrices will be used exclusively to represent an object's rotation. The two forms of rotation for an object are world rotation and local rotation. A world rotation is where the object's rotation is around the world axis, while a local rotation involves rotating the object around its own local axis. Rotating an object around its local axis involves moving it back to the world origin, applying the rotation, and then moving it back to its previous position. The world rotation does not involve the translation but does result in the object's position changing. The vector dot product and cross product are not compatible with matrices.

It is recommended that for further understanding of the mathematical structures the reader should refer to books dedicated to vector math such as Lay's, "*Linear algebra and it's applications*" (Lay, 2006).

INTERSECTION TESTS

Axially Aligned Bounding Boxes

This section will focus on the first of the two collision volumes which are trivial to implement, the Axially Aligned Bounding Box (AABB). The AABB data structure allows for inexpensive collision checks at the cost of precision. An AABB does not always allow for a tight fit around the game object, especially if that game object is rotating. This results in empty space registering as a collision when the game object is not rectangular. An AABB has a couple of defining qualities in that it cannot be rotated and its edges must be parallel to one of the world axes.

Figure 1 shows two boxes with their bounding boxes displayed in dark grey. This feature allows for very inexpensive intersection tests. An overlap test between two AABBs in two dimensions is, at most, four bound checks; checking the minimum and maximums of both boxes. Concerning AABBs, Eberly notes that, "In two dimensions, Boolean operations are applied to axis-aligned rectangles by decomposing the problem into one-dimensional operations" (Eberly 2008, pp. 5).

For an example see Code 1.

The three dimensional bound check includes the above checks but has an additional two checks for the z-axis test. Figure 2 shows the two AABBs colliding at their corners. The collision test is fairly simple as is the data maintenance of the AABB. During an object's update, if its position should change then the AABB encapsulating it should be updated as well. The only collision data that needs to be updated during the object's update is the AABB's minimum and maximum corners. As shown above, these two vectors are the only data structures needed to check for a collision. Additionally, these two corners allow for the entire AABB to be reconstructed as needed based on combinations of the minimum and maximum vectors.

The AABB method can be the main intersection test used in a game if the imprecision of the collision check is acceptable. The only other drawback would be the quantity of collision checks done as the number of game objects increases. In the best case there are $O(n(n-1)/2)$ checks where n is the number of collidable objects in the game's world. This can be mitigated by using one or more of the organizational techniques discussed later in this chapter. For most games, AABBs are not precise enough to use as the only intersection test.

Radial Intersections

This method of collision detection involves using the object's center point and a radius to perform intersection tests. The collision area in two dimensions is a circle, while in three dimensions it is a sphere. In the same way that AABB's are more

Figure 1. This figure displays two boxes with their axially aligned bounding boxes shown around them. The world axis is represented as x-axis in dark grey, y-axis in grey and z-axis in light grey

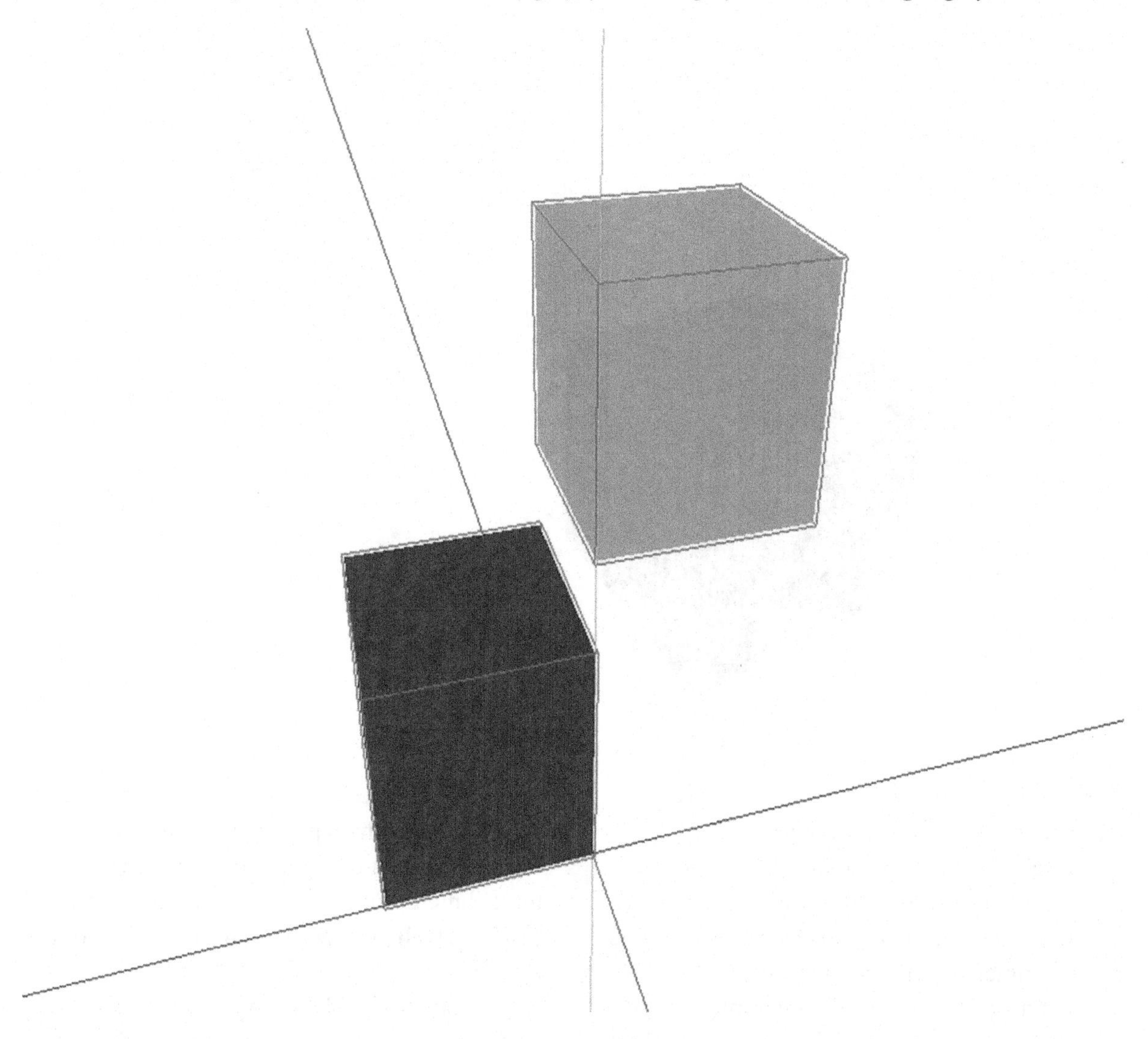

Code 1. The code above shows the AABB overlap test for the x-axis only, but provides enough detail to extend this code to test additional axes

```
bool CheckAABB(AABB box01, AABB box02){
check if box 1's maximum x is less than box 2's minimum x;
     return false if it is;
check if box 1's minimum x is less than box 2's maximum x;
     return false if it is;
/* Do the same as above for the y axis and z axis */

return true if the checks above fail
}
```

Figure 2. In this image, the light grey box moved towards the dark grey box and collided with it at a corner

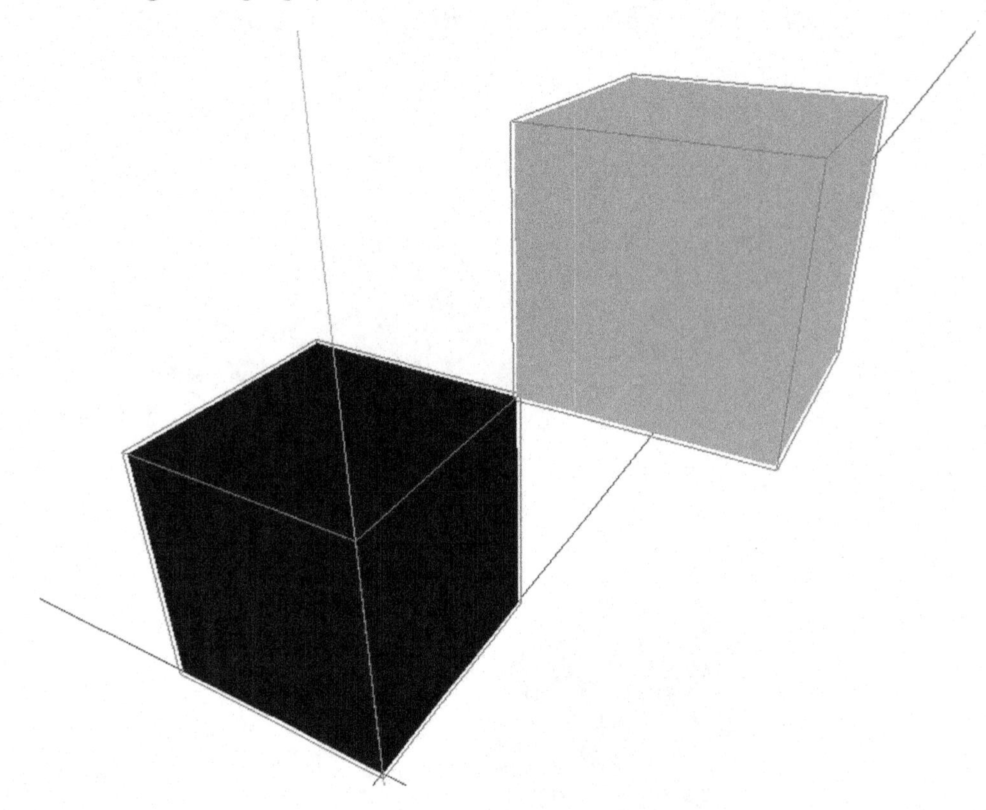

precise with rectangular objects, radial checks are more precise with spherical objects.

The collision test for radial objects is inexpensive in that it is just checking the distance between the two objects' centers and their radii. Here is the two dimensional check in Code 2.

This distance check can be optimized by removing the square root and squaring the result of adding the radii together. (Code 3)

The CPU takes extra time to perform a square root which is why squaring the right side of the equation results in a performance increase. The three dimensional check includes the above example as well as an additional calculation for the z position.

The use of AABBs and radial volumes in conjunction with one another involves a few conditional checks and a distance check. The

Code 2. This logic will indicate whether or not an intersection has occurred by returning true for overlap and false otherwise

```
float flDistOnX = (object01's x - object02's x)^2;
float flDistOnY = (object01's y - object02's y)^2;
if (SquareRoot(flDistOnX + flDistOnY) <
                    (object01's radius + object02's radius)) {
return true;
}
     return false;
```

Code 3. This logic improves Code 2's logic by removing the need to perform a costly square root

```
if ((flDistOnX + flDistOnY) <
                    (object01's radius + object02's radius)^2)
return true;
     return false;
```

first step is to find the closest point on the AABB volume to the radial volume's center point. Next, a distance check, much like the one presented above, is performed to see if the radius of the sphere is larger than the distance from the point on the AABB to the origin of the radial volume. Here is some example code in Code 4.

Now that the point has been located on the outside of the AABB, the slightly modified collision check for radial objects can be used to see if the AABB and radial object are overlapping. As before, the code can be optimized by squaring the radius and not using a square root for the distance calculation. (Code 5)

The same issues and benefits mentioned previously for using AABBs as the sole intersection checks exist for radial checks as well. Their use in high profile modern games has a similar function to AABBs and will be mentioned in the next section.

Bounding Volume Hierarchies

Bounding Volume Hierarchies (BVHs) are an alternate approach to using one or two types of intersection volumes exclusively. This method makes use of the binary tree data structure. Binary trees are explained in the key terms section at the end of this chapter. By using a binary tree representation of bounding volumes, the precision of an object's intersection tests can be improved without increasing the complexity of the intersection tests. This is useful in both two and three dimensional environments even if more precise collision checks are needed. This is because these two approaches can be used to cull the number of expensive collision checks needed. An AABB can be used to test whether an object has come within close enough proximity for the more expensive collision tests.

Code 4. The function, FindPointOnAABB(...) will locate a point on an AABB that is closest to the radial object and return that point

```
Vector3 FindPointOnAABB(AABB object01, Vector3 position){

Vector3 point; /* [0]: x, [1]: y, [2]: z */
if (object01's min x < position[0]) {
point[0] = object01's min x;
} else if (object01's max x > position[0]) {
point[0] = object01's max x;
} else {
point[0] = position[0];
     }
/*  Do the same above for y and z as needed */
return point;
}
```

Code 5. This CheckCollision(...) function makes use of the logic in Code 4 to find a point to test whether or not the radial object is overlapping that point

```
bool CheckCollision(AABB object01, Radial object02){
Vector3 point = FindPointOnAABB(object01, object02.position);
float flDistOnX = (point's x - object02's x)^2;
float flDistOnY = (point's y - object02's y)^2;
float flDistOnZ = (point's z - object02's z)^2;

if ((flDistOnX + flDistOnY + flDistOnZ) < (object02's radius)^2)
{
    return true;
}
return false;
}
```

One way of using bounding volume hierarchies is to have AABB or radial collision volumes at each level of the binary tree. At the top level of the tree, the largest bounding volume restricts access to its children based on whether or not anything has collided with it. Bounding checks are done first at root of the tree, as Ericson notes, "With a hierarchy in place, during collision testing children do not have to be examined if their parent volume is not intersected" (Ericson, 2005). The tree is traversed by taking the branches that indicate a potential intersection until a node has no children or until the intersection test fails on both children. If the intersection tests fail, then there is no overlap, but if there are no children, then there is an intersection. This hierarchy is useful because it allows for inexpensive detection methods to be used as often as possible.

Object Oriented Boxes

Object Oriented Boxes (OOB) are similar to AABBs in that they are both rectangular bounding volumes. They differ in everything else. The first major difference between the two is that OOBs rotate with the object that they are encapsulating. This allows for more precise overlap checks with fewer false collision events. The second difference between the two bounding boxes is that all of the OOB's corners must be maintained at all times. The third difference is concerned with the intersection test which, "amounts to processing each axis of the 15 potential separating axes. If a separating axis is found, the remaining ones of course are not processed" (Eberly, 2008, pp. 3). The AABB was able to take advantage of its axially aligned property to do its collision checks. The OOB does not have that luxury. As seen in Figure 3, the way to check for overlap isn't as apparent as with the AABB volumes. OOBs intersection test is much more complicated and makes use of the Separate Axis Theorem in order to accomplish the overlap test.

The Separate Axis Theorem states that for two convex shapes to not be overlapping there must exist a plane where one object is completely on one side of the plane and the other object is completely on the opposite side of the plane. This plane is found by projecting all of the vertices of the OOBs onto a plane that is perpendicular to this separating plane, and thus, detecting whether or not there is a gap between the points. If a gap is found then there must exist a plane that separates the two objects resulting

Figure 3. This image shows the use of BVHs to encapsulate a stick figure and a bar bell using multiple bounding boxes and multiple radial spheres. An inner bounding volume is not tested unless an intersection was found on the outer volume

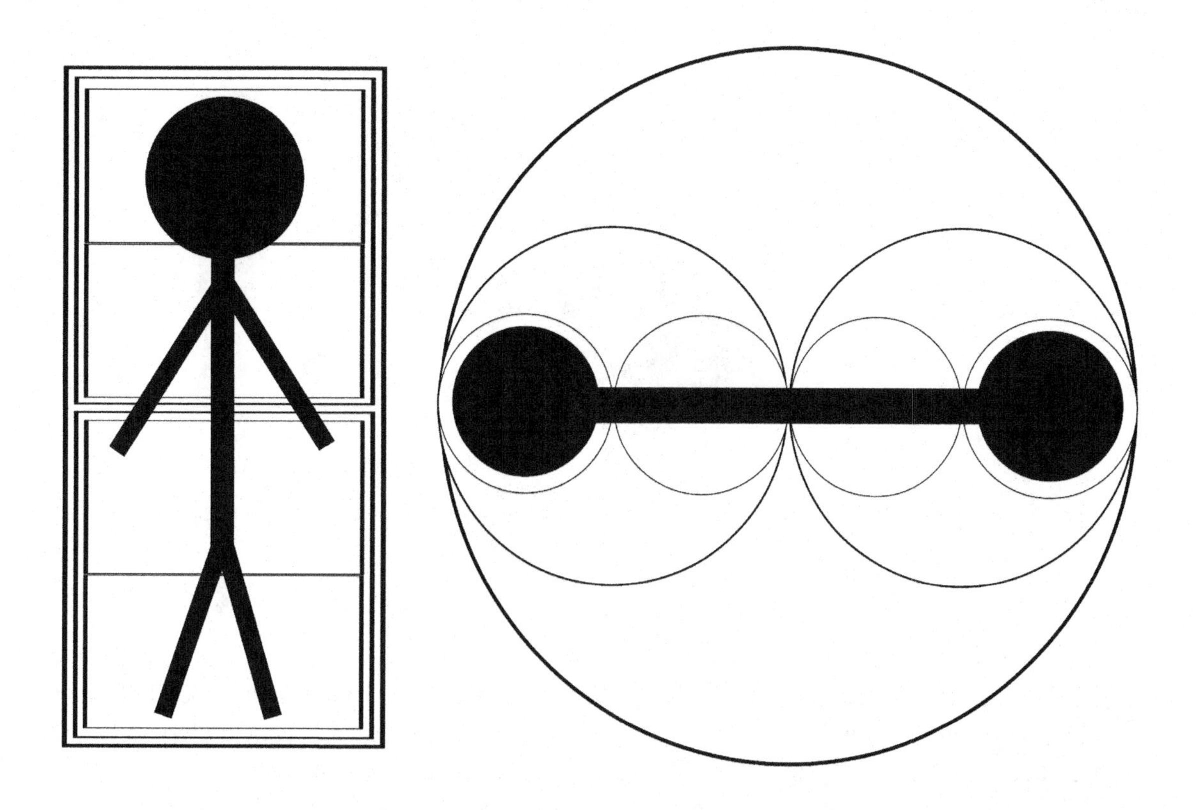

in no overlap. Figure 4 shows two object oriented boxes colliding. It is important to note that this intersection test will not work with concave shapes. If a concave shape were to fit around any shape such that only a curved plane could separate the two objects, then a false collision would be dedicated.

The Theorem applies in the same way to two dimensional objects as it does to three dimensional objects. The difference between the two is that the additional dimension requires additional axis checks. There are an infinite number of planes that could be checked, but fortunately, the number of planes to check can be reduced to checking a finite set based off of the rotation matrices of the two objects. The local axes of both objects are the first to be checked and only after they all indicate that there isn't a gap do the other planes get checked. There are d^2, where d is the number of dimensions, additional planes to check.

In order to perform the separate axis check with OOBs, several data structures are necessary. Each corner, represented as a vector, of the OOBs is needed, as well as each object's rotation matrix. As previously stated, the first planes to project onto, via the dot product, and test are the local x, y and z axes of the first object and then the local x, y and z axes of the second object. If a gap is not found on these six planes, then the cross product is used to find the planes perpendicular to both of the object's local axes. The same method for checking that was performed on the local axes is then performed on each of these newly calculated

Figure 4. Two object oriented boxes positioned so as to not overlap each other. The local axis for each box can be seen by the rectangles that are parallel to the edges of the boxes at each corner

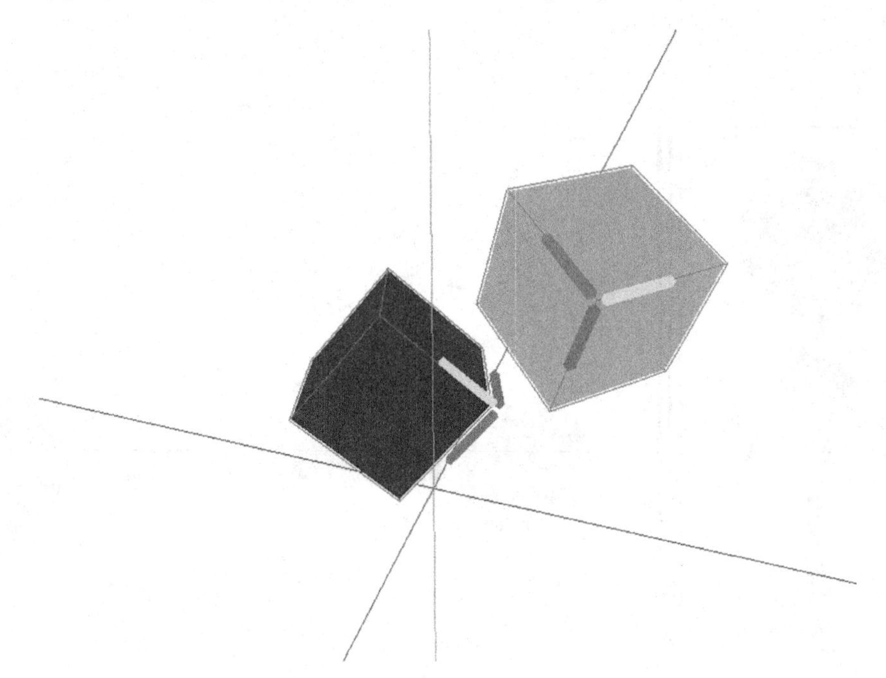

planes. Only after all of these planes have been checked does the algorithm indicate that there is a collision between the two objects. If the algorithm finds a gap during its intersection tests, it can stop looking and indicate that the two objects do not overlap.

Pseudo code has been provided in Code 6.

The OOBTester object in Code 6 has three publically accessible functions. It has a default constructor and destructor used for setting up and cleaning up the object. The TestForOverlap(...) object is the callable function that will

Code 6. This logic detects whether or not objects with oriented bounding boxes are overlapping

```
class OOBTester{
    public:

        OOBTester ();
        ~OOBTester();
        bool TestForOverlap(OrientedObject oobObj01,
                            OrientedObject oobObj02);
    private:
        bool CheckLocalAxis(OrientedObject oobObj01,
                            OrientedObject oobObj02);
        bool CheckCrossedAxis(OrientedObject oobObj01,
                              OrientedObject oobObj02);
};
```

return whether two OOBs are overlapping. (Code 7)

True is returned at the end of the CheckLocal-Axis(...) function indicating that a gap was not found between the two objects on their local axes. (Code 8)

True is returned for the same reason as Check-LocalAxes(...) indicating that the objects are overlapping. For additional information on the Separating Axis Theorem and optimized example code refer to Ericson's "*Real-time Collision Detection*" (Ericson, 2005).

Gilbert-Johnson-Keerthi Distance Algorithm

The Gilbert-Johnson-Keerthi distance algorithm is the last intersection test covered in this chapter. This algorithm is more versatile than the previously described intersection tests because it can be used to do an intersection check between any

Code 7. The code section above gives more insight into the overall logic that is going on in Code 6

```
TestForOverlap(...)
    if CheckLocalAxis(...) returns false
        return false for TestForOverlap(...) since a gap
            was found

    otherwise, if CheckCrossedAxis(...) returns true
        return true for TestForOverlap(...)
        return false otherwise indicating a gap was
            found
CheckLocalAxis(...)
    for each local axis in oobObj01
        for each point in oobObj01
            Find the minimum and maximum point along the current
            axis by doing a dot product
        for each point in oobObj02
            Find the minimum and maximum point along the current
            axis by doing a dot product
        compare the minimum and maximum points found in the above
            two loops

        if a gap exists between the set of points then
            return false
            else do nothing

    Repeat the loop above with oobObj01's local axes

    return true
```

Code 8. The CheckCrossedAxis(...) function takes care of the additional axes that need to be checked

```
CheckCrossedAxis(...)
    for each unique cross product between the local axes of
        oobObj01 and oobObj02
        for each point in oobObj01
            Find the minimum and maximum point along the current
            axis by doing a dot product
        for each point in oobObj02
            Find the minimum and maximum point along the current
            axis by doing a dot product
        if a gap exists between the set of points then
            return false
            else do nothing

    return true
```

two convex shapes. An additional benefit of the GJK algorithm is that it works independently of an object's rotation. The algorithm has these two characteristics through the use of a derivation of the Minkowsky Sum.

The Minkowsky Sum is an important part of the GJK algorithm. It is the process of creating a set of points that contains every element in the set of A added to every element in the set of B where sets A and B contain the two object's bounding vertices respectively. The summation is not what is useful in detecting a collision. The variation of the Minkowsky Sum, the Minkowsky Difference, is what makes it useful to the GJK algorithm. Figure 5 shows two boxes with their bounding boxes and local axis as well as the Minkowsky Sum between the two objects' bounding vertices. The difference tells us whether or not the objects are overlapping in two or three dimensions. An example of this is when two objects have a vertex occupying the same position; one of the vertices in the Minkowsky Difference between those two objects will be the origin. Additionally, the distance between the two objects, when they are not colliding, can be determined by doing a distance calculation between the closest point to the origin in the Minkowsky set and the origin itself.

The GJK algorithm does not calculate all of the Minkowsky vertices because not only would it be time consuming, but only a small portion of the set is useful for intersection detection. Instead it iteratively samples the vertices of the two objects by using a support function for each of the objects. Jovanoski notices that, "the versatility of GJK is a result of the fact that it relies solely on support mappings for reading the geometry of an object. A support mapping fully describes the geometry of a convex object and can be viewed as an implicit representation of the object" (Jovanoski, 2008, pp. 5). An iteration of the GJK algorithm involves maintaining a list of vertices that make up a simplex. For our purposes a simplex is a convex geometrical shape ranging from a single point to a four point tetrahedron for three dimensional intersection testing. The algorithm manages this list by adding or removing vertices as it is attempting to build a tetrahedron, using the Minkowsky Difference, containing the origin. A two dimensional GJK algorithm will be concerned with building a simplex of three points containing the origin.

Figure 5. The two objects have collided at light grey's bottom corner into the dark grey box's edge

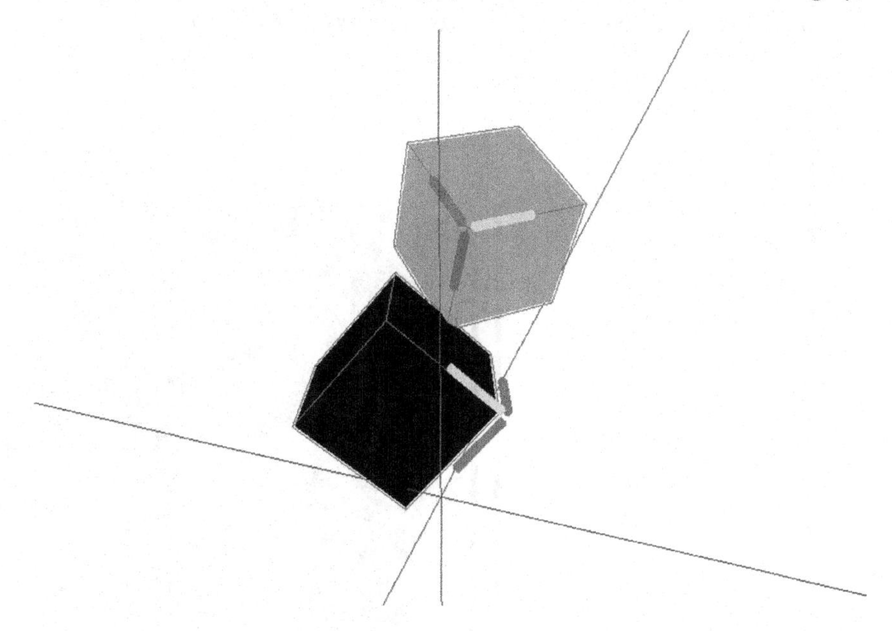

An object's support function finds the furthest vertex of its bounding volume along the direction that is passed in. The direction is updated each iteration of the algorithm to indicate where the origin is in relation to the current progress of the simplex. Figure 6 shows the iterative building process that the algorithm goes through when attempting to detect an overlap. The example code in Code 9 is one way to structure an object to perform the algorithm.

The GJKTester class in Code 9 performs the GJK algorithm on any two convex shapes that have a publically accessible internal support function. The first two functions handle the initialization, allocation, and deallocation of the object. The only publically available function in this class other than the constructor and destructor is the CheckGJK(...) function. CheckGJK(...) controls the main loop of the algorithm. (Code 10)

If the loop fails to find a simplex containing the origin, then the objects are not colliding. The 'return false' will end the overlap test for these two objects. (Code11)

Hill Climbing is a method that can be used to iterate over the vertices in a bounding volume. This method is useful for the GJK algorithm when the algorithm needs the most extreme point of a large set of points of a bounding volume in a specified direction. For this to work, each vertex needs to have references to its immediate neighboring vertices. The focus of the method is now on comparing the current vertex with that of its neighbors in order to find the most extreme current point. This process is repeated for each current point until there are no neighbors more extreme than the current vertex. This vertex is then returned as the most extreme point in a given direction. The Hill Climbing, "approach is very efficient, as it explores only a small corridor of vertices as it moves toward the extreme vertex" (Ericson, 2005, pp. 406).

OVERLAP TEST REDUCTION

Dynamic and Static Lists

This section examines the benefit of creating two subsets, dynamic and static, from the set of all collidable objects. The purpose is to reduce intersection tests from object versus every other object

Figure 6. Shows how two objects are compared on a plane. The objects' points are projected onto the axis. The minimum and maximum points for each box are marked with crosses

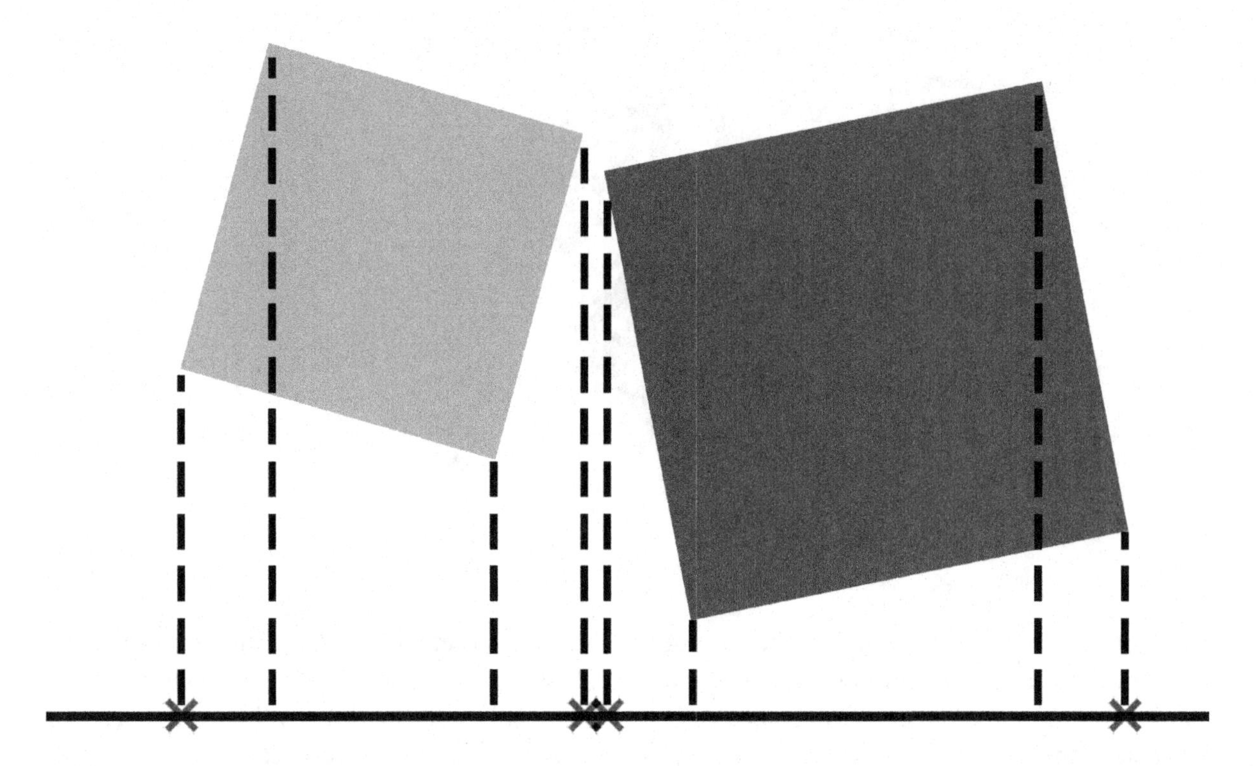

to the set of dynamic objects versus every other dynamic object and every static object. The set of dynamic objects will only contain objects that are moving or will move in the near future. The set of static objects will only contain objects that are not moving unless acted upon by an outside force or are intended to never move. Depending on the mechanics of the game, objects can move between these two sets but must never be in both sets at the same time. This removes the need to check if an object is doing a collision check versus itself. The performance improvement comes from not checking collisions of static objects, such as walls, with other static objects. These objects are not moving and therefore do not require collision checks during their update process. A wall should never have to check for a collision with a wall that is parallel to it unless it were acted on by outside forces that would cause it to move, turning it into a dynamic object. An added benefit is that it allows for dynamic objects that have ceased moving for a period of time to transfer themselves to the static object set to remove its collision check versus all other objects. For example, the number of initial collision checks using a static and dynamic set is shown in Code 12.

This loop compares every object in the dynamic set to every other object in this set once. It then compares every object in the dynamic set to every object in the static set. The static set is never compared with itself as these objects are not moving. A limiting factor on this method is when the number of dynamic objects far exceeds the number of static objects. At this point, the performance gain is negligible compared to the total number of checks that would need to be performed.

Sort and Sweep Algorithm

The Sort & Sweep (S&S) algorithm is an alternate way to reduce the number of object to object collision checks by culling the set of possible collisions. The algorithm identifies objects that are near enough to each other to need an expensive intersection test performed on them. This ensures that two objects on opposite ends of the game space are not degrading performance with unnecessary intersection tests between them. Terdiman explains that the, "basic idea behind the algorithm is this: two AABBs overlap if and only if their projections on the X, Y and Z coordinate axes overlap" (Terdiman, 2007, pp. 1). The algorithm works by comparing bounds of all collidable bounding volumes in each dimension of the game space. This process makes use of one-dimensional testing to perform the culling. The algorithm then performs collision checks on the objects that overlapped each other on all of the examined lists if they require a more precise overlap test other than an AABB.

The algorithm starts by sorting the minimum and maximum bounding values for each collidable into separate lists based on the dimensions of the game space, i.e. minimum and maximum x values for the x-axis on one list. This is done for each axis that is deemed necessary. An insertion sort should be used for the sorting process because it performs well on nearly sorted lists. The lists will be mostly sorted due to the game objects not having moved very much during one pass of the game loop. If the majority of the objects are moving very quickly between each frame then this algorithm will spend an excessive amount of time sorting. It then steps through the lists one at a time examining whether or not a minimum or a maximum value is seen. This is the sweep phase of the algorithm because it is identifying objects that may have the potential to intersect based on the current axis it is checking. When a minimum is found the object is put into a separate list. This

Code 9. The GJKTester class tests the two passed in objects for overlap using the GJK algorithm

```
class GJKTester{

    public:
        GJKTester ();
        ~GJKTester ();
        bool CheckGJK(ConvexObject cobObj01,
                      ConvexObject cobObj02);
    private:
        Vector3 Mediator(ConvexObject cobObj01,
                         ConvexObject cobObj02);
        void CheckSimplex1();
        void CheckSimplex2();
        void CheckSimplex3();
        bool CheckSimplex4();

        Vector3 vecDir;
        list< Vector3 > lisMinkowsky;
};
```

Code 10. CheckGJK(...) is the outer most layer of the GJK layer. It iteratively builds GJK simplex by calling the private functions inside of the GJKTester class described in Code 9

```
CheckGJK(...) does the following:
    initialize vecDir to a random direction
    initialize lisMinkowsky to an empty list

    Call the Mediator(...) and append the
        returned Vector3 object onto the end of
        lisMinkowsky

    Set vecDir to point from the one point in
        lisMinkowsky to the origin

    Loop until some specified count
        call Mediator(...) and append the
            returned Vector3 object on the end of
            lisMinkowsky

        ensure that the newest object crosses the origin
            going from the previous point to the new
            point.
                If it does not return false as a simplex
                    cannot be made around the origin
                else continue

        Based on the number of objects in lisMinkowsky
            call one of the 4 CheckSimplex<#>()
            functions where <#> is the number of objects
            in the lisMinkowsky list.

            if CheckSimplex4() returns true then
                CheckGJK(...) should return true as a
                simplex was made that contained the
                origin

                Otherwise, do nothing
    return false
```

Code 11. These four functions handle the different aspects of the GJK algorithm. The first function, Mediator(...), finds a new point to add to the current list of Minkowsky points. CheckSimplex<#>() validates the current simplex by testing it against the origin

```
Mediator(...)
    each object's support function is called
        object01Support is assigned the result of calling
            object01's Support Function with vecDir as the
            parameter
        object02Support is assigned the result of calling
            object02's Support Function with -1*vecDir as the
            parameter

    return (object01Support - object02Support)

CheckSimplex1()
    this is the first step of the iterative simplex
        validation.
    this function updates vecDir to point from the only
        point in lisMinkowsky to the origin

CheckSimplex2()
    this check does not need to validate that the
        the origin is between the two points because
        the previous early check in CheckGJK(...)
        ensures that the newest point crossed the origin

    vecDir is updated to point from the edge made by
        the two points in lisMinkowsky to the origin

CheckSimplex3()
    this function checks to see if the triangle made
         by three points in lisMinkowsky contains the
         origin.

        if the origin is not contained within the
            triangle then one of the newest edges are
            checked to see if they contain the origin.
            The two vertices that make up that edge
            are kept in lisMinkowsky while the other
            vertex is removed.
```

continued on following page

Code 11. Continued

```
        vecDir is updated to point from the face of the
            triangle to the origin.

    CheckSimplex4()
        this function checks to see if the tetrahedron made
            by four points in lisMinkowsky contains the
            origin.

            because CheckSimplex3() verified the three
                oldest vertices in lisMinkowsky the
                algorithm only has to check if the sides
                created by the newest point are valid.

                this function takes advantage of the
                    ordering of vertices in lisMinkowsky to
                    test the three new sides of the
                    tetrahedron.  By building a face of each
                    triangle to face away from the origin a
                    dot product between the direction of
                    the face and the direction from the
                    newest vertex to the origin should be
                    less than 0 if the origin is within the
                    tetrahedron.

                    If each dot product between one of these
                    faces and the direction from the newest
                    point to the origin is less than 0 then
                    CheckSimplex4() should return true.

                    Otherwise, lisMinkowsky needs to remove
                    the least useful vertices by checking
                    the edges of the triangle face that
                    failed.  It will then remove the least
                    useful vertices and update vecDir to be
                    from the edge of the remaining two
                    vertices or from the only remaining
                    point to the origin.  False is returned
                    to CheckGJK(...) to keep it going.
```

Code 12. This looping example ensures that all objects are compared once and only objects that are actively moving should be checked against other objects. The second loop containing, 'for each object02... ' is intended to start at the next object in lisDynamic and iterate from there till the end of the list

```
void LoopExample(list<object*> lisDynamic,
                 list<object*> lisStatic){
    for each object01 in lisDynamic
        for each object02 starting one past object01 in lisDynamic
            CheckForOverlap(object01, object02)

        for each staticObject in lisStatic
            CheckForOverlap(object01, staticObject)
}
```

separate list indicates that its maximum has not yet been seen and that any other object in the list, or that is put into the list, is a potential intersection event. The object is removed from this separate list when its maximum is seen in the current sorted list. This indicates that the object does not have the potential to intersect any objects further along the axis as seen in Figure 9 with the first object.

There are multiple methods of handling the situation in the sweep phase where an object's minimum is found before the previous object's maximum is found. The method described here is to add all objects currently in the watch list to an internal list inside of the current bounding object's data structure. As each dimension is swept and the object's minimum is seen, its internal list is updated to only include objects currently seen. The objects are then iterated over in order to identify the objects that appear on all three internal lists. These identified objects are then tested between the current object for overlap with more precise detection algorithms. (Code13)

The BoundaryType enumator is used by the SortObject to indicate whether or not it is a minimum value or a maximum value on whichever axis list it is on. (Code 14)

This object is used by a sorting method to sort objects within their dynamics lists. It should make use of the enumator defined in Code.# to indicate its boundary type. (Code 15)

The SortSweep class in Code 15 provides the structure for a sort and sweep implementation. The Add(...) and Remove(...) procedures allow for objects to be inserted and erased from the algorithm's private containers, lisAxes and lisObjects. (Code 16)

A disadvantage of using this algorithm is when the majority of the game objects have lined up on one or more of the axes it is checking. This results in a larger number of objects being referenced in the internal lists of the objects.

Spatial Partitioning

Spatial Partitioning is another way of culling out unnecessary intersection tests in detection systems. The two methods described here are quadtrees and octrees. These two methods make use of modified versions of the binary tree data structure previously mentioned. The differences from a binary tree is that a parent node in a quadtree will have four children whereas a parent node in an octree will have eight children in addition to, "each node also has a finite volume associated with it" (Ericson, 2005, pp. 308). Figure 7 shows the difference between these three trees. Quadtrees are primarily used for two or three dimensional partitioning while octrees are for three dimensional partitioning. The main reason for partitioning the game space into cells

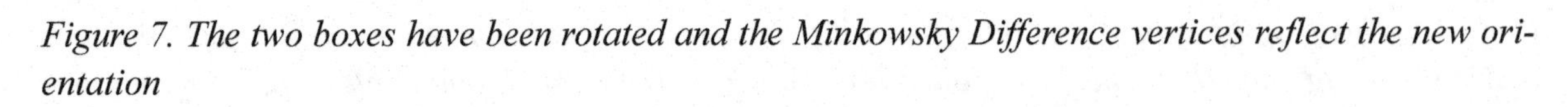

Figure 7. The two boxes have been rotated and the Minkowsky Difference vertices reflect the new orientation

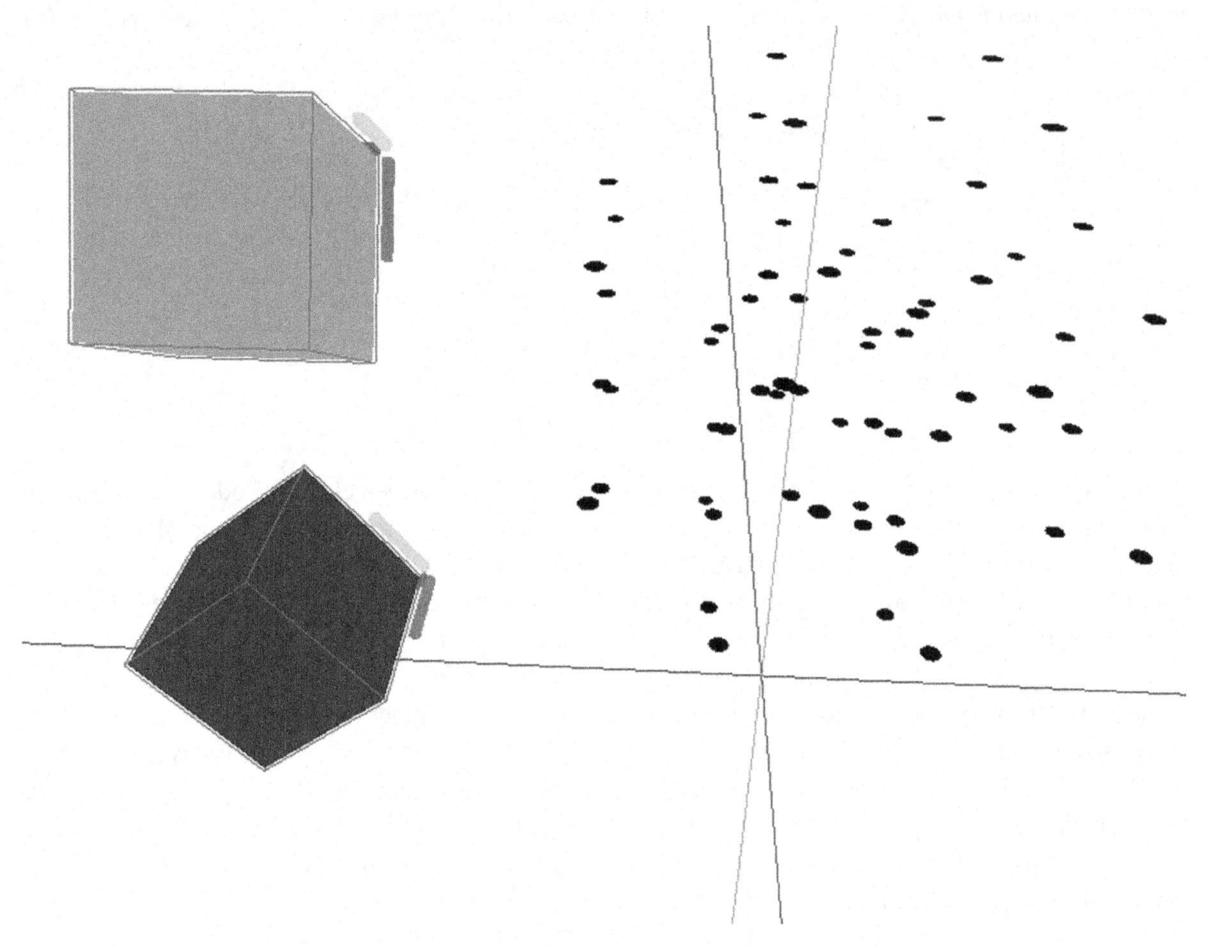

Code 13. This enumerated type should contain a minimum bound and a maximum bound

```
enum BoundaryType{
    int minimum, maximum;
};
```

Code 14. This object should have a unique identifier for the object it represents in intObject as well as whether or not it represents a minimum boundary or a maximum boundary. The position along whichever axis it is associated with should be stored in floValue

```
struct SortObject {
    int intObject;
    BoundaryType btyCurrent;
    float floValue;
};
```

Figure 8. This graphic shows six different stages of the GJK algorithm as it iteratively builds a simplex to contain the origin. It also shows all of the Minkowsky Sum points as black ovals. #1 shows the algorithm finding two extreme points of the Minkowsky Sum between the two objects and is represented as a line connecting two Minkowsky points. #2 shows the algorithm at the 2-simplex stage, the triangle. #3 shows that the algorithm has decided that the triangle face is not useful in finding the origin and therefore uses the closest edge to the origin to continue its search. #4 is a closer view of stage #3. #5 shows a triangle while #6 shows a tetrahedron containing the origin

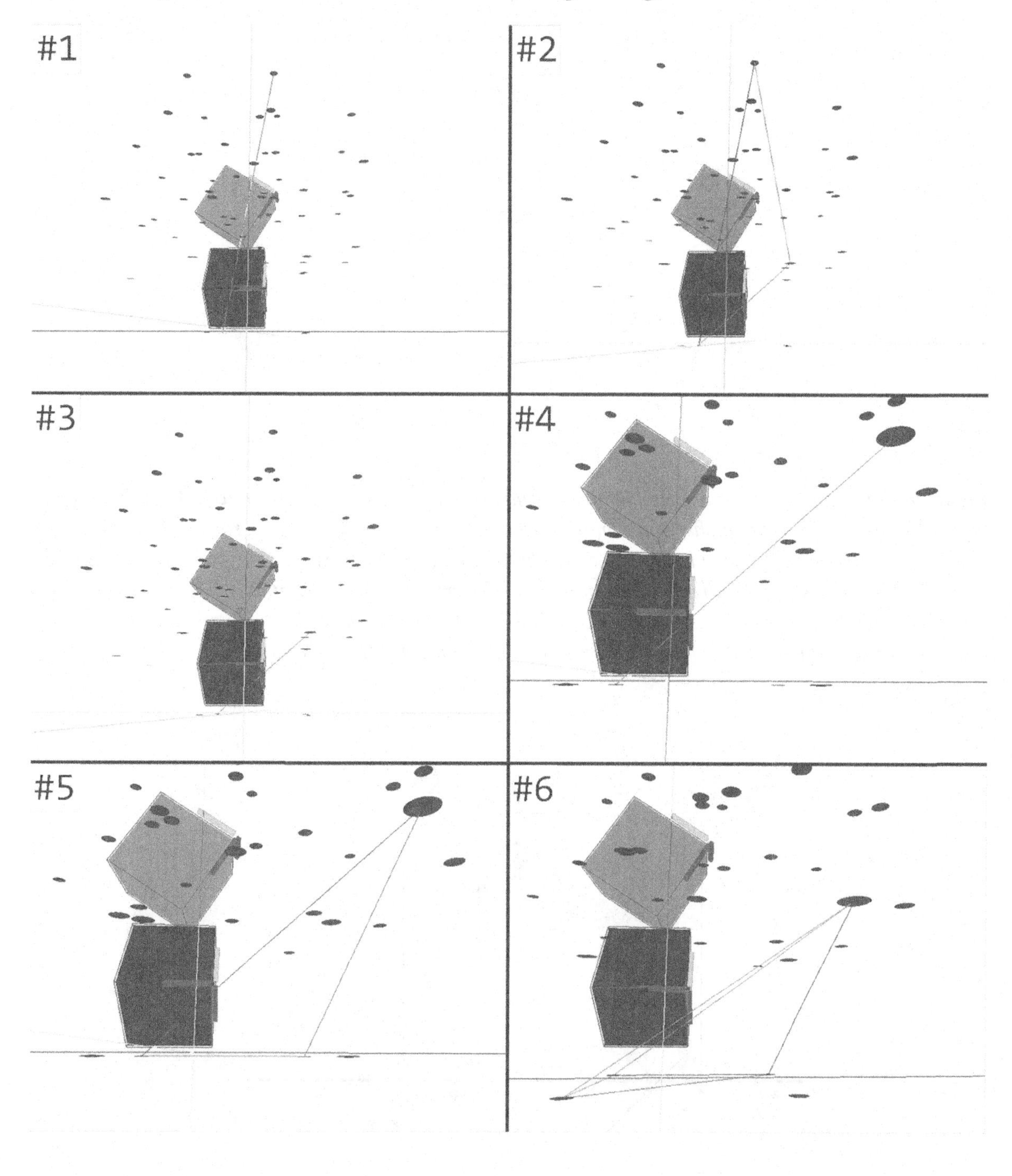

Code 15. This SortSweep object is one way to implement the sort and sweep algorithm

```
class SortSweep {
    public:
        SortSweep();
        ~SortSweep();
        void Add(Object *objObjectIn);
        void Remove(Object *objObjectOut);
        void CheckObjects()

    private:
        void SortList();
        void TestLists();
        void OverlapTesting();
        list< SortObject > lisAxes[3], /* [0] for x,
                                          [1] for y
                                          [2] for z */
                           lisSeen;

        list< Object* > lisObjects;

};
```

Figure 9. This image shows the sorted minimum and maximum points for a given axis. By following the algorithm, min.1 is put on the list and removed when max.1 is seen. The next iteration of the loop sees min.2 on the list. Min.3 is seen before max.2 and therefore the object associated with min.3 is notified that the object associated with min.2 has potential to overlap on the current axis. The end result from sweeping over the sorted lists is that the object's associated with min.3 will have min.2 as a potential intersection test candidate and min.4 will have a min.3's object as its candidate

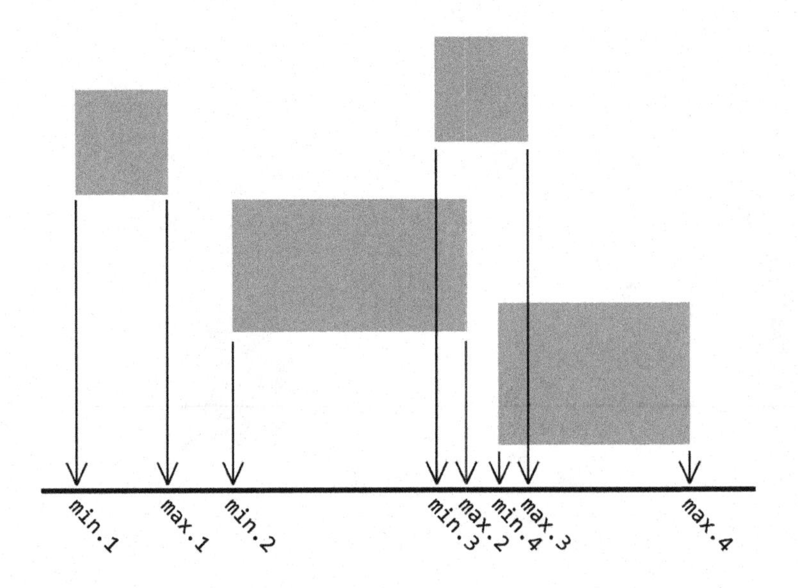

Code 16. CheckObjects() shows the basic ordering of the sort and sweep's function calls. SortList() sorts the objects in the three lists while TestLists() traverses the three lists identifying the objects that are likely overlapping. OverlapTesting() performs any additional intersection testing that is required

```
CheckObjects()
    call the SortList() function
    call the TestLists() function
    call the OverlapTesting() function
SortList()
    for each index of lisAxes
        Sort the list using insertion sort

TestLists()
    for each index of lisAxes
        for every object in lisAxes at the current index
            if currentObject is a minimum then
                add every object in lisSeen to the
                    currentObject's internal list
                    corresponding to the current axis.

                append the currentObject to the end of
                    lisSeen

            else currentObject is a maximum then
                locate the currentObject's corresponding
                minimum in lisSeen and remove it

OverlapTesting()
    for each object in the SortSweep object
        for each object that appears on all of the
            outerObject's internal lists

            perform a more precise overlap test between
                the outerObject and the innerObject

                if there is an overlap then notify both
                    objects
                else do nothing
```

to reduce the number of collision checks needed each pass of the game loop. This is useful because any two objects need not have an intersection test performed on them unless they are both contained within the same cell. At most, any object will need an intersection test performed between objects in its own cell and those cells adjacent to it.

Similarly to the dynamic and static lists described earlier, the same separation concept can be applied with the use of these two tree structures. The separation of the two types of objects in the game allows for one of the tree structures to be pre-generated in order to improve performance. The dynamic structure does not have this performance luxury and must be maintained dynamically. The dynamic tree structure can make use of recursion in its maintenance of the tree data. Each child node is itself an octree and therefore needs to maintain its own children in much the same way as the root node does.

Generating and maintaining either of these tree structures for dynamic objects involves using dynamically linked lists at each node in the tree. Each node should have a pointer to its parent node for use in moving objects from one cell of the tree to another as they move around. An array of pointers to the children of a node should also be maintained for use in allocating new child trees when the number of objects contained within a cell reaches a threshold. Data should be maintained in each node to ensure the node knows which section of space it references. The size of a cell should never become so small that it cannot contain an object's bounding volume.

As objects move around the game space, the cells will need to be updated accordingly. A cell that is empty should notify its parent that it can be cleaned up and its resources can be returned back to the system. An object that occupies more than one cell should exist in the list of all cells it overlaps, i.e. if it is straddling the border of two or more cells. This multiple occupancy results in situations where all cells will need to be updated according to whether or not the object moves in or out of a cell. A way of handling this is to maintain a data structure which keeps references to which octree cells the object is contained in.

Collision checking would involve accessing a node's occupancy list and iterating over it in a manner similar to the one presented in Dynamic and Statics list section specifically, the loop that had O(n*(n-1)/2) iterations. The intersection test would then be based on whichever intersection testing methods that are available between the offending objects. This would be done recursively for each node in the tree. The example header code (Code 17) should provide some insight into the layout of an Octree class:

The TreeNode object from Code 17 is used as each node in the Octree class in Code 18.

The class in Code 18 has the functionality to create, maintain, and destroy an Octree. The traversal through the tree for intersection tests is left out but is trivial to add in during implementation. (Code 19)

For a more robust example code and ways to optimize this data structure refer to Ericson's "*Real-time Collision Detection*" (Ericson, 2005).

FUTURE RESEARCH DIRECTIONS

This chapter focused on presenting a few intersection volumes, tests on those volumes and methods for culling the number of checks needed. A few additional research topics are: kd-trees, fast movement, penetration depth, and multi-threading detection. Kd-trees are similar to octrees except that they allow for a more balanced distribution of objects per cell. Unlike the octree structure the kd-tree's cells are not rigidly partitioned but are created as needed to split the quantity of objects in the current cell evenly.

There are situations where objects are moving too fast between time steps for the collision detection methods described in this chapter to detect. These scenarios can be avoided by ensuring the

Figure 10. #1 shows a binary tree. #2 shows a quadtree. #3 shows an octree. These trees have a depth of three but can extend to any reasonable depth as necessary. A reasonable depth insures that cell's boundaries do not intersect an object's AABB

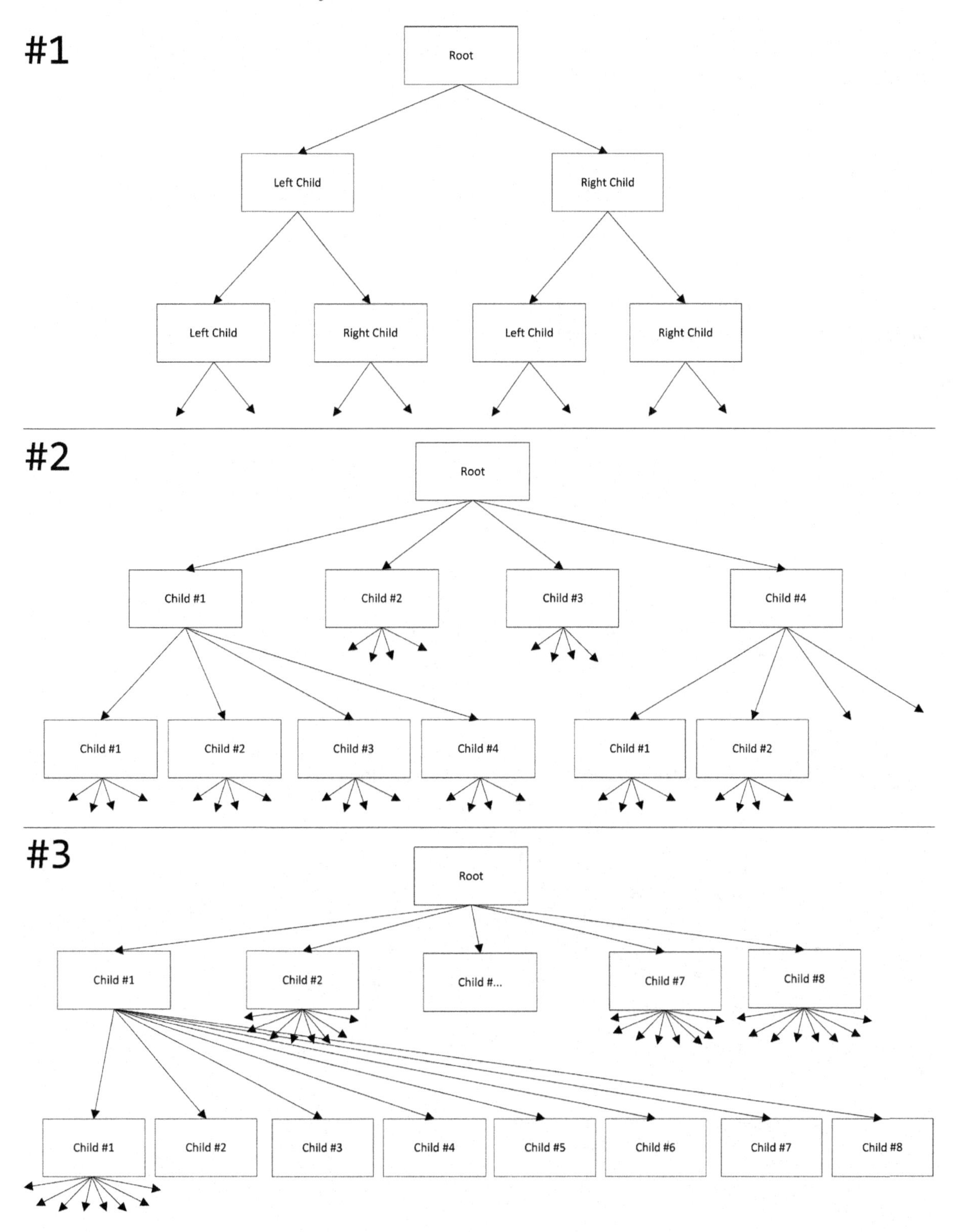

Code 17. This object is used by Octree in Code 18 to build and maintain the Octree data structure

```
struct TreeNode {

    Vector3 vecMinCorner, vecMaxCorner, vecCenter;
    list< GameObject* > lstObjects;
    TreeNode *trnChildNodes[8], *trnParentNode;
};
```

Code 18. The class above is set up to manage an Octree data structure for the purpose of grouping objects based on their position in the game

```
class Octree {
    public:

        Octree(int intTreeDepth,
               Vector3 vecMinCorner,
               Vector3 vecMaxCorner);
        ~Octree();

        void AddObject(GameObject * gobObjIn);
        void RemoveObject(GameObject * gobObjOut);
        void UpdateObject(GameObject * gobObjUp);

    private:
        void ConstructTree(TreeNode * tnoCurrentNode,
                           int intTreeDepth,
                           Vector3 vecMinCorner,
                           Vector3 vecMaxCorner);
        void AddObject(GameObject * gobObjIn,
                       TreeNode *trnNode);
        TreeNode * InsertObject(GameObject * gobInObj,
                                TreeNode *trnNode);
        TreeNode * trnRoot;
        unordered_map< int, list< TreeNode* > > umaContents;
};
```

Code 19. The above logic descriptions correlate to the functions in Code 18

```
Octree(...)
    create a new TreeNode object and assign it to
        trnRoot

    call ConstructTree(...) passing in trnRoot along
        with the three parameters passed to this
        constructor.  Decrement intTreeDepth by 1.

~Octree(...)
    clean up the dynamically allocated memory by
        traversing tree nodes until all nodes have been
        deallocated.

AddObject(...)
    calls the private recursive function,
        AddObject(...,...)

RemoveObject(...)
    the object's id is used as the key into umaContents
        to get the list of TreeNode's that this
        object is contained in

        for each TreeNode in the list
            search through this TreeNode's object list
                and remove the current object from it

    remove the object from umaContents

UpdateObject(...)
    the object's id is used as the key into umaContents
        to get the list of TreeNode's that this object
        is contained in

        for each TreeNode in the contained list
            check if the object is still contained
                within it or is intersecting it

                if it is then
                    do nothing
                else
                    remove the object from the TreeNode
```

continued on following page

Code 19. Continued

```
                        remove the TreeNode object from the
                            list in umaContents

                        call the private AddObject(..., ...)
                            function with the current object
                            and the current TreeNode's
                            parent

ConstructTree(...)
    assign vecMinCorner, vecMaxCorner, and vecCenter
        based off of the passed in values

    if intTreeNode is greater than or equal to 0
        for eight iterations
            create a new TreeNode object inside of the
                tnoCurrentNode object at
                trnChildNodes[currentIteration]

            calculate the newly created TreeNode object's
                minimum and maximum bounds

            call ConstructTree (
                        trnChildNodes[currentIteration],
                        intTreeDepth - 1,
                        minimumBound,
                        maximumBound
                       )
    else
        do nothing

AddObject(..., ...)
    for each element in the array trnChildNodes
        Does the passed in object fit completely within
        the this child tree node?

            if it does then call AddObject(..., ...)
                with the current child node

        otherwise do nothing

    if the passed in object, gobObjIn, did not fit within
        any of the elements in the array trnChildNodes
```

continued on following page

Code 19. Continued

```
        then call InsertObject(...) passing it gobObjIn
        and the current TreeNode.

InsertObject(...)
    if gobInObj is not already in trnNode's list of
        objects

        insert gobInObj into the list

        add trnNode to umaContents based on gobInObj's
            key

    else the object is already in this node's object
        list therefore do nothing
```

movement speed of an object is capped at a reasonable amount or have appropriate algorithms in place. If fast moving objects are necessary in terms of game play then here are a couple of suggestions:

- Encapsulate an object's previous position and next position with a bounding volume and perform collision checks with it.
- Iteratively check for collisions during one frame of movement by reducing the time step to counter fast movement

There are situations when the penetration depth of a collision event is needed for an appropriate response. The penetration depth is the smallest amount of movement that would move the two colliding objects out of their overlap. There is a variation of the GJK algorithm that is explained by Ganjugunte on getting the penetration depth in, "*A Survey on Techniques for Computing Penetration Depth*" (Ganjugunte, 2011). The last topic mentioned for further research, multithreading detection, is concerned with removing the intersection testing from the game loop so as to improve the game's overall performance. Having the collision system in its own thread has an obvious advantage of being able to perform intersection tests more frequently and not bottleneck the game loop. There are numerous complexities to consider when attempting to multithread any system, and research into the subject should be done prior to an attempt. An example research prior to building a multithreaded collision system can be found in Refenes, "*Sponsored Feature: Multi-Threading Goo!: A Programmer's Diary*" (Refenes, 2011).

CONCLUSION

Collision detection's performance in video games can be anywhere from the dastardly villain to the unsung hero. The end user attempts to relate himself to a video game by applying their real life experiences to those of the game. Unless there is a game mechanic that is supposed to allow objects to visually overlap the player expects objects to behave as they normally would in reality. Objects that collide in game should have a response that is triggered by an accurate detection system. A system that fails at detecting obvious overlap can result in a frustrating user experience.

The use of AABBs or radial volumes as the only detection method, allows for easy implementation and fast overlap testing. Using these two intersection testing methods as the first phase of a collision system makes for increased performance over OOB or GJK for detection testing. Object type filtering is the next way to improve performance. If the quantity of objects starts causing a performance bottleneck then spatial partitioning should be considered.

The easiest and fastest collision detection is performed using AABBs and radii tests. The precision of these two testing methods can be increased by using BVHs. BVHs can also be used with the next two intersection tests, preferably with an AABB check or radius check as the first overlap test. The OOB separating axis theorem method of intersection testing is the next system to use. This method has better performance than the GJK algorithm when dealing with collision volumes with a large number of vertices. If the quantity of vertices is small then the GJK algorithm is far superior to the OOB method.

Partitioning objects into dynamic and static sets is an inexpensive strategy when dealing with a relatively small set of objects. The sort and sweep algorithm is the next consideration for collision culling when performance is degrading due to an increasing quantity of collidable objects. The last method of culling the number of intersection tests resides in the quadtree and octree. The concepts and structures presented in this chapter should give a good starting point for designing an accurate collision detection system.

REFERENCES

Eberly, D. (2008). Intersection of Convex Objects: The Method of Separating Axes. Retrieved from www.geometrictools.com/Documentation/MethodOfSeparatingAxes.pdf

Eberly, D. (2008a). Boolean operations on intervals and axis-aligned rectangles. Retrieved from www.geometrictools.com/Documentation/BooleanIntervalRectangle.pdf

Eberly, D. (2008b). Dynamic Collision Detecting using Oriented Bounding Boxes. Retrieved from www.geometrictools.com/Documentation/DynamicCollisionDetection.pdf

Ericson, C. (2005). *Real-time collision detection: Christer Ericson.* Amsterdam

Jovanoski, D. (2008). Bachelor seminar: The Gilbert -Johnson -Keerthi (GJK) algorithm. Retrieved from reference.kfupm.edu.sa/content/b/a/bachelor_seminar__the_80305.pdf

Muratori, C. (n.d.). *Implementing GJK.* Retrieved from http://mollyrocket.com/849

Terdimann, P. (2007). Sweep-and-prune. Retrieved from www.codercorner.com/SAP.pdf

ADDITIONAL READING

Berg, M. D. (2000). *Computational geometry: Algorithms and applications* (2nd ed.). Berlin, Germany: Springer.

Bergen, G. J. (2004). *Collision detection in interactive 3D environments.* San Francisco, CA: Morgan Kaufman.

Bourg, D. M. (2002). *Physics for game developers.* Sebastopol, CA: O'Reilly Media.

DeLoura, M. A. (2000). *Game programming gems. Newton Center.* MA: Charles River Media.

DeLoura, M. A. (2001). *Game programming gems 2. Newton Center.* MA: Charles River Media.

Dunn, F., & Parberry, I. (2002). *3D math primer for graphics and game development.* Plano, TX: Wordware Pub.

Eberly, D. H. (2005). *3D game engine architecture: Engineering real-time applications with Wild Magic.* San Francisco, CA: Morgan Kaufmann Publishers.

Eberly, D. H. (2007). *3D game engine design a practical approach to real-time computer graphics* (2nd ed.). San Francisco, CA: Morgan Kaufmann Publishers.

Eberly, D. H. (2010). *Game physics* (2nd ed.). San Francisco, CA: Morgan Kaufmann Publishers.

Ericson, C. (2005). *Real-time collision detection: Christer Ericson*. San Francisco, CA: Morgan Kaufmann Publishers.

Ganjugunte, S. K. (n.d.). *A survey on techniques for computing penetration depth.* Retrieved from www.cs.duke.edu/courses/spring07/cps296.2/course_projects/shashi_proj.pdf

Gould, H., Tobochnik, J., & Christian, W. (2007). *An introduction to computer simulation methods: applications to physical systems* (3rd ed.). Reading, MA: Addison-Wesley.

Horn, R. A., & Johnson, C. R. (2010). *Matrix analysis* (23rd ed.). Cambridge, UK: Cambridge University Press.

Jurgen, R. K. (2007). *Object detection, collision warning & avoidance systems*. Warrendale, PA: SAE International.

Kim, Y. J. (n.d.). DEEP - Dual space-expansion for estimating penetration depth between convex polytopes. *Geometric Algorithms for Modeling, Motion, andAnimation - GAMMA UNC*. Retrieved from http://gamma.cs.unc.edu/DEEP/

Knuth, D. E. (1998). *The art of computer programming* (3rd ed.). Reading, MA: Addison-Wesley.

Lass, H. (1950). *Vector and tensor analysis*. New York, NY: McGraw-Hill.

Lay, D. C. (2006). *Linear algebra and its applications* (3rd ed.). Reading, MA: Addison-Wesley.

Millington, I. (2007). *Game physics engine development*. San Francisco, CA: Morgan Kaufmann.

Millington, I. (2010). *Game physics engine development: How to build a robust commercial-grade physics engine for your game* (2nd ed.). San Francisco, CA: Morgan Kaufmann.

Moller, T., Haines, E., & Hoffman, N. (2008). *Real-time rendering* (3rd ed.). Wellesley, MA: A.K. Peters. doi:10.1201/b10644

N Tutorial B - Broad-Phase Collision. (2011). *Welcome to metanet software inc*. Retrieved from http://www.metanetsoftware.com/technique/tutorialB.html

Palmer, G. (2005). *Physics for game programmers*. New York, NY: Apress.

Refenes, T. (2008). Multi-threading goo!: A programmer's diary. *Gamasutra - The Art & Business of Making Games*. Retrieved from http://www.gamasutra.com/view/feature/3782/sponsored_feature_multithreading_.php

Schneider, P. J., & Eberly, D. H. (2003). *Geometric tools for computer graphics*. San Francisco, CA: Morgan Kaufmann.

Standish, T. A. (1980). *Data structure techniques*. Reading, MA: Addison-Wesley.

Treglia, D. (2002). *Game programming GEMS 3. Newton Center*. MA: Charles River Media.

Tutorial, N. A - Collision Detection and Response. (2011). *Welcome to metanet software inc*. Retrieved from http://www.metanetsoftware.com/technique/tutorialA.html

Verth, J. M., & Bishop, L. M. (2004). *Essential mathematics for games and interactive applications: A programmers guide*. San Francisco, CA: Morgan Kaufmann Publishers.

KEY TERMS AND DEFINITIONS

Big O Notation: A way to categorize the way an algorithm will perform in terms of cycles based on the input size. This method of categorization explains the bounds of the algorithm in terms of best case, worst case and average case. This book chapter focused on the average case.

Binary Trees: Binary trees are a computer science data structure. They are used to store data in a structured way. It has a parent to child structuring where in which each parent has exactly two children or none at all. Each node in the binary tree structure is a binary tree.

Concave Shape: A shape is concave if any point on its surface cannot be reached by a straight line from all other points without the line leaving the internal area of the shape. Common examples of concave shapes are crescent moons and the torus.

Convex Shape: A shape is convex if any point along its surface can be reached by a straight line from every other point on the shape and the line never leaves the internal area of the shape. Common convex shapes are triangles, boxes and circles.

Culling: Culling is the process of reducing the quantity of a set based on a given criteria. For collision detection this usually means reducing the number of intersection tests by organizing the collidable objects in some way.

Set Theory: A branch of mathematics is dedicated to set theory. A set is a grouping of objects that share one or more characteristics. An example of a set is the integer. An integer can be any whole number from negative infinity to positive infinite. Each number in the set of integer is an integer.

Simplex: Simplex is a way to describe geometrical shapes in terms of their quantity of vertices. A 0-simplex is a point whereas a 1-simplex is a line. A triangle is a 2-simplex and so on.

Chapter 11
Collision Detection Using the GJK Algorithm

William N. Bittle
dyn4j.org, USA

ABSTRACT

GJK is a fast and elegant collision detection algorithm. Originally designed to determine the distance between two convex shapes, it has been adapted to collision detection, continuous collision detection, and ray casting. Its versatility, speed, and compactness have allowed GJK to be one of the top choices of collision detection algorithms in a number of fields.

INTRODUCTION

Collision detection is the process of detecting when two objects intersect, overlap, or collide. It's found in simulations, video games, robotics, and even basic GUI applications. The mouse, a standard user interface device, uses a simplified form of collision detection to achieve tasks like issuing a click command on a button or moving a window. The mouse's current position must be found inside the rectangle that forms the bounds of the button or window to perform the respective action. This type of collision detection is often referred to as hit-testing.

On the other hand, robotics, simulations, and video games have more challenging requirements. For applications like these it's essential to know when a player attempts to exit the playable area, when two objects occupy the same space, or when to avoid a robotic arm crashing through a wall. However, unlike the interaction between the mouse and GUI, objects within these environments can have any shape and can move and rotate freely. Early applications used simple geometric constructs, like circles and non-rotating axis-aligned rectangles to make detection easy. Detecting if two circles are colliding can be accomplished by checking if the distance between their centers is greater than the sum of their radii. In addition, detecting if two non-rotating axis-aligned rect-

DOI: 10.4018/978-1-4666-1634-9.ch011

angles are colliding can be done by comparing the extents, usually the top-left and bottom-right vertices, against each other with simple less-than greater-than comparisons. Yet, as technology has advanced, objects within the environment have become more complex, requiring the use of less specialized collision detection routines. One solution might be to use a separate collision detection algorithm for each shape type pair; circle-circle, circle-axis-aligned rectangle, and so on. This can be a good solution for applications that require only a few different shape types, but it can quickly become unmanageable as the number of shape types increase. An ideal solution would be able to handle any moving/rotating shape efficiently and accurately with one algorithm. The GJK algorithm is one such solution.

In addition to the increased complexity, the size of the environments themselves has increased. It's unlikely to find an environment that includes only a handful of objects. Likewise, it's difficult to find objects that are represented with only *one* simple shape. As such, collision detection software has evolved into a phased approach, usually including two or three phases, which attempt to reduce the number of collision tests that must be performed by expensive algorithms. The first phase is referred to as the *broad-phase*. The broad-phase performs the n x n collision tests, where n is the number of objects in the scene, using one simple geometric shape that encloses the entire object. This phase is designed to be extremely fast but at the same time conservative; meaning it should not miss any collisions, but may report false collisions. Sweep and Prune, BSP Trees, Uniform Grids, and Hierarchical Grids are some examples of broad-phase collision detection algorithms. The pairs found in the broad-phase are then passed to either an intermediate phase, called the *mid-phase*, or directly onto the *narrow-phase*. The mid-phase is used to reduce the number of collision tests for objects that are represented by a combination of simple shapes, often using the same algorithms as the broad-phase or variations thereof. Finally, the narrow-phase uses the exact geometry of the objects to perform an exact collision test. The narrow-phase is typically the most expensive algorithm. GJK and the Separating Axis Theorem are some examples of narrow-phase collision detection algorithms. The phased approach is not required but can improve performance significantly. This chapter focuses solely on the GJK algorithm.

Beginning with some background information we will cover the basic principles of the algorithm along with common pitfalls using concrete examples and pseudo code. After which, we will cover a simplified version of the original method for obtaining the distance between two convex shapes and the closest points. Next, we will cover a supplementary algorithm to obtain collision information useful in collision resolution. Finally, we will touch on how the algorithm has been applied to continuous collision detection and ray casting.

BACKGROUND

To begin detecting collisions between objects within an environment, they must have a representation of their shape. The most common shapes used are circles, rectangles, and polygons because of their simplicity. In three dimensions common shapes include spheres, rectangular boxes, and cylinders. An object, like a chair, can be represented by one or more of these simple shapes depending on the accuracy required. The closer the simple shapes fit to the original object the higher the accuracy but the lower the performance. We could use a rectangular box that encloses the entirety of the chair or we could create multiple rectangular boxes that enclose each leg, arm, and other features. Either way, the representation of these simple shapes is required for a collision detection algorithm. For the remainder of the chapter we will focus on two types of shapes; circles and

Figure 1. a) A convex shape with a sample line segment test. b) A non-convex shape with a line segment test that passes. c) A non-convex shape with a line segment test that does not pass.

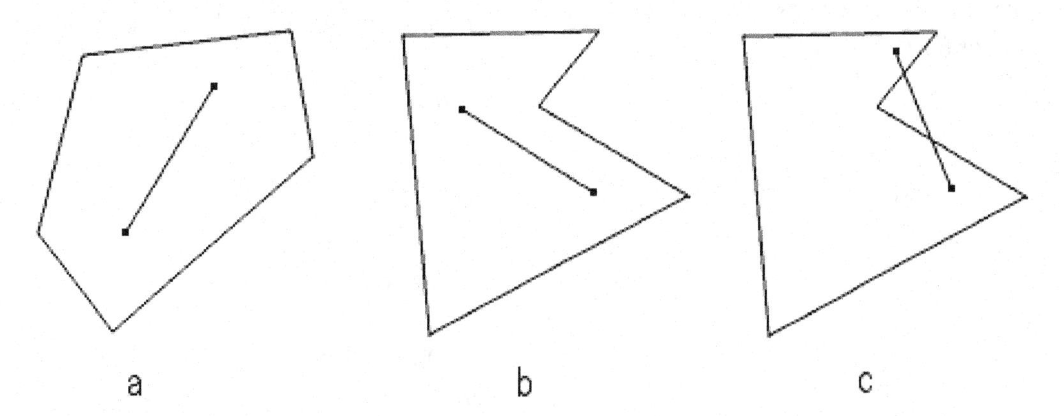

convex polygons. Our representation of a circle shape will be a radius and center point and our representation of a convex polygon will be a list or array of vertices.

To fully understand the GJK algorithm we must cover the basic principles on which the algorithm is based. The first principle is convexity, as it applies to shapes, with the Minkowski Sum being the second.

Convexity

A shape is convex "if it contains all the line segments connecting any pair of its points" (Weisstein para. 1). In other words, if you can draw a line segment between any two points within the shape and the entire line segment is *not* contained within the shape, then the shape is *not* convex.

Figure 1a shows a test against a shape that is convex. Any line segment created from the points inside the shape will always be completely contained within the shape. Figure 1b is a non-convex shape but we see that this shape passes the first test that we perform. However, if we choose a different set of points, as shown in Figure 1c, we see that the shape does not pass. Since we were able to find one line segment that that is not fully contained within the shape, then the shape is not convex.

Performing this test works great on paper, but what if we wanted our application to do this? Testing using this method isn't practical because there exists an infinite number of combinations of points that our application could choose from. Instead we will use a different method involving the vector cross product.

The cross product of two vectors returns another vector, more precisely a *pseudo vector*, which is perpendicular to the other two. In two dimensions the cross product yields a signed scalar. The cross product can produce two different perpendicular vectors depending on the "handedness" of the coordinate system. Figure 2 shows the results of the cross product of two vectors; the first from a right-handed system and the second from a left-handed system. We will use a right-handed coordinate system for the remainder of the chapter.

Using the result of the cross product of each adjacent edge we can determine if the shape is convex. If the direction or sign of the cross product of the current edge is different than the previous cross product then the shape is not convex (Bourke, 1998).

This implementation could contain a number of improvements but has been left as follows for simplicity and understanding. A x B represents the cross product of the vectors A and B.

Figure 2. The cross product of two vectors, A and B, for a right handed coordinate system and the cross product of two vectors, A and B, for a left handed coordinate system.

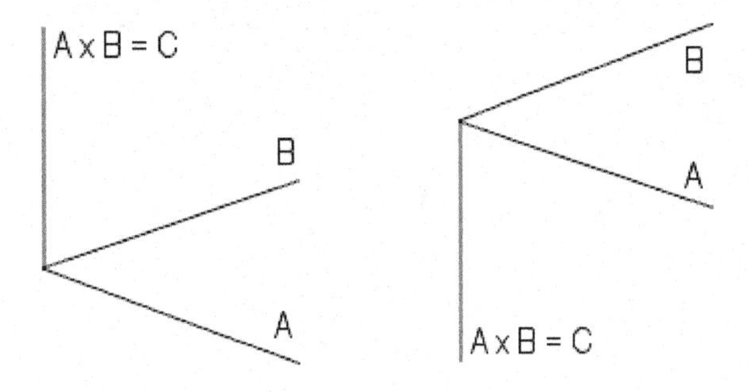

Listing 1. The isConvex function

```
isConvex(shape)
  prev = 0
  for each point in shape
    prev = point.prev
    next = point.next
    A = point - prev
    B = next - point
    result = A x B
    if first iteration
      prev = result
    else
      if result * prev < 0
        return false
      end if
    end if
  end for
  return true
```

Using Figure 3a we can test our method on a convex shape:

Listing 2. The isConvex function applied to Figure 3a

```
Iteration 1: result = +46
Iteration 2: result = +33
Iteration 3: result = +37
Iteration 4: result = +50
```

As we can see, all the cross products of the adjacent edges are the same sign, therefore the shape is convex. If we do the same for Figure 3b:

Listing 3. The isConvex function applied to Figure 3b

```
Iteration 1: +70
Iteration 2: +30
Iteration 3: -11
```

On the third iteration we see that one of the cross products yielded a negative, so we conclude that the shape is not convex. Comparing the result from the current cross product with the last allows the code to work for clockwise or anti-clockwise point ordering (winding), however, we must skip the check if it's the first iteration so that the previous result can be set. The cross product can also return zero, indicating the adjacent edge points are collinear. In this case *result * prev* will yield zero and the function will continue normally. If we decide that all shapes within our application have anti-clockwise winding then the check can be changed from *result * prev < 0* to *result < 0* rather than performing a multiply and storing the previous result. Likewise if we use clockwise winding then the check can be changed from *result * prev < 0* to *result > 0.*

This algorithm will easily handle any polygonal shape without holes. Circles and triangles will always be convex and therefore do not need to be tested.

Convex shapes are key to an efficient collision detection algorithm. *The GJK algorithm requires that all shapes are convex.* The algorithm may fail on a shape like Figure 3b returning a false positive or a false negative. Non-convex shapes are common however, so we must have a way to handle them. To allow the GJK algorithm to work with non-convex shapes, we must decompose the convex shape into multiple convex shapes and test each one. This is called convex decomposition. Figure 3c shows an example decomposition of the

Figure 3. a) A sample convex shape. b) A sample non-convex shape. c) Shape b decomposed into two convex shapes

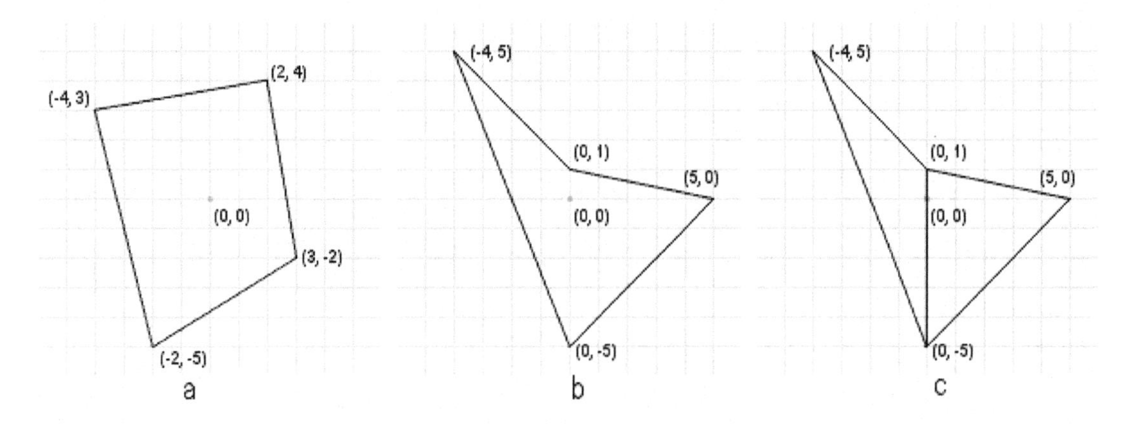

shape in 3b. Convex decomposition is a topic in and of itself and will not be covered in this chapter.

Minkowski Sum

The Minkowski Sum is the backbone of the GJK algorithm. Also called Minkowski Addition, the Minkowski Sum is an operation that can be performed on two shapes resulting in a new shape. The new shape is computed by adding *each* point in one shape to *all* the points in the other shape. This operation can be performed on both convex and non-convex shapes. When performed on two convex shapes, the resulting shape is guaranteed to be convex (Cameron, 1997).

Though each shape has an infinite number of points in its interior, using convex polygons allows us to use the *vertices* of the polygons and take the convex hull of the resulting point set instead. For example, Figure 4a and 4b show two convex shapes and their Minkowski Sum. Notice that only the polygon vertices were added together and that a convex hull operation was performed to give the final shape.

A convex hull is simply a convex shape that encloses all the points of a point set. As we see in Figure 4b, all the points in the point set are contained within the shape, some of which make up the vertices of the hull.

The Minkowski Sum uses the addition operator to create the resulting point set. What would happen if we changed the operation from addition to subtraction; i.e. subtract *each* point in one shape from *all* the points in the other shape? We'll call using the difference operator the Minkowski Difference for clarity. Figure 5b shows the Minkowski Difference of the *non-colliding* shapes in Figure 5a. Figures 5c and 5d show the same shapes *in collision* and their Minkowski Difference.

Notice in Figure 5d that the origin is contained within the hull and in Figure 5b the origin is *not* contained in the hull. If we performed this operation on a number of different convex shapes, both colliding and non-colliding, an extremely useful property is revealed: *if the Minkowski Difference contains the origin, then the two shapes are colliding (Ericson, 2004).*

THE GJK ALGORITHM

In the last section we revealed that if two convex shapes are colliding, then the resulting Minkowski Difference will contain the origin. This is the basis of the GJK algorithm. A naïve approach would

Figure 4. a) Two example convex shapes. b) The Minkowski Sum of the two convex shapes in 4a

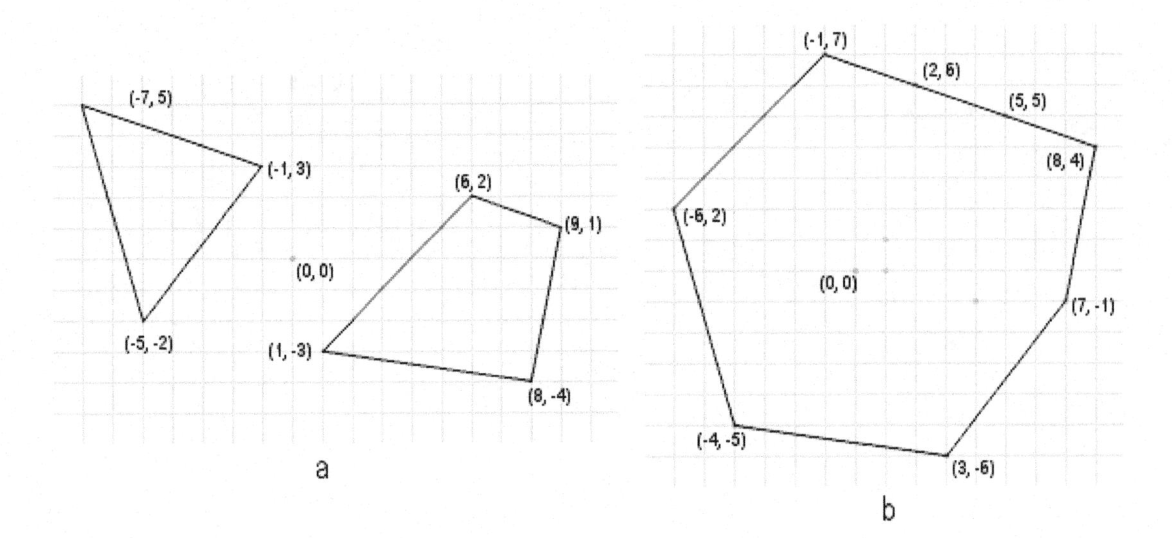

be to compute the Minkowski Difference directly using the polygon vertices, take the convex hull of the resulting point set, and finally perform a point in polygon test to see if the Minkowski Difference contains the origin.

However, there are some problems with this approach. First, computing the Minkowski Difference directly requires *m* x *n* number of vector subtractions; where *m* and *n* are the number of vertices in the two shapes. Secondly, the previous examples did not include curved shapes. Circles and other curved shapes prevent us from calculating the Minkowski Difference directly because there are an infinite number of points that make up curved edges.

Instead of calculating the Minkowski Difference directly, GJK uses a different approach. Recall that the only criterion to find whether two convex shapes are colliding is that the Minkowski Difference of the two shapes contains the origin. *If we can create a convex shape within the Minkowski Difference that contains the origin, then we can conclude that the Minkowski Difference contains the origin.* The convex shape that we create inside the Minkowski Difference is called the *simplex*. The simplex is convex, so naturally the simplest convex shapes are used; a triangle for two dimensions and a tetrahedron for three dimensions. Consequently, the simplex will never have more than three vertices in two dimensions and no more than four vertices in three dimensions. Figure 6 illustrates an example simplex inside the Minkowski Difference from Figure 5d.

The GJK algorithm creates and modifies the simplex iteratively. Each iteration will attempt to modify the simplex so that it encloses the origin. If the algorithm detects that the simplex cannot enclose the origin, then it returns that the shapes are not colliding.

Listing 4. The GJK collision detection algorithm

```
gjk(A, B)
  d = some initial direction vector
  a = support(A, B, d)
  simplex.add(a)
  while (the last point we added
  is past the   origin in the
  direction of d)
    if checkSimplex(simplex, d)
      return true
    else
```

Figure 5. a) The two shapes from Figure 4a. b) The Minkowski Difference of the shapes in 5a. c) The shapes in 5a moved into collision. d) The Minkowski Difference of the shapes in 5c

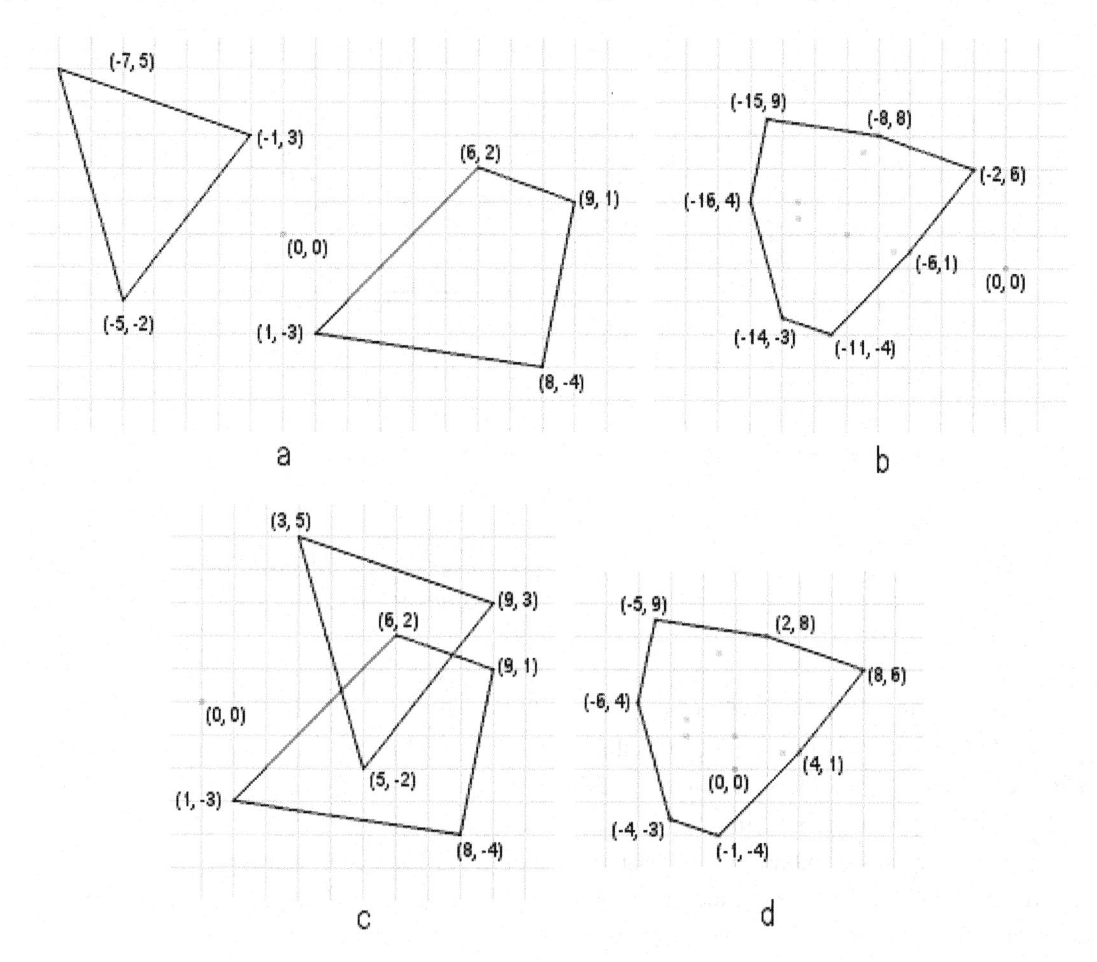


```
      p = support(A, B, d)
      simplex.add(p)
    end if
  end while
  return false
```

The algorithm, as described in this section has been referred to as the GJK separating-axis algorithm in literature because of the lack of the distance sub-algorithm and the use of a separating-axis termination condition (Bergen, 1999, pg. 10).

Support Function

To understand the algorithm above we must understand the *support* function. It was stated earlier that computing the Minkowski Difference directly is not a viable solution to the problem and that our goal instead should be to create a simplex within the Minkowski Difference. *The support function is used to obtain points on the edge of the Minkowski Difference to create the simplex.* It can be defined simply as: the point farthest in shape A along d minus the point farthest in shape B along –d (Bergen, 2004).

Figure 6. The Minkowski Difference from Figure 5d with an example simplex

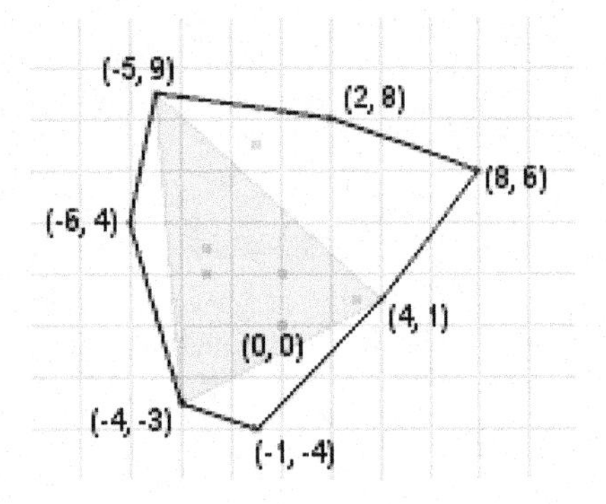

Listing 5. The support function

```
support(A, B, d)
  a = A.farthest(d)
  b = B.farthest(-d)
  return a - b
```

The *support* function is performing the Minkowski Difference, but only on the points that are of interest. Since GJK requires that both shapes are convex, thereby making the Minkowski Difference convex, we can guarantee that the points generated this way will always lie on the edge of the Minkowski Difference. This is important for keeping the simplex contained within the Minkowski Difference and for determining non-collision cases.

The *farthest* function is dependent on the shape. A circle and polygon will have different implementations. This allows for good object oriented design. A new shape can be supported by implementing the *farthest* function only. For better understanding, an implementation for a circle shape and polygon shape are provided (see Listings 6 and 7).

Listing 6. The farthest function for a circle shape

```
farthest(d)
  c = the circle center
  r = the radius
  d = normalize(d)
  d = d * r
  return c + r
```

Figure 7. All the points of the shape projected onto a vector

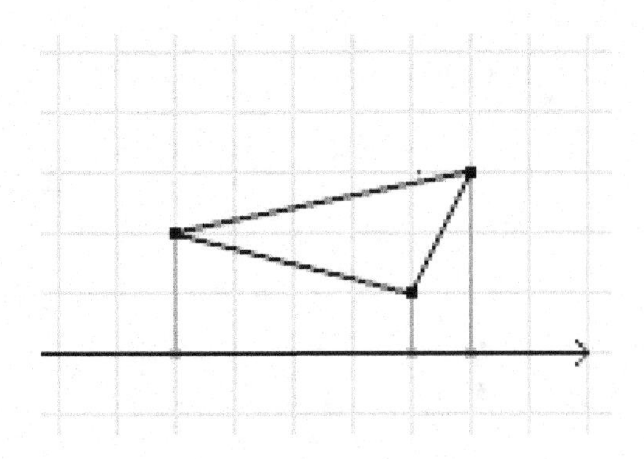

Listing 7. The farthest function for a convex polygon shape

```
farthest(d)
  l = the number of vertices
  p = the farthest point;
  initially null
  proj = the largest projection;
  initially -(largest value)
  for each vertex, v, in the polygon
    test = dot(v, d)
    if (test > proj)
      proj = test
      p = v
    end if
  end for
  return p
```

For a circle we extend the center point along *d*, *r* distance, to obtain the farthest point. For a polygon we find the vertex of the polygon whose projection along *d* is the largest (see Figure 7). See Bergen 1999 pg. 7-8 for a list of methods for common three dimensional shapes.

The while Condition

The while condition from listing 4 checks if the origin could be contained within the Minkowski Difference. This is achieved by projecting the last point added to the simplex onto the direction vector and verifying that it's non-negative.

Listing 8. The while loop condition for listing 4

```
dot(simplex.last, d) > 0
```

Failing this test means that there is no possible way to enclose the origin. Figure 8 shows an example of a failure case on the second iteration of the GJK algorithm. The point *b* is not passed the origin in the direction of *d*. If the new point passes the origin then it's possible that we could enclose the origin, but if the new point does not pass the origin, then it's guaranteed that we cannot enclose the origin (Muratori 2006). This is a result of the Minkowski Difference being convex and the fact that the new point is the farthest point in the Minkowski Difference along *d*.

Figure 8. An example failure case for the while loop condition

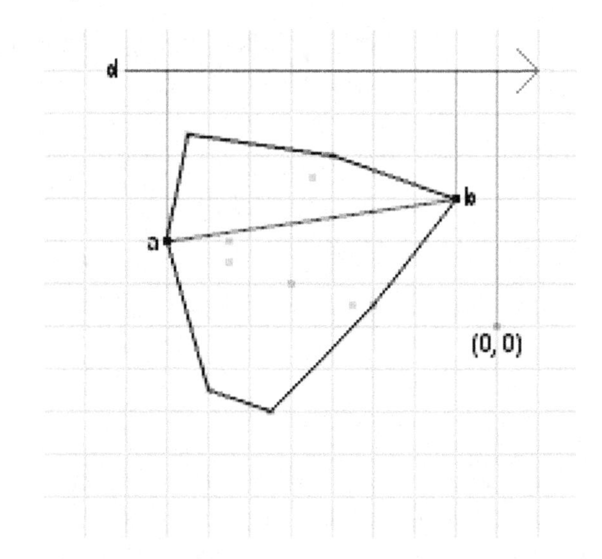

The checkSimplex Function

The next function we must understand is the *checkSimplex* function. This function performs double duty, both returning true if the simplex contains the origin and changing the direction, *d*, to a new direction if the simplex does not contain the origin.

The *checkSimplex* function must be able to handle all the simplex cases: a point, a line segment, and a triangle. This is a result of the iterative nature of the algorithm, with the simplex starting as a single point. Adding another point creates a line segment, and adding a third point creates a triangle. Once the triangle case has been reached the simplex will be a triangle from thereon. For three dimensions the check simplex function must be able to handle a tetrahedron case in addition to the other three cases.

Finding the new search direction and determining whether the origin is contained in the simplex can be done together using the concept of Voronoi regions (Muratori 2006). A Voronoi region, in relation to a shape, is a subdivision of space in which all points contained in a region are closer to one feature than any other feature. The feature can be a vertex, edge or face. Figure 9 shows a line segment and triangle in two dimensions with the respective Voronoi regions. In Figure 9b, region 7 could have been further subdivided; however, this isn't necessary for our purposes as we will see later.

From the pseudo code in listing 4, the first iteration will pass the *point* simplex case to the *checkSimplex* function. A point cannot contain the origin so we only need to return a new search direction. For this case we simply return *-d*.

Listing 9. The checkSimplex function's point case

```
d = -d
return false
```

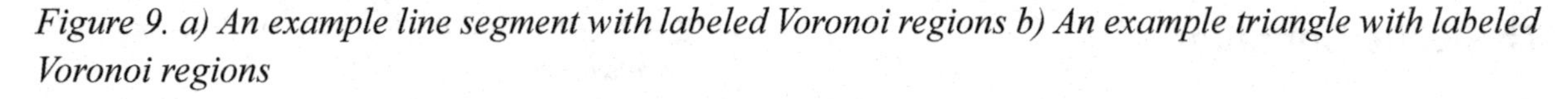

Figure 9. a) An example line segment with labeled Voronoi regions b) An example triangle with labeled Voronoi regions

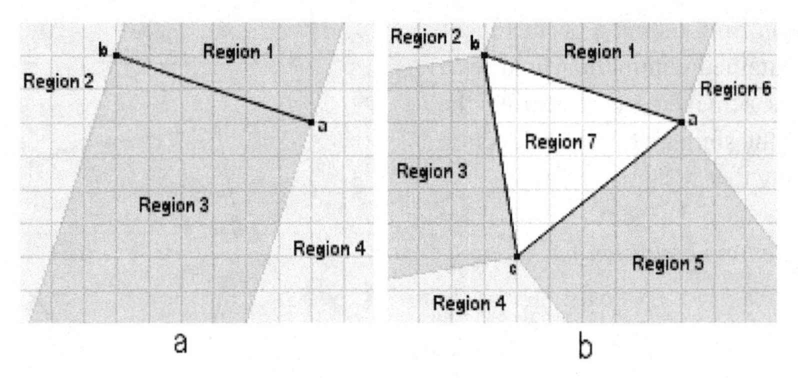

The second iteration passes a *line segment* to the *checkSimplex* function. Again, a line segment cannot contain the origin, so we only need to return a new search direction. The new search direction should always be a normal of an edge or face of the simplex in the direction of the origin. Figure 9a shows the Voronoi regions that could contain the origin. The origin cannot be past point *a* since it's the farthest point in that direction and it's past the origin (determined by the while loop condition) and the origin cannot be past point *b* since it's the farthest point in the opposite direction and it's also past the origin (Muratori 2006). Therefore we can eliminate regions 2 and 4 and easily obtain a new search direction using the vector triple product:

Listing 10. The checkSimplex function's line segment case

```
ab = b - a
ao = o - a (or just -a)
d = ab x ao x ab
return false
```

Listing 10 obtains the vector from point *a* to point *b*. Then it performs the cross product with *ab* and *ao* to obtain a vector perpendicular to *ab* and *ao*. Lastly, it performs the cross product between the last result and *ab* again to obtain the normal of the edge *ab* in the direction of the origin.

All other iterations will pass a *triangle* to the *checkSimplex* function. Figure 9b shows the Voronoi regions of a triangle. Since we know that point *c* was the last point added to the simplex we can eliminate regions in which the origin cannot be just like we did in the line segment case. Regions 2, 4, and 6 cannot contain the origin because of the while loop condition. Region 1 can also be eliminated because the origin was found to be in the opposite direction as shown by the new point *c*. This leaves only three regions to test; 3, 5, and 7 (Muratori, 2006). See Figure 9b for an illustration.

We can find the region that contains the origin by using the triple product, and then use the result of the triple product as the new search direction. First we obtain the normal vectors of the simplex edges *cb* and *ca* then we can use these to test where the origin is located.

Listing 11. The checkSimplex function's triangle case

```
cb = b - c
ca = a - c
cbNormal = ca x cb x cb
caNormal = cb x ca x ca
co = o - c (or just -c)
if dot(cbNormal, co) >= 0
  remove point a from the simplex
  d = cbNormal
```

```
  return false
else
  if dot(caNormal, co) >= 0
    remove point b from the simplex
    d = caNormal
    return false
  else
    return true
  end if
end if
```

The first check determines if the origin lies in region 3, and if so removes point *a*, since we will replace it with a new point later, and sets the new direction to edge *cb*'s normal. If this check fails, then we know that the origin is either in region 5 or region 7. Next, we check if the origin lies in region 5, and if so we remove point *b* and set the new search direction to edge *ca*'s normal. If both of these checks fail then we can guarantee that the origin lies within the simplex and therefore in the Minkowski Difference, finally concluding that the two shapes are colliding.

Listing 12. The checkSimplex function combined from listings 9, 10, and 11

```
checkSimplex(simplex, d)
  if simplex is a point
    d = -d
    return false
  else if simplex is a line segment
    ab = b - a
    ao = o - a (or just -a)
    d = ab x ao x ab
    return false
  else
    cb = b - c
    ca = a - c
    cbNormal = ca x cb x cb
    caNormal = cb x ca x ca
    co = o - c (or just -c)
    if dot(cbNormal, co) >= 0
      remove point a from the simplex
      d = cbNormal
      return false
    else
      if dot(caNormal, co) >= 0
        remove point b from
        the simplex
        d = caNormal
        return false
      else
        return true
      end if
    end if
end if
```

Each iteration of the triangle case attempts to find the origin relative to the simplex, remove a point, obtain a new search direction and then add another point and start the process over. *Storing the simplex points in the order in which the points were added allows listing 12 to be simple and straightforward.* If they were not stored this way we wouldn't be able to identify the Voronoi regions that have already been tested and would have to test all the regions every iteration (Muratori, 2006).

Example Collision Case

The first example consists of two convex shapes, A and B, in collision (see Figure 10a). We will use the previous pseudo code listings to determine whether these two shapes are colliding.

Listing 13. The vertices of shape A and B

```
A = {(0, 2), (5, 4), (1, 6)}
B = {(5, 2), (8, 3), (7, 5), (4, 6)}
```

For a visual aid see the Minkowski Difference in Figure 10b.

Following listing 4 we must begin by choosing an initial direction. We'll use the vector from the center of A to the center of B.

Figure 10. a) An example collision between two convex shapes. b) The Minkowski Difference of the two convex shapes in 10a. c) The simplex inside the Minkowski Difference before entering the while loop. d) The simplex inside the Minkowski Difference during the second iteration along with the new direction. e) The simplex inside the Minkowski Difference during the third iteration along with the new direction (point b gets removed in this iteration). f) The simplex inside the Minkowski Difference before the fourth iteration

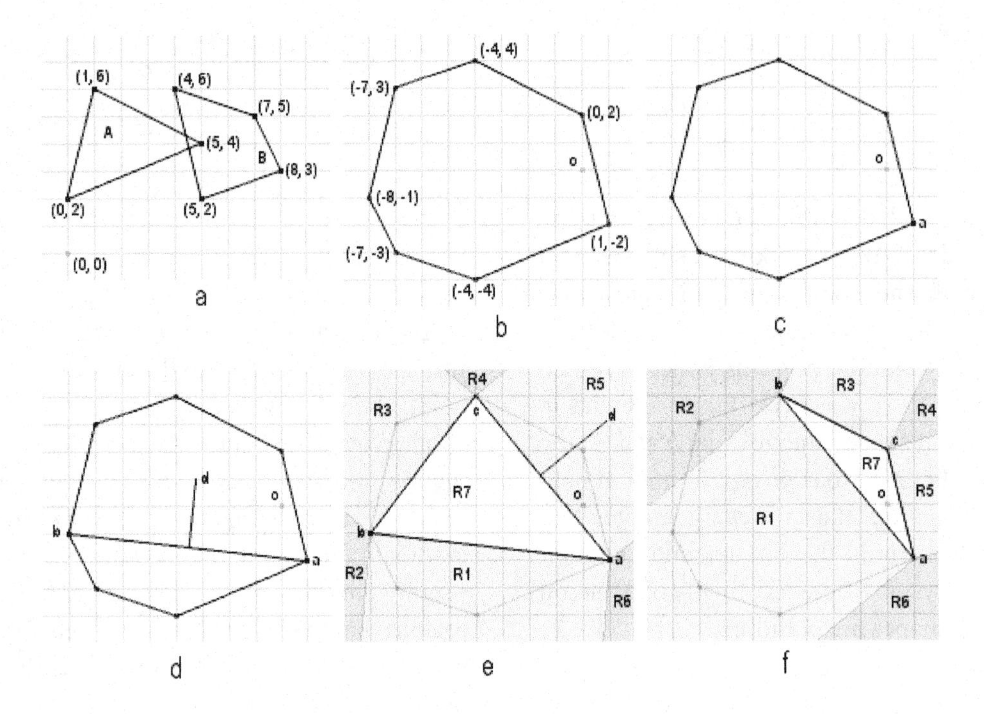

Listing 14. Computing the initial direction

```
A_center = (2, 4)
B_center = (6, 4)
d = (6, 4) - (2, 4) = (4, 0)
```

Then we obtain our first support point using listings 5 and 7 and add it to the simplex (Listing 15).

See Figure 10c for an illustration of the simplex before entering the while loop (it's simply point *a*). Then we perform the while loop check to see if the point we just added to the simplex is past the origin along *d*. (Listing 16)

Since the dot product of the last point added along *d* is greater than zero we can continue to the *checkSimplex* function. Since this is the single point case of the *checkSimplex* function, it returns false and a direction of *–d*. Since the *checkSimplex* function returned false we move to the else block and obtain another support point. (Listing 17)

Then we check if the last point added is past the origin. (Listing 18)

The dot product passes so we move on to the *checkSimplex* function. This time the simplex is a line segment. Therefore we follow listing 10 to obtain the new direction and return false since a line segment cannot contain the origin. (Listing 19)

See Figure 10d for an illustration of the simplex during the second iteration after obtaining a new direction from the *checkSimplex* function.

Listing 15. Using the support function we obtain our first simplex point

```
dot((0, 2), (4, 0)) = 0 * 4 + 2 * 0 = 0
dot((5, 4), (4, 0)) = 5 * 4 + 4 * 0 = 20
dot((1, 6), (4, 0)) = 1 * 4 + 6 * 0 = 4
A_farthest(d) = (5, 4)
dot((5, 2), (-4, 0)) = 5 * -4 + 2 * 0 = -20
dot((8, 3), (-4, 0)) = 8 * -4 + 3 * 0 = -32
dot((7, 5), (-4, 0)) = 7 * -4 + 5 * 0 = -28
dot((4, 6), (-4, 0)) = 4 * -4 + 6 * 0 = -16
B_farthest(-d) = (4, 6)
MD_point = (5, 4) - (4, 6) = (1, -2)
Simplex = {(1, -2)}
```

Listing 16. Making sure the last point added is past the origin along d

```
dot((1, -2), (4, 0)) = 1 * 4 + -2 * 0 = 4
```

Listing 17. Using the support function to obtain the second simplex point

```
-d = (-4, 0)
dot((0, 2), (-4, 0)) = 0 * -4 + 2 * 0 = 0
dot((5, 4), (-4, 0)) = 5 * -4 + 4 * 0 = -20
dot((1, 6), (-4, 0)) = 1 * -4 + 6 * 0 = -4
A_farthest(d) = (0, 2)
dot((5, 2), (4, 0)) = 5 * 4 + 2 * 0 = 20
dot((8, 3), (4, 0)) = 8 * 4 + 3 * 0 = 32
dot((7, 5), (4, 0)) = 7 * 4 + 5 * 0 = 28
dot((4, 6), (4, 0)) = 4 * 4 + 6 * 0 = 16
B_farthest(-d) = (8, 3)
MD_point = (0, 2) - (8, 3) = (-8, -1)
Simplex = {(1, -2), (-8, -1)}
```

Listing 18. Making sure the last point added is past the origin along d

```
dot((-8, -1), (-4, 0)) = -8 * -4 + -1 * 0 = 32
```

Listing 19. Using the line segment case to find the next direction

```
ab = (-8, -1) - (1, -2) = (-9, 1)
ao = -a = (-1, 2)
  (ab x ao) = -9 * 2 - 1 * -1 = -18 + 1 = -17
  (ab x ao) x ab = (-1 * -17 * 1, -17 * -9) = (1, 9)
d = (1, 9)
```

Listing 20. Using the support function to obtain the third simplex point

```
dot((0, 2), (1, 9)) = 0 * 1 + 2 * 9 = 18
dot((5, 4), (1, 9)) = 5 * 1 + 4 * 9 = 41
dot((1, 6), (1, 9)) = 1 * 1 + 6 * 9 = 55
A_farthest(d) = (1, 6)
dot((5, 2), (-1, -9)) = 5 * -1 + 2 * -9 = -23
dot((8, 3), (-1, -9)) = 8 * -1 + 3 * -9 = -35
dot((7, 5), (-1, -9)) = 7 * -1 + 5 * -9 = -52
dot((4, 6), (-1, -9)) = 4 * -1 + 6 * -9 = -58
B_farthest(-d) = (5, 2)
MD_point = (1, 6) - (5, 2) = (-4, 4)
Simplex = {(1, -2), (-8, -1), (-4, 4)}
```

Listing 21. Making sure the last point added is past the origin along d

```
dot((-4, 4), (1, 9)) = -4 * 1 + 4 * 9 = 32
```

Since the line segment case also returns false, we enter the else block and are tasked to find another support point. (Listing 20)

To remain in the while loop the last point must be past the origin along *d*. (Listing 21)

Next we check if the origin is contained within the simplex. In this iteration the simplex is a triangle so we use listing 11 to test if the origin is contained within the simplex and to find the new direction. (Listing 22)

As we can see from the first if statement (which uses *cbNormal*), the origin is not in region 3. Then we move to the else block and to another if statement. The result of this if statement tells us that the origin is in region 5, therefore we remove point *b* and let *caNormal* be the new direction. See Figure 10e for an illustration of the simplex inside the Minkowski Difference during the third iteration. Point *b* is left in the figure for clarity.

Since we didn't enclose the origin in the *checkSimplex* function, we enter the else block where we perform another support function call. (Listing 23)

Remember that the ordering of the points allows the *checkSimplex* function to be as simple as possible. *The points should be stored in the order in which they are added (Muratori 2006).*

Now that we have added another point we must check if we should remain in the while loop. (Listing 24)

Listing 22. Using listing 11 to determine if the origin is in the simplex and to obtain a new direction

```
cb = (-8, -1) - (-4, 4) = (-4, -5)
ca = (1, -2) - (-4, 4) = (5, -6)
  (ca x cb) = (5, -6) x (-4, -5) = 5 * -5 - -6 * -4 = -49
  (ca x cb) x cb = (-1 * -5 * -49, -4 * -49) = (-5, 4)
cbNormal = (-5, 4)
  (cb x ca) = (-4, -5) x (5, -6) = -4 * -6 - -5 * 5 = 49
  (cb x ca) x ca = (-1 * -6 * 49, 5 * 49) = (6, 5)
caNormal = (6, 5)
co = -c = (4, -4)
dot(cbNormal, co) = -5 * 4 + 4 * -4 = -36
dot(caNormal, co) = 6 * 4 + 5 * -4 = 4
```

Listing 23. Using the support function to obtain a new third simplex point

```
dot((0, 2), (6, 5)) = 0 * 6 + 2 * 5 = 10
dot((5, 4), (6, 5)) = 5 * 6 + 4 * 5 = 50
dot((1, 6), (6, 5)) = 1 * 6 + 6 * 5 = 36
A_farthest(d) = (5, 4)
dot((5, 2), (-6, -5)) = 5 * -6 + 2 * -5 = -40
dot((8, 3), (-6, -5)) = 8 * -6 + 3 * -5 = -63
dot((7, 5), (-6, -5)) = 7 * -6 + 5 * -5 = -67
dot((4, 6), (-6, -5)) = 4 * -6 + 6 * -5 = -54
B_farthest(-d) = (5, 2)
MD_point = (5, 4) - (5, 2) = (0, 2)
Simplex = {(1, -2), (-4, 4), (0, 2)}
```

Listing 24. Making sure the last point added is past the origin along d

```
dot((0, 2), (6, 5)) = 0 * 6 + 2 * 5 = 10
```

See Figure 10f for an illustration of the simplex at the beginning of the fourth iteration.

Since the new point is past the origin along *d* we continue to the *checkSimplex* function where again we have a triangle simplex. Using listing 11 we attempt to find whether the origin is inside the simplex and find the new direction if it's not. (Listing 25)

The first *if* statement tells us the origin is not in region 3 and the second *if* statement tells us that the origin is not in region 5, therefore the origin must be in region 7. Since the simplex contains the origin, the Minkowski Difference contains the origin, from which we can conclude that the two convex shapes A and B are colliding.

Example Non-Collision Case

The next example demonstrates the algorithm in a non-collision case. As was done in the

Listing 25. Using listing 11 to determine if the simplex contains the origin

```
cb = (-4, 4) - (0, 2) = (-4, 2)
ca = (1, -2) - (0, 2) = (1, -4)
  (ca x cb) = (1, -4) x (-4, 2) = 1 * 2 - -4 * -4 = -14
  (ca x cb) x cb = (-1 * 2 * -14, -4 * -14) = (2, 4)
cbNormal = (2, 4)
  (cb x ca) = (-4, 2) x (1, -4) = -4 * -4 - 2 * 1 = 14
  (cb x ca) x ca = (-1 * -4 * 14, 1 * 14) = (4, 1)
caNormal = (4, 1)
co = -c = (0, -2)
dot(cbNormal, co) = 2 * 0 + 4 * -2 = -8
dot(caNormal, co) = 4 * 0 + 1 * -2 = -2
```

previous examples, the pseudo code listings provided before will be used to determine whether the two shapes, A and B from Figure 11a are colliding.

Listing 26. The vertices of shape A and B

```
A = {(0, 2), (5, 4), (1, 6)}
B = {(6, 2), (9, 3), (8, 5), (5, 6)}
```

For a visual aid refer to Figure 11b.

Following listing 4 we must begin by choosing an initial direction. Again, we'll use the vector from the center of A to the center of B.

Listing 27. Computing the initial direction

A_{center} = (2, 4)
B_{center} = (7, 4)
d = (7, 4) - (2, 4) = (5, 0)

Then we obtain our first support point using listings 5 and 7 and add it to the simplex (see Listing 28).

Then we perform the while loop check to see if the point we just added to the simplex is past the origin along *d*. (Listing 29)

Since the dot product is not greater than zero the check fails. The algorithm exits the loop and returns false. In this case we were able to detect that the two shapes are not colliding nearly immediately. The fact that the loop continually checks for non-intersection allows the GJK collision detection algorithm to exit early.

Robustness

The GJK algorithm is easy to understand and, when implemented correctly, is fast and robust. However, there are a few pitfalls to consider during implementation.

A touching collision is where two convex shapes are touching rather than overlapping. A decision must be made as to whether this is a collision or not. A touching collision creates a Minkowski Difference whose edge passes through the origin. The pseudo code from listing 8 uses a greater than sign to verify that each point added to the simplex is past the origin. This means that if we obtain a new point from the *support* function that is on the edge that passes through the origin, the dot product will yield zero (or close to it), returning that there is no collision. To allow touching to be considered a collision we need to make a few modifications to our pseudo code. First, the while loop condition must be modified to allow for zero by using a greater than or equal sign instead of

Figure 11. a) An example of two shapes not colliding b) The Minkowski Difference of the two convex shapes in 11a

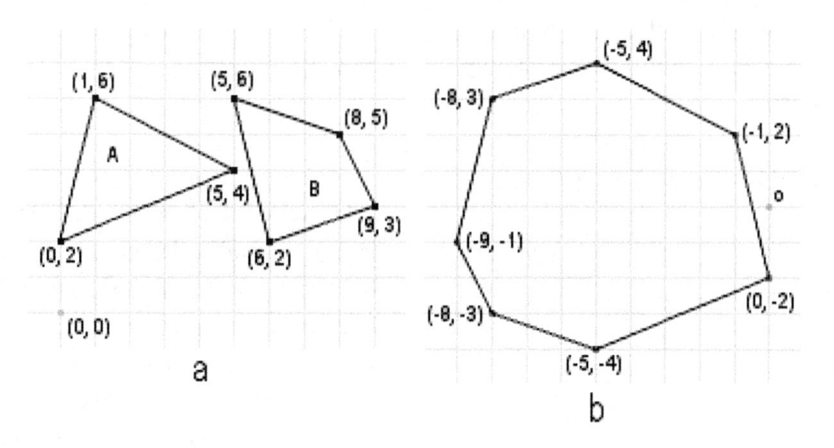

Listing 28. Using the support function to obtain the first simplex point

```
dot((0, 2), (5, 0)) = 0 * 5 + 2 * 0 = 0
dot((5, 4), (5, 0)) = 5 * 5 + 4 * 0 = 25
dot((1, 6), (5, 0)) = 1 * 5 + 6 * 0 = 5
A_farthest(d) = (5, 4)
dot((6, 2), (-5, 0)) = 6 * -5 + 2 * 0 = -30
dot((9, 3), (-5, 0)) = 9 * -5 + 3 * 0 = -45
dot((8, 5), (-5, 0)) = 8 * -5 + 5 * 0 = -40
dot((5, 6), (-5, 0)) = 5 * -5 + 6 * 0 = -25
B_farthest(-d) = (5, 6)
MD_point = (5, 4) - (5, 6) = (0, -2)
Simplex = {(0, -2)}
```

Listing 29. Making sure the last point added to the simplex is past the origin along d

```
dot((0, -2), (5, 0)) = 0 * 5 + -2 * 0 = 0
```

a greater than sign. Next, the line segment case of the *checkSimplex* function must be modified when the triple product returns zero (see listing 30). This indicates that the origin is on the line segment simplex. Finally, the triangle case of the *checkSimplex* function must be modified. Removing the equals from the less than or equals signs in listing 11/12 allows these cases to be considered a collision.

In touching contact cases the origin can lie on the edge of the Minkowski Difference, but the origin can also lie on an edge of the simplex within the Minkowski Difference. This is a problem for the line segment case since the triple product would yield the zero vector. The *farthest* functions for both the circle and polygon will return incorrect Minkowski Difference points if the direction vector is zero. In this case a backup process should be used.

Listing 30. A revised version of listing 10

```
ab = b - a
ao = o - a or just -a
d = ab x ao x ab
if isZero(d)
  d = perpendicular(ab)
end if
```

In two dimensions we can get the perpendicular vector of another vector by switching the coordinates and negating one coordinate. There are two possible perpendicular vectors to use, depending on the coordinate negated, however, it doesn't matter which one is used.

Listing 31. The perpendicular function

```
perpendicular(v)
  perp.x = -v.y
  perp.y = v.x
  return perp
```

The next important pitfall is the choice of the initial search direction. Listing 4 didn't define what the initial search direction could be. The answer is that it can be any direction, but some directions are more optimal for performance and robustness.

A good choice might be to use the vector from the center of the first shape to the center of the second shape as was used in the examples. This is a great choice and can rule out non-collision cases quickly, however, there is a problem: what happens if both shape centers are the same point? This makes the first search direction the zero vector. Since this is a possibility, we should check for a zero initial search direction and choose an arbitrary initial direction instead.

Listing 32. A revised portion of listing 4

```
d = B.center - A.center
if isZero(d)
  d = (1, 0)
end if
```

However, the best choice for the initial search direction would be the search direction from the last GJK test. If we test two shapes and find that they were not colliding we could save the search direction for the next time we perform the test. Then, if the two shapes didn't move or moved very little then it's possible that on the first iteration we can determine that the shapes are not colliding by using the saved search direction as the initial direction. This takes advantage of temporal coherence, the idea that most objects in a scene don't move very far in between tests for collisions (Bergen, 2001).

Next we should consider the environment the code is running on. Computing devices have finite precision and cannot represent all floating point numbers exactly. Operations that you would expect to return a whole number or zero may return a number that is instead really close to that number. In addition, accuracy can be lost through calculations. A naïve approach to the *isZero* function used above implementation might be:

Listing 33. The isZero function

```
isZero(v)
  return v.x == 0 and v.y == 0
```

A better implementation would be to check the values against a value that is considered zero.

Listing 34. The revised version of listing 33

```
isZero(v)
  return abs(v.x) < e
  and abs(v.y) < e
```

This implementation allows a choice of e to be anything close to the value of zero, 1×10^{-9} for example.

Finally, we should add another bit of code to our while loop to terminate if a certain number of iterations has been reached. This prevents infinite looping in cases where numerical error occurs. This can also be used to manage the accuracy to performance tradeoff. Polygon vs. polygon cases will always complete within a finite number of iterations (Bergen, 1999), whereas curved shapes like circles may not.

DISTANCE AND CLOSEST POINTS

In the beginning of this chapter we explained that the GJK algorithm was first created to obtain the distance between two convex shapes and was adapted to test for collisions. As you might imagine, the collision detection algorithm and the distance algorithm work similarly. Both require convex shapes, both use the Minkowski Difference, and both are iterative.

To review, if two convex shapes are colliding then the Minkowski Difference of those shapes contains the origin. If they are not colliding then the Minkowski Difference does not contain the origin. This means that we cannot enclose the origin with a simplex for the distance function. Instead, the distance function keeps a line segment simplex and attempts to find the simplex contained in the Minkowski Difference that is closest to the origin.

Listing 35. The GJK distance algorithm

```
gjk_distance(A, B)
  d = some initial direction vector
  a = support(A, B, d)
  d = -d
  b = support(A, B, d)
  d = closestToOrigin(a, b)
  while (true)
    d = -d
    c = support(A, B, d)
    pc = dot(c, d)
    pa = dot(a, d)
    if (pc - pa < e)
      d = normalize(d)
      normal = d
      distance = -dot(c, d)
      return true
    else
      p1 = closestToOrigin(a, c)
      p2 = closestToOrigin(c, b)
      if (magnitudeSquared(p1) <
          magnitudeSquared(p2))
        b = c
        d = p1
      else
        a = c
        d = p2
      end if
    end if
  end while
```

Listing 35 and listing 4 are very similar. Both make use of the *support* function, are iterative, create and modify a simplex, and must be primed with an initial direction. There are a few differences however:

The first difference is the while loop condition. We cannot use the while loop condition from listing 4 because the simplex points may not meet that condition (that they are past the origin in the direction of *d*). In addition, we cannot use the opposite logic because a new point may validly pass the origin in the direction of *d. As a result, listing 35 assumes that the shapes are separated.*

Since the while loop condition is lacking a termination expression we must add it somewhere inside the while loop. The difference between *pa* and *pc* is the new termination condition. If the new support point is the same distance from the origin as the current simplex then we know that we cannot get any closer and can exit the loop. We use point *a* because both *a* and *b* produce the same projection *if*

Figure 12. a) Two shapes that are not colliding b) The Minkowski Difference of the shapes in 12a c) The simplex inside the Minkowski Difference before starting the first iteration d) The simplex inside the Minkowski Difference after the first iteration e) The simplex inside the Minkowski Difference after the second iteration

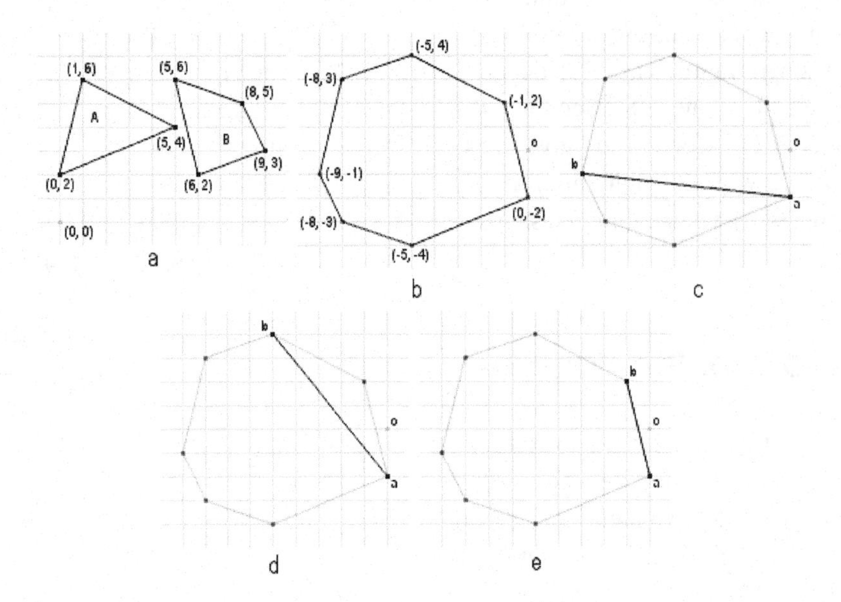

d is normal to the line *ab*. See the robustness section for an explanation of the case when *d* is not normal to the line *ab*.

Next, a new function has been added: *closestToOrigin*. This function obtains the closest point on the simplex to the origin.

The closestToOrigin Function

The *closestToOrigin* function is a simplified version of getting the closest point on a line segment to another point. We use a parameterized version to easily get the closest point on the line segment instead of the closest point on the line.

Listing 36. The closestToOrigin function

```
closestToOrigin(a, b)
  ao = -a
  ab = b - a
  ab2 = dot(ab, ab)
  aoab = dot(ao, ab)
  t = aoab / ab2
  t = clamp(t, 0.0, 1.0)
  return ab * t + a
```

If t is greater than 1 or less than 0, then we know that the closest point is on the line but not the line segment. Therefore we clamp t to get either end point of the line segment.

Example

The distance algorithm only needs to maintain the closest line segment to the origin, unlike the triangle that was maintained in the collision detection algorithm. Each iteration attempts to move the line segment closer to the origin by replacing either of the end points with a new support point until we cannot get any closer. Because many of the concepts are the same between the distance and collision detection algorithms we can jump right into an example (see Figure 12a for the non-colliding shapes and 12b for their respective Minkowski Difference).

Listing 37. Computing the initial simplex points using the support function

```
A_center = (2, 4)
B_center = (7, 4)
d = (7, 4) - (2, 4) = (5, 0)
dot((0, 2), (5, 0)) = 0 * 5 + 2 * 0 = 0
dot((5, 4), (5, 0)) = 5 * 5 + 4 * 0 = 25
dot((1, 6), (5, 0)) = 1 * 5 + 6 * 0 = 5
A_farthest(d) = (5, 4)
dot((6, 2), (-5, 0)) = 6 * -5 + 2 * 0 = -30
dot((9, 3), (-5, 0)) = 9 * -5 + 3 * 0 = -45
dot((8, 5), (-5, 0)) = 8 * -5 + 5 * 0 = -40
dot((5, 6), (-5, 0)) = 5 * -5 + 6 * 0 = -25
B_farthest(-d) = (5, 6)
a = (5, 4) - (5, 6) = (0, -2)
d = -d = (-5, 0)
dot((0, 2), (-5, 0)) = 0 * -5 + 2 * 0 = 0
dot((5, 4), (-5, 0)) = 5 * -5 + 4 * 0 = -25
dot((1, 6), (-5, 0)) = 1 * -5 + 6 * 0 = -5
A_farthest(d) = (0, 2)
dot((6, 2), (5, 0)) = 6 * 5 + 2 * 0 = 30
dot((9, 3), (5, 0)) = 9 * 5 + 3 * 0 = 45
dot((8, 5), (5, 0)) = 8 * 5 + 5 * 0 = 40
dot((5, 6), (5, 0)) = 5 * 5 + 6 * 0 = 25
B_farthest(-d) = (9, 3)
b = (0, 2) - (9, 3) = (-9, -1)
Simplex = {(0, -2), (-9, -1)}
```

Listing 38. Obtaining the new direction before entering the while loop

```
ao = -a = (0, 2)
ab = (-9, -1) - (0, -2) = (-9, 1)
ab2 = -9 * -9 + 1 * 1 = 82
aoab = 0 * -9 + 2 * 1 = 2
t = 2/82 = 1/41
t = clamp(1/41, 0.0, 1.0) = 1/41
d = (-9/41, 1/41) + (0, -2) = (-9/41, -81/41) ≈ (-0.22, -1.98)
```

Listing 39. Using the support function to obtain a new point for the simplex

```
d = -d ≈ (0.22, 1.98)
dot((0, 2), (0.22, 1.98)) = 0 * 0.22 + 2 * 1.98 = 3.96
dot((5, 4), (0.22, 1.98)) = 5 * 0.22 + 4 * 1.98 = 9.02
dot((1, 6), (0.22, 1.98)) = 1 * 0.22 + 6 * 1.98 = 12.1
A_farthest(d) = (1, 6)
dot((6, 2), (-0.22, -1.98)) = 6 * -0.22 + 2 * -1.98 = -5.28
dot((9, 3), (-0.22, -1.98)) = 9 * -0.22 + 3 * -1.98 = -7.92
dot((8, 5), (-0.22, -1.98)) = 8 * -0.22 + 5 * -1.98 = -11.66
dot((5, 6), (-0.22, -1.98)) = 5 * -0.22 + 6 * -1.98 = -12.98
B_farthest(-d) = (6, 2)
MD_point = (1, 6) - (6, 2) = (-5, 4)
c = (-5, 4)
```

The first step before entering the while loop is to create our initial simplex by obtaining two points on the edge of the Minkowski Difference using the support function. Our initial direction will be the vector pointing from the center of A to the center of B. (Listing 37)

See Figure 12c for an illustration of the current simplex. Once point *a* and point *b* have been found we need to use the *closestToOrigin* function to obtain the new direction. (Listing 38)

Immediately after entering the loop we get a new support point using the negative of the search direction. (Listing 39)

Next we compare the distance of point *c* and point *a* along *d* to see if we have moved closer to the origin. Remember that we could have used point *a* or point *b* to compare against point *c as long as d is normal to the line ab.*

Listing 40. Testing the termination condition in the first iteration

```
-d ≈ (0.22, 1.98)
dot(c, d)
  = -5 * 0.22 + 4 * 1.98 = 6.82
dot(a, d)
  = 0 * 0.22 + -2 * 1.98 = -3.96
pc - pa = 6.82 - -3.96 = 10.78
```

Since *pc* – *pa* is not small enough we must continue to loop (see Listing 41).

Since *p1*'s magnitude is smaller than *p2*'s magnitude, we can conclude that the line segment *ac* is closer to the origin than *cb*. Therefore, we replace point *b* in the simplex with the new point *c* giving us the new simplex shown in Figure 12d.

Then we loop and begin the process over again. (Listing 42)

Now make sure we have moved far enough.

Listing 43. Testing the termination condition in the second iteration

```
d ≈ (0.98, 0.82)
dot(c, d)
  = -1 * 0.98 + 2 * 0.82 = 0.66
dot(a, d)
  = 0 * 0.98 + -2 * 0.82 = -1.64
pc - pa = 0.66 - -1.64 = 2.3
```

Since *pc* – *pa* is not small enough we must continue to loop (see Listing 44).

Since *p1*'s magnitude is smaller than *p2*'s magnitude, we can conclude that the line segment *ac* is closer to the origin than *cb*. Therefore, we replace point *b* in the simplex with the new

Listing 41. Computing the closest to origin for the line segments ac and cb

```
p1 = closestToOrigin(a, c)
ao = -a = (0, 2)
ab = c - a = (-5, 4) - (0, -2) = (-5, 6)
ab2 = -5 * -5 + 6 * 6 = 61
aoab = 0 * -5 + 2 * 6 = 12
t = 12/61
t = clamp(12/61, 0.0, 1.0) = 12/61
p1 = (-60/61, 72/61) + (0, -2) = (-60/61, -50/61) ≈ (-0.98, -0.82)
p2 = closestToOrigin(c, b) ≈ (-5.64, 3.2)
ao = -c = (5, -4)
ab = b - c = (-9, -1) - (-5, 4) = (-4, -5)
ab2 = -4 * -4 + -5 * -5 = 41
aoab = 5 * -4 + -4 * -5 = 0
t = 0/41 = 0
t = clamp(0, 0.0, 1.0) = 0
p2 = (0, 0) + (-5, 4) = (-5, 4)
magnitudeSquared(p1) ≈ 1.63
magnitudeSquared(p2) = 41
b = c = (-5, 4)
d = p1 ≈ (-0.98, -0.82)
Simplex = {(0, -2), (-5, 4)}
```

Listing 42. Using the support function to obtain a new point for the simplex

```
-d ≈ (0.98, 0.82)
dot((0, 2), (0.98, 0.82)) = 0 * 0.98 + 2 * 0.82 = 1.64
dot((5, 4), (0.98, 0.82)) = 5 * 0.98 + 4 * 0.82 = 8.18
dot((1, 6), (0.98, 0.82)) = 1 * 0.98 + 6 * 0.82 = 5.9
A_farthest(d) = (5, 4)
dot((6, 2), (-0.98, -0.82)) = 6 * -0.98 + 2 * -0.82 = -7.52
dot((9, 3), (-0.98, -0.82)) = 9 * -0.98 + 3 * -0.82 = -11.28
dot((8, 5), (-0.98, -0.82)) = 8 * -0.98 + 5 * -0.82 = -11.94
dot((5, 6), (-0.98, -0.82)) = 5 * -0.98 + 6 * -0.82 = -9.82
B_farthest(-d) = (6, 2)
MD_point = (5, 4) - (6, 2) = (-1, 2)
c = (-1, 2)
```

Listing 44. Computing the closest to origin for the line segments ac and cb

```
p1 = closestToOrigin(a, c)
ao = -a = (0, 2)
ab = c - a = (-1, 2) - (0, -2) = (-1, 4)
ab2 = -1 * -1 + 4 * 4 = 17
aoab = 0 * -1 + 2 * 4 = 8
t = 8/17
t = clamp(8/17, 0.0, 1.0) = 8/17
p1 = (-8/17, 32/17) + (0, -2) ≈ (-0.47, -0.12)
p2 = closestToOrigin(c, b)
ao = -c = (1, -2)
ab = b - c = (-5, 4) - (-1, 2) = (-4, 2)
ab2 = -4 * -4 + 2 * 2 = 20
aoab = 1 * -4 + -2 * 2 = -8
t = -8/20
t = clamp(-8/20, 0.0, 1.0) = 0
p2 = (0, 0) + (-1, 2) = (-1, 2)
magnitudeSquared(p1) ≈ 0.24
magnitudeSquared(p2) = 5
b = c = (-1, 2)
d = p1 ≈ (-0.47, -0.12)
Simplex = {(0, -2), (-1, 2)}
```

Listing 45. Using the support function to obtain a new point for the simplex

```
-d ≈ (0.47, 0.12)
dot((0, 2), (0.47, 0.12)) = 0 * 0.47 + 2 * 0.12 = 0.24
dot((5, 4), (0.47, 0.12)) = 5 * 0.47 + 4 * 0.12 = 2.83
dot((1, 6), (0.47, 0.12)) = 1 * 0.47 + 6 * 0.12 = 1.19
A_farthest(d) = (5, 4)
dot((6, 2), (-0.47, -0.12)) = 6 * -0.47 + 2 * -0.12 = -3.06
dot((9, 3), (-0.47, -0.12)) = 9 * -0.47 + 3 * -0.12 = -4.59
dot((8, 5), (-0.47, -0.12)) = 8 * -0.47 + 5 * -0.12 = -4.36
dot((5, 6), (-0.47, -0.12)) = 5 * -0.47 + 6 * -0.12 = -3.07
B_farthest(-d) = (6, 2)
MD_point = (5, 4) - (6, 2) = (-1, 2)
c = (-1, 2)
```

point c giving us the new simplex shown in Figure 12e.

Then we loop and begin the process over again. (Listing 45)

Now make sure we have moved far enough.

Listing 46. Testing the termination condition in the third iteration

```
d ≈ (0.47, 0.12)
dot(c, d)
  = -1 * 0.47 + 2 * 0.12 = -0.23
dot(a, d)
  = 0 * 0.47 + -2 * 0.12 = -0.24
pc - pa = -0.23 - -0.24 = 0.01
```

Since the difference in distance along d for point a and point c is nearly zero then we can conclude that we are as close to the origin as possible which is confirmed by Figure 12e. Therefore we return d normalized as the separation normal and the projection of c along the normalized d as the separation distance. In this example the $pc - pa$ result would have been much closer to zero if we used better precision in our computations.

Closest Points

The GJK algorithm is also suited to obtain the closest points between the two shapes. We can solve for the closest points using a system of equations (Bergen, 1999). If we store the farthest points that make up each support point we can use these to find the closest points on each shape. Before returning true in listing 35 we can perform a quick calculation to get the closest points using the current simplex.

Let Q be the closest point to the origin on the final simplex. The closest point on a line to a given point is the point which creates a perpendicular line to the given point. We can represent this by the following equation:

Listing 47. The closest point Q on a line created by A and B to a point P has a dot product of zero

```
dot(Q - P, A - B) = 0
```

Where P is the point we are trying to get close to, Q is the closest point, and A and B are the control points of the line. In our case the point P is the origin. This simplifies $Q - P$ to Q. In addition, we'll let $L = A - B$ to achieve:

Listing 48. A simplified version of listing 47

```
dot(Q, L) = 0
```

We also know that the closest point, Q, must be a convex combination of A and B (Bergen, 1999, pg. 7):

Listing 49. A convex combination of a pair of points (Hedegaard para. 1)

```
Q = aA + bB
a + b = 1
```

Let $a = 1 - b$ and substitute $Q = (1 - b)A + bB$ into listing 48 and solve for b:

Listing 50. Solving for the b coefficient

```
dot((1 - b)A + bB, L) = 0
dot(A - bA + bB, L) = 0
dot(A - b(A - B), L) = 0
dot(A - bL, L) = 0
dot(A, L) - dot(bL, L) = 0
dot(A, L) - b * dot(L, L) = 0
-b * dot(L, L) = -dot(A, L)
b = dot(A, L) / dot(L, L)
```

Then we substitute b back into the equation in listing 49 to obtain a. Once the a and b coefficients are found, they can be used to find the closest points on the shapes by using the stored support points.

Listing 51. Finding the closest points on each shape

```
Aclosest = aAs1 + bBs1
Bclosest = aAs2 + bBs2
```

Where A_{s1}, A_{s2}, B_{s1}, and B_{s2} are the points used to create the Minkowski Difference points A and B of the final simplex.

Example

Using the previous example we can solve for the closest points. The final simplex of the last example was:

Listing 52. The final simplex of the example from the GJK distance algorithm section

```
Simplex = {(0, -2), (-1, 2)}
```

Let $A = (0, -2)$ and $B = (-1, 2)$. By looking at the points on the shapes that created the Minkowski Difference points we can define:

Listing 53. The points on the shapes that made up the simplex points

```
As1 = (5, 4)
Bs1 = (5, 4)
As2 = (5, 6)
Bs2 = (6, 2)
```

From this information we can solve for the closest points:

Listing 54. Solving for the closest points

```
b = dot(A, L) / dot(L, L)
where L = A - B
a = 1 - b
L = (0, -2) - (-1, 2) = (1, -4)
dot(A, L) = 0 * 1 + -2 * -4 = 8
dot(L, L) = 1 * 1 + -4 * -4 = 17
b = 8/17
```

Figure 13. The closest points between shapes A and B

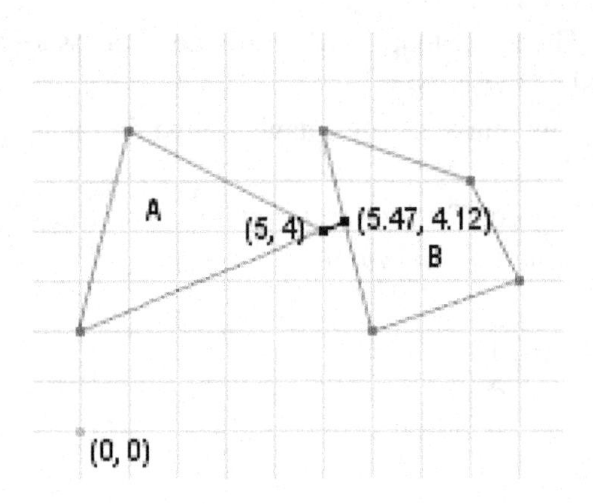


```
a = 1 - b = 9/17
Aclosest
  = (9/17)(5, 4) + (8/17)(5, 4)
  = (5, 4)
Bclosest
  = (9/17)(5, 6) + (8/17)(6, 2)
  = (45/17, 54/17) + (48/17, 16/17)
  = (93/17, 70/17)
  ≈ (5.47, 4.12)
```

Robustness

The distance function, just like the collision detection function requires some robustness enhancements. The first enhancement is the $pc - pa$ condition which compares against e; some small value that is considered zero. This is necessary for curved shapes since the pseudo code can infinitely get closer and closer to the exact answer. Comparing against e will ensure that we get a good enough result.

Next, we need to handle the case where d is not normal to the line ab. This happens when the origin lies in a vertex Voronoi region of the simplex, in which case the closest point is either end point of the simplex (see Figure 14). This would seemingly require a change to the termination condition; however, if we inspect the two cases we can see that the termination case is still valid.

Figure 14. a) Case 1: if the origin lies in a's Voronoi region of the simplex b) Case 2: if the origin lies in b's Voronoi region of the simplex

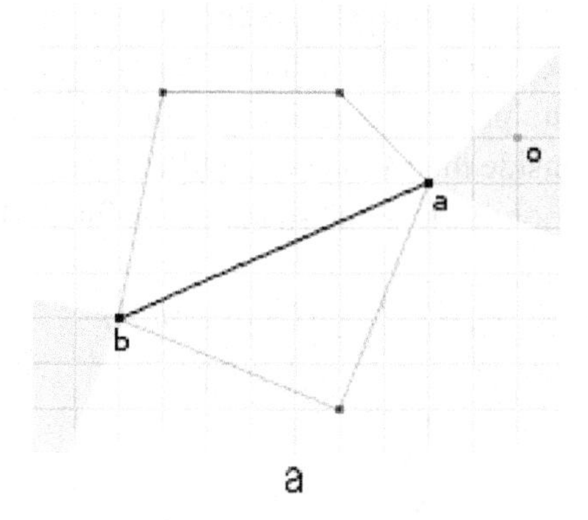

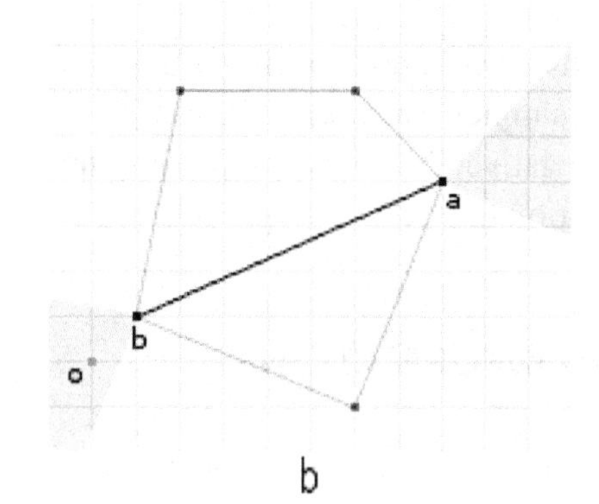

Case 1: The closest point is point a on the simplex ab. We test the projection of point a against point a which will always return zero and terminate. This isn't a problem because point a is the closest feature of the simplex to the origin because the origin lies in the Voronoi region of point a.

Case 2: The closest point is point b on the simplex ab. We test the projection of point a against point b which will always return different projections since they can only return the same projection when the closest point is on the line. Because of this we move onto comparing which edge is closer to the origin. The two closest points, $p1$ and $p2$, will be identical since both edges contain the closest point c. As a result, point a is replaced with point c. Now point a and point b are the same point. When the next iteration is performed, the same point will be found again and will terminate by Case 1.

The closest points function contains a few pitfalls as well. The first is if $L = A - B = 0$, which means that the points A and B are the same point (in other words, the end simplex from the distance function is a vertex on the Minkowski Difference). Later we divide by L's squared magnitude to obtain b causing a divide by zero.

Listing 55. Fix for divide by zero problems when $A - B = 0$

```
if dot(L, L) < e
  A_closest = A_s1
  B_closest = A_s2
end if
```

We compare against e again to avoid close to zero values. The use of A_{s1} and A_{s2} is arbitrary since $A = B$.

The second problem with the closest points function is that the coefficients a or b can be less than zero. But, according to the definition of a convex combination, the coefficients a and b must be greater than or equal to zero (Hedegaard para. 1). This happens when the closest point on the line is outside the range of the line segment simplex which is possible if Case 1 is reached. To fix this we add:

Listing 56. Fix for negative a or b coefficients

```
if a < 0
  A_closest = B_s1
  B_closest = B_s2
else if b < 0
  A_closest = A_s1
  B_closest = A_s2
else
```

```
  listing 50 & 51
end if
```

In these cases we must use the points that come from the simplex point that has a coefficient which is not negative. Otherwise we perform the computation normally.

COLLISION INFORMATION

Whenever a collision detection algorithm is discussed, it typically returns true or false for simplicity. However, if we wanted to *do something* with the colliding shapes, to resolve the collision for instance, we need more information. This extra information is called the Minimum Translation Vector or MTV. The MTV is the vector of minimum magnitude used to translate the shapes out of collision. The normalized MTV is simply called the normal and the magnitude of the MTV is called the depth.

Unfortunately, the GJK collision detection algorithm cannot give a collision normal and depth directly, and must be supplemented with another algorithm.

Expanding Polytope Algorithm

The Expanding Polytope Algorithm, or EPA, is an iterative method which uses the Minkowski Difference to obtain the MTV from a collision of two shapes (Bergen, 2001). The algorithm assumes that the origin has already been found inside the Minkowski Difference and that the shapes are convex. Then, it finds the closest face or edge on the Minkowski Difference to the origin. The normal of the closest face or edge is the collision normal of the MTV and the projected distance from that face or edge to the origin is the depth.

EPA fits well with the GJK algorithm because it can use the final simplex from a detected collision (Bergen, 2001). Using the final simplex from the GJK collision detection algorithm, we iteratively expand it by adding more support points until the closest feature of the simplex to the origin cannot be expanded any more. A good way to visualize the process is to imagine blowing up a balloon inside the Minkowski Difference, where the balloon starts in the shape of the final GJK simplex.

Listing 57. The EPA algorithm

```
epa(simplex)
  while true
    n, d, and
    i = findClosestEdge(simplex)
    p = support(n)
    t = dot(p, n)
    if t - d > e
      simplex.add(i, p)
    else
      return n and -t
    end if
  end while
```

Where n is the normal and d is the depth of the closest edge of the simplex to the origin. The variable i represents the index, in the list of points that makes up the simplex, where we need to insert the new support point, p, if we need to continue expanding.

Finding a support point in the direction of the normal of the closest edge of the simplex will yield a point on the edge of the Minkowski Difference just like any other direction we have used in the collision detection and distance methods. But, if the new point is *on* the closest edge of the simplex, then we know that we cannot expand anymore and terminate (the *else* block of the *if* statement). If the new point is not on the closest edge of the simplex, then we know we need to add the new point to the simplex at index i. This in effect splits the closest edge of the simplex into two edges. This also means that unlike GJK, the EPA simplex will grow continually until the algorithm terminates, instead of having a maximum of three or four points.

Since EPA uses the Minkowski Difference, we can use the same support function that both the GJK distance and collision detection methods use. However, we do have to define a new function that finds the edge closest to the origin:

Listing 58. The function to find the closest edge on the simplex to the origin

```
findClosestEdge(simplex)
  n, d, i
  min
  for each edge, e, in simplex
    ab = e.b - e.a
    ao = -e.a
    normal = -(ab x ao x ab)
    normal = normalize(normal)
    distance = dot(e.a, normal)
    if distance < min
      min = distance
      n = normal
      d = distance
      i = b.index
    end if
  end for
  return n, d, and i
```

Where each edge of the simplex is defined by two points; a and b. The projection of both point a and point b on the normal produces the same distances; we chose to use point a in listing 58 but either can be used. As in listing 57, n is the normal and d is the depth of the closet edge of the simplex to the origin. We use the index of the end point of the closest edge of the simplex so that when we insert the new support point into the simplex it splits the closest edge.

Example

We will use the same shapes as the previous examples; however, their overlap in this example will be greater to fully illustrate the algorithm. The shapes are shown in Figure 15a with their Minkowski Difference in Figure 15b. Figure 15c is the final simplex from the GJK collision detection algorithm that will be passed to EPA:

Listing 59. The end simplex from GJK illustrated in Figure 15c

```
Simplex = {(2, -2), (-7, -1), (-3, 4)}
```

Following listing 57 we must first find the closest edge to the origin. (Listing 60)

Since *edge3* was the closest, we use its normal to obtain a new support point (see Listing 61).

Next, we test if the new point is on the edge by testing the difference in projection between the new point and point a (that projection was saved in d from the *findClosestEdge* function) (see Listing 62).

Because the difference is greater than zero, meaning adding the new point will indeed expand the simplex, we add the new point to the simplex at the i-th position, in effect splitting the edge creating two new edges; see figure 15d. Then we continue and find the next closest edge (see Listing 63).

In this iteration the closest edge is *edge4*; *p1* to a. One thing to notice here is that *edge1* and *edge2* didn't have to be recomputed again (even though we did for clarity) since their normals and distances will remain the same. Only the new edges, c to *p1* and *p1* to a, must be computed since these are new. This is not implemented in the pseudo code above but should be trivial to add.

Once we have the closest edge we compute another support point (see Listing 64).

Next we check the projection against the edge's distance from the origin (see Listing 65).

Since the projection is nearly zero we know that we cannot expand the simplex at the closest edge and therefore can exit the loop and return the MTV (see Figure 15e).

Figure 15. a) Two colliding shapes. b) The Minkowski Difference of the two colliding shapes from 15a. c) The final simplex after the GJK collision detection algorithm has detected a collision. d) The expanded simplex after the first iteration of the EPA algorithm. e) The MTV found in the second iteration of the EPA algorithm, with the length being the depth, and the direction being the normal

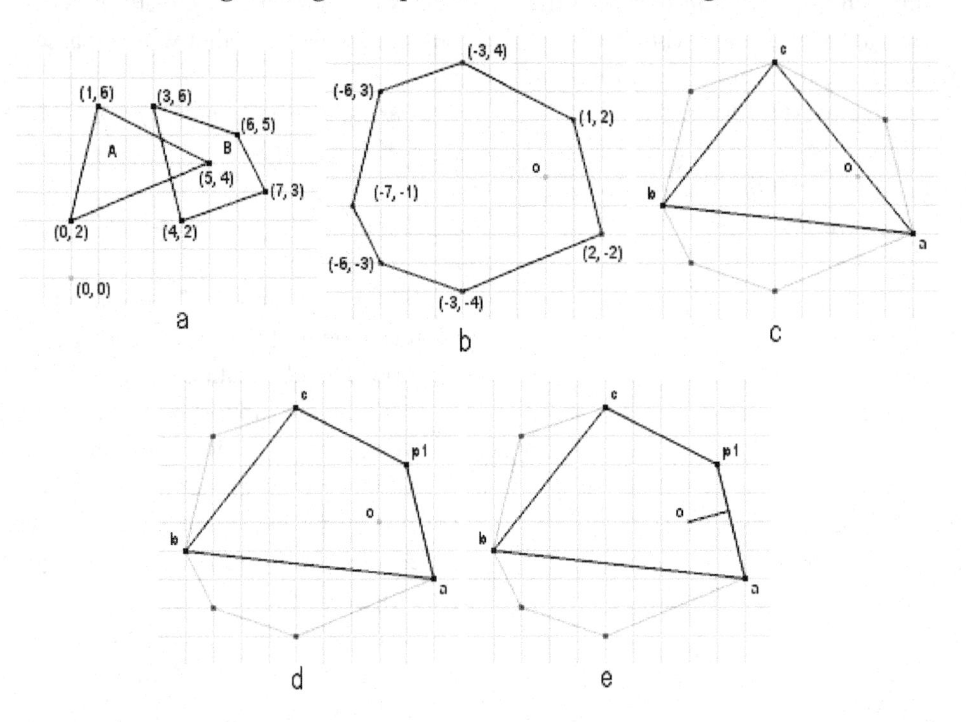

From listing 64 we can see that B's farthest point could be either (4, 2) or (3, 6) if we had used more precision. This makes sense since n is normal to the line created by (4, 2) and (3, 6). Depending on which is selected, $p1$ or a will be returned. If we use (4, 2) instead of (3, 6) we will get $p = (1, 2)$. Using (1, 2) in listing 65 instead of (2, -2) still results in 1.45 which means that it doesn't matter which one is selected.

Robustness

As with the GJK collision detection and distance algorithms we need to improve some areas of the pseudo code. First is the use of e, a number close to zero, in our termination conditions just like the other algorithms in this chapter to handle the loss of accuracy.

The previous example ended in two iterations, where the collision example in the GJK section terminates in one. The EPA algorithm does perform rather quickly; however, if we examine the second iteration of the previous example, we can see that each iteration adds another point/edge to the simplex that will have to be tested in the *findClosestEdge* function. In addition to this, we must perform a normalization of the normal to correctly choose the closest edge. This can create a performance problem for curved shapes since we can infinitely get closer to the true edge while the simplex edges get shorter and shorter. To combat this, it's recommended that a maximum number of iterations condition be placed on the while loop.

In addition, curved shapes can cause numerical problems when the edges of the simplex get small enough. Listing 58 uses the triple product to find

Listing 60. Computing the closest edge to the origin in the first iteration

```
edge1 = a to b = (2, -2) to (-7, -1)
ab = (-7, -1) - (2, -2) = (-9, 1)
ao = (-2, 2)
(ab x ao) = (-9, 1) x (-2, 2) = -9 * 2 - 1 * -2 = -18 + 2 = -16
(ab x ao) x ab = (-1 * 1 * -16, -9 * -16) = (16, 144)
normal = -(ab x ao x ab) = (-16, -144) ≈ (-0.11, -0.99)
distance = dot((2, -2), (-0.11, -0.99)) = 2 * -0.11 + -2 * -0.99
         = 1.76
edge2 = b to c = (-7, -1) to (-3, 4)
ab = (-3, 4) - (-7, -1) = (4, 5)
ao = (7, 1)
(ab x ao) = (4, 5) x (7, 1) = 4 * 1 - 5 * 7 = 4 - 35 = -31
(ab x ao) x ab = (-1 * 5 * -31, 4 * -31) = (155, -124)
normal = -(ab x ao x ab) = (-155, 124) ≈ (-0.78, 0.62)
distance = dot((-7, -1), (-0.78, 0.62)) = -7 * -0.78 + -1 * 0.62
         = 4.84
edge3 = c to a = (-3, 4) to (2, -2)
ab = (2, -2) - (-3, 4) = (5, -6)
ao = (3, -4)
(ab x ao) = (5, -6) x (3, -4) = 5 * -4 - -6 * 3 = -20 + 18 = -2
(ab x ao) x ab = (-1 * -6 * -2, 5 * -2) = (-12, -10)
normal = -(ab x ao x ab) = (12, 10) ≈ (0.77, 0.64)
distance = dot((-3, 4), (0.77, 0.64)) = -3 * 0.77 + 4 * 0.64
         = 0.25
n = (0.77, 0.64)
d = 0.25
i = 2
```

Listing 61. Obtaining a new support point to add to the simplex

```
n = (0.77, 0.64)
dot((0, 2), (0.77, 0.64)) = 0 * 0.77 + 2 * 0.64 = 1.28
dot((5, 4), (0.77, 0.64)) = 5 * 0.77 + 4 * 0.64 = 6.41
dot((1, 6), (0.77, 0.64)) = 1 * 0.77 + 6 * 0.64 = 4.61
A_farthest(d) = (5, 4)
dot((4, 2), (-0.77, -0.64)) = 4 * -0.77 + 2 * -0.64 = -4.36
dot((7, 3), (-0.77, -0.64)) = 7 * -0.77 + 3 * -0.64 = -7.31
dot((6, 5), (-0.77, -0.64)) = 6 * -0.77 + 5 * -0.64 = -7.82
dot((3, 6), (-0.77, -0.64)) = 3 * -0.77 + 6 * -0.64 = -6.15
B_farthest(-d) = (4, 2)
MD_point = (5, 4) - (4, 2) = (1, 2)
p = (1, 2)
```

Listing 62. Testing whether we expanded any more using the new simplex point

```
dot((1, 2), (0.77, 0.64)) = 1 * 0.77 + 2 * 0.64 = 2.05
2.05 - 0.25 = 1.8
```

Listing 63. Computing the closest edge to the simplex in the second iteration

```
edge1 = a to b = (2, -2) to (-7, -1)
ab = (-7, -1) - (2, -2) = (-9, 1)
ao = (-2, 2)
(ab x ao) = (-9, 1) x (-2, 2) = -9 * 2 - 1 * -2 = -18 + 2 = -16
(ab x ao) x ab = (-1 * 1 * -16, -9 * -16) = (16, 144)
normal = -(ab x ao x ab) = (-16, -144) ≈ (-0.11, -0.99)
distance = dot((2, -2), (-0.11, -0.99)) = 2 * -0.11 + -2 * -0.99
         = 1.76
edge2 = b to c = (-7, -1) to (-3, 4)
ab = (-3, 4) - (-7, -1) = (4, 5)
ao = (7, 1)
(ab x ao) = (4, 5) x (7, 1) = 4 * 1 - 5 * 7 = 4 - 35 = -31
(ab x ao) x ab = (-1 * 5 * -31, 4 * -31) = (155, -124)
normal = -(ab x ao x ab) = (-155, 124) ≈ (-0.78, 0.62)
distance = dot((-7, -1), (-0.78, 0.62)) = -7 * -0.78 + -1 * 0.62
         = 4.84
edge3 = c to p1 = (-3, 4) to (1, 2)
ab = (1, 2) - (-3, 4) = (4, -2)
ao = (3, -4)
(ab x ao) = (4, -2) x (3, -4) = 4 * -4 - -2 * 3 = -16 + 6 = -10
(ab x ao) x ab = (-1 * -2 * -10, 4 * -10) = (-20, -40)
normal = -(ab x ao x ab) = (20, 40) ≈ (0.45, 0.89)
distance = dot((-3, 4), (0.45, 0.89)) = -3 * 0.45 + 4 * 0.89
         = 2.21
edge4 = p1 to a = (1, 2) to (2, -2)
ab = (2, -2) - (1, 2) = (1, -4)
ao = (-1, -2)
(ab x ao) = (1, -4) x (-1, -2) = 1 * -2 - -4 * -1 = -2 - 4 = -6
(ab x ao) x ab = (-1 * -4 * -6, 1 * -6) = (-24, -6)
normal = -(ab x ao x ab) = (24, 6) ≈ (0.97, 0.24)
distance = dot((1, 2), (0.97, 0.24)) = 1 * 0.97 + 2 * 0.24
         = 1.45
n = (0.97, 0.24)
d = 1.45
i = 3
```

Listing 64. Obtaining another support point to expand the simplex

```
n = (0.97, 0.24)
dot((0, 2), (0.97, 0.24)) = 0 * 0.97 + 2 * 0.24 = 0.48
dot((5, 4), (0.97, 0.24)) = 5 * 0.97 + 4 * 0.24 = 5.81
dot((1, 6), (0.97, 0.24)) = 1 * 0.97 + 6 * 0.24 = 2.41
A_farthest(d) = (5, 4)
dot((4, 2), (-0.97, -0.24)) = 4 * -0.97 + 2 * -0.24 = -4.36
dot((7, 3), (-0.97, -0.24)) = 7 * -0.97 + 3 * -0.24 = -7.51
dot((6, 5), (-0.97, -0.24)) = 6 * -0.97 + 5 * -0.24 = -7.02
dot((3, 6), (-0.97, -0.24)) = 3 * -0.97 + 6 * -0.24 = -4.35
B_farthest(-d) = (3, 6)
MD_point = (5, 4) - (3, 6) = (2, -2)
p = (2, -2)
```

Listing 65. Testing whether the new point expanded the simplex

```
dot((2, -2), (0.97, 0.24)) = 2 * 0.97 + -2 * 0.24 = 1.46
1.46 - 1.45 = 0.01
MTV_normal = (0.97, 0.24)
MTV_depth = 1.46
```

the normal of the edge. In cases where the origin lies on the edge of the Minkowski Difference or on the edge of the simplex, or if the points *a* and *b* that make up an edge are equal, the triple product will return the zero vector. To fix this we use the same fix from the GJK algorithm outlined in listing 30. The difference is that we must choose the normal that points away from the simplex, not just either one. To do this, we must know the winding direction of the simplex.

Listing 66. The function to return the winding of the simplex (Bourke, 1998)

```
getWinding(simplex)
  for each edge, e, in simplex
    en = the next edge
    e1 = e.b - e.a
    e2 = en.b - en.a
    cross = e1 x e2
    if cross > 0
      return 1
    else if cross < 0
      return -1;
    end if
  end for
  return 0
```

We can easily find the winding of the simplex by performing the cross product of just one set of edges in the simplex. If one set of edges returns zero then we continue and test the next pair of edges. Once we know the winding of the simplex we can easily determine which direction to use:

Listing 67. A revised portion of listing 58 using the winding as a fall back

```
if isZero(normal)
  if winding < 0
    normal = (normal.y, -normal.x)
```

```
    else
      normal = (-normal.y, normal.x)
    end if
  end if
```

Only the winding of the initial simplex must be computed because the winding is retained throughout the algorithm as each new point is inserted into the simplex.

Because EPA can be a performance bottle neck in some cases, it's mainly used in deep penetration cases, where shapes significantly overlap. In the cases where the overlap is small, the GJK distance algorithm is used. This may seem strange since the distance algorithm, as we described before, only works when the shapes are separated. The trick here is to represent the real shapes with smaller "core" shapes. When the real shapes overlap a little, the core shapes won't. If we perform a GJK distance check on the core shapes, then incorporate the radial expansion amount from both shapes we can get the penetration depth. The normal is the normal from the distance check. If the core shapes overlap then we use the EPA algorithm. This also suffers from some problems. First is how to choose the radial expansion. The second is the additional complexity of creating a core shape rather than using the original shape.

Another alternative to finding the penetration normal and depth is sampling. Sampling is the process of taking a number of vectors around the unit circle and projecting the Minkowski Difference onto them to see which has the minimum penetration depth from the origin. This method may not produce the exact depth and normal but may be good enough depending on the application. The accuracy can be improved by using the surface and/or edge normals of the shapes as well.

FUTURE RESEARCH DIRECTIONS

GJK can be used in what's called continuous collision detection. Continuous collision detection (CCD), is the process of detecting collisions that were missed as a result of the discrete nature of collision detection. Typically, collision detection is performed once each time step of an application. If the time step is sufficiently long or an object moves fast enough (or both), the object can pass through another object without the collision being detected. This is referred to as tunneling. Conservative advancement is an algorithm that can use GJK to test for distances between objects and predict when a collision will or has been missed.

GJK can be used in ray casting as well. Ray casting is the process of shooting a ray and detecting the objects that are intersected by that ray. This is useful in a number of areas such as AI visibility tests. GJK can be adapted to using a ray and a convex shape for fast ray casting using the same support function that is used in the original algorithm.

CONCLUSION

Using the concepts of convexity and the Minkowski Difference we have covered and implemented the GJK algorithm for collision detection. In addition, we have covered a simplified version of the GJK distance method used to obtain the distance and closest points between two convex shapes. We also covered the EPA algorithm to obtain the MTV. All algorithms were supplemented with examples that point out key concepts. GJK is a great choice for collision detection in demanding applications because of its speed, robustness, simple implementation and versatility.

REFERENCES

Bergen, G. (1999). A fast and robust gjk implementation for collision detection of convex objects. Retrieved from http://www.win.tue.nl/~gino/solid/jgt98convex.pdf

Bergen, G. (2001). Proximity queries and penetration depth computation on 3d game objects. Presented at *Game Developers Conference, 2001*. San Jose, CA. Retrieved from http://www.win.tue.nl/~gino/solid/gdc2001depth.pdf

Bergen, G. (2004). *Collision detection in interactive 3d environments*. San Francisco, CA: Morgan Kaufmann Publishers.

Bourke, P. (1998). *Determining whether or not a polygon (2D) has its vertices ordered clockwise or counterclockwise*. Retrieved from http://paulbourke.net/geometry/clockwise/index.html

Cameron, S. (1997). Enhancing GJK: Computing minimum and penetration distances between convex polyhedra. Proceedings from *IEEE International Conference on Robotics and Automation*. Albuquerque, NM.

Ericson, C. (2004, August). The gilbert-johnson-keerthi (gjk) algorithm. Presented at *SIGGRAPH 2004*, Los Angeles, CA. Sony Computer Entertainment America. Retrieved from http://realtimecollisiondetection.net/pubs/

Hedegaard, R. (n.d.). Convex combination. MathWorld--A Wolfram Web Resource created by Eric W. Weisstein. Retrieved from http://mathworld.wolfram.com/ConvexCombination.html

Muratori, C. (2006). *Implementing GJK* [MP4]. Molly Rocket Nebula. Retrieved from http://mollyrocket.com/849

Weisstein, E. W. (n.d.) *Convex. MathWorld*--A Wolfram Web Resource created by Eric W. Weisstein. Retrieved from http://mathworld.wolfram.com/Convex.html

ADDITIONAL READING

Bergen, G. (2004). *Ray casting against general convex objects with application to continuous collision detection*. Breda, Netherlands: Playlogic Game Factory.

Coumans, E. (2005). *Continuous collision detection and physics*. Sony Computer Entertainment.

Ericson, C. (2005). *Real-time collision detection*. San Francisco, CA: Morgan Kaufmann.

Gilbert, E. G., Johnson, D. W., & Keerthi, S. S. (1988). A fast procedure for computing the distance between complex objects in three-dimensional space. *IEEE Journal on Robotics and Automation*, *4*(2), 193–203. doi:10.1109/56.2083

Weisstein, E. W. (n.d.). *Cross product*. MathWorld--A Wolfram Web Resource created by Eric W. Weisstein. Retrieved from http://mathworld.wolfram.com/CrossProduct.html

Weisstein, E. W. (n.d.). *Dot Product*. MathWorld--A Wolfram Web Resource created by Eric W. Weisstein. Retrieved http://mathworld.wolfram.com/DotProduct.html

KEY TERMS AND DEFINITIONS

Collision Detection: The process of detecting whether two objects have collided or overlapped.

Convex Shape: A shape, without holes, where any combination of two interior points creates a line segment that is fully contained within the shape.

Expanding Polytope Algorithm: (EPA): Used in conjunction with GJK to find the MTV of a collision.

Gilbert-Johnson-Keerthi (GJK) Algorithm: For determining the distance between two convex shapes and adapted to collision detection.

Minimum Translation Vector: (MTV): Is the vector of minimum magnitude that can be used to push two colliding shapes out of collision.

Minkowski Difference: The Minkowski Sum using the subtraction operator instead of the addition operator.

Minkowski Sum: An operation performed on shapes in which a resulting shape is obtained, defined by adding all points in the first shape to all the points in the second shape.

Simplex: A convex polygon that's created inside the Minkowski Difference used to enclose the origin; a triangle in two dimensions and a tetrahedron in three dimensions.

Support Function: A subroutine used to obtain a point on the edge of the Minkowski Difference during the creation of the simplex.

Voronoi Region: A region of a shape in which all points in that region are closer to some feature of the shape than any other feature.

Section 4

Game Models and Implementation

Chapter 12
Designing Multiplayer Online Games Using the Real-Time Framework

Sergei Gorlatch
University of Muenster, Germany

Frank Glinka
University of Muenster, Germany

Alexander Ploss
University of Muenster, Germany

Dominik Meiländer
University of Muenster, Germany

ABSTRACT

This chapter describes a novel, high-level approach to designing and executing online computer games. The approach is based on our Real-Time Framework (RTF) and suits a wide spectrum of online games including Massively Multiplayer Online Games (MMOG) and First-Person Shooters (FPS). The authors address major design issues like data structures and Area of Interest (AoI), with a special focus on the scalability of games implemented on multiple servers, including distribution of the game state, inter-server communication, object serialization and migration, etc. The chapter illustrates the approach with two case studies: the design of a new multi-player online game and bringing the single-server commercial game Quake 3 to multiple servers in order to increase the number of simultaneous players. The authors show the place of their approach in the taxonomy of game development approaches, and they report experimental results on the performance of games developed using RTF.

DOI: 10.4018/978-1-4666-1634-9.ch012

INTRODUCTION

The development of *Massively Multiplayer Online Games* (MMOG) is significantly more complex as compared to games with small user numbers. Higher user numbers cause more events to be processed and data to be exchanged while larger game worlds accommodating multiple users increase the amount of maintained environmental data. Although hardware performance advancements have increased the number of users that a single server can serve, going beyond these limits requires a multi-server design which is able to employ multiple resources for the game processing and data distribution tasks. This leads to a number of new, challenging design issues, including inter-server communication, object migration between servers, synchronization of data across servers, load balancing, latency hiding, etc. The current practice of game development usually relies on the know-how of developers and employs low-level programming and networking tools, which makes it quite an expensive, time-consuming and risky endeavor.

This chapter describes a novel, high-level approach for developing real-time interactive MMOG and their efficient execution. Our approach is based on the new software platform – Real-Time Framework (RTF) – developed at the University of Muenster for a wide spectrum of online games: from traditional single-server games, over first-person shooters to large-scale multi-server MMOG, with an additional possibility to enhance an initially single-server design to a multi-server solution in order to support an increasing number of players. The RTF, whose architecture and the supported development methodology was introduced in (Ploss, Glinka, Gorlatch, & Müller-Iden, 2007), provides integrated solutions for both game development and their run-time execution. In the chapter, we focus on the following topics: the object-oriented design and efficient transmission of game data structures; interfaces and services that allow the developer to efficiently process an MMOG on multiple servers; integrated monitoring and controlling functionality; management mechanisms for the multi-server distribution of MMOG; and the trade-offs for employing the proposed development approach and its multi-server design.

The contributions and the structure of the chapter are as follows. We start by describing the contemporary design model of online games based on a real-time loop and present a comprehensive taxonomy of the current game development approaches, with respect to their complexity and flexibility. We then outline our high-level game development approach and describe the concepts of the supporting RTF middleware as a development tool. We continue with a practical case study on how a single-server multiplayer game is designed using RTF. The following section describes the multi-server distribution mechanisms in RTF and demonstrates how they allow to realize a seamless virtual environment across zone borders and a transparent transfer of game parts between resources. Then, we show how the originally single-server commercial shooter game Quake 3 was ported to a multi-server design using RTF, and we experimentally evaluate the RTF-based version against the original Quake 3 with respect to the most important performance and quality metrics: the achieved maximal client numbers, resource consumption, and client-server response time. The conclusion summarizes our contributions in the context of related work.

GAME DEVELOPMENT APPROACHES

The majority of today's online games simulate a spatial virtual world which is conceptually separated into a static part and a dynamic part. The static part covers, e.g., environmental properties like the landscape, buildings and other non-changeable objects. Since the static part is pre-known, no information exchange about its state is required between servers and players. The dynamic part covers *entities* like avatars, non-playing characters (NPCs) controlled by the computer, items that can be collected by players or, generally, objects that can change their state. Both parts, together, build the game state which represents the game world at a certain point of time.

In a continuously progressing game, the game state is repeatedly updated in real time in an endless loop, called *real-time loop*. Figure 1 shows the structure of an iteration of the real-time loop for a multiplayer game based on the client-server architecture. The figure shows one server; in a multi-server scenario, there is a group of server processes distributed among several machines. A loop iteration consists of three major steps: At first, the clients process the users' inputs and transmit them to the server (step 1 in the figure). The server then calculates a new game state by applying the received user actions and the game logic, including the artificial intelligence (AI) of NPC and the environmental simulation, to the current game state (step 2). As the result of this calculation, the states of several dynamic entities may change. The final step 3 transfers the new, updated game state to the clients.

When designing the real-time loop for a particular game, the developer deals with several tasks. In steps 1 and 3, the developer has to organize the network transfer of the data structures that realize user actions and entities. If the server is distributed among multiple machines, then step 2 also requires the developer to organize the distributed computation of the game state and necessary communications for its update across different servers.

In order to compare different approaches to the game development, we build a taxonomy which is based on the following three major aspects that must be designed in an online game:

1. Game logic, including entities, events (implemented as data structures), and processing rules for the virtual environment;
2. Game engine which continuously processes user actions and events in the real-time loop to compute a new game state, according to the rules of the game logic;
3. Game distribution, including logical partitioning of the game world among multiple servers, and dynamic management of distributed computation and communication.

Figure 1. One iteration of the server real-time loop

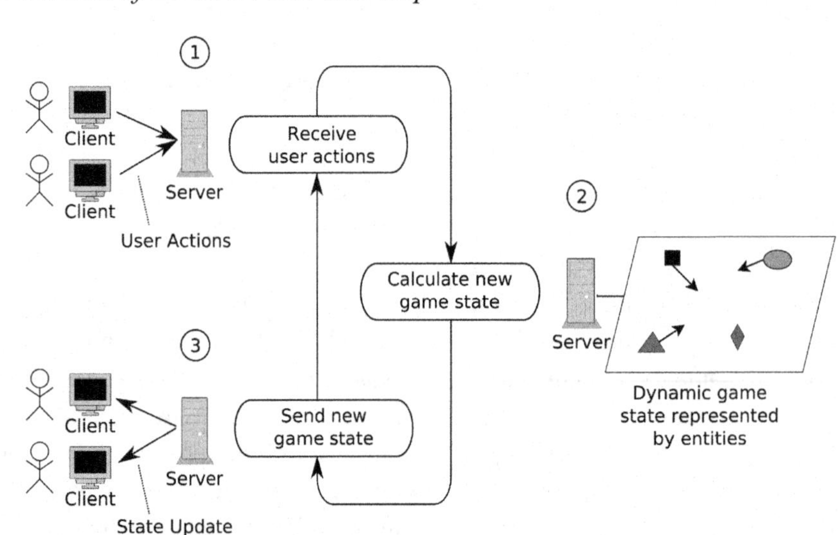

These three aspects are treated differently, depending on the requirements and properties of a particular game genre. For example, fast-paced action games rely on an efficient communication and engine implementation while using a relatively simple game logic. The complexity of the game distribution aspect usually increases with the number of avatars and the density of the participating users and is thus particularly challenging for MMOG.

In Figure 2, we distinguish four contemporary approaches a) – d) to game development, according to how they treat these three aspects, while the fifth approach e) is our alternative approach based on RTF. In each approach, the aspects shown in white are managed by the human developer, whereas the shaded areas are supported by the development system.

(a) Custom development: The most direct approach used for game development is to design and implement the entire software of the new game individually. The development team designs and implements from scratch all three major aspects of the game software system: game logic, game engine and game distribution. This allows the developers to have full control over their code and to optimize the implementation with focus on the individual performance needs of the game. While the custom development of an entire game is very complicated, cost-intensive and error-prone, it is sometimes the only way to achieve the particular objectives of the game design due to its flexibility.

(b) Game communication middleware: This approach uses special communication libraries and middleware systems - like Quazal Net-Z (Quazal) - for game development. As shown in Figure 2b), the game developer employs the middleware to realize the communication between clients and servers while implementing the game engine and logic on his own. Using this approach, the developer has enough flexibility to design and implement various aspects of game logic and game engine while the middleware deals with the

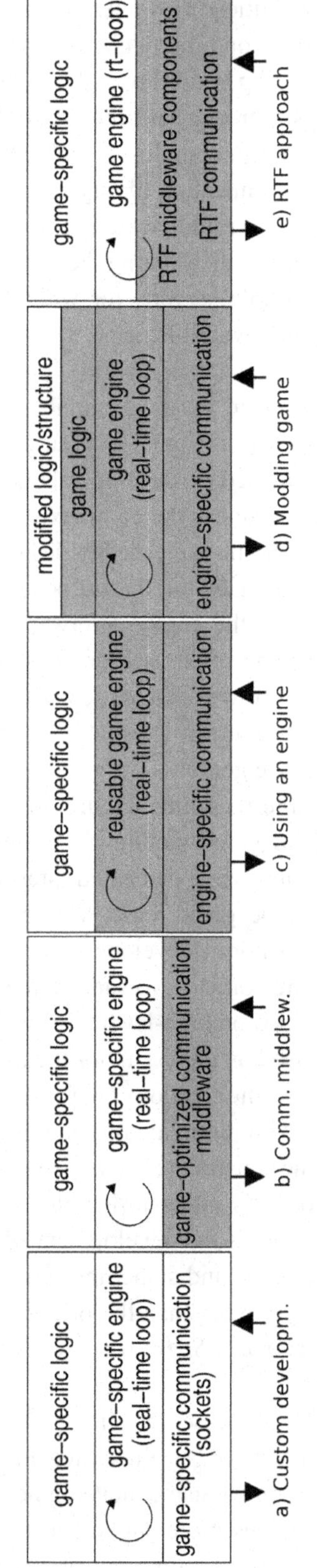

Figure 2. Main approaches to game development. White: self-developed; Gray: using existing components

game distribution. However, available libraries usually focus on a particular architecture setup, thus decreasing flexibility of the engine development. This approach has been used only rarely for the development of multi-server MMOG because a pure communication library is insufficient for these games: a middleware has to deal with the difficult task of distributing the game processing among multiple servers, for which only a few systems are available, e.g., (Emergent Game Technologies) and (BigWorld).

(c) Using an existing engine: with this approach, shown in Figure 2c), an existing game engine is reused to develop a completely new game. This reduces the complexity of game development. Some game studios design their game engines primarily for the purpose of reselling and licensing the engine afterwards. Examples of popular and often used engines are *Quake 3* and *Unreal*. However, a particular engine is quite inflexible because it is usually closely tied to a specific game genre.

(d) Game modding: Figure 2d) outlines the approach of game modding, i.e. modifying an existing game via a dedicated interface for programming the game logic. This approach was first used by hobby developers who modified the actual game content. Nowadays, the creation of mods is based on high-level tools which are created and used by game studios themselves. Such tools support the creation of new game content by designers with minimal programming effort; because of the constraints of an existing game logic, the approach is rather inflexible. Nevertheless, modding allows the development of innovative game concepts, and sometimes a mod becomes even more popular than the original game, e.g., the mod *Counter-Strike* based on Valve's game *Half-Life*.

(e) RTF Multi-Server Middleware: Our Real-Time Framework, as illustrated in Figure 2e), enables a novel game development approach which provides more support to the developer than using only a communication middleware, but does not constitute a complete game engine, thus allowing higher flexibility. Therefore, RTF can be positioned in between the approaches b) and c). The characteristics and usage of the RTF-based approach, justifying this classification, are discussed in the next sections.

Figure 3 illustrates our taxonomy of the five development approaches a) - e) with respect to their flexibility and complexity. The most simplicity in terms of distributed software infrastructure is offered by existing game engines c) and modding toolkits d). However, these approaches have the serious drawback of being quite inflexible. Obviously, the fully customized development a) offers most flexibility while being rather complex. The use of special middleware b) is a promising alternative for particular tasks: its use reduces the complexity of game development. Communication libraries are not enough for MMOG: for such large distributed systems, the multi-server management is quite extensive and increases the development complexity. As indicated in the taxonomy, RTF is designed to provide the developer with the highest possible flexibility in game design while freeing him from complex low-level implementation tasks in the game development process.

AN OVERVIEW OF RTF

The Real-Time Framework provides a high-level development API together with a communication and computation middleware for the execution of single-server and multi-server online games. RTF supports both the server-side and client-side processing in an online game. Figure 4 shows a generic example of a multi-server game developed on top of RTF, with 2 clients and 2 servers.

The role distribution between the framework and the developer is as follows. RTF deals with the entity and event handling in the real-time client loop and with the continuous game state processing in the real-time server loop, as well as

Figure 3. Taxonomy of game development approaches

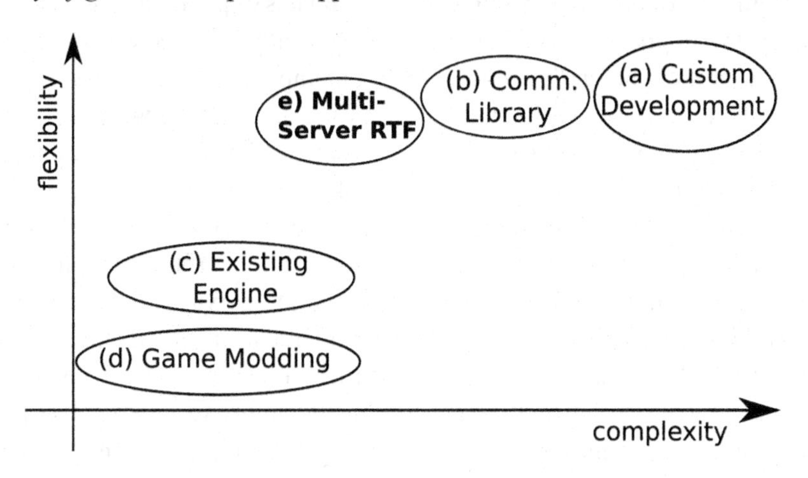

with the distribution of the game state processing across multiple servers. The developer implements the game-specific real-time loop on client and server, as well as the game logic, using RTF to automatically exchange information between the processes.

The architecture of RTF follows a modular approach and provides components for various aspects of distributed games in addition to the game processing itself. Figure 5 shows the RTF components, the central of which is the *Communication and Computation Parallelization (CCP) module* which handles the game processing. The

Figure 4. RTF-based multi-server game

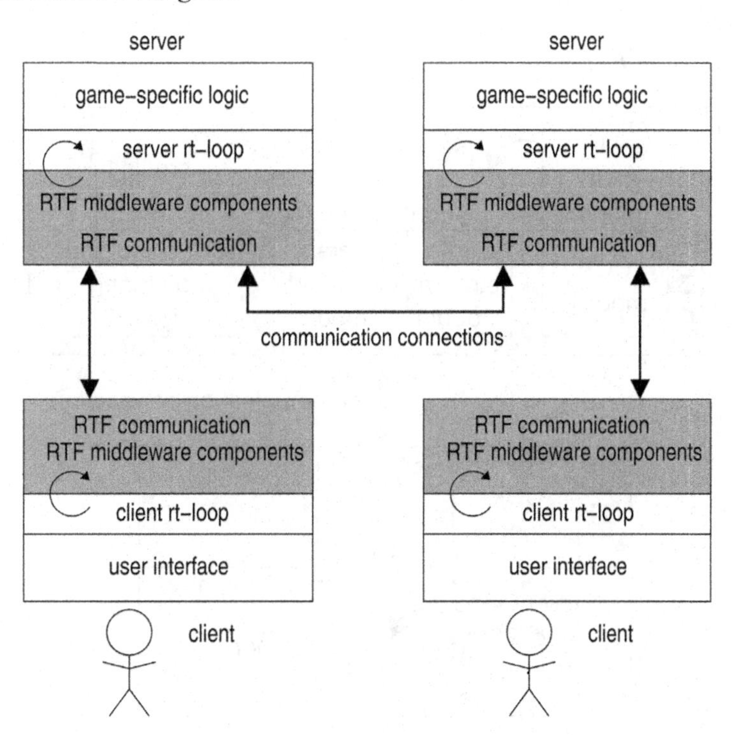

persistency module allows for storing/retrieving entities specified by the application developer to/from a relational database. The *audio streaming module* supports Voice over IP (VoIP) audio communication over RTP (Real-Time Protocol) and provides an interface for setting up channels, and switching users between them. The modular architecture makes RTF extensible without a major impact on the existing parts of the middleware.

The *controlling and monitoring module* is the middleware-endpoint that allows the game developer to issue commands for steering the game application at runtime and to report the application status. The developer can define a particular profile on top of this module, which reflects the game-specific monitoring characteristics the developer is interested in. There also exist predefined profiles for typical monitoring metrics and controlling tasks in games: RTF can report communication characteristics (bandwidth consumption, latency, packet rate) and distribution characteristics (number of clients, entities, and exchanged events), such that the application developer is not required to request this information explicitly.

RTF is implemented as a C++ library, because C++ is arguably the language of choice in most contemporary MMOG designs. Developing an online game consists of several tasks shown in Figure 6. The developer deals with three general tasks - AoI management, game state processing and data-structure design - when designing a game on top of RTF. If the game uses multiple servers, then multi-server parallelization and distribution also have to be taken care of. To simplify these tasks for the developer, RTF provides a variety of functionality like optimized serialization and communication, management and processing (possibly distributed) of the game state. This separation of roles between the developer and RTF reflects the overall idea of the RTF-based approach: providing high-level game functionality on top of an optimized communication middleware.

Figure 5. Modular design of RTF

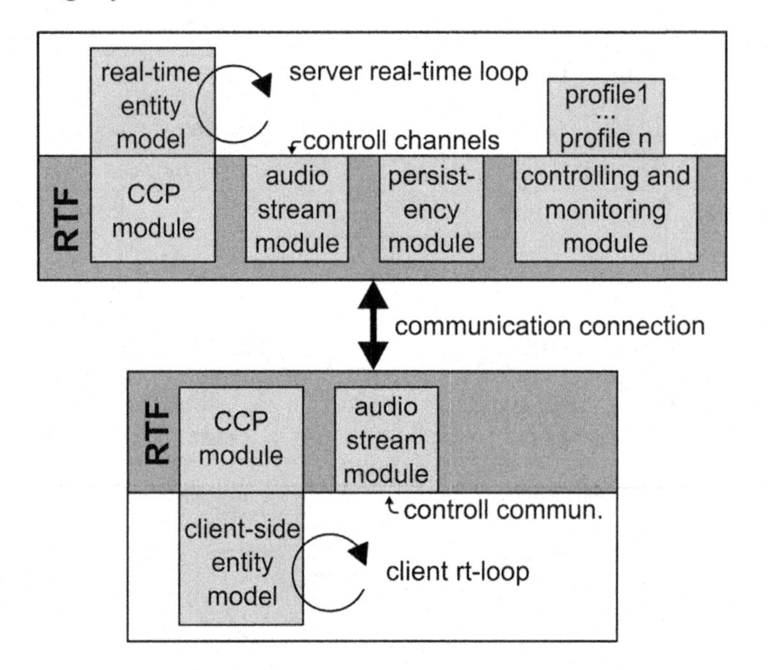

Figure 6. Development tasks for a multi-server game: role distribution between RTF and developer

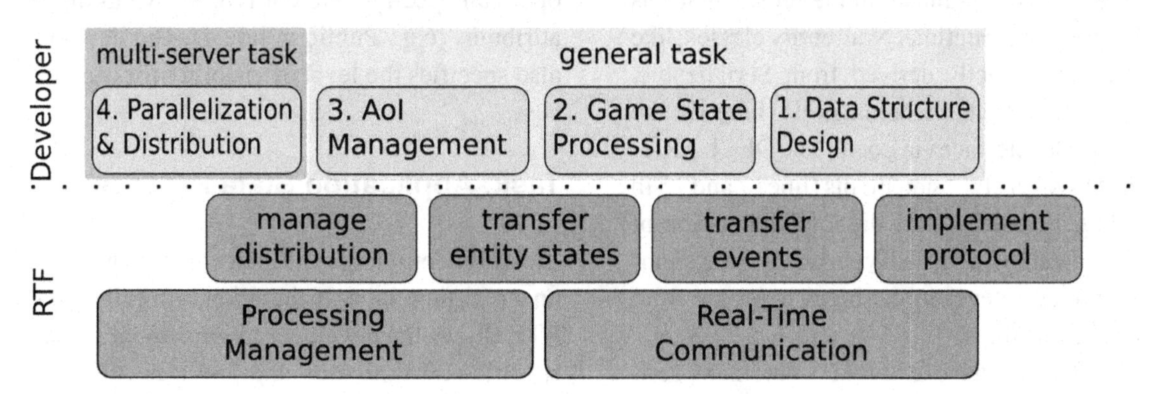

CASE STUDY: USING RTF FOR DEVELOPING A MULTI-PLAYER ONLINE GAME

In this section, we demonstrate the practical use of RTF for game development by designing a simple but still quite challenging multi-player online computer game.

Task: Data Structure Design

The dynamic state of a game is usually described as a set of *entities* which represent avatars and non-player characters in the game world. Besides entities, *events* are other important objects in a virtual environment used for representing user inputs and game world actions. Hierarchical data structures for events and entities in complex virtual worlds have to be serializable for efficient network communication.

Describing the Entity State When using RTF, entities and events are implemented as C++ classes. The developer defines the semantics of the data structures according to the game logic. The only semantics of entities that are predetermined by RTF is the information about their position in the virtual world. Entities, therefore, are derived from a particular base class Local of RTF that defines the representation of an entity position. This is necessary since the distribution of the state processing across multiple servers is based upon the entity location. Besides the requirement of inheriting from Local, the design of the data structures is completely customizable to the particular game logic.

We start developing our example game from a class to model the state of a player's avatar: Box 1 shows the class Avatar inheriting from Local, in which the position and dimension of the entity are described. Other attributes describe the game-dependent state of the avatar (lines 9-14). Attributes can be primitive types (health, maxHealth) including enumerations (AvatarType in line 9), classes (velocity and dimension), or even more complex containers of classes (annotations). Methods think and move (lines 5 and 6) implement the modifications of the avatar state.

RTF Serialization RTF provides automatic serialization of the entities and events defined in C++, implements marshaling and unmarshaling of data types, and optimizes the bandwidth consumption of the messages. While the developer specifies entities and events as usual C++ classes, RTF provides a generic communication protocol implementation for all data structures following a special class hierarchy. All network-transmittable classes inherit from the base class Serializable of RTF. The Serializable interface can be either implemented by the developer or automatically implemented using the serialization mechanism provided by RTF. For all entities and events implemented in this manner, RTF automatically generates network-transmittable representations and uses them at runtime. Non-entity classes, like actions, are directly derived from Serializable, whereas the Avatar automatically inherits the Serializable interface via Local. The DECLARE_SERIALIZABLE_* statements (lines 7 and 15 in Box 1) generate code for the implementation of the Serializable interface. TypeAvatar is a system-wide unique integer to distinguish Avatar from other Serializables.

Box 2 shows the use of RTF's automatic serialization. The IMPLEMENT_SERIALIZABLE statement (line 3) generates the implementation of the Serializable interface. The developer needs to describe attributes that should be transmitted over the network. For example, the ADD_ATTRIBUTE statement (line 4) adds the velocity attribute to the description of the serialized form of the Avatar. The automatic serialization mechanism can handle delta updates, i.e., only transmitting information that indeed changed, and differentiated updates for different processes. In order to use delta updates, the developer tracks the modification of attributes in a mask provided by RTF. To use differentiated updates, the developer can specify different types of visibility for attributes (e.g., Public in line 4). The developer also specifies the level of visibility for each process.

Task: Application State Processing

Most contemporary multiplayer games are based on a real-time loop; its iterations are called *ticks*. RTF allows the developer to implement a game-specific real-time loop and provides substantial support for implementing and running this loop on both the server and client side: the client determines the user actions and displays the current game state; the server performs the processing of the events and the updating of the game state. We describe the design of these steps next.

Client: Determine User Actions

At first, the client reads the user's input (from keyboard/mouse). Then it determines the desired action and sends it to the server, using the ClientCCPModule (Box 3, line 5). Serialization and transmission are done by RTF transparently.

Box 1. Class avatar models the state of a player's avatar

```
class Avatar: public RTF::Local {
public:
 /* process a new avatar state */
 void think(const double& passedSec);
 void move(RTF::Vector movement);
 [..]
 DECLARE_SERIALIZABLE_PUBLIC(Avatar, TypeAvatar)
private:
 AvatarType avatarType; // type of the Avatar (enum)
 RTF::Vector velocity; // movement
 RTF::Vector orientation; // direction
 rtf_uint16 health; // cur hitpoints
 rtf_uint16 maxHealth; // max hitpoints
 RTF::Annotation annotations; // State changes, Actions
 DECLARE_SERIALIZABLE_PRIVATE(Avatar) };
```

Box 2. Implementation of the avatar

```
[..] // application-specific code goes here
#include <rtf/GenericSerializerImpl.cpp>
IMPLEMENT_SERIALIZABLE_DERIVED(Avatar, RTF::Local,
     ADD_ATTRIBUTE(Avatar, velocity, Unreliable, Public)
[..] // dito for all network-transmittable attributes
     ADD_ATTRIBUTE_DEFAULT(Avatar, annotations))
```

Box 3. Sending a user action via ClientCCPModule from client to server

```
void ClientActionFactory::sendActionMove(){
 serverPos = mAvatar->getLocation().getPos();
 RTF::Vector newPos = mGraphicManager->getPlayersPosition();
 ActionMove moveEvent(newPos-serverPos);
 mClientCCP->sendEvent(moveEvent); }
```

Server: Process User Actions

On the server side, events are automatically received by RTF and appended to the event queue. The server processes these events and calculates the new game state as shown in Box 4. It accesses RTF's event queue via the EventManager (line 4) and uses the RTF type identification to determine the correct class of the event (lines 5 and 6). The move action refers to a specific entity (line 7) which can be retrieved from RTF via the ClientManager, which allows the developer to determine the client which has sent the event and to access its avatar (line 8). After that, the action is applied to the game state (line 10).

Server: Compute the New Game State

At this step (Box 5), the active entities are updated according to the game rules and game logic. As our implementation applies move actions directly to the entity state, we do not have to move play-

Box 4. Processing of game

```
// RTF::EventManager& em = ccpModule.getEventManager();
// RTF::ClientManager& cm = ccpModule.getClientManager();
void Server::processEvents() {
for(RTF::Event* e=em.popEvent(); e!=NULL; e=em.popEvent()) {
 switch(e->getEvent().getType()) {
 case ActionMove::TYPE: {
 Avatar &actor = (Avatar&)
      cm.findClient(e->getSender())->getAvatar());
 ActionMove& actionMove = (ActionMove&)e->getEvent();
 actor.move(actionMove.getMovement());
} break; } } }
```

Box 5. Update all entities

```
void Server::updateAllEntities() {
 std::map<RTF::DGObjectID, RTF::Local*>::const_iterator it
     = om.getActiveObjects().begin();
 for(; it != om.getActiveObjects().end(); it++) {
 switch(it->second->getType()) {
 case Avatar::TYPE:
 Avatar& avatar = (Avatar&) *it->second;
 avatar.think(ticklength); // let every active
[..] } } } // process other types of entities etc.
```

ers' avatars at this step anymore. But we have to update the rest of the game state, e.g., to move the non-player characters.

Server Real-Time Loop

The complete processing cycle for the server is shown in Box 6. During the onBeforeTick (line 2) and onFinishedTick (line 6) calls, RTF fills the event queue and sends the state updates to the clients. Therefore, the game application should not modify the game state concurrently to these calls. The call in line 6 deals with the AoI Management as described below.

Client Real-Time Loop

The real-time loop on the client side (Box 7) is similar to the one on the server side.

After determining user actions and sending them to the server (line 5), newly arrived updates from the server are processed (line 6) and the new game state is displayed on the screen (line 7). The client loop is completed by the onFinishedTick (line 8) and onBeforeTick (line 3) calls, during which incoming events sent by the server are enqueued and game state updates are applied.

AoI Management

The *Area of Interest (AoI)* concept assigns a specific area to each avatar in the game world where dynamic game information is relevant for this avatar and thus has to be transmitted to the avatar's client. AoI optimizes network bandwidth by omitting irrelevant information in the communication. A common approach considers entities and, correspondingly, their updates to be relevant for a particular player if they are within a certain distance to the player's avatar, typically the view range. More advanced approaches combine multiple techniques in order to reduce the

Box 6. Server real-time loop

```
while(!serverQuit) {
 ccpModule.onBeforeTick(); // inform RTF about begin of tick
 processEvents();
 updateAllEntities();
 interestManagement.update();
 ccpModule.onFinishedTick(); // inform RTF about end of tick
 [..] } // sleep, check for server quit etc.
```

Box 7. Client real-time loop

```
while(mInputProcessor->mContinue) {
 // sleep, calculate time since last tick
 mClientCCP.onBeforeTick();
 // Capture input and send actions (e.g., sendActionMove)
 mInputProcessor->capture(inputTimer.getTicklength());
 updateEntities(timeSinceLastTick); // apply state updates
mGraphicManager->renderFrame(); // render a frame
 mClientCCP.onFinishedTick(); }
```

amount of transmitted information even further, e.g., exploit the viewing angle and attenuate the update frequency for distanced entities for a given avatar (Bezerra, Cecin, & Geyer, 2008).

RTF supports the custom implementation of arbitrary AoI concepts, with the exception of update frequency attenuation, by offering a generic publish/subscribe interface. The custom interest management continuously determines which entity is relevant for a client and notifies RTF about each change of the "interested" relation through client.subscribe and client.unsubscribe calls. RTF guarantees that the entity is available and updated at the client or removed from it and automatically manages the replicas at the other processes (clients or servers) according to the designed AoI. Clients are informed about (dis-)appearing entities via the ClientCCPModuleListener interface.

Box 8 shows the implementation of the objectAppeared callback, during which the application performs procedures to handle the newly appeared object, e.g., preparing the entity for introduction to the graphics engine (line 5).

General Task: Client Connection and Entity Creation

A general task that occurs independently of the continuous state updating is introducing new entities to the application state when they are created. A typical example is a new client connecting to the session or a newly spawned NPC. RTF informs the application about the connecting clients by the clientConnected callback of the ClientListener which allows creating, initializing and registering a client's avatar. The steps for creating an NPC are shown in Box 9.

DISTRIBUTION CONCEPTS SUPPORTED BY RTF

RTF currently supports three distribution concepts and their combinations for scaling multiplayer games: zoning, instancing and replication, all three shown in Figure 7. A novel feature of RTF is the possibility to arbitrarily combine the three

Box 8. Notification about new entities (client-side)

```
// ClientCCPModuleListener interface implementation
void Client::objectAppeared(RTF::Local& obj) {
switch(obj.getType()) {
 case gcf::TypeAvatar: {
 Avatar& avatar = static_cast<Avatar&>(obj);
 mGraphicManager->AvatarAppeared(avatar); break; } } }
```

Box 9. Introducing new entities to the application state

```
// Place a walking NPC in the world.
Avatar& npc = *new Avatar(Avatar::ZONE_TRAVELER,
RTF::Space(2400, FLAT_HEIGHT, 750, 40.0f, 85.0f, 40.0f),
RTF::Vector(50,0,0), RTF::Vector(1,0,0));
ccpModule.getObjectManager().registerActive(npc);
```

orthogonal distribution approaches depending on the requirements of a particular game design. Figure 7 illustrates a possible combination of these approaches in a single game. This is an improvement compared to contemporary MMORPG which combine zoning and instancing, whereas replication is currently not available for a combination with either of them.

Zoning (Cai, Xavier, Turner, & Lee, 2002; Macedonia, Zyda, Pratt, Barham, & Zeswitz, 1994; Rosedale & Ondrejka, 2003) partitions the virtual world into disjoint parts, called zones, and assigns each zone to one server. Figure 7 shows a virtual world with four zones on four servers; clients and entities move and interact within these zones. Although clients can move between zones, no interaction between clients in different zones is possible which is a potential drawback for some game designs. Zoning is widely used in contemporary MMORPG and allows large player numbers in such games. The zoning approach fits best for games where the players are reasonably distributed in a very large virtual world. It requires that clients are always connected to their responsible server, i.e., the one processing the client's zone. RTF supports zoning by performing run-time checks, for all clients, that monitor whether this condition is met and by transferring clients automatically to their responsible server. For the developer, such a transfer happens completely transparently on the client side and only causes a notification on affected servers.

Instancing creates multiple copies of special parts of the game world. Figure 7 shows how a small area (the grey rectangle) is processed in separate copies by two different servers. Instancing in online games comes in two flavors. Firstly, complete zones may have several independent copies which can be accessed by any player. Players that enter an instanced zone have to choose one particular copy or the game logic chooses it on behalf of the user. The second flavor is to have

Figure 7. Combination of zoning, instancing, and replication for a single-game world in RTF

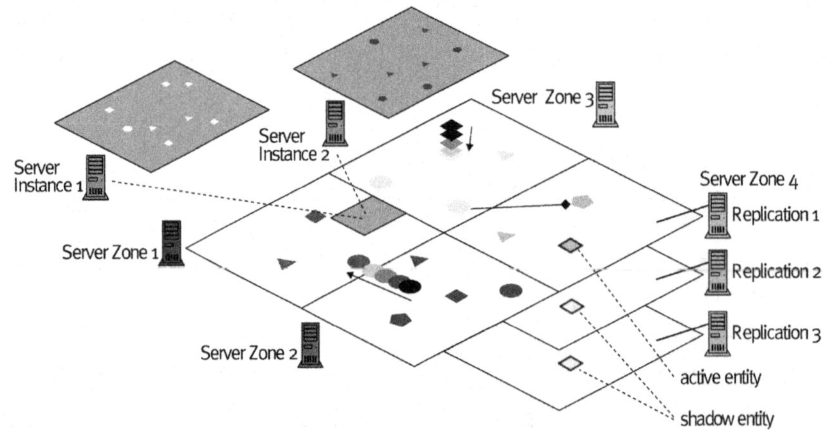

instances for smaller areas in the game world, such that an instance is created on demand by the players or groups of players, upon entering such an area. Both flavors of instancing are supported within RTF. A client that enters an instanced area - depending on the instancing flavor - either triggers an automatic instance creation within RTF or an existing instance must be specified as the transfer target. Subsequently, the client is automatically transferred by RTF to the server that is responsible for the new or the specified instance.

Replication (Bharambe, Pang, & Seshan, 2006; Müller-Iden, 2007) is an alternative distribution approach recently discussed in academia. Figure 7 shows three servers which cooperatively process the same zone. Each of the servers replicates its data in a symmetric manner and each server is responsible for some disjoint part of the overall data. RTF supports the replication concept by allowing the addition of entities to a zone that is managed by multiple servers. A server which creates a new entity in a zone automatically becomes the responsible server for this entity which is then called *active entity* on this server. RTF automatically starts to replicate the active entity on all the remaining servers of the zone; these replicas are called *shadow entities* on the remaining servers.

All three orthogonal distribution concepts aim at different scalability dimensions: zoning allows high user numbers in large MMORPG worlds, instancing runs a large number of game world areas independently in parallel, and replication targets high user density for action and player-vs-player-oriented games. If the players are regularly distributed across the game world, then zoning is the best choice as it scales linearly with the number of zones. But if players tend to crowd at certain places, then instancing or replication become necessary. Instancing scales linearly with the number of instances, but it introduces multiple copies and their conceptual drawbacks. If the creation of multiple copies of certain areas does not fit into the game logic, then only replication allows coping with increased player density while providing a single, seamless world. The game logic usually determines the best possible combination of these concepts. For example, zoning can be used for a huge game world, while certain dungeons in this game world are instanced for groups of players, and cities in this game world are replicated to allow for an increased player density.

The integration of all three concepts enables RTF to provide new functionalities for game developers, without the need for a specific application support. The next paragraphs explain how the combination of zoning and replication enables RTF to transfer clients from one zone into another without a noticeable interruption on the client side. This combination also allows for interactions across zone boarders – which must otherwise be implemented separately. The combination of instancing and replication allows for the reassignment of zones and instances during run-time to new machines as a reaction to the increase of load in certain zones instances. Adding an additional server to a zone or instance redistributes the load and thus raises the supported number of clients for this zone or instance.

Traditionally, using zoning as a distribution concept is subject to two restrictions: a) entities and clients must be transferred between the participating servers if they move across the zones, and b) no interactions are allowed across zone borders. Therefore, the game developer usually has to implement the required transfer by explicitly establishing a connection to the new server and communicating the entity's view from this server to the client. To allow interactions, e.g., attacking a remote entity across the zone border, special synchronization and inter-server communication are required, typically with two negative effects: increasing the overall complexity of the game architecture and reducing its scalability.

Let us describe how RTF provides a transparent solution for these problems and, furthermore, allows the movement of zones between servers. RTF supports a seamless inter-zone migration and interaction by creating an overlapping area

Figure 8. Using overlappings for a seamless migration of an entity between two zones

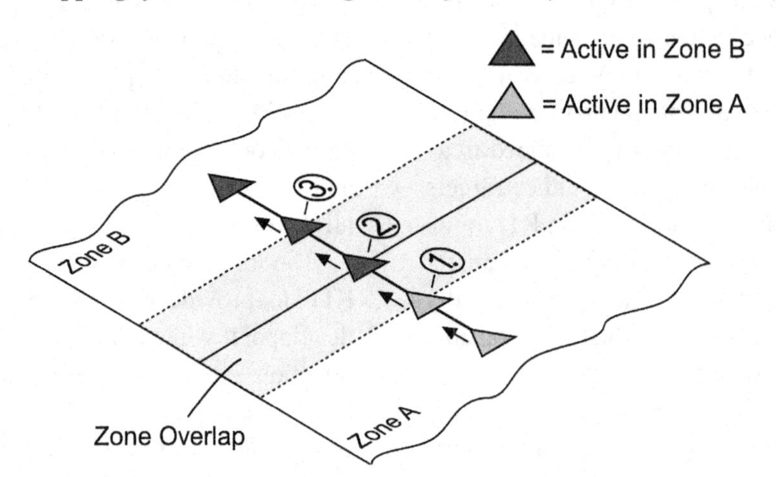

between two or more adjacent zones. Since RTF allows to freely combine zoning with replication, this overlap is replicated across the servers. Figure 8 illustrates how two zones are overlapped in the 2D case, thus creating a seamless game world. The movement of an entity between these zones is handled as follows:

1. The entity moves within zone A into the overlap and is replicated as a shadow entity on the server of zone B (step 1 in the figure).
2. As soon as the entity has moved half the distance of the overlap, RTF automatically changes its status in A from active to shadow and vice versa in B (step 2).
3. As soon as the entity leaves the overlap, RTF removes it from zone A (step 3).

If the entity is a client's avatar, then the communication connection of the client to the server of zone A must be transferred during step 2 to the server of zone B. RTF manages this seamlessly if the developer makes the overlap bigger than the area of interest of the client: in this case, both servers responsible for A and B have the same view of the game world within the replicated area, such that no initial communication between the new server and the client is necessary. Furthermore, interactions across the border of the two zones are now possible because they take place within the replicated overlap area and the client is located in both overlapping zones at the same time. In summary, this leads to a completely seamless game world for the clients.

RTF also supports a migration of a zone to another server during runtime, by using a replication mechanism similar to the one shown in Figure 8. Since the migration is performed over an extended period of time, no interrupts are necessary, such that the players enjoy a smooth game flow. In addition, this allows dynamically re-assigning of servers depending on the current system load, which is particularly relevant for the utilization of Grid or Cloud resources.

With respect to the supported load-balancing concepts for zoning, RTF follows the microcell and region approach presented in (Ng, Si, Lau, & Li, 2002). Here, the game world is divided into relatively small *microcells* of fixed size and position. These microcells can be grouped to form a contiguous area called *region* or *macrocell* and each region is assigned to one server. The microcells can then be moved dynamically between different macrocells, balancing the load among the available resources, following the dynamic player distribution within the game world. Load-balancing schemes like (Bezerra & Geyer,

2009; de Vleeschauwer *et al.*, 2005) use the client/entity/message per microcell metric and the number of interactions between clients located in different microcells for an optimal grouping of microcells. Obviously, microcells assigned to the same server do not generate additional synchronization overhead and, therefore, these algorithms aim at an optimal microcell/region grouping which minimizes the overhead between two regions as the result of the dynamic player distribution and interaction.

The load-balancing algorithms are not built directly into RTF. Instead, a rich, standardized monitoring and controlling interface is offered which allows an external load-balancing and resource management component to monitor load-indicative metrics and issue load-balancing commands, e.g., moving microcells between servers, creating instances or replicating highly crowded microcells on multiple servers. Keeping the load-balancing and resource management as external components proved to be suitable, especially for complex resource management scenarios based on Grid and Cloud infrastructures (Meiländer, Ploss, Glinka, & Gorlatch, 2011). So far, a neural network-based approach is used within the resource management system (Nae, Prodan, & Fahringer, 2008).

FROM A SINGLE TO MULTI-SERVER ONLINE GAME: A QUAKE 3 CASE STUDY

In this section, the development of multi-server games using RTF is illustrated using a real-world case study – the popular Quake 3 Engine, which was originally designed for a single server. We test the *responsiveness* (time for the user to perceive the result of his action) and *scalability* (increasing player numbers by using additional servers) of the RTF-based multi-server version of Quake 3 against the original single-server Quake.

The Quake 3 Engine

The Quake 3 Arena game engine (*Quake* in the sequel) was first released by ID Software in 1999 and has been one of the most popular and successful FPS games since then.

All main challenges of multiplayer action games can be observed for Quake: fast-paced gameplay, short response time, very high rate of interaction, and high density of players (Kwok & Yeung, 2005). Quake requires players to continuously move around and interact with the game environment, e.g., picking up powerups or attacking other players. The game is designed to support constant action where no player is likely to stay alone for a long time.

Like other action games, Quake traditionally uses a single-server architecture for the processing of the game state, see Figure 9: processing steps are executed in a continuously running real-time loop (RTL). This computation of the new state for all participants of a game session is handled by a single process which ensures a consistent game state for all participants. In Quake, this processing server can run either as part of a graphical user client or as a dedicated server on a separate machine. A client in a Quake session mostly implements the interaction with the player, i.e., it reads the user's input and displays the current game state (graphics rendering and sound output) received from the server. Therefore, user actions and game state updates have to be continuously transmitted between client and server. A game session is operated by the server in a continuously running loop of frames (Quake refers to one processing cycle as *frame*, RTF as *tick)*, the pseudo-code of which is shown in Box 10.

In order to incorporate multi-server processing, the update of the game state (lines 6-7 in Box 10), implemented in the game logic, needs to be parallelized. The challenge is to identify the parallelization criteria and implement the algorithms of the game logic in a parallel manner. The entity management has to deal with the distribution

Figure 9. Quake 3 client-server architecture

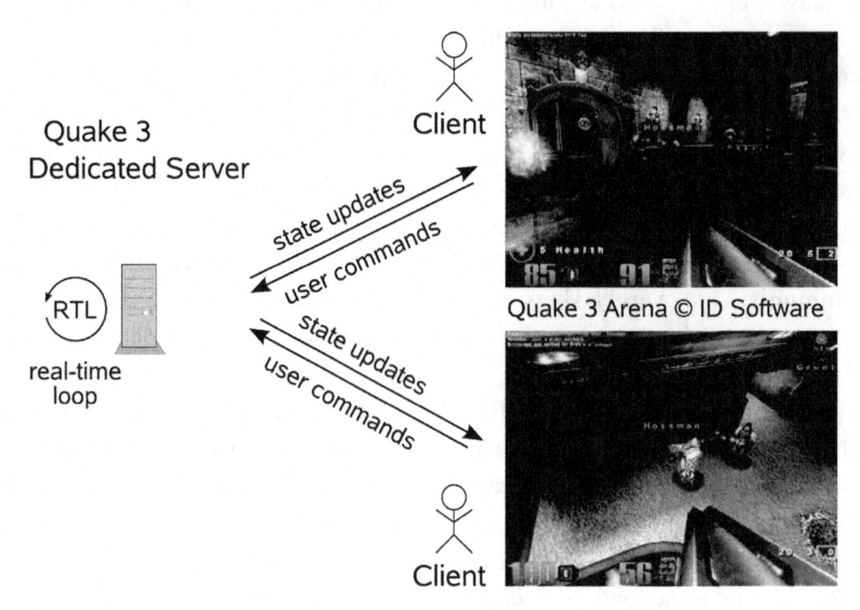

of the game state. The processing algorithm for the gameplay rules has to be coordinated between the participating servers. This requires additional synchronization and communication between multiple servers, which should be implemented by a suitable network communication component.

Multi-Server RTF Architecture for Quake

We employ the replication concept of RTF, in which the decision about the entity responsibility is made depending on the game rules and particular game play: entities may have relations with each other, which influence their interaction or require special coordination. Therefore, RTF leaves the task of entity distribution to the application developer who possesses the corresponding knowledge.

For Quake, this leads to a simple entity assignment: the server which creates an entity becomes responsible for it. This is the case for all player entities and also for dynamically created items (e.g., those dropped by players). The items that are a fixed part of the game world (powerups, weapons) are created and handled by one server throughout the game. The rate of interaction in Quake is very high and entities are constantly moving around, so we decided not to change the

Box 10. Quake server frame (three main steps)

```
void SV_Frame(int msec) {
 // NET_Sleep gives the OS time slices until either
 // get a packet or time for a server frame has gone
 NET_Sleep(frameMsec - sv.timeResidual);
// let everything in the world think and move
 G_RunFrame(sv.time);
// send messages back to the clients
 SV_SendClientMessages();
```

method of distributing entities based on their current location.

Figure 10 shows our multi-server architecture for distributing the Quake game state processing and the flow of information within the architecture between distributed processes. Each client is connected to one dedicated server: the client sends the user commands to this server and receives from it the state updates for all (visible) entities. Thus, the distributed state processing is transparent for the clients. The use of multiple servers requires communication between them: state updates are sent to shadow copies of entities, and the interaction between shadow and active copies needs to be coordinated. The next sections describe how we implement this for Quake using RTF.

The Quake Game State Processing

Figure 11 shows the main steps of the Quake processing cycle in one server when using RTF. In addition to the three main steps of the original Quake server frame (NET_Sleep, G_RunFrame, SV_SendClientMessages in the figure), the engine has to inform the RTF about the begin (onBeforeTick) and the end (onFinishedTick) of the processing cycle. The clients and other servers send *events*, e.g., user commands, via RTF, which enqueues them to the list of pending events. The processing of pending events is explicitly invoked from the engine (*processIncomingServerEvents*), and accesses the event queue (2.1 in the figure). To process the game state (3. G_RunFrame), the engine requires an access to the distribution management of the game state (3.1). RTF provides a list of active and shadow entities (getObjects) that allows the register to determine if a certain entity is active or shadow and also to create and introduce new entities (registerActive). The transmission of state updates is automatically handled by RTF during the onFinishedTick call, including the creation and transmission of messages. The original Quake sends the state updates explicitly in step 4., SV_SendClienMessages. With RTF, this step is reduced to a simple light-weight notification message that broadcasts the current server frame time which is required by the client engine.

The assignment of an entity to a server is done by registering the entity as active with RTF (registerActive). RTF automatically handles the replication of entities: it provides shadow copies of all entities to the remaining servers. These

Figure 10. The multi-server architecture for Quake

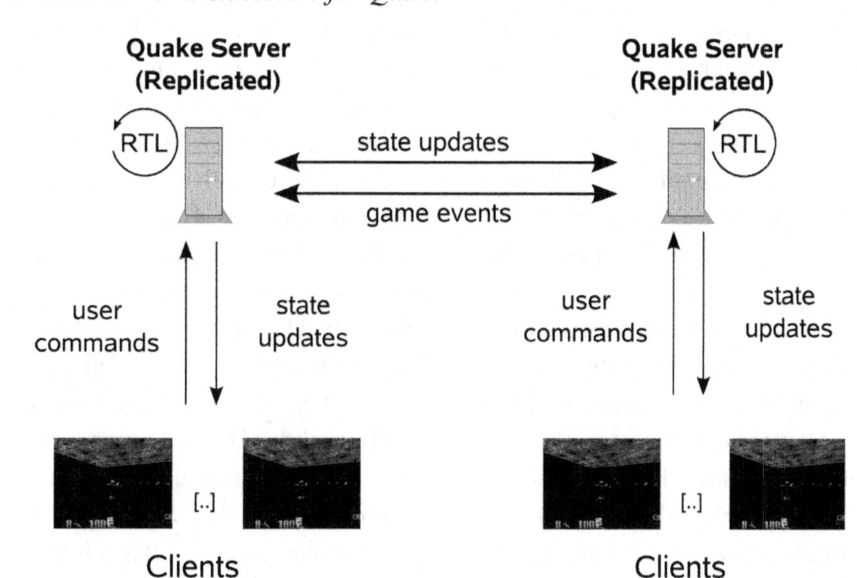

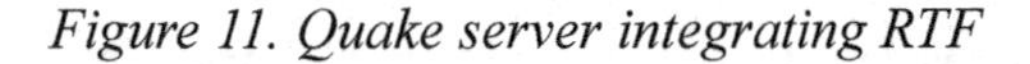
Figure 11. Quake server integrating RTF

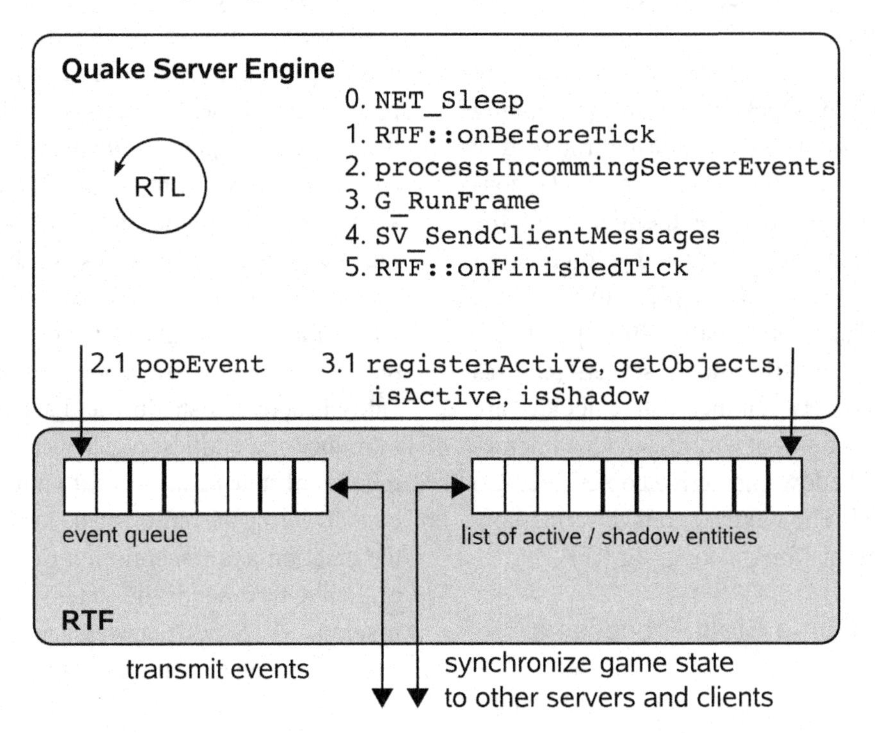

shadow copies are automatically updated with changes from their server. Moreover, RTF automatically informs the game developer about appearing and disappearing objects.

Distributed Game Logic Processing

Other than movement, interaction between entities (attacking, touching items or teleports and jump pads) is the main content of one game frame (3. G_RunFrame in Figure 11). When such an interaction is detected, e.g., entities collide with each other, the original Quake engine invokes a procedure to handle the interaction. For example, a combat interaction, e.g., a rocket hitting a player, is handled by procedure G_Damage. This procedure receives as arguments the target (hit player), inflictor (rocket), and attacker (player who fired the rocket) and the caused damage. It applies the damage, creates new entities like explosions, manages scores or issues splash damage for other entities. For the single-server case, this algorithm is quite simple. The algorithm is more complicated for the distributed case. If the involved entities are all active on the current server, then the procedure can be applied as usual. If the game entity is a shadow copy, then the interaction has to be handled by the server responsible for this entity.

Figure 12 illustrates how the remote interaction is organized for the multi-server Quake. If a server detects a damage interaction, the original G_Damage procedure is invoked (1. in the figure). The procedure determines, with the support of RTF, whether the involved entities are shadow or active entities (2.1). The result of the interaction is directly applied to the active entity (3). For the shadow entity, an event (4.1, DamageEvent) describing the interaction is sent through RTF (4.2) to the server responsible for the entity RTF automatically determines this server. On this server, the event is enqueued by RTF until the Quake engine processes it during step 5.2 of the real-time loop cycle. The remote server invokes the same G_Damage procedure (6.) with an additional flag which denotes the call as triggered remotely. This prevents a loop for

Figure 12. Distribution of the G_Damage procedure in multi-server Quake

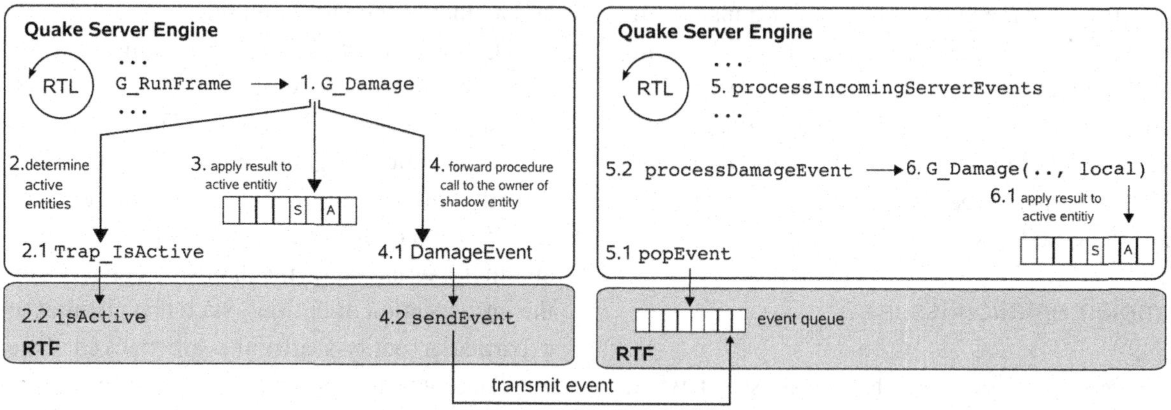

Quake Server Engine
RTL
...
G_RunFrame → 1. G_Damage
...
2.determine active entities
3. apply result to active entitiy
4. forward procedure call to the owner of shadow entity
2.1 Trap_IsActive
4.1 DamageEvent
2.2 isActive
4.2 sendEvent
RTF
transmit event
Quake Server Engine
RTL
...
5. processIncomingServerEvents
...
5.2 processDamageEvent → 6. G_Damage(.., local)
6.1 apply result to active entitiy
5.1 popEvent
event queue
RTF

shadow interaction handling, which is necessary because the active/shadow roles are non-swapped. This pragmatic approach is suitable for the Quake engine, since the interaction between entities is deterministic: each server will apply the same rules based on the arguments of the procedure. Hence, it is not necessary to coordinate the interaction between the servers.

Quake Engine Modifications

To use the described multi-server architecture for Quake with RTF, we modified and/or exchanged some components of the original Quake (Figure 13). The classes realizing the entities were simplified as they now rely on the automatic (de)serialization mechanism of RTF. The Entity Management component was modified to use RTF for the distribution of the game state. Since RTF automatically handles the transmission of events and state updates, the Network Communication component of the original Quake has become obsolete and, therefore, was removed.

Replacing Network Communication

In the original Quake, communication must be explicitly invoked whenever processing steps create messages and use the network communication component to send them. The concept of RTF is different. While the sending of (user) commands is also invoked explicitly by the application, serializing and sending state updates is automatically handled by RTF. The developer still has to define the usual C++ objects and describe, directly in the game code, the attributes that should be transmit-

Figure 13. Quake 3 Arena main software components

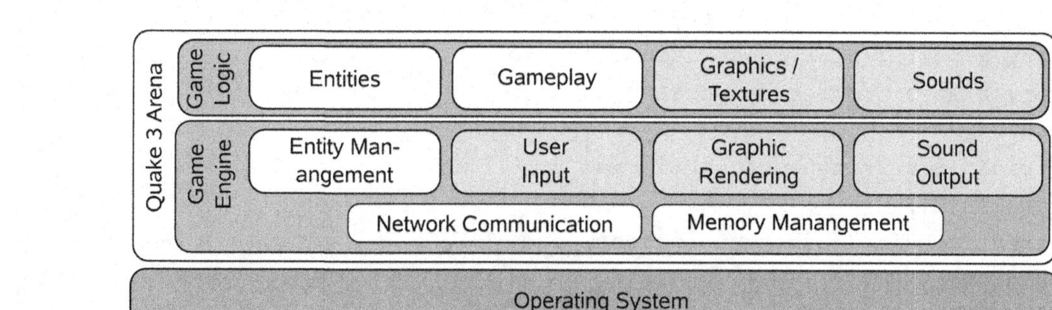

ted over the network. The ported Quake engine also uses this mechanism of RTF for transmitting user commands. The synchronization of the game state between servers and clients is now handled automatically by the RTF. Like the original Quake -- and like almost every online game RTF uses UDP as transport protocol for transmitting events and state updates.

Implementation Issues

We had to solve some technical issues when combining RTF with Quake. Quake was originally written in plain C while the RTF is designed for C++ which is used by most game companies nowadays. After some minor modifications that did not change the semantics of Quake, it was possible to compile the Quake sources with a C++ compiler. However, the separation of game engine and game logic, with the game logic compiled as binary code executed by the Quake Virtual Machine (QVM), could not be preserved. To circumvent this problem, we decided to compile game logic and engine together. Thereby, we lost the ability to use arbitrary modifications (mods) through pre-compiled plugins. However, the Quake modding community has also decided recently to give up pre-compiled binary game logic modules: mods like Urban Terror are delivered as a binary package including the Quake engine.

The Quake entities forming the game state are implemented as plain C structs (Box 12) which we mapped onto RTF classes. The engine accesses and modifies the fields of these structs directly throughout the code. To support the transmission of application-level objects with an automatic serialization mechanism, RTF relies on the introspection of entities, such that changes to a particular entity's attribute are tracked. One challenge was to find a feasible solution for the modification tracking without changing all accesses to an entity attribute in the original Quake code. Our solution is the template class ManagedType that can be used with several primitive types like int; it overloads all necessary operators, such that an instance of ManagedType<int> can be used like a usual int variable with tracked changes (Box 12).

Response Time Model

Fast-paced action games like Quake are highly sensitive to the system's responsiveness (Beigbeder et al., 2004; Pantel & Wolf): it is critical for players to observe the state changes caused by their own actions or actions of other players as soon as possible. The *response time* (t_{resp}) is

Box 11. Original Quake entity struct

```
public:
  int number; // entity index
  entiyType_t eType; // enum entityType_t
  trajectory_t pos; // for calculating position
  trajectory_t apos; // for calculating angles
  int time;
  vec3_t origin[3]; // typedef int[3] vec_3
  vec3_t angles[3];
  int otherEntityNum; // shotgun sources, etc.
  int groundEntityNum; // -1 = in air
[...]} entityState_t;
```

Box 12. Quake entity rewritten for RTF

```
class EntityState: public RTF::Local {
public:
 RTF::ManagedType<int> number; // entity index
 RTF::ManagedType<int> eType; // entityType_t
 Trajectory pos; // for calculating position
 Trajectory apos; // for calculating angles
 RTF::ManagedType<int> time;
 float origin[3];
 float angles[3];
 RTF::ManagedType<int> otherEntityNum; //shotgun, etc
 RTF::ManagedType<int> groundEntityNum; //-1 = in air
[...]};
```

the timespan between the moment when the user triggers a particular action (e.g., a shot with a rocket launcher) and the moment when the result is displayed on the client after one iteration of the real-time loop (RTL) has been executed (e.g., the fired rocket becomes visible).

The absolute response time t_{resp} consists of several phases accomplished in the game processing system: sending of the user command, processing of the command by the game logic in the server real-time loop (RTL), preparing and transmitting the information over the system network, and notifying the clients of the result. In order to analyze the game responsiveness as precisely as possible, we study the duration of these processing phases. In order to analyze the performance of the software independently of the effects caused by the network communication, we also consider the so-called *network-adjusted response time*, described in the sequel.

Our model of response time splits one iteration of the game's real-time loop in several phases. These phases are separated by nine so-called *breakpoints,* shown in Figure 14 for the original Quake engine and in Figure 15 for the RTF-port of Quake. Both sides of the figure consist of two parts: client and server. The breakpoints correspond to the method calls performed in the real-time loop; the methods' names are different for the two implementations, but we enumerate them according to their semantics, so that we can compare the phases of the two implementations. Figure 15 demonstrates that RTF automatically takes care of important parts of the network communication in the game, which are traditionally managed "manually" by the game developer. For example, at breakpoint 2., the user command is sent to the server using the CL_WritePacket method in the original Quake client engine; in RTF, this is done automatically by sendEvent.

In the following, we briefly comment on the main breakpoints and phases in the model. The start of a user action is marked by creating the user command in Quake (1. CL_CreateUmd); it takes time t_{cmd} for creating a command message which includes wrapping of the command in a network-transmittable form; it ends with dispatching the message to the operating system network stack (2.). The next phase – transmitting the command message over the network - ends when the message is unwrapped after it is received from the OS network stack (3.). This phase starts from (3.), takes time t_{event} and ends with the parsing of the user command (4.). The time for processing the command, denoted by t_{create}, ends when the rocket entity is created (5.). Afterwards, the update

Figure 14. Response time model of original Quake

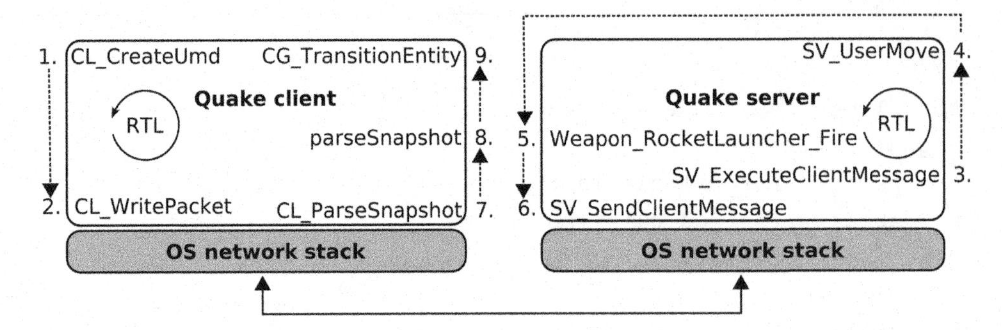

Figure 15. Response time model for Quake using RTF

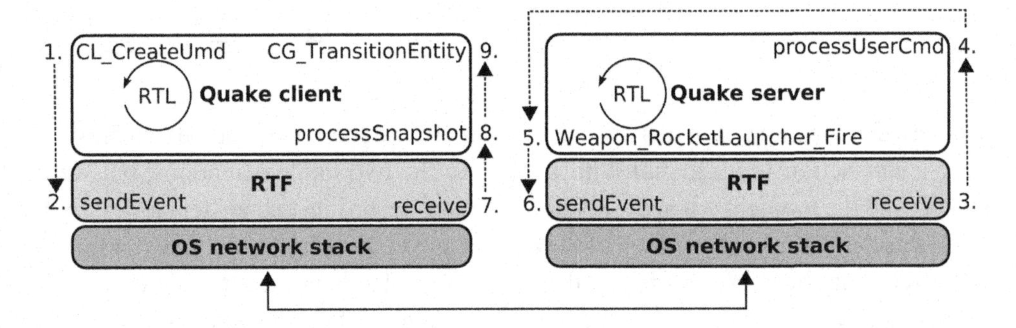

of the game state for the client is prepared, taking time t_{update}, and finishes by dispatching the update message to the OS network stack (6.). On the client side, the update phase t_{snap} begins upon the receipt of the update message (7.) and ends with applying the content of the update message to the client's game state (8.). The final breakpoint is the appearing of the rocket entity on the client display (9.).

To calculate the network-adjusted response time, we measure the time at breakpoints: T_1, ..., T_9, and compute the time of the described phases as follows:

$$t_{cmd}=T_2-T_1,\ t_{snap}=T_8-T_7,\ t_{disp}=T_9-T_8$$

$$t_{event}=T_4-T_3,\ t_{create}=T_5-T_4,\ t_{update}=T_6-T_5$$

$$t_{client}=t_{cmd}+t_{snap}+t_{disp},\ t_{server}=t_{event}+t_{create}+t_{update}$$

$$t_{process}=t_{server}+t_{client}$$

Here, t_{client} denotes the client processing portion and t_{server} the server processing portion, which together constitute the network-adjusted response time $t_{process}$. The total response time t_{resp} can obviously be measured as the difference between T_1 and T_9. The relation between t_{resp} and $t_{process}$ can be approximated as doubled latency between client and server, denoted by *RTT* (Round Trip Time):

$$t_{resp}=T_9-T_1\approx t_{process}+RTT$$

Experiments on Responsiveness

To compare the responsiveness of the original Quake with the RTF-based multi-server port of Quake, we examine the described use case of a player firing a rocket. We measure the times T_1 till T_9 which constitute our model. We conducted two kinds of experiments: 1) server and client running on the same machine, and 2) server and client running on different machines connected via LAN or via Internet. We used the first experiment to compare RTF and original Quake with respect to the network-adjusted response time ($t_{process}$). The second experiment allows us to study the absolute response time (t_{resp}) in comparison to the network-adjusted time.

Figure 16 shows the measurement results for the different phases of network-adjusted response time for client and server running on the same machine. For both implementations – original Quake and RTF-based, multi-server Quake - the largest amount of time is spent in phase t_{disp} on the client side, where the client integrates the appeared entity to the graphics engine. For the server part, we observe a difference between the two implementations: the phase t_{update} is comparatively long for the original Quake, whereas for the multi-server version the t_{event} phase takes most time. This is due to the fact that the length of one processing frame in Quake is fixed (50ms): an update for the game state is sent at the end of the frame, and hence is independent from when the event actually arrived within the server tick. For the two implementations, the point when the event is processed is different (RTF: at the end of the frame, original: directly upon arrival). However, for both versions the processing time on the server was always within one tick.

Absolute Response Time

To evaluate the impact of the network performance, we conducted experiments where we measured the absolute response time over different network con-

Figure 16. Response time, RTF-based vs. original Quake

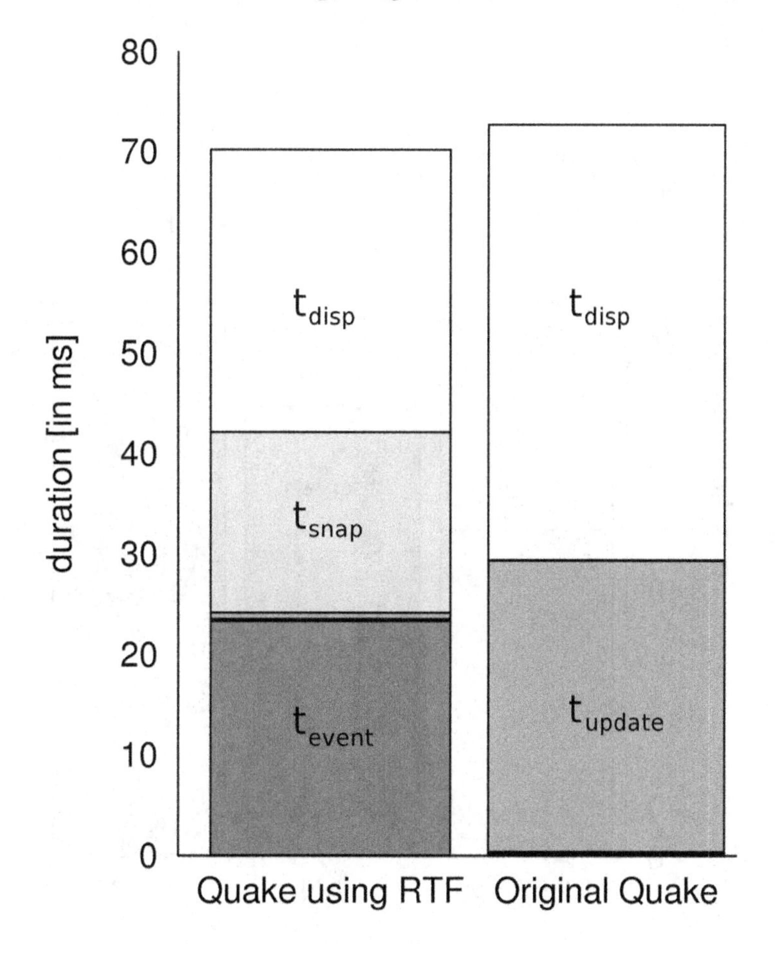

nections. Figure 17 shows the measured response time t_{resp} in comparison to the accumulated client and server processing time (t_{client}, t_{server}) obtained using our adjusted model and the round trip time (RTT) between client and server when using RTF. The RTT was measured using the Linux ping tool. The results show that for the Quake port using RTF, the measured response time is equal to the sum of the network-adjusted response time and the RTT, which confirms the estimation of our analytical model.

Figure 18 shows our measurements of the real-time loop saturation while connecting more and more clients. For this scalability test, most of the players were simulated using aimbots which automatically target other players, move their avatars and issue attack commands. The bots were configured to approximate typical Quake game play. We used P4 1.7 GHz processors with 500MB of RAM as servers, connected via LAN (100 MBit/s). In Figure 18, the saturation of the real-time loop is plotted against the number of clients per session. A saturation over 100% means that the server's active time exceeded the pre-defined frame time of 50 ms. The curves in the figure show the average saturation over several measurements. For one server using the RTF version, we also show the actual measured values, each as a single dot. For the multi-server case, we accumulated the average saturation over the 2, 3, and 4 servers while adding the total number of clients. The clients were evenly distributed among the servers.

Figure 17. t_{resp} for Quake using RTF over different networks

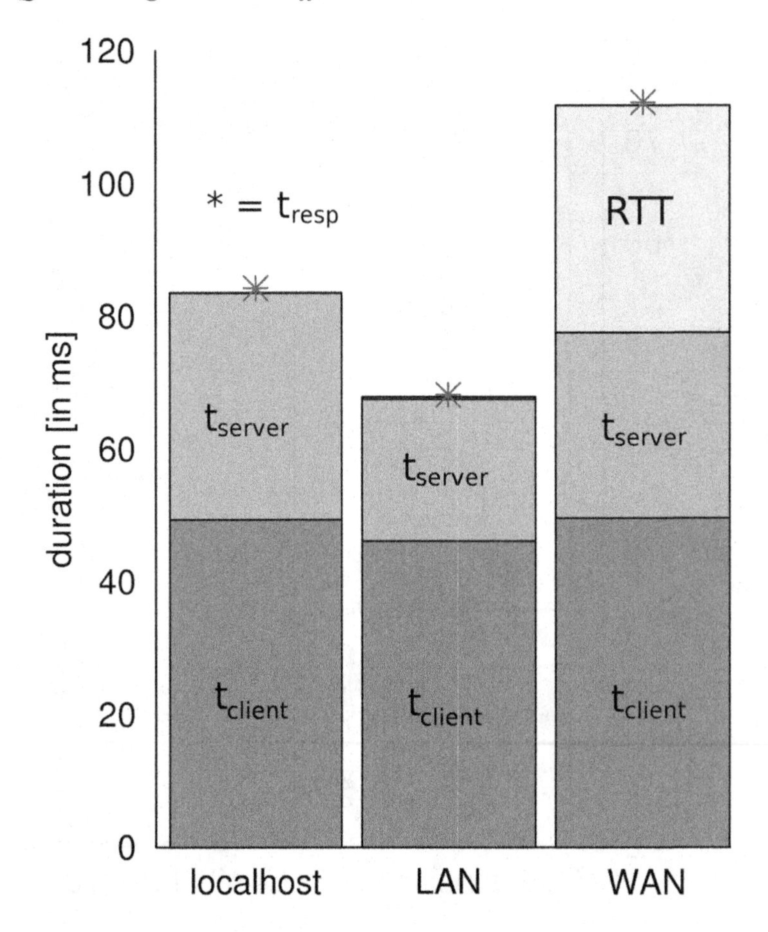

Figure 18. Real-time loop saturation vs. number of clients

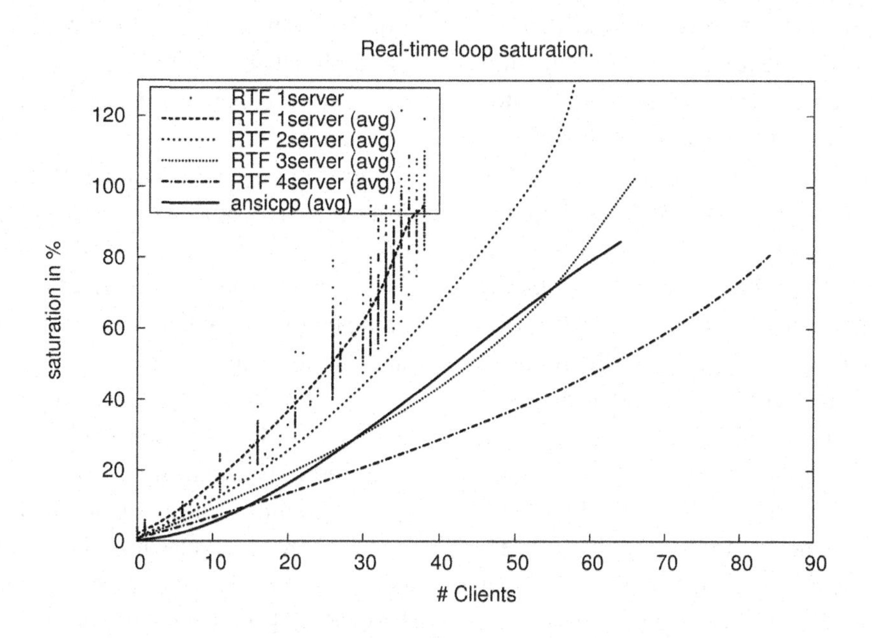

The experiments show that, with the ported RTF version, it is possible to handle up to 38 clients using a single node until the real-time loop becomes saturated, and about 54 clients with 2 servers, 66 and 84 clients with 3 and 4 servers, correspondingly. With the original Quake engine, a single server handling 64 clients was about 90% saturated. The maximum number of participating clients is technically limited to 64 for the original Quake engine. In our ported RTF-version, each server can handle more than 64 clients, so we extended the technical limit for the overall maximum number of clients per session. With 4 servers it is possible to support more clients than with the original single-server version.

CONCLUSION AND RELATED WORK

In this paper, we have studied and analyzed the contemporary development methods for massively multiplayer online games and demonstrated to what extent the traditional low-level custom development can be substituted by a high-level approach using game middleware, in particular for scalable multi-server engines. Our particular new contributions are as follows:

1. We develop a comprehensive taxonomy of contemporary game development approaches with respect to their flexibility and level of abstraction. Based on this taxonomy, we describe a new game development approach based on the Real-Time Framework (RTF) that aims both at single-server and multi-server settings and still provides a high degree of flexibility. Game developers are liberated from the low-level communication and distribution management tasks.
2. We describe in detail our RTF middleware system which is used to support the human developer in the game development process and to enable an efficient game execution. RTF's comprehensive distribution capabilities enable a smooth transition from a single-server to a multi-server game design. Since RTF focuses on the processing part of games, it puts no constraints on the remaining development tasks as, e.g., graphics or game logic implementation.

3. Using case studies, we demonstrated that:
   a. Game data structures are specified as plain C++ entities. The serialization for the network transfer is done automatically by RTF, and the underlying communication implementation is optimized with delta updates to reduce the amount of data sent over the network; the game logic and entities are implemented in an object-oriented way.
   b. The proven multi-server distribution concepts - zoning, instancing and replication - as well as their combinations and parallelization of the game state processing are fully handled by RTF. Zones can be reallocated to new servers during runtime and inter-server client migrations are realized in a seamless way.

In comparison to existing approaches in the field of basic communication middleware like Net-Z (Quazal) or RakNet (Rakkarsoft) RTF provides a higher level of abstraction: it includes automatic entity serialization and hides nearly all of the technical network communication aspects. On the other hand, RTF is significantly more flexible than reusable game engines like the Quake or Unreal engines, because it is not bound to predefined entities and events or a specific graphics engine and leaves the real-time loop implementation to the developer, who is now supported by the high-level mechanisms of RTF for entity and event handling. RTF incorporates three different parallelization and distribution approaches and is open to be extended in future game designs. This flexible support of different parallelization concepts allows RTF to be usable for a broader range of MMOG concepts than existing multi-server middleware like the Emergent Server Engine (Emergent Game Technologies) or BigWorld (BigWorld) which focus mostly on the concept of zoning.

Zoning (Assiotis & Tzanov, 2006) and replication (Bharambe *et al.*, 2006; Müller-Iden, 2007) have already been investigated as independent distribution concepts for scalable multiplayer online games. Our high-level development approach integrates both concepts, together with instancing, in a seamless way. Previous high-level game development approaches, e.g., (BinSubaih & Maddock, 2008) investigated the abstraction of the overall game engine, making it possible to exchange the underlying game engine without modifying the game. This development approach represents an even higher level of abstraction than ours, but it requires implementing various network-related issues, graphics-rendering, input-processing, etc. In our approach, RTF supports an easy realization of these issues in the context of online games. The Project Darkstar (Sun Microsystems) proposed a separation of all game-related processing into task objects that are freely distributable across multiple servers. An underlying run-time and global object store distributes the tasks and manages the distributed object access. However, a global object store and random distributed access might be difficult to manage efficiently in a very responsive and highly interactive game.

RTF follows the multi-server architecture model, establishing a setup of fully connected servers, while client connections are managed by the middleware according to their location in the game world. Due to a trusted and controlled environment at the server, with server-side processing and input checks, RTF obviously is quite resistant to various techniques of cheating that peer-to-peer systems (P2P) typically suffer from or must address specifically (Kabus, Terpstra, Cilia, & Buchmann, 2005). Furthermore, adding communication encryption, authentication and message signing support prevents man-in-the-middle attacks and repudiation (Ploss, 2011). With respect to performance and scalability, P2P solutions aim at exploiting the network and computational resources being present at each client.

In (Hampel, Bopp, & Hinn, 2006), a highly distributed P2P gaming architecture uses an overlay network based on distributed hash tables for the input and update exchange between the peers and exploits locality by segmenting the game world into several regions, each one being managed by the peers residing in the region. Despite the cost and scalability advantages, using an overlay network introduces additional hops and latency between the peers for input and state update transmissions, which is not suitable for fast-paced and latency-sensitive interactive games. Furthermore, fault-tolerance is an issue for persistent games running 24/7 as clients may drop out of sessions in an ad-hoc manner. Therefore, game developers must choose between the tradeoff of building their own resource infrastructure with predictable latency and persistency characteristics versus reducing the infrastructure burden by using P2P-based systems.

Our proposed high-level development approach addresses a large variety of online game types, ranging from fast-paced and small action games and first-person shooters to large-scale MMOGs and is efficiently supported by the current RTF implementation. It both provides a high level of abstraction and preserves design flexibility for single- and multi-server game engines. We conducted several case studies which showed that RTF is indeed easy to use and it successfully shields the developer from the low-level tasks of online game implementation.

REFERENCES

Assiotis, M., & Tzanov, V. (2006). A distributed architecture for MMORPG. *NetGames '06:Proceedings of the 5th ACM SIGCOMM workshop on Network and system support for games* (pp. 4). Singapore: ACM.

Beigbeder, T., Coughlan, R., Lusher, C., Plunkett, J., Agu, E., & Claypool, M. (2004). The effects of loss and latency on user performance in unreal tournament. *NetGames '03:Proceedings of the 2nd ACM SIGCOMM Workshop on Network and System Support for Games*. Portland, OR.

Bezerra, C. E., Cecin, F. R., & Geyer, C. F. R. (2008). A3: A novel interest management algorithm for distributed simulations of MMOGs. Proceedings from the *2008 12th IEEE/ACM International Symposium on Distributed Simulation and Real-Time Applications.* Vancouver, Canada.

Bezerra, C. E., & Geyer, C. F. (2009). A load balancing scheme for massively multiplayer online games. *Multimedia Tools and Applications, 45*(1-3), 263–289. doi:10.1007/s11042-009-0302-z

Bharambe, A., Pang, J., & Seshan, S. (2006). Colyseus: A distributed architecture for online multiplayer games. *NSDI'06: Proceedings of the 3rd Conference on 3rd Symposium on Networked Systems Design & Implementation* (pp. 12-12). San Jose, CA: USENIX Association.

BigWorld Technology. (2011). Retrieved from http://www.bigworldtech.com

BinSubaih,, A., & Maddock, S. C. (2008). Game portability using a service-oriented approach. *International Journal of Computer Games Technology, 7*. doi:10.1155/2008/378485

Cai, W., Xavier, P., Turner, S. J., & Lee, B.-S. (2002). A scalable architecture for supporting interactive games on the internet. *Proceedings of the 16th Workshop on Parallel and Distributed Simulation* (pp. 60-67). Washington, D.C.

de Vleeschauwer, B., van den Bossche, B., Verdickt, T., de Turck, F., Dhoedt, B., & Demeester, P. (2005). Dynamic microcell assignment for massively multiplayer online gaming. Paper presented at the *Proceedings of 4th ACM SIGCOMM Workshop on Network and System Support for Games.* New York, NY.

Emergent Game Technologies. (2007). Retrieved from http://www.emergent.net

Hampel, T., Bopp, T., & Hinn, R. (2006). A peer-to-peer architecture for massive multiplayer online games. Paper presented at the *Proceedings of 5th ACM SIGCOMM workshop on Network and System Support for Games*. Singapore.

Kabus, P., Terpstra, W. W., Cilia, M., & Buchmann, A. P. (2005). *Addressing cheating in distributed MMOGs*. Paper presented at the *Proceedings of 4th ACM SIGCOMM Workshop on Network and System Support for Games*. Hawthorne, NY.

Kwok, M., & Yeung, G. (2005). Characterization of user behavior in a multi-player online game. *ACE '05: Proceedings of the 2005 ACM SIGCHI International Conference on Advances in Fomputer Entertainment Ttechnology* (pp. 69-74). Valencia, Spain: ACM.

Macedonia, M. R., Zyda, M. J., Pratt, D. R., Barham, P. T., & Zeswitz, S. (1994). NPSNET: A network software architecture for large-scale virtual environments. *Presence (Cambridge, Mass.)*, *3*(4), 265–287.

Meiländer, D., Ploss, A., Glinka, F., & Gorlatch, S. (2011). Software development for real-time online interactive applications on clouds. *Frontiers in Artificial Intelligence and Applications*, *231*. do i:doi:10.3233/978-1-60750-831-1-81

Müller-Iden, J. (2007). Replication-based scalable parallelization of virtual environments. Doctoral dissertation. Retreived from http://pvs.uni-muenster.de/jmueller/jmi_thesis.pdf. Universität Münster, Germany.

Nae, V., Prodan, R., & Fahringer, T. (2008). Neural network-based load prediction for highly dynamic distributed online games. Paper presented at *The 14th International Euro-Par Conference European Conference on Parallel and Distributed Computing*. Las Palmas de Gran Canaria, Spain.

Ng, B., Si, A., Lau, R. W. H., & Li, F. W. B. (2002). A multi-server architecture for distributed virtual walkthrough. Paper presented at the *Proceedings of the ACM symposium on Virtual Reality Software and Technology*. Hong Kong, China.

Pantel, L., & Wolf, L. C. (2002). On the impact of delay on real-time multiplayer games. *NOSSDAV '02: Proceedings of the 12th International Workshop on Network and Operating Systems Support for Digital Audio and Video* (pp. 23-29). Miami, Florida, USA: ACM.

Ploss, A. (2011). *On Efficient Dynamic Communication for Real-Time Online Interactive Applications in Heterogeneous Environments.* Doctoral dissertation. University of Müenster, Germany.

Ploss, A., Glinka, F., Gorlatch, S., & Müller-Iden, J. (2007). Towards a high-level design approach for multi-server online games. *GAMEON'2007: Proceedings of the 8th International Conference on Intelligent Games and Simulation* (pp. 10-17). Bologna, Italy.

Quazal. (2011). Quazal Net-Z. Retrieved from http://www.quazal.com

Rakkarsoft. (2011).RakNet. Retrieved from http://www.rakkarsoft.com

Rosedale, P., & Ondrejka, C. (2003). *Enabling player-created online worlds with grid computing and streaming*. Retrieved from http://www.gamasutra.com/resource_guide/20030916/rosedale_01.shtml

Sun Microsystems. (2009). I. Project Darkstar. Retrieved from http://www.projectdarkstar.com

ADDITIONAL READING

Armitage, G. (2006). *Networking and Online Games (Vol. Grenville Armitage and Mark Claypool and Philip Branch)*. Chichester, UK: John Wiley & Sons.

Armitage, G. J. (2003). An experimental estimation of latency sensitivity in multiplayer quake 3. Paper presented at the *11th IEEE International Conference on Networks (ICON 2003)*. Sydney, Australia.

Bharambe, A., Douceur, J. R., Lorch, J. R., Moscibroda, T., Pang, J., Seshan, S., & Zhuang, X. (2008). Donnybrook: enabling large-scale, high-speed, peer-to-peer games. *SIGCOMM Computer Communication Review*, *38*, 389–400. doi:10.1145/1402946.1403002

Boulanger, J.-S., Kienzle, J., & Verbrugge, C. (2006). Comparing interest management algorithms for massively multiplayer games. Paper presented at the NetGames *'06: Proceedings of 5th ACM SIGCOMM Workshop on Network and System Support for Games*. New York, NY.

Chen, J., Wu, B., Delap, M., Knutsson, B., Lu, H., & Amza, C. (2005). Locality aware dynamic load management for massively multiplayer games. Paper presented at the *Proceedings of the 10th ACM SIGPLAN Symposium on Principles and Practice of Parallel Programming*. New York, NY.

Claypool, M., & Claypool, K. (2006). Latency and player actions in online games. *Communications of the ACM*, *49*(11), 40–45. doi:10.1145/1167838.1167860

Coulouris, G., Dollimore, J., & Kindberg, T. (2005). *Distributed systems: Concepts and design* (4th ed.). Essex, England: Addison Wesley.

Fischer, T., Daum, M., Irmert, F., Neumann, C., & Lenz, R. (2010). Exploitation of event-semantics for distributed publish/subscribe systems in massively multiuser virtual environments. Paper presented at the Proceedings *of the Fourteenth International Database Engineering & Applications Symposium.* New York, NY.

Fritsch, T., Ritter, H., & Schiller, J. (2005). The effect of latency and network limitations on MMORPGs: A field study of Everquest 2. Paper presented at the *NetGames '05: Proceedings of 4th ACM SIGCOMM Workshop on Network and System Support for Games*. New York, NY.

GauthierDickey. C., Zappala, D., Lo, V., & Marr, J. (2004). Low latency and cheat-proof event ordering for peer-to-peer games. Paper presented at the *Proceedings of the 14th International Workshop on Network and Operating Systems Support for Digital Audio and Video*. New York, NY.

Ghosh, C., Wiegand, R. P., Goldiez, B., & Dere, T. (2010). An architecture supporting large scale MMOGs. Paper presented at the *Proceedings of the 3rd International ICST Conference on Simulation Tools and Techniques, ICST.* Brussels, Belgium.

Henning, M. (2004). Massively multiplayer middleware. *Queue*, *1*(10), 38–45. doi:10.1145/971564.971591

Hsiao, T.-Y., & Yuan, S.-M. (2005). Practical middleware for massively multiplayer online games. *IEEE Internet Computing*, *9*(5), 47–54. doi:10.1109/MIC.2005.106

Iimura, T., Hazeyama, H., & Kadobayashi, Y. (2004). Zoned federation of game servers: A peer-to-peer approach to scalable multi-player online games. NetGames '04*: Proceedings of 3rd ACM SIGCOMM Workshop on Network and System Support for Games* (pp. 116-120). New York, NY: ACM Press.

Joselli, M., Zamith, M., Clua, E., Leal-Toledo, R., Montenegro, A., & Valente, L. (2010). An architecture with automatic load balancing for real-time simulation and visualization systems. *Journal of Computational Interdisciplinary Sciences*, *1*(3), 207–224. doi:10.6062/jcis.2010.01.03.0023

Kienzle, J., Verbrugge, C., Kemme, B., Denault, A., & Hawker, M. (2009). Mammoth: A massively multiplayer game research framework. Paper presented at the *FDG '09: Proceedings of the 4th International Conference on Foundations of Digital Games*. New York, NY.

Knutsson, B., Lu, H., Xu, W., & Hopkins, B. (2004). Peer-to-peer support for massively multiplayer games. Paper presented at the *INFOCOM 2004: 23rd Annual Joint Conference of the IEEE Computer and Communications Societies.* Hong Kong, China.

Mauve, M., Vogel, J., Hilt, V., & Effelsberg, W. (2004). Local-lag and timewarp: Providing consistency for replicated continuous applications. *IEEE Transactions on Multimedia, 6*(1), 47–57. doi:10.1109/TMM.2003.819751

Miller, R. (2009). WoW's back end: 10 Data centers, 75,000 cores. Retrieved from http://www.datacenterknowledge.com/archives/2009/11/25/wows-back-end-10-data-centers-75000-cores/

Morillo, P., Orduna, J. M., Fernandez, M., & Duato, J. (2005). Improving the performance of distributed virtual environment systems. *IEEE Transactions on Parallel and Distributed Systems, 16*, 637–649. doi:10.1109/TPDS.2005.83

Morse, K. L., Bic, L., & Dillencourt, M. (2000). Interest management in large-scale virtual environments. *Presence (Cambridge, Mass.), 9*, 52–68. doi:10.1162/105474600566619

Müller, J., & Gorlatch, S. (2005). GSM: A game scalability model for multiplayer real-time games. Paper presented at the *Proceedings of the IEEE 24th Annual Joint Conference of the IEEE Computer and Communications Societies INFOCOM 2005*. Miami, FL.

Narayanasamy, V., Wong, K.-W., & Fung, C. C. (2006). Complex systems-based high-level architecture for massively multiplayer games. In M. Dickheiser (Ed.), *Game Programming Gems 6* (pp. 607-622). Newton Center, MA: Charles River Media.

Scharf, O., Gorlatch, S., Blanke, F., Hemker, C., Westerheide, S., Priebs, T., et al. (2010). Scalable distributed simulation of large dense crowds using the real-time framework (RTF). Paper presented at the *Proceedings of the 16th international Euro-Par conference on Parallel processing: Part I.* Berlin, Germany.

Shaikh, A., Sahu, S., Rosu, M.-C., Shea, M., & Saha, D. (2006). On demand platform for online games. *IBM Systems Journal, 45*(1), 7–19. doi:10.1147/sj.451.0007

Sheldon, N., Girard, E., Borg, S., Claypool, M., & Agu, E. (2003). The effect of latency on user performance in warcraft III. Paper presented at the *NetGames '03: Proceedings of the 2nd workshop on Network and system support for games*. New York, NY.

Shelley, G., & Katchabaw, M. (2005). Patterns of optimism for reducing the effects of latency in networked multiplayer games. Paper presented at the *Proceedings of FuturePlay 2005*. East Lansing, MI.

Smed, J., & Hakonen, H. (2006). *Algorithms and Networking for Computer Games*. Chichester, UK: John Wiley & Sons. doi:10.1002/0470029757

Wang, T., Wang, C.-L., & Lau, F. C. (2006). An architecture to support scalable distributed virtual environment systems on grid. *The Journal of Supercomputing, 36*(3), 249–264. doi:10.1007/s11227-006-8296-z

Waveren, J. M. P. v. (2001). *The quake III arena bot.* Doctoral Dissertation. University of Technology Delft, The Netherlands.

Zhang, L., & Lin, Q. (2006). Support dynamic network architecture for large-scale collaborative virtual environment. *International Journal of Information Technology*, *12*(1), 26–36.

KEY TERMS AND DEFINITIONS

Event: An input triggered by a player or a computer initiated in-game action which needs to be processed.

Entity: A dynamic, in-game object, e.g., a non-player character or item. An entity is 'active' at a server if the server is responsible for the entity's update and event processing, and 'shadow' otherwise.

Micro-/Macrocells: The partitioning of the game world into small areas (microcells) which are grouped along a load-balancing scheme to macrocells before they are assigned to a particular processing resource.

Real-Time Framework (RTF): A framework for the high-level development of scalable real-time online games and their efficient execution through a variety of parallelization and distribution techniques.

Real-Time Loop: The periodic execution of state update and event processing tasks under soft real-time constraints in a central loop.

Response Time: Timespan between the moment when the user triggers a particular action and the moment when the result of the action is displayed on the client.

Scalability: An application's capability to support an increasing number of users or higher workload by adding resources, e.g., by distributing the game processing among multiple servers.

Chapter 13
Modular Game Engine Design

Aaron Boudreaux
University of Louisiana at Lafayette, USA

Brandon Primeaux
University of Louisiana at Lafayette, USA

ABSTRACT

The usage of software engineering principles in designing a game engine is discussed in this chapter using a simple tower defense game implemented using C# and the XNA Framework to illustrate usage of the engine. Essential functions, such as collision detection, input/output, graphics, object management, state management, and sound, will be implemented as independent units called managers. Because each manager is independent from the rest, essential development tasks such as implementing each manager and isolating bugs are much simpler.

INTRODUCTION: WHAT IS A GAME ENGINE?

A game engine is a system designed to facilitate rapid development of video games. A game engine generally acts as a layer of abstraction between hardware or low level code and the engine interface to reduce the amount of time spent designing or coding a game. Additionally, game engines may utilize third party libraries to provide desired functionality, such as physics, sound, and artificial intelligence, rather than having to "reinvent the wheel". There exist many libraries on each of these topics which can both quickly add new features to a game engine and reduce the overall time required for engine development.

"Game Engine" is a generic term and does not define any specific features which must be included other than aiding in the game development process. However, many game engines include, but are not limited to, any number of the following standard features:

- 2D or 3D rendering
- Audio
- Physics
- Input Detection
- Scripting

DOI: 10.4018/978-1-4666-1634-9.ch013

- Entity Management
- Networking
- Modeling and Animation
- Graphical User Interface Management
- Particle Systems

In addition to these core features, it is also desirable to develop the engine to provide these features using a component based architecture. Ideally, each of the previously listed features is implemented as an individual, independent component. Realistically, some inter-component dependencies may be required in order for some libraries to function correctly, or to prevent duplication of work. For example, the GUI system requires the input system in order to detect GUI interactions. While a GUI library may implement its own basic input detection, it would be completely unnecessary to use both the basic GUI library input detection, as well as a more fully-featured input library which may support many other input libraries.

Component based systems tend to be more modular. This allows engine developers to either swap out one implementation of a module or component for another, or provide multiple component implementations, allowing a game developer to have a choice of implementation. Additionally, a game engine may allow a game developer to either add new components or replace the default components with their own custom components. A great example would be a component based 3D rendering system which allows the developer to choose a target platform such as the Microsoft Xbox 360, Sony Playstation 3, or Nintendo Wii. It is even possible to compile the game using each of these systems to allow game developers to focus on multiple platforms.

This chapter illustrates the concept of modular design through actual design and implementation of a simple game engine. It features many of the components listed above, including 2D rendering, audio, input detection, and graphical user interface management.

BACKGROUND

This section will cover the basics of the XNA framework and C# before delving into our engine.

C# Keywords and Other Functionality

The game engine uses several keywords and other features of C# that may be unfamiliar to those without experience in the language. A brief overview will be provided here. For more information, please refer to the C# books listed in the additional reading section at the end of the chapter.

- **Override:** Used to extend or change an abstract or virtual member of a base class. The override keyword must be used in the declaration of the derived class or member (Override, 2011).
- **Sealed:** The sealed keyword can be applied to class members, methods, fields, properties, or events that override a virtual member of the bass class. Other classes or members cannot override a sealed object (Abstract and Sealed Classes and Class Members, 2011).
- **Internal:** A method or class member that is declared to be internal is only accessible to other classes within the same namespace (internal, 2011).
- **Delegate:** The delegate keyword functions similarly to a function pointer in C or C++. It is used to specify the signature of the events which will be thrown by events specified using this delegate. Methods registered for the event must also match the signature of this delegate (i.e., a void method with a single parameter of type string) (Delegates Tutorial, 2011).
- **Event:** The event keyword is used to specify an object which can be used to store a list of specific methods to be called. An

event must be declared using a delegate type.
- **accessor methods:** C# provides properties for auto-implementation of get and set methods (Auto-Implemented Properties, 2011.).

XNA Framework

The 2D game engine described in this chapter was created using Microsoft's XNA Framework. This framework, which is implemented in C#, provides many of the core and low level features required by games such as the game loop (discussed shortly), pre-processing, and loading game resources (textures, sounds, etc.), graphics rendering, raw input detection, and many other features which are not covered by this chapter. Several classes will be frequently referenced throughout the chapter and are discussed below.

The GameServicesContainer provides a means of registering game services while maintaining loose coupling between the service provider and the code using the service consumer. This service is created by instantiating a class object and adding that object to the services container. When adding the object, the type for the class object to be registered under must also be included. As long as the service is not removed from the container, any block of code with access to the GameServicesContainer can retrieve the service provider by requesting the specific type. All of the managers described in this chapter are registered as services. Additionally, all of the managers are registered using their interface type, rather than the manager type. This allows the service consumer the freedom to use any (of possibly many) implementations of the service without requiring any code changes to the service consumer.

Below is an example of adding a service. The GameEngine is registered as type Game.

```
Services.Add(typeof(Game), new Ga-
meEngine());
```

A service is retrieved via the Get method. Type casting is used because Get returns the service as type "object," which all classes inherit by default.

```
((GameEngine)Services.
Get(typeof(Game))).Exit();
```

The GameComponentsCollection class provides a way to add content to the game while keeping the code modular. A class can inherit either the GameComponent class or the DrawableGameComponent class; instances of that class object can be registered with the GameComponentsCollection in order to have its Update and Draw methods automatically called each frame. Initially, all of the manager classes described in this chapter were intended to put this feature to use, however, it was decided against in favor of an implementation which would not be dependent upon any specific framework or library. Thus, the GameEngine class was designed to provide this feature to the managers, which will be described in more detail later in this chapter. Regardless, the GameComponentCollection can still be accessed via the static accessors within the GameEngine class if compatibility with any third party XNA plugins or libraries is required.

Other Game Engines

Examples of open source engines include Radium, OGRE, and Crystal Space.

Raydium is an open source game engine which touts a quick and flexible system to rapidly develop games (Raydium Game Engine, 2011). The Raydium engine includes an OpenGL renderer, physics support, a networking api, OpenAL audio, php scripting, a GUI system, input detection with force feedback on supported devices, and a particle engine. The Raydium engine provides developers with most of the desired game engine components and abstracts away much of the lower level coding which communicates directly with the hardware. While there is a separation of the individual components, the engine is not designed to be able to replace or add custom components. For example, the engine is designed to use a specific rendering engine, MyGLUT, which provides a small subset of functionality found the OpenGL library, GLUT. Without a thorough understanding of the the specific implementation it would be very difficult to expand this engine to support an alternative rendering library like DirextX.

OGRE (Object-Oriented Graphics Rendering Engine) is not a game engine, but many games, both commercial and non-commercial, have used the engine (OGRE, 2011). The OGRE libraries abstract away the implementation details of OpenGL and DirectX and provide an interface based on a hierarchy of objects. OGRE does not provide native support for audio, networking, physics, collision detection, and many other features required of a game engine. However, it does use a component based architecture, thus providing easy integration for outside libraries that do provide these functions. This allows the user to pick and choose the libraries that are most appropriate to the needs of his or her game.

Crystal Space is a cross-platform, modular 3D game engine that uses OpenGL and provides support for shaders, physics, audio, animation, terrain, and more (Crystal Space, 2011). A plugin system is also available to allow users to extend or customize existing features or add new ones. CEL is another component supported by the developers of Crystal Space. It supplies additional game specific functionality such as an event system, camera management, and artificial intelligence. CEL also adds entity management, which is missing from the base Crystal Space engine.

About the Game

A simple tower defense game will be used to illustrate salient features of the game engine. In this type of game, the player places defenses to attack computer controlled enemy creatures that move along a set path. Defeating enemies earns gold, which can then be used to buy or upgrade more defenses. The game is over if enough enemies reach the player's base at the end of the path and destroy it. A screenshot of the game is shown in Figure 1.

Commercial games of this genre include Crystal Defenders (2008), PixelJunk Monsters (2011), Savage Moon (2011), and Comet Crash (2011).

Figure 1. Screenshot of sample game

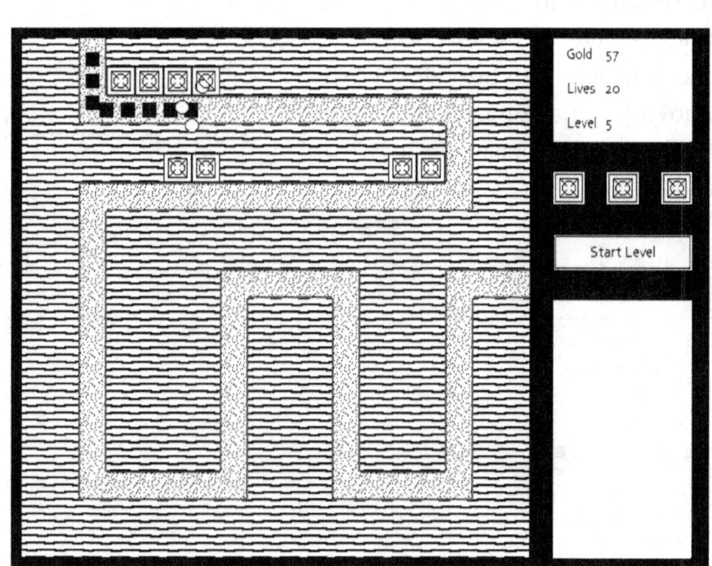

GAME ENGINE OVERVIEW

The Game Loop and Game Engine Class

The game loop refers to the process by which a game repeatedly gathers input, modifies the state of the game, and renders the new state. This process is at the core of all modern day games, and happens many times per second. The engine described in this chapter was also designed around this idea. However, this loop is created behind the scenes by the XNA framework and is not visible. Within the XNA framework, the game loop is started at the time of the Game.Run method call. The game loop is ended at the time of the Game.Exit call. The GameEngine class inherits the XNA Game class, which is where the game loop is implemented. As an example (Box 1), a simple game loop can be created with very few lines of code.

Components of the game, such as sound, collision, object management, etc. are implemented as separate classes called managers and registered to the GameEngine class. Each manager implements their own versions of the Initialize, Update, Draw, and Cleanup functions. During the game loop, the GameEngine's Update and Draw call the same functions in each respective manager.

The GameEngine class inherits from the XNA Game class. It additionally adds the ability to maintain a list of IGameManagers. This provides support for adding, removing, initializing, updating, drawing, and cleaning up the managers. It adds the following fields:

- **s_dqManagers** is a double ended queue (Deque) of IGameManager objects. The deque was chosen for its performance over the generic list provided by C#. The various managers, such as sound, collision, logging, game states, etc. are added to this list. Only managers in this list will be updated/drawn as necessary.
- Each of these fields are included solely to provide static access to objects which are already part of the XNA Game class:
  - Spritebatch s_spritebatch
  - GameServicesContainer s_services
  - GameComponentsCollection s_components
  - ContentManager s_content
  - GraphicsDeviceManager s_graphicsDeviceManager
  - GameWindow s_window

Figure 2. The game loop

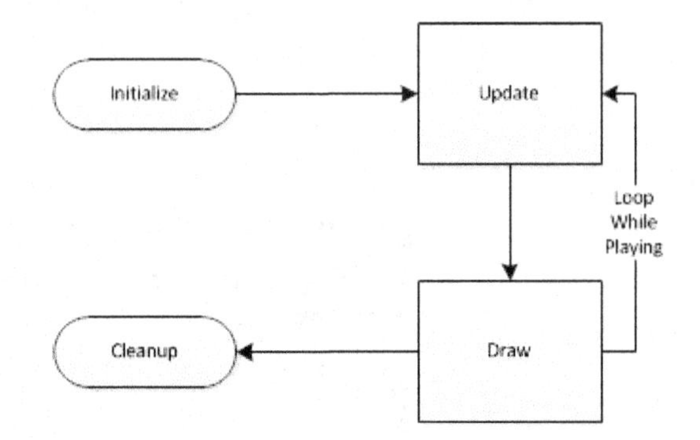

Box 1. Simple game loop

```
initialize game engine
loop until game ends
{
     process input, update the world, etc.
     draw the game
}
unload resources and deallocate memory
```

Figure 3. Abstract view of the GameEngine

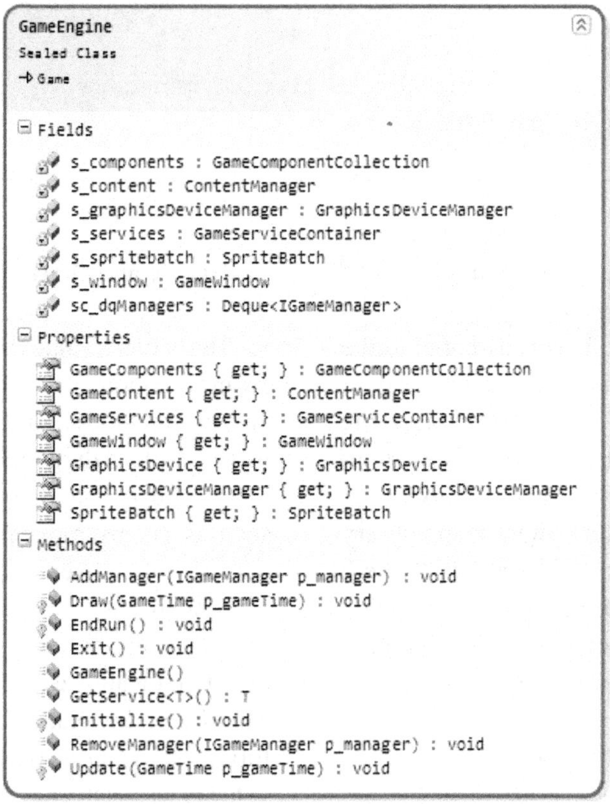

```
GameEngine
Sealed Class
-> Game

Fields
  s_components : GameComponentCollection
  s_content : ContentManager
  s_graphicsDeviceManager : GraphicsDeviceManager
  s_services : GameServiceContainer
  s_spritebatch : SpriteBatch
  s_window : GameWindow
  sc_dqManagers : Deque<IGameManager>
Properties
  GameComponents { get; } : GameComponentCollection
  GameContent { get; } : ContentManager
  GameServices { get; } : GameServiceContainer
  GameWindow { get; } : GameWindow
  GraphicsDevice { get; } : GraphicsDevice
  GraphicsDeviceManager { get; } : GraphicsDeviceManager
  SpriteBatch { get; } : SpriteBatch
Methods
  AddManager(IGameManager p_manager) : void
  Draw(GameTime p_gameTime) : void
  EndRun() : void
  Exit() : void
  GameEngine()
  GetService<T>() : T
  Initialize() : void
  RemoveManager(IGameManager p_manager) : void
  Update(GameTime p_gameTime) : void
```

Additionally, the defined properties within the GameEngine class are all used to provide access to the static fields.

The constructor assigns the static fields to reference their corresponding local instances which are automatically instantiated by the XNA framework via the Game class. The Initialize, Update and Draw methods iterate over each registered manager and call their respective methods. To exit the game, the Exit method is called. This triggers the EndRun method to call the Cleanup method for each registered manager. The AddManager method adds the manager to the end of the deque. In RemoveManager, the Remove method will delete the specified manager from the deque.

The power of this methodology is that the GameEngine doesn't know what types of managers are given. It only knows that it needs to call Update and Draw on each manager each frame. If a manager needs to be temporarily disabled for any reason its SkipUpdate and SkipDraw properties may be set to true.

The following example is a sample main function, which serves as the entry point for the game. This method instantiates the GameEngine, registers all the managers as services, and adds the services to the list of managers. The initial game state is set to the menu state and the game is started (Box 2).

Interfaces

An interface separates the signature of a class from its implementation. By using an interface, one can easily change the manager that inherits from the interface without altering the rest of the code base. Interface classes are abstract by definition, so they do not provide any implementation.

Box 2. A sample main function that registers managers as services and sets the initial state

```
create new GameEngine
repeat for each desired manager
     register as a service and add to list of managers
set starting state to the menu state
run game
```

However, they are useful for defining a set of methods that must be implemented by all classes that inherit the interface. This is essential to the modular design of the game engine. In addition, using an interface allows the user of the game engine to switch plugins or libraries without having to recode the entire game. For example, if a user was first using the FMOD sound library then switched to OpenAL later on, as long as there was an interface between the sound library and the rest of the code, the other classes would not know what sound library was being used.

Example: IGameManager

IGameManager is an interface class that all managers must inherit in order to be registered in the game engine. Each class that inherits IGameManager (i.e. all the managers) must implement the methods Initialize, Update, Draw, and Cleanup. As shown in the game loop section above, the list of managers can be iterated over each frame and the Update and Draw methods for each manager can be called.

The boolean values SkipUpdate and SkipDraw are used to indicate if a specific manager's Update or Draw method should be skipped, respectively. For example, if the game is currently paused, there is no need for anything exclusive to gameplay, such as enemy movement, to be updated. The pause menu, however, would need to be updated and drawn.

The LoggingLevel enumerator property is used to indicate the log verbosity of the manager. Modifying this value changes the amount of information which is output to the log file. If this value is less than or equal to the value within the ManagerLogs, the manager will write its log data to the logfile.

Again, because interface classes are abstract, the methods are only declared here. The implementation of the following four methods will be specific to the needs of each individual manager. In general:

- Initialize handles the required setup, such as setting the initial game state or registering a service.
- Update does any processing that is needed each frame. For example, the collision manager will check if there are collisions between game objects during this step.
- Draw will specify what needs to be drawn to the screen. Because the input manager has no graphical representation, this manager has an empty Draw method. The game object manager, on the other hand, is

Figure 4. Abstract view of the IGameManager

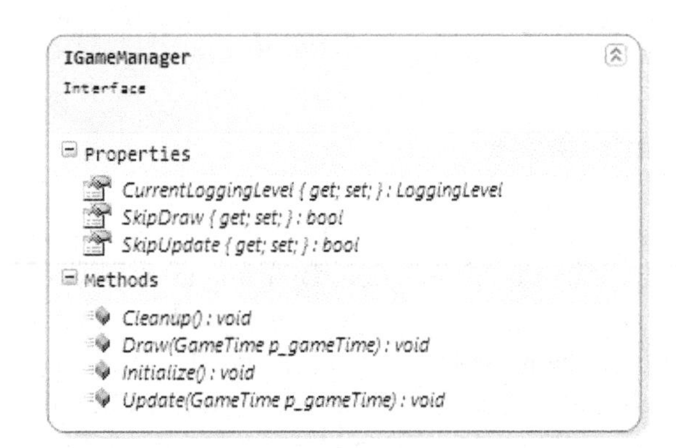

responsible for calling the drawing methods of all objects currently being used in the game.
- Cleanup defines what actions are to be taken when the manager is destroyed.

MANAGERS

Managers oversee many objects with similar functionality or behavior within the game. Managers implement the plugin/library/module specific code and provide access via the manager's interface. This decouples the rest of the code from the specific plugin/library/module and promotes maximum flexibility. It also allows multiple programmers to work on separate managers without interfering with each other's efforts and makes isolating bugs a simpler task.

This section will cover the various types of managers used in the game engine. Each subsection will cover a manager implementation as well as an example of using the manager in the game.

Game State Manager

Each specific section of a game is a game state. States common to many games include the intro video, menu screen, gameplay, pause, game over screen, and credits screen. The Game State Manager handles the management of transitioning from one game state to another based on user input and game events. A stack of GameState objects is used to keep track of which state is currently active. While any number of states may exist on the stack, only the top state is active.

The m_sStackStack object is the underlying stack of GameState objects which manages the current active state. The stack may be manipulated via the interface methods PushState, PopState, and ChangeState. Both push and pop are common stack operations, while change is a

Figure 5. Abstract view of the ManagerGameState

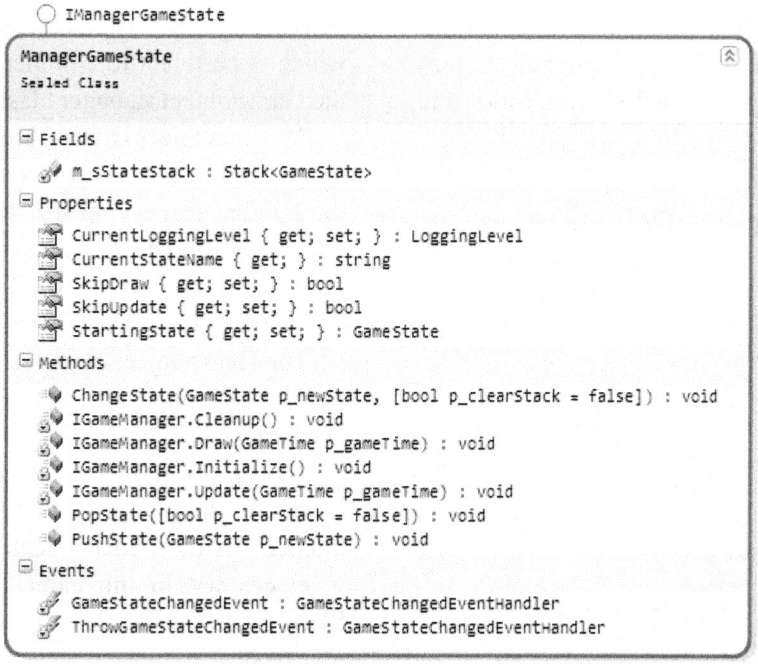

```
IManagerGameState
ManagerGameState
Sealed Class

Fields
  m_sStateStack : Stack<GameState>
Properties
  CurrentLoggingLevel { get; set; } : LoggingLevel
  CurrentStateName { get; } : string
  SkipDraw { get; set; } : bool
  SkipUpdate { get; set; } : bool
  StartingState { get; set; } : GameState
Methods
  ChangeState(GameState p_newState, [bool p_clearStack = false]) : void
  IGameManager.Cleanup() : void
  IGameManager.Draw(GameTime p_gameTime) : void
  IGameManager.Initialize() : void
  IGameManager.Update(GameTime p_gameTime) : void
  PopState([bool p_clearStack = false]) : void
  PushState(GameState p_newState) : void
Events
  GameStateChangedEvent : GameStateChangedEventHandler
  ThrowGameStateChangedEvent : GameStateChangedEventHandler
```

convenience method which first pops the stack, then pushes the new state onto the stack.

The StartingState property must be assigned before the game engine is initialized, or an exception will be thrown. This property defines a GameState object to use when the game first starts. The CurrentStateName property returns the name of the state which is currently on the top of the stack.

The Initialize method sets the logging level and calls the ChangeState method with the StartingState as a parameter. SkipUpdate and SkipDraw are initialized to false. The Update and Draw methods call their respective methods on the top state on the stack. The Cleanup method pops all states from the stack and calls the Cleanup method of each state popped.

The ManagerGameState class also includes an event which allows an object to register to receive notifications when the state of the game changes. ThrowGameStateChangedEvent is the event object which maintains the list of registered objects, while the GameStateChangedEvent property is defined in the interface through which the object may register.

The GameState class is an abstract class which defines the minimal characteristics of a game state.

The _Initialize, _Update, _Draw, and _Cleanup methods follow the game loop convention and serve the same purpose within a game state. Like their counterparts in IGameManager, these methods will be called in the same situations.

_OnFocusGained and _OnFocusLost are called automatically by ManagerGameState whenever a state becomes the top state on the stack, or is no longer the top state on the stack, respectively. This allows for special conditions to take place, such as pausing the game. The pause state is pushed on top of the play state, which can notify the game object manager to stop updating objects. When the pause state is popped from the stack, the play state then notifies the game object manger to resume updating objects. Popping the play state before pushing the pause state would erase all the user's progress in the game, as when the play state is pushed back on, it's Initialize method is called.

Figure 6. Abstract view of the GameState class

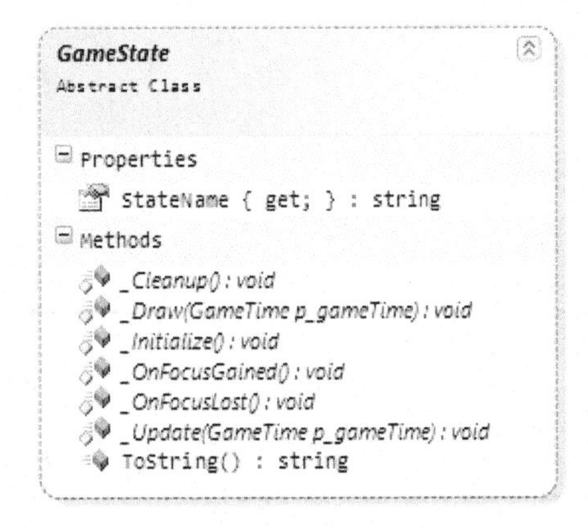

Game Object Manager

This manager handles the creation, updating, and destruction of objects within the game. It maintains a list of all the objects which will be updated each frame. This includes player controlled objects, enemy AI objects, objects to trigger certain events, etc. The GameObject class will first be discussed as this class contains information which is pertinent to completely understanding the GameObjectManager class.

A GameObject is any object which exists in the game world. The class is designed to follow the familiar game loop structure containing the Initialize, Update, Draw, and Cleanup methods. Each of these methods fulfill these previously described roles for the GameObject class.

The GameObject class contains a single integer property which is used to uniquely identify a GameObject. When the object is added to ManagerGameObject, it is assigned a unique integer identifier which is stored in the ID property. This integer can be used to identify specific object instances within the game. Additionally, the SkipUpdate and SkipDraw boolean properties may be set to toggle invocation of their respective

Figure 7. Abstract view of the GameObject class

methods. The ID property may not be set within classes which inherit GameObject. This is by design to prevent an object from having its ID changed after it is added to the manager.

The GameObject class also contains an event when may be thrown to notify registered listeners of the destruction of a specific object. This event is automatically thrown by the ManagerGameObject as soon as the manager is notified to destroy the object. The Destroy method is implemented as a convenient way to destroy a object. This method simply calls the DestroyObjectByID method within the ManagerGameObject and passes in its own unique identifier.

After having looked at the design of the GameObject class, the ManagerGameObject implementation is easier to follow. The local members of this class consist of the following:

- Dictionary<uint, GameObject> m_dGameObjects - A dictionary which maps an object's unique identifier to the GameObject. This data structure was chosen for its efficiency in object lookups from an integer value which may not be sequential. The dictionary is the storage structure for active objects within the game.
- List<GameObject> m_lNewObjects - A list to store objects which are waiting to be added to the manager. When objects are added to the manager, the object is added to this list. Once each frame, all objects within this list will be moved to the dictionary.
- List<GameObject> m_lDestroyedObjects - A list to store objects which are waiting to be destroyed. When objects are destroyed, they are added to this list. Once each frame, all objects within this list will be removed from the dictionary and cleaned up.
- uint m_nextObjectId - An unsigned integer to provide unique object identification values. This integer begins at zero, and is incremented each time an object is added to the manager. Upon reaching the limit of an unsigned 32-bit integer value, this value is reset back to zero. Duplicate entry checks are performed to prevent multiple objects from having the same unique identifier.

The GameObjectManager implements the IGameManager interface. The Initialize method ensures the local data members contain no entries and resets the unique object identifier generator back to zero. The Update method has three loops. The first loop iterates through m_lDestroyedObjects and removes each object from the dictionary and calls the Cleanup method on the object. The destroyed objects list is cleared after this loop. The second loop initializes and adds to the dictionary all objects in the new object list. The new object list is cleared after this loop. The third loop iterates over each object stored in the dictionary, calling each objects' Update method if its SkipUpdate property is not set to true (Box3).

Three separate lists were chosen specifically to prevent problems which arise from an object removing itself (or other objects) from the GameObjectManager while it is in the process of updating. Traditional iteration would skip over

Figure 8. Abstract view of the ManagerGameObjects

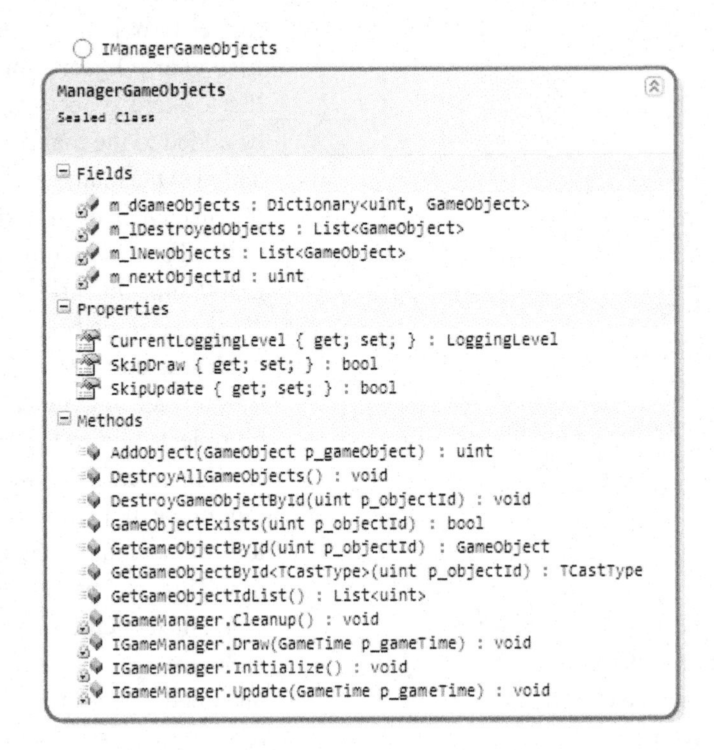

an object within a list if an object were to remove itself from the list during iteration.

The Draw method iterates over the GameObjects contained in m_dGameObjects, calling the Draw method for each GameObject that does not have its SkipDraw property set to true. Cleanup removes all objects from the game by calling DestroyAllGameObjects, which consists of iterating through the lists of game objects, new objects, and destroyed objects and calling the Cleanup method for each object.

Box 3. Update function for the GameObjectManager.

```
Update()
          for each object in the destroyed objects list
                    call the objects' cleanup method
                    remove from the list of active objects collection
          clear the destroyed objects list

          for each object in the new objects list
               call the objects' initialize method
                    add the object to the active objects collection
          clear the new objects list
          for each object in the active objects collection
           call the objects' update method
```

To destroy a specific game object, the method DestroyGameObjectById is provided. The integer input parameter is checked to determine if the dictionary contains a GameObject which is mapped to the integer. If the check passes, the object specified by the integer input parameter object is added to m_lDestroyedObjects and will be removed when Update or DestroyAllGameObjects is next called.

Adding a new object to the manager is handled by the AddObject method. The first step to adding an object is to assign a unique identifier. The value m_nextObjectId is incremented until its value does not exist in the keys stored in the dictionary. The GameObject is assigned this value for its id. Next, the object is appended to m_lNewObjects to be added to m_dGameObjects during the next Update call. Finally, the Initialize function is called and the object's id value is returned to provide a way for objects to be retrieved via ID value.

The GetGameObjectByID method receives the ID as an integer parameter. This value is used to retrieve the GameObject, if it exists, from the dictionary. An additional method overload is provided to accept a template parameter to specify a type to which the GameObject is type-casted and returned.

The pseudocode in Box 4 is from a class that is responsible for creating other game objects on a set interval. The Initialize method sets up an alarm with a countdown of one second and a message that indicates this alarm is to trigger enemy spawns. Each time Update is called, the alarm is updated. When the required time has passed, an alarm event method is called. This method adds enemies to the list of objects in play so that the player has something to fight.

Because the sample game is in 2D, an additional class, GameObject2D, was derived from GameObject. This class contains parameters common to all objects, such as position, angle, scale, texture, and depth.

Box 4. Pseudocode to spawn enemies on a set interval

```
Initialize()
          create an instance of a class which inherits from GameObject
          pass this object to the AddObject method of GameObjectManager
          initialize an alarm to trigger the creation of another game object
after one second. The alarm message is specified as "enemy spawn"
Update()
          update the alarm using the parameter time elapsed since last frame
          if alarm time remaining is less than 0, call the alarm event meth-
od, passing the alarm message
Alarm event()
          receive alarm message parameter
          if alarm message is enemy spawn
               create new instance of an enemy
                    add the enemy to the GameObjectManager
               if there are more enemies to spawn
                    reset the alarm to start the countdown for the next enemy
               else
                    destroy enemy spawner
```

Collision Detection Manager

The collision detection manager handles detection and notification of collisions between objects. Objects which require collision detection must inherit and implement the ICollidable2D interface. An object may then be registered with the manager. Collision checks for all registered objects will be automatically performed each frame, with colliding objects receiving collision notifications when appropriate. This section will only cover basic code not specific to any implementation, as collision detection algorithms and methods are beyond the scope of this chapter. Due to the modular design of the engine, any algorithm can be implemented in the collision detection manager. For this project a simple collision algorithm was chosen. For detailed discussion on collision detection methods, please see Chapter XX (Ben's chapter).

The collision manager makes use of two hash tables, one of mobile objects and one of stationary objects. Because stationary objects have zero chance of colliding with each other the number of comparisons is greatly reduced by separating them out into a different list.

The Update method performs the collision checks between all of the mobile objects and each mobile object with each stationary object for a total of n^2 checks. The code specific to the user's chosen collision detection algorithm is contained within this method. When a collision is detected both collidable objects are notified of the collision and are passed the id of the offending object.

RegisterCollidable is used to add a game object to the collision manager. The method requires both the object's id value and if it is static (stationary) or not. The object will be added to either the mobile or stationary object list as appropriate. When an object no longer needs to have collisions checked on it, the UnRegisterCollidable method will remove it from the proper list.

Input Manager

The input manager handles the detection of raw input from the device and allows for the registration of objects to be notified of input events. This manager makes use of the Nuclex framework, an XNA library which provides many utilities that are not provided by the standard XNA framework such as particle systems, graphical user interface

Figure 9. Abstract view of class ManagerCollision2D

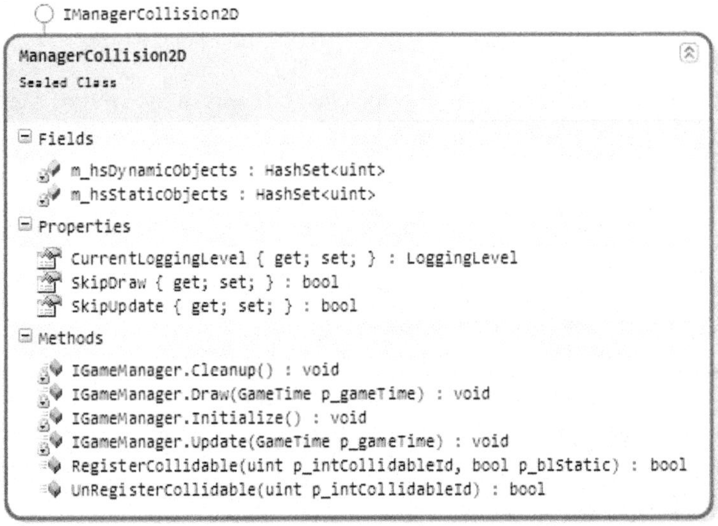

```
IManagerCollision2D

ManagerCollision2D
Sealed Class

Fields
  m_hsDynamicObjects : HashSet<uint>
  m_hsStaticObjects : HashSet<uint>
Properties
  CurrentLoggingLevel { get; set; } : LoggingLevel
  SkipDraw { get; set; } : bool
  SkipUpdate { get; set; } : bool
Methods
  IGameManager.Cleanup() : void
  IGameManager.Draw(GameTime p_gameTime) : void
  IGameManager.Initialize() : void
  IGameManager.Update(GameTime p_gameTime) : void
  RegisterCollidable(uint p_intCollidableId, bool p_blStatic) : bool
  UnRegisterCollidable(uint p_intCollidableId) : bool
```

Figure 10. Abstract view of class ManagerInput

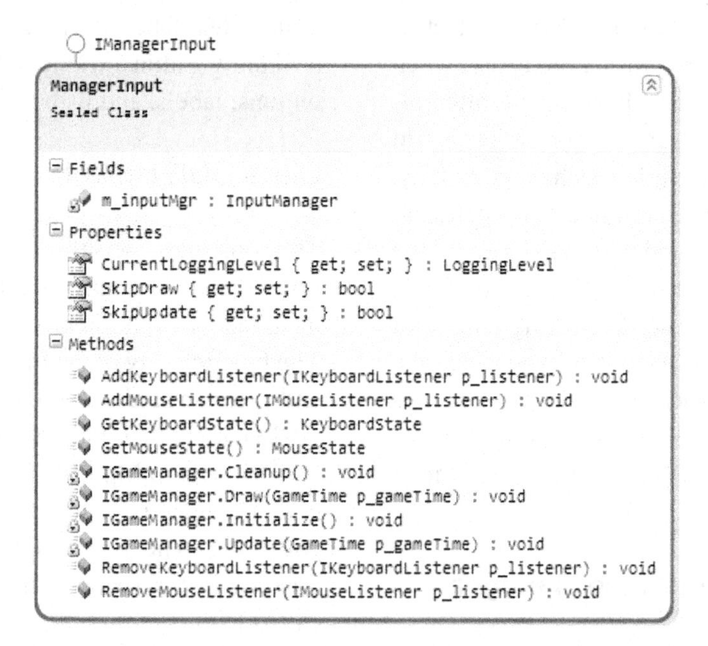

(GUI) support, and LZMA zip file compression ("Nuclex Framework", 2011). Our ManagerInput (not to be confused with Nuclex's InputManager class) uses Nuclex to abstract away much of the redundant code required for detecting input from the system, such as mouse events and button presses.

The m_inputMgr field allows direct access to the Nuclex input manager object. The Initialize method simply registers m_inputMgr as a service. As the GUI manager for this engine both implements and builds upon the Nuclex GUI framework, it is optimized to look for this service, which detects mouse interaction with the GUI.

The Update method is where the actual input detection takes place. The engine calls Nuclex's InputManager's Update method. In this method, the previous state for all input devices is saved then the input devices are polled and compared to their previous states. If there are any discrepancies between the two states of the device, then some kind of input event has taken place and the appropriate events will be triggered. Unlike the other managers seen thus far, the input manager does not allow for skipping of its Update method. An exception is thrown if an attempt is made to change SkipUpdate.

Because the input manager has no graphical representation, there is no reason to ever call the Draw method. As a result, we implement our own version of the get method that always returns true when asked for the value of SkipDraw. The Draw method must still be defined, but it does nothing.

The functions AddKeyboardListener, RemoveKeyboardListener, AddMouseListener, and RemoveMouseListener register or unregister the appropriate device for receiving input from the Nuclex InputManager. GetKeyboardState and GetMouseState retrieve the current state of the appropriate device from Nuclex and pass it to the requesting class.

Example: Using the input manager

The play state makes use of the input manager to determine if the player is interacting with towers or other GUI elements via the keyboard or the mouse. As part of the initialization process, the state registers for keyboard and mouse events so that interaction with GUI elements can be detected. MouseMoved specifies what actions to take when the user moves the mouse. If the user is attempting to place a tower, this method snaps it to a grid so that towers cannot overlap each other. Finally, pressing the left mouse button has two potential actions, shown in method MousePressed. If the user is not currently holding a tower to be placed, it checks to see if the user is clicking on one of the tower types. Otherwise it tries to place the currently selected tower wherever the mouse is currently located. Clicking the right mouse button (method not shown) cancels tower placement mode (Box 5).

Graphics User Interface Manager

The GUI manager handles displaying the user interface. As previously stated, the GUI manager makes use of the Nuclex GUI system. The GUI manager class maintains a list of GUIElement objects which are currently displayed. Any number of GUIElement objects may be active at one time. A GUIElement object may be as simple as a single text label on the screen, or as complex as a window within a window which contains many buttons, labels, and listboxes.

Class GUIElement

The constructor sets this GUIElement to be notified whenever a control is added or removed from its collection of children, m_hsDisplayedChildren. The functionality of the Initialize, Update, Draw, and Cleanup methods will be dependent upon the GUI element that inherits from this class.

The _OnGuiModified method is called whenever a control is added to or removed from the collection. The RefreshGUI method, which is covered in the section on the GUI manager, is called to ensure that the changes to the GUI are reflected on the screen.

Class ManagerGUI

Do not confuse the engine's manager for the user interface with the one used by Nuclex, which is

Box 5. The Initialize function of ManagerInput and sample mouse functions.

```
Initialize()
          ...
          load the GUI
          turn on mouse notifications for GUI interaction
          turn on keyboard notifications for GUI interaction
          ...
MouseMoved()
          if the user is attempting to place a tower
                    if the mouse is in a valid position to place a tower
                    snap the tower image to the grid
MousePressed()
          case LeftMouseButton
               if user is not placing a tower
                    check if user is selecting a tower
               else
                    attempt to place tower at current mouse position
```

Figure 11. Abstract view of class GUIElement

```
GUIElement
Abstract Class

Fields
  m_hsDisplayedChildren : ObservableHashSet<Control>
Properties
  _Children { get; } : ObservableHashSet<Control>
Methods
  _Cleanup() : void
  _Draw(GameTime p_gameTime) : void
  _Initialize() : void
  _OnGuiModified(object p_sender, NotifyCollectionChangedEventArgs p_e) : void
  _Update(GameTime p_gameTime) : void
  GUIElement()
  ThrowGuiEvent(string p_guiEvent) : void
  ToString() : string
Events
  GuiEvent : GUIMessageEventHandler
```

named GuiManager. Our engine makes use of the latter through the variable m_guiMgr.

The currently loaded GUI elements are stored in a linked list called m_lElements.

Initialize sets both SkipUpdate and SkipDraw to false. Setting SkipUpdate to true would prevent any changes to the GUI from being shown to the user, and setting SkipDraw to true would mean that the user interface would never be shown. Next, the actual GUI screen is created using the width and height given from the user's graphics device. Finally, the Nuclex GuiManager is initialized with the screen settings specified.

The Update method loops through the list of GUI elements and calls the Update method for each. When that is completed, the Nuclex Update is called. The Draw method draws the GUIElements from m_lElements to the screen. Cleaning up the manager consists of calling Cleanup on each GUIElement in the linked list then removing the element from the list. The screen is then cleared so the removed GUIElements are no longer displayed.

The RefreshGUI method forces an update, to ensure that changes made to any GUIElement are reflected on the screen.

When adding an element to the GUI, the linked list is checked to ensure that the element doesn't already appear in the list. The GUIElement is then

Figure 12. Abstract view of class ManagerGUI

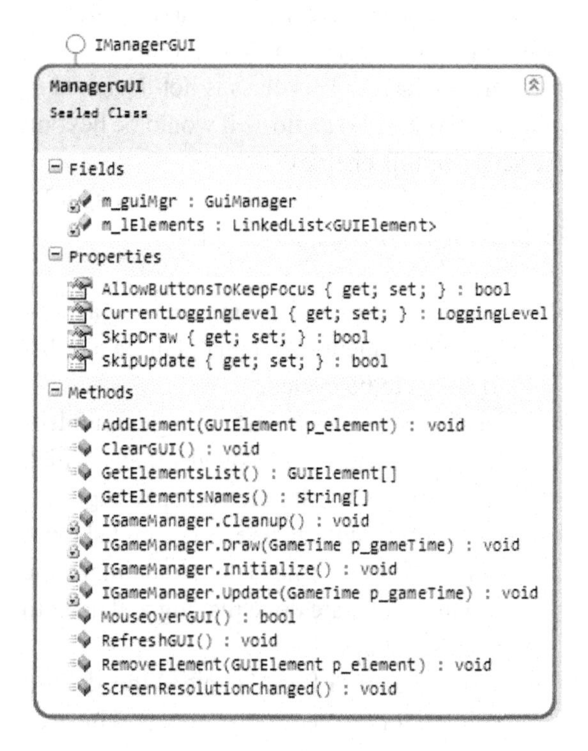

initialized and added to the linked list. After the element is added, the RefreshGUI method from above is called so the user can see and interact with the new element. The RemoveElement method also makes use of RefreshGUI, but after removing and cleaning up the GUIElement which is passed to the method.

ClearGUI is used in situations where one layout is being exchanged for another, such as transitioning from the menu state to the play state. Each element must be cleaned up and removed from the screen before the new layout can be displayed.

Other methods included in this class are GetElementsList, which returns the contents of the linked list of GUIElements, GetElementsNames, which returns the string names of all currently visible GUIElements, and MouseOverGUI, which checks if the mouse cursor is currently over some aspect of the GUI.

Audio Manager

This manager handles playing sound effects and music. This manager makes use of the Mediaplayer class, a static class designed to provide basic functions common to most desktop media software packages. This class is not designed to play positional (3D) audio as it would be beyond the scope of this chapter.

The local fields within the AudioManager2D include the following:

- Deque<Song> m_dqSongQueue - a double-ended queue which maintains a playlist of songs to be cycled.
- HashSet<SoundEffects> m_hsSoundEffects - a hash-set of sound effects which are currently instantiated and playing.
- HashSet<SoundEfects> m_hsDoneSoundEffects - a hash-set to maintain the sounds which are done playing and need to be cleaned up.
- float m_soundEffectVolume - the default volume for created sound effects. This value is only used if the volume is not passed in via the PlaySoundEffect method.

The class also contains the following properties which are inherited from the interface:

- bool IsLooping - is the currently playing song looping
- float SoundEffectVolume - get or set the default volume level for sound effects
- float SongVolume - get or set the song volume level

The Initialize method registers the two listener methods to receive notifications for the song and player state being changed. When either of these events happen, the appropriate function will be called automatically and information will be written to the logs. The Update function iterates through all of the sound effects in the set, attempting to find sound effects which are finished playing. Any finished sound effects are added to the alternate set m_hsDoneSoundEffects. Afterwards, the m_hsDoneSoundEffects set is iterated over, and each sound effect is disposed of and removed from the primary active sound effect set. After the iteration, the m_hsDoneSoundEffects set is cleared.

The sound manager is another instance of a manager that has no graphical representation and has an empty Draw method. The Cleanup function calls StopAllSoundEffects followed by ClearPlaylist to stop all audio, followed by loops to iterate over both sets of sound effect collections to cleanup all memory allocated for sounds. Finally, the event listener methods are unregistered from the MediaPlayer notification of media player state changes and song switches.

The default C# property get and set methods are overridden to allow SongVolume and SoundEffectVolume to perform validation checks on the value being passed in for the volume. The remaining methods defined in this class are largely self-explanatory, performing functions such as adding/

Figure 13. Abstract view of class ManagerAudio2D

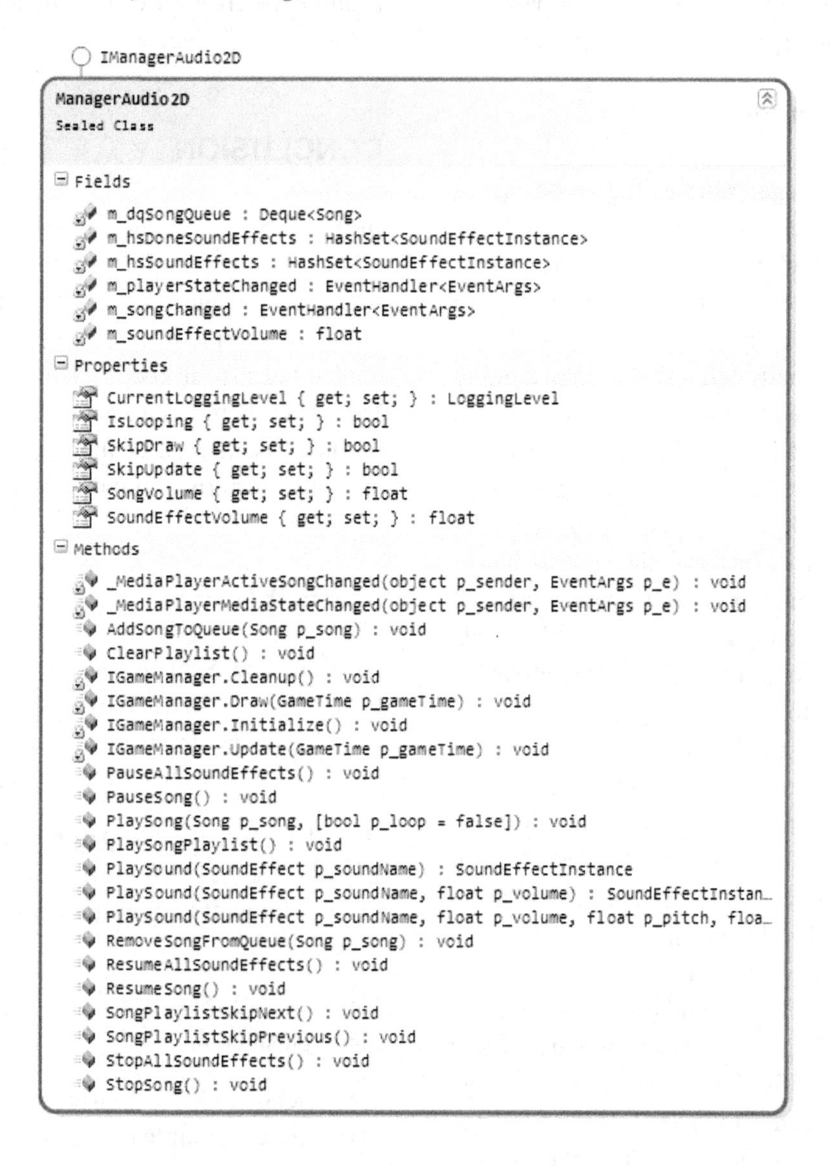

Figure 14. Abstract view of class ManagerLogs

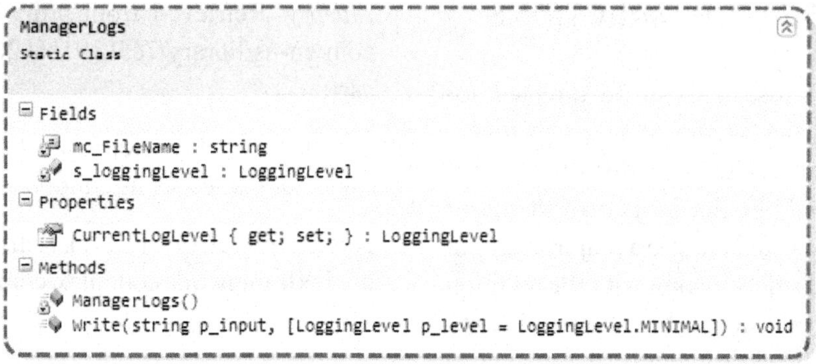

```
ManagerLogs
Static Class

Fields
  mc_FileName : string
  s_loggingLevel : LoggingLevel
Properties
  CurrentLogLevel { get; set; } : LoggingLevel
Methods
  ManagerLogs()
  Write(string p_input, [LoggingLevel p_level = LoggingLevel.MINIMAL]) : void
```

removing songs from the song queue or pausing/unpausing/stopping songs or sound effects.

Logging Manager

The logging manager handles the recording of information to files. Note that the ManagerLogs class is static which nullifies the need to instantiate it as a service. Varying degrees of logging detail can be specified via the CurrentLogLevel property. This property defines the current logging level for the engine. Whenever the Write method is called, a logging level is either specified at the time of the call or automatically assigned to the minimum level. The message is only logged if the logging level received in the method is less-than or equal-to the logging level stored in ManagerLogs. This method of logging allows multiple levels of logging without requiring log level validation before each call to the log manager.

The static constructor is called the first time an object attempts to access either the Write method or modify the logging level. The constructor deletes the logfile if it already exists from a previous iteration of the program, then sets the default logging level to minimal.

The Write method takes in a string and writes it to the default file. The second parameter defaults to minimal logging level, but an alternate, higher level can be specified. If the file already exists, the string is appended to the file. If the file does not exist, it is created before writing the string.

FUTURE RESEARCH DIRECTIONS

Future points of research can be directed towards rewriting the engine to take advantage of multi-threaded architectures. Ideally, each of the managers could be run in its own thread of execution which would require a major overhaul to enforce thread safety. Along with the addition of multi-threading, a more robust event subscriptions could be added to many of the managers to improve cross-thread notification of specific actions taking place.

CONCLUSION

In this chapter, a modular game engine was discussed. Using a simple tower defense game as an example, managers such as collision, sound, and input were implemented. This method is a great choice because it keeps each manager separate from the others. As a result, the implementation of these essential game functions can be changed without interfering with the rest of the code base.

REFERENCES

Abstract and Sealed Classes and Class Members. (2011) In *MSDN Library*. Retrieved from http://msdn.microsoft.com/en-us/library/ms173150%28v=VS.100%29.aspx

Auto-Implemented Properties. (2011) In *MSDN Library*. Retrieved August 10, 2011 from, http://msdn.microsoft.com/en-us/library/bb384054.aspx

Crash, C. (2011). Retrieved from http://www.cometcrash.com/

Defenders, C. (2008). Retrieved from http://www.crystaldefenders.jp/na/index.html

Gregory, J. (2009). *Game engine architecture*. Boca Raton, FL: AK Peters Ltd. internal. (2011) In *MSDN Library*. Retrieved from http://msdn.microsoft.com/en-us/library/7c5ka91b%28v=VS.100%29.aspx

Moon, S. (2011). Retrieved http://www.fluffy-logic.net/games/savage-moon/

Nuclex framework. (2011). Retrieved from http://nuclexframework.codeplex.com/

OGRE. (2011). Retrieved from http://www.ogre3d.org/

Override. (2011) In *MSDN Library*. Retrieved from http://msdn.microsoft.com/en-us/library/ebca9ah3%28v=VS.100%29.aspx

PixelJunk Monsters. (2011). Retrieved from http://pixeljunk.jp/

Raydium Game Engine. (2011). Retrieved August 10, 2011 from, http://radium.org

Space, C., III. (2011). Retrieved from http://www.crystalspace3d.org/main/Main_Page

Tutorial, D. (2011) In *MSDN Library*. Retrieved from http://msdn.microsoft.com/en-us/library/aa288459%28v=vs.71%29.aspx

What Is a Game Loop? (2011) In *MSDN Library*. Retrieved August 10, 2011 from, http://msdn.microsoft.com/en-us/library/bb203873.aspx

ADDITIONAL READING

C# Programming Guide. (2011). In *MSDN Library*. Retrieved from http://msdn.microsoft.com/en-US/library/67ef8sbd.aspx

Dalmau, D. (2004). *Core techniques and algorithms in game programming*. Indianapolis, IN: New Riders.

Finney, K. (2007). *3D game programming all in one* (2nd ed.). Boston, MA: Thomson Course Technology.

Gamma, E., Helm, R., Johnson, R., & Vlissides, J. (1995). *Design patterns: Elements of reusable object-oriented software*. Reading, MA: Addison-Wesley.

Gittleman, A. (2012). *Computing with c# and the. net framework*. Sudbury, MA: Jones & Bartlett Learning.

Morns, D., & Hartas, L. (2004). *Strategy games*. Boston, MA: Thomson Course Technology.

Pedersen, R. (2009). *Game Design Foundations* (2nd ed.). Plano, TX: Wordware Publishing.

Penton, R. (2005). *Beginning c# game programming*. Boston, MA: Thomson Course Technology.

Rabin, S. (2005). *Introduction to game development. Newton Center*. MA: Charles River Media.

Reed, A. (2008). *Learning xna 3.0*. Sebastopol, CA: O'Reilly.

Rollings, A., & Morris, D. (2000). *Game architecture and design*. Scottsdate, AZ: Coriolis Group.

Troelsen, A. (2007). *Pro c# 2008 and the. NET 3.5 Platform* (4th ed.). New York, NY: Apress. doi:10.1007/978-1-4302-0422-0

KEY TERMS AND DEFINITIONS

Collision Detection: The process of determining if two game objects are sharing space in the game world.

Game Loop: The process by which a game repeatedly gathers input, modifies the state of the game, and renders the new state.

Game Object: Any object that exists within a game world. This includes characters, terrain objects, enemies, weapons, etc.

Game State: A distinct section of a game. Common states include menu, play, pause, credits, and game over.

Interface: An abstract class that lists the methods a class inheriting the interface must implement. Keeps other classes from seeing the implementation of the child classes.

Modularity: The division of a software project into separate, independent pieces.

Manager: A class that implements code specific to a plugin/library/module and provides access to the rest of the code base via an interface.

Chapter 14
A Gameplay Model for Understanding and Designing Games

Erik Hebisch
Institute of Human Factors and Technology Management IAT, University of Stuttgart, Germany

Ulrich Wechselberger
Institute for Computational Visualistics, University of Koblenz-Landau, Germany

ABSTRACT

The study of video games involves many characteristics such as story, artwork and sound design. While it is possible to describe these qualities of a game in great detail, the same can generally not be said for the interactive qualities. This article argues that gameplay is the fundamental characteristic of a game and can be studied independently from the other qualities. The authors present a definition of gameplay and its components that enable the study of gameplay on an abstract level. They show how to use their definition for designing the interactions of a simple casual and a serious game, for analyzing the game-play mechanics of an existing game and how this impacts on the study of interactivity in video games.

INTRODUCTION

Every game designer has a vision of which elements he would like to have in a game. Whether he works backward from a story to design the scenes a player encounters during play or forward from an idea of what would be interesting for a player to interact with, a game designer evaluates how each element he employs adds or subtracts from the vision of how a player plays the game. The designer chooses story elements such as characters, plot lines and twists, visual elements such as character and monster appearance, or aural elements such as the game's score and the game's sound effects. And, of course, he designs how a player sees the game and interacts with it. This has a great effect on the experience of the game but it is often rather undefined what the ele-

DOI: 10.4018/978-1-4666-1634-9.ch014

ments of the interactive experience are and how a good experience can be achieved. However, it is the responsibility of the game designer to decide what plays well in a video game.

Every player can say whether he enjoys a game or not. Usually, however, he cannot fully describe what exactly makes a particular game fun to play. Is it the engaging story? Is it the crisp graphics? Is it the responsive controls? The critic who reviews a game or compares it to other games has an equally hard time to qualify the elements that make a particular game fun to play. While visual and audio quality are important factors to how a game is perceived, a crucial aspect of games is that they are interactive. How does he describe how a game works? What exactly are the elements that make it interesting and engaging? How do the interactive elements impact the perception of a game? What makes one game better than another?

These questions apply to academics, too. Be it for media impact studies or the design of a serious game, on some occasions it is crucial for a scientist to describe how a game plays. With this, he can model the interaction between a player and a game system (and thus the cognitive activity of a player), either with regard to game analysis or game design.

What is missing are concepts for talking about the most fundamental aspects in a game. In section 2 we discuss several theoretical foundations for describing games. We use these foundations to find the characteristic that is independent from presentational aspects such as artwork or sound design, and find it in the concept of gameplay. In section 3 we present a definition of gameplay that aims to strike a balance between formal validity and practical applicability. We define the element of gameplay in detail and put together a collection of building blocks for gameplay mechanics. In section 4 we apply our definition to the design process of both a simple casual and a serious game as well as to the analysis of an existing game. We show how gameplay can be created and studied independently from presentational aspects. In section 5 we conclude our findings and ponder the consequences for game studies in general.

BACKGROUND

In search of a workable definition for what makes a game what it is we turn to the theorists.

According to Lindley (2005), games can be seen as systems consisting of three components: Simulation, Narration and Gameplay, each having a different tradition, language and methodology. Basic features and functions of the game world are part of the simulation layer. It is "the level at which the authored logic and parameters of a game system together with the specific interactive choices of the player determine an (implied) diegetic (i.e. represented) world" (Lindley, 2005). The *simulation layer* is based on an underlying set of rules, functions and constraints that have been developed by the game designers. It describes non-player character behavior (e.g. friendly/aggressive, the choices they make when certain things happen etc.), physics of the game world (e.g. the speed of a car, the distance a player can jump), the mechanics that govern all kinds of interaction between game elements (e.g. bonuses/penalties for fighting units based on terrain), and others. Some video games focus more, some focus less on the simulation layer. For example, simulation games like SimCity (Maxis 1989) lay stress on this layer while many other games do not contain many simulation elements. The *game layer* consists of a "framework of agreed rules" (Lindley, 2005). It defines what a player can and cannot do in a game, what he has to do in order to win, and the consequences of the players' actions. Although rules are a fundamental part of any computer game, some games stress this layer more than others. For example, abstract games like Tetris (Pajitnov, 1984) concentrate on this layer, while adventure games like Heavy Rain (Quantic Dream, 2010) pay more attention to the other levels like the narrative layer. The third level of semiotic encoding, the *narrative layer*, is the "representation of the causally interconnected events of a story" (Lindley, 2005). It comprises the efforts of creating a believable narrative for the game. According to Lindley, a common narrative structure of computer games consists of three acts: establishment of conflict, playing out its implication and solving the conflict. Games like hypertext adventures and game books focus on the narrative layer of game systems.

An alternative distinction between different game layers is made by Mäyrä (2008). Thus, games consist of a core and a shell. The *core* represents both all the actions a player can take in a game and the rules that regulate these actions. It is the layer on which the gameplay takes place. The *shell*, on the other hand, is the sign system that represents all the aspects of the game. While the core of the game is abstract, the shell is the contextualization that brings the core elements to life. When the shell is changed, the game play stays the same, and so does the identity of the game - mostly. However, there is a certain impact of the shell's semiotic representation on the game. Just as the identity of a chess game depends on its texture and workmanship (e.g. cheap plastic vs. premium ivory), a video game is influenced by its shell. For example, on the core layer both video games The Elder Scrolls IV: Oblivion (Bethesda Softworks, 2006) and Fallout 3 (Bethesda Softworks, 2006) are seemingly very different. However, there are great differences regarding the semiotic representation of each game (and its narrative). While few people who just *see* both games might find them to be quite similar on the shell layer, most people who *play* them might notice the similarities on the core layer.

These two approaches come from a media studies background and are used for game analysis. A different perspective, taken from a game design standpoint, is found in the MDA framework by Hunicke, LeBlanc and Zubek (2004). It serves as a guide for game design and conceptualizes games as dynamic systems on three levels: mechanics, dynamics and aesthetics. *Mechanics* represent the game's components at a very abstract level (data representation and algorithms). They include actions, behaviors and control mechanisms the player encounters within a game. They are crucial for game play and the basis on which dynamics can emerge. For example, game mechanics like highly accurate weapons and rifle scopes can produce game play dynamics like sniping. *Dynamics* represent the "run-time behavior of the mechanics" (Hunicke, LeBlanc and Zubek, 2004), including the mechanic's reaction to the player's input. They create an experience for the player. Lastly, *aesthetics* describe the player's affective reactions to the game system. The authors mention reactions like sensation, fantasy, fellowship, challenge and others, which might occur in different combinations (depending on the game system). The MDA framework is quite abstract and therefore provides only "touchstones" for the game design process (Sellers, 2006).

None of the abovementioned approaches offer a model of gameplay that is concrete enough to be easily used for analyzing a game's interactive components or designing specific interactions. However, each of them provides some valuable input for this article:

1. Gameplay has to be isolated from narrative and simulative layers of the game.
2. Gameplay has to be understood on an abstract level and to be separated from its semiotic representation.
3. Gameplay can be seen as an interaction of a player's input (actions) the rule-based response system of the game (rules, challenges).

In this article we argue that the defining aspect of a game, the game's identity, is the gameplay. Gameplay is what really differentiates interactions between different games. It is therefore crucial to study gameplay and explicitly define its components.

GAMEPLAY

We build on the definition of gameplay defined by Hebisch and Jeong (2009). They define gameplay as a combination of challenges and methods. Methods comprise those that are afforded to the player by the game to overcome the challenges and those that a designer uses to represent the challenges in the game (Hebisch & Jeong, 2009). The definition of the challenges is based on Adams and Rollings (2006), in which fundamenal challenges are defined in a pragmatic way. Hebisch and Jeong complete their definition of gameplay by compiling a list of fundamental methods.

We decided to revisit this definition and describe it and its components as precisely as possible. Our revised definition of gameplay is as follows:

Gameplay is a combination of challenges and methods. Challenges are the basic underlying concepts a player has to deal with in a game. Methods are the building blocks of the interaction in a game and describe how a player accesses the game and influences the outcome. Gameplay is achieved when a player approaches a challenge by means of the methods a game offers to him.

In the remainder of this section we will define our challenges and methods.

Challenges

Challenges are the foundation of every game. Without them there would be no need to do anything in a game and thus nothing that intrigues the player. They are the driving purpose of a game. The following list of challenges builds on Adams and Rollings (2006) as well as preliminary work by Hebisch and Jeong (2009):

- coordination,
- logic and mathematical puzzles,
- race and time pressure,
- trivia,
- recall,
- pattern recognition,
- exploration and discovery,
- conflict,
- economic challenges,
- and reasoning.

In the following we define these challenges precisely.

Coordination challenges require precise input handling of the player. They can be divided into different categories. Speed challenges require the player to rapidly input commands over a certain time while reaction time challenges demand quick input decisions within a certain time frame. Accuracy and precision challenges require the player to guide an element in the game as close as possible to a predefined goal, usually within a limited time frame. This can include aiming a gun as well as steering a car. Physics challenges require the player to understand and adapt to the game's underlying physics system to successfully control elements in the game. Examples are the physics in driving simulations or jumping physics in any game. Timing and rhythm challenges require the player to input commands at recurring intervals. Combination challenges require the player to input commands in predefined sequences to succeed.

Logic and mathematical challenges require the player to reason about the current state of the game to decide his next step. In formal logic puzzles there is usually one solution that can be reached by applying deductive reasoning. The player has all the information readily available and all the time he needs for these conclusions. Mathematical challenges are mostly concerned with estimating and calculating probabilities of outcomes that are beneficial to the player. Purely algorithmic or computing challenges are rare.

Races and time pressure challenges require things to be done in a limited amount of time. In races the player has to accomplish something first in competition with other players. In time pressure challenges the player races alone against the clock.

Factual knowledge challenges are challenges that require knowledge from outside the game world. They present the player with questions about the real world[1] to solve a task in the game. The player has to have this knowledge prior to playing the game; he cannot acquire it during play (other than trying all answers and remembering the correct ones for the next session).

Recall challenges build on knowledge that the player acquires while playing the game. These are facts about the game world, the story, its characters and other aspects. Anything that the player can experience in the game is subject to a recall challenge later on.

Pattern recognition challenges demand that the player recognizes repetition(s) during play. The repetition in the behavior of the game's elements, such as movement patterns of obstacles, or in the game world itself has to recognized and properly dealt with by the player. There is usually a certain strategy involved with every pattern.

Exploration challenges are about the game world the player navigates through. The player discovers the world, his position and path relative to it and learns how the world fits together. The discovery of the game world is challenge and reward in itself. A popular example of this can be found in Castlevania: Symphony of the Night (KCET, 1997), that tracks the player's progress of visiting every space in the game's world on a map accessible in-game. Games that focus on exploration usually employ elements such as maze-like layouts, teleporters or illogical spaces to make the discovery more interesting.

Conflict challenges are competitions for limited resources. The player is confronted with situations where he has to defend his own resources and/or diminish the resources of others. Combat is not necessarily involved. Instead, the conflict arises from the competition itself. This challenge requires strategy and tactics to come up with a plan and executing it subsequently.

Economic challenges are about resources, too, but the handling of resources follows a built-in economy. Resources have an abstract value and can be exchanged accordingly. Economic challenges build on the idea of accumulating resources and achieving an economic balance of the different resources.

Reasoning challenges build on knowledge that comes from outside of the actual game. In conceptual reasoning challenges the behavior of a game element or the consequence of an action is identical to its real-life counterpart. Factual knowledge and physics challenges are similar if the physics in the game world simulate those in the real world. Lateral thinking challenges on the other hand eschew the most obvious solutions or make them impossible to achieve so that the player has to think of and improvise an alternative. The solutions are often obscure and convoluted, making their discovery much more satisfying.

Methods

The second part of our definition of gameplay covers the methods for expressing and implementing the challenges. A game based on coordination challenges may let the player take control of an avatar directly and make him react to obstacles, for example by avoiding them or by hitting them in some fashion. A game based on logic challenges may display all the information a player needs to discover the solution, but the player can only influence the needed elements indirectly. Ultimately, methods describe how a player accesses the game and influences the outcome. This section lists potential methods for presenting challenges in a game.

We build on the list of methods in Hebisch and Jeong (2009) and divide them into the following categories:

- obstacles,
- manipulation,
- degrees of freedom,

- player representation,
- abilities,
- game progression,
- event progression,
- resources,
- playing field overview,
- and opponents.

Obstacles hinder the player's progress in the game. They can be passive or active. Passive obstacles are non-moving objects such as hills or trenches the player has to overcome. They are fixed on the playing field. Active obstacles on the other hand can move. Among others, they include moving platforms and opponents.

Manipulation allows the player to interact with the world. It can be direct or indirect. With direct interaction, the player controls objects in the game world, e.g. the avatar or a puzzle piece. With indirect interaction, the player does not affect a particular object itself, but the environment it is embedded in. In the first instance, an object is manipulated directly, whereas in the other instance its conditions (and thereby indirectly its behavior) are changed. Direct manipulation is arguably the most used form of manipulation: The player controls an avatar on-screen by pressing a button or other forms of interaction. An example for indirect manipulation is the game Loco Roco (Japan Studio 2006), in which the player has to rotate the whole game world to make the avatar move in a certain direction. Indirect manipulation requires the player to understand the cause and effect of his actions.

Degrees of freedom affect how much control the player has over the positioning of the game's elements he controls. In a two-dimensional game with all degrees of freedom, the player can freely position the game elements on the game field. As an example, consider a battlefield simulation where the player can order his troops to move to any location on the field. An example for a three-dimensional game with all degrees of freedom is the game Descent (Parallax Software 1995), a game with a science fiction theme, that gives players complete freedom to move and orient their spaceships within the geometry of the game level. However, in most games the degrees of freedom are restricted to some extent. In the game R-Type (Irem 1987), the player can move his ship to any location on the screen but not in the level itself, as it is scrolling by automatically. In Super Mario Bros (Nintendo 1985), the player cannot freely move the avatar to any position on-screen. He is limited to the level's geometry and the positions he is able to reach by jumping with the avatar. This method also affects the control of the virtual camera that defines what the player can see on-screen. In most two-dimensional games, the player can only see a section of the complete level at a time and cannot move the virtual camera to inspect the rest of it. In three-dimensional games it is often possible to move the virtual camera to get a better overview of the game world. In games that use the first-person perspective, for example, the movement of the virtual camera corresponds to the movement of the avatar's head, allowing the player to experience the game from the perspective of his in-game alter ego.

Player representation regards the visual manifestation of the player within the game world. If there is a representational element that is under the player's control or has to be manipulated by the player, we call it an avatar. The player controls the avatar directly from either a first-person or a third-person perspective. Examples for avatars can be found in Nintendo's (1985) Super Mario Bros (third-person perspective) or Bungie's (2001) Halo: Combat Evolved (first-person perspective). If there is no single element that is controlled by the player but instead multiple objects, we call the player representation unit representation. In Tetris (Pajitnov, 1984), the player controls a single tetromino at any given point of time but a lot of them over the course of a session. In StarCraft (Blizzard Entertainment, 1998), the player gives orders to a subset of units that he selects according to his strategy. If there is no visual manifestation of the player in the game at all we call it god representation. For example, in games like Loco Roco (Japan Studio, 2006, or SimCity (Maxis, 1989), the player manipulates the game elements by changing the rotation of the game world or by setting up the structures that make up the simulated world.

Abilities affect the actions a player can invoke in the game by granting or removing certain capabilities during play. Power-ups allow the player to temporarily use a different set of abilities during play. They can be acquired with items or by a series of key combinations. Permanent changes to the player's abilities are usually acquired by a development process. The development process is often based on the acquisition of some sort of resource, e.g. experience points, time, money or equipment.

Game progression is the concept of giving the player information on how to proceed in the game. In a linear game progression the player experiences the different events of the game in a predefined order. A non-linear game progression lets the player choose the next event.

Event progression determines the order of actions while playing the game. If the game is real-time, the player can influence the game at any point during the event. In turn-based games on the other hand, the player and his opponents perform their actions one after the other in a predetermined order.

Resources are elements that the player has to acquire or preserve during play. Running out of a specific resource, e.g. time, energy, bullets or lives, has consequences for the game progress. Additive resources grant the player additional abilities and thus make game play easier (power-ups). Limiting resources diminish the player's abilities or a critical resource; but nevertheless he can continue the game, albeit under more difficult conditions (energy, ammunition, time). The loss of a critical resource usually makes the player lose the game and resets his progress back to a certain point (lifes, continues, tries etc.).

Overview regards the information the player has at any moment during the game. Overview can be complete, letting the player acquire all the information he needs to decide on his next action. It can also be limited so that the player is not fully informed but has to react spontaneously to the events.

Opponents are related to active obstacles. While active obstacles are defined as simply being obstacles that move according to a pattern, opponents move and react to the player's actions during the game. Opponents can be controlled by the computer or by other human players in a multiplayer game. This method has several characteristics. The attitude towards the player determines how other players react to the player: In a competitive setting they actively try to hinder the player; in a cooperative setting they help the player; in a team-based setting the player has opponents from both sides. The player and his opponents might have equal or non-equal abilities. A third characteristic is the spacial setting: Player and opponents can share the playing field or play on separate ones. An accompanying characteristic is the possibility of interference between the opponents and the player: It determines whether the opponents can actively hinder the player's progress on the playing field (e.g. by blocking his path) or if there is no interference allowed.

Discussion

The above collections are not complete. However, they cover the most basic challenges and methods and can already be used to create gameplay combinations. This offers us the possibility to create a raw experience, the core of a game, which can later be augmented by the context of a game, i.e. the theme, the story or other things attributed to the shell of a game. This abstract view of gameplay has several advantages and implications.

We can use our definition of gameplay and its components to analyze a game based on its gameplay characteristics. We give an example of such an analysis in the next section. This gives us a solid basis for comparing the gameplay of different games and makes it much clearer to understand how a game plays, furthermore giving us the possibility to talk about gameplay the same way we talk about the other aspects that make a game, e.g. the story, the graphics or the sound. The ability to view gameplay as independent from the other properties makes it possible to define combinations that are not bound by their context and thus might not have been thought of yet.

Hebisch and Jeong (2009) arrange the challenges and methods in a matrix, where the challenges are listed vertically and the methods horizontally. They use this matrix to classify the gameplay of well-received games in order to find out which combinations are most popular. Beyond that, the matrix can be used to research which combinations appear less often and create new experiences. It is a stepping stone to create fundamentally new experiences according to a methodical approach.

We can use the above definition for the creation of games, for example serious games, where a way to connect the educational content to the gameplay is needed. By having the building blocks for gameplay available we can evaluate which combinations are used best when we want to get a certain point across. We can select the methods that accompany the underlying challenge of the learning task best.

Ultimately, we can use the above collections to systematically research which combinations work and how they affect the experience of the player during play with regard to experience, playability and enjoyment.

APPLICATION

Challenges and methods can be used as building blocks for choosing gameplay combinations or creating new ones. In a scientific context, this is important, for instance, when designing a serious game ("synthesis"). Here it is essential to create game structures that represent the curricular content and learning objectives. However, guides addressing this issue are often vague and leave many questions unanswered. Moretti and Dondi (2003) describe features required in relation to certain learning objectives (e.g. real time game play or accurate description of the problem for the learning objective decision maker). However, these features are still somewhat abstract. Furthermore, the authors' game recommendations are based on video game genres (adventure games, simulation games etc.) which are far too imprecise to systematically create specific game play experiences fostering the desired learning processes. Our matrix of challenges and methods can provide a pragmatic tool to systematically choose interactions in order to create game play experiences that can easily be linked to learning outcomes. Also, one could use our matrix for game analysis. By identifying game play profiles and interactions one could describe cognitive experiences that might be of relevance within media impact studies. The next sections show how the matrix of challenges and methods can be used for synthesis of two games (a simple casual and a serious game) and analysis.

Synthesis: A Simple Game

In this section we show an example of how the above definitions can be used to design the gameplay of a simple game starting from scratch.

We start by choosing which challenge should form the foundation of the game. While this could be achieved by simply picking one from the list at random, we start with the idea of having the player use swinging, momentum and trajectories to move, i.e. we start with the idea of a game based on a physics challenge. We find it to be easier to start with an idea of an interaction, however vague, and use the list of challenges to find out which of them is the most dominant in it. With an understanding of what the dominant challenges in the game should be, we can evaluate ideas during the design process and keep the gameplay mechanic focused. After defining the core challenge of the mechanic we have to choose the methods to construct an interesting gameplay mechanic.

First we pick the *obstacles*. We want obstacles that obstruct the player's progress and that have to be overcome by arranging a trajectory in some fashion. We have the option of having fixed and moving obstacles. We choose only fixed obstacles. Moving obstacles would require the player to observe two things at once, the swinging motion and the motion of the moving obstacles. In order to keep the gameplay tight we want the player to concentrate only on one thing at a time.

With the same reasoning, we choose to let the player only *control* a single element in the game, i.e. an avatar. We go a step further and decide that he is going to control a character that moves around by swinging.

We want to let the player *manipulate* the avatar directly. However, swinging does not appear to be a purely direct manipulation. Usually, something swings while being affixed to something, e.g. a rope or a cable. But having the player control a rope to control the movement of the avatar would make the manipulation indirect. Therefore, we decide to make the avatar use his arms to swing around a suitable object. This decision could be used to create the backstory of the game by telling why he uses his arms and not a rope. He might be a gymnast in the Olympics or a thief trying to escape.

Next we have to decide the *degrees of freedom* we want to give the player. We decide to keep it as simple as possible. The player uses only one button to control the avatar; it is not possible for him to position the avatar anywhere he likes. Since we decided that the avatar uses his arms to swing around, we need to determine how the button interacts with the avatar. If the player presses the button the avatar reaches for an object to swing around. As long as the player keeps the button pressed, the avatar rotates around the object with the same momentum he had when grabbing it. When the player releases the button the avatar lets go of the object and launches into the air according to the trajectory simulated by the physics engine of the game. In order to succeed in the game, the player has to find the right moment to launch the avatar and make him fly to the next object he can rotate around, all while avoiding the obstacles in his way. We realize that what we have designed here is a kind of obstacle course that only allows flying as transportation method.

In order to correctly estimate when to let go of an object to achieve the correct trajectory in order to move on, the player has to see the avatar, the obstacles and the objects that he can use to rotate. Therefore, we decide to *present* the game from a third-person perspective and in 2D. This enables the player to estimate the distance between objects, the position of obstacles and the speed of his avatar better than in a three-dimensional setting or from a first-person perspective.

In the above we decided on an obstacle course, now we have to determine whether we give the player a complete *overview* over it or not. We could give the player a complete overview, however, then we might only reach a certain level of complexity, i.e. the more complex the obstacle course gets, the smaller the individual elements have to be to fit them all on screen. To us, it seems to make sense to sacrifice overview for visibility of crucial game elements. Therefore, we decided to only allow partial overview over the obstacle course. The player can only see a segment of the obstacle course; the game takes care of always displaying the relevant segment of the course by scrolling horizontally or vertically as needed.

We mitigate the partial overview by choosing that the game consists of a series of levels that the player can traverse as often as he likes. This enables him to memorize the placement of obstacles and objects. We think that the effect of the memorization *challenge* will be negligible in regard to the resulting gameplay mechanic. An additional challenge that we might however want to consider is the time pressure challenge. We could measure

Figure 1. A sketch of the playing field for the simple game. The avatar has traversed part of the level by means of hanging onto the objects, swinging around and flying between them.

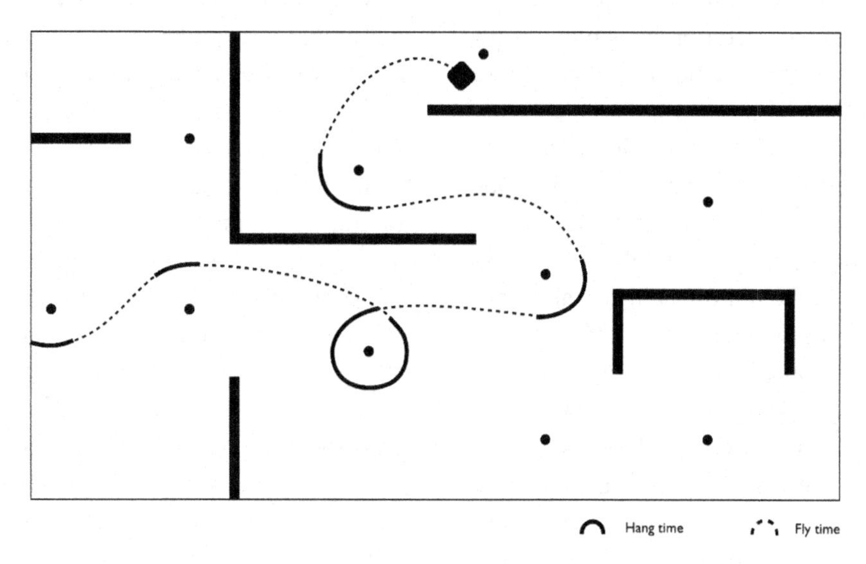

the time the player needs to traverse the obstacle course in each level so that he can improve it on his next try. A popular technique is to show the player the replay of his best attempt of an earlier run as a kind of ghost so that he can race it to the end of the level.

Other than the replay we decided that there will be no other *opponents* in the game.

In order to increase the tension of the game mechanic we decided that there should be one *resource* in the game that is critical to the success of the player. Restricting the number of retries does not make sense for an obstacle course; we have to find a more critical resource. We find it in the time in which the player has to decide his next move. We put a limit on the "hang time", i.e. the time a player can hang on to an object. There is even a suitable analogy: The avatar's arms are getting more tired the longer he rotates around an object. If the player holds on for too long, the avatar lets go and might launch in an undesired direction.

We decided that there are no *abilities* that the player can gain throughout the game. The only resource in the game is the hang time. We could make special items that prolong it for a certain time or set up a system where the player gains hang time by leveling up his arms. But this would unnecessarily complicate the game.

The last method to be considered from our list is the *event progression*. It is clear to us from the progress we made so far that we want a real-time game. While we can envision a round-based system where in each round the player sets the angle and speed to move from object to object we have made other decisions earlier that disagree with this option. We bring this up here to show that the possible options for a certain method might influence already made decisions to the point that one might rethink the game from another perspective.

It is now time to summarize the game. In the game the player has to traverse an obstacle course by making an avatar swing around certain objects and launching him over the obstacles that make up the course. In order to do that the player presses a button to grab a suitable object, keeps the button pressed to rotate around it and releases it to launch the character in the desired direction. The trajectory is simulated correctly. However, the avatar keeps the momentum he had when grabbing an object while swinging around it. The player has to pay attention to the arm strength of his avatar. He can try a level as often as he likes and there are no opponents. Once he finishes a level he can try to improve his time from start to finish.

We have shown a comprehensible design process that leads from an idea to a workable description of the mechanics of a simple game. Note that there are still lots of decisions that must be made and questions to be answered. How does the avatar gain momentum in the first place? What happens if there is no object to hold on to when the player presses the button? What exactly are these objects that the avatar can rotate around? How are they displayed on the screen?

These are valid questions. However, we think answering them in detail is out of the scope of this article. The core gameplay mechanic is solid; here we only wanted to show how to construct it using the definitions we presented earlier. We have listed some of the alternatives when choosing a particular method and encourage the reader to think about which alternatives he would have chosen and how it would have affected the gameplay.

We now apply the same process to the design of the mechanics of a serious game.

Synthesis: A Serious Game

When designing a serious game, one is well advised to link game structure and educational content as tightly as possible (Wechselberger, 2009). While this task might sound easy, it could post a challenge to one's creativity. Our characterization of gameplay as a combination of challenges and methods can provide some starting points that ease the conceptualization process.

During a research project on the effectiveness of game-based learning in an industrial context, a serious game teaching the start-up procedure of a testing plant had to be created (Wechselberger, 2011). Due to didactical considerations derived from a constructivist learning perspective, we used very few instructional methods and a lot of interactive gameplay episodes that were modeled using challenges and methods.

Challenges were supposed to initiate as much self-directed behavior on the part of the learners as possible, fostering learning as an active process. *Exploration challenges* played an important role in the serious game. The player had to get familiar with the premises and figure out how to eliminate obstacles like disabled plant modules etc. Furthermore, we informed the player that he had to activate several power switches. However, in order to engage him to actively deal with the environment, we refrained from telling them where these switches were located. Instead, this information was hidden within a *logic challenge*: The player had to decrypt an abstract schematic drawing of the plant's layout, identify the switches and transfer this information to the three-dimensional model of the plant (Figure 2). *Logic riddles* also played an important role for teaching the player how to activate a plant module using the attached touch screens. Instead of providing immediate information regarding which buttons had to be pushed in what order, every button played a different tone. Tones and buttons were linked in a way that as soon as the player hit the tones in the order of the major scale, the module was successfully activated. In order to foster the information the player had gained by his/her exploration, we added some practice episodes. These were implemented as "boss battles" (which are very suitable for mastering a skill, Gee, 2007) as a combination of the *challenges recall and time pressure*. For example, after exploring the activation sequence of a plant module, the player had to activate all remaining plant modules within a given timeframe. This was supposed to consolidate the player's newly acquired knowledge in a suspenseful and therefore entertaining manner (Vorderer, Hartmann, & Klimmt, 2003).

Regarding methods, the player could only indirectly *manipulate* the plant by activating switches and buttons which were then processed by the dynamic object system of the technical framework (Rilling & Wechselberger, 2011). This was supposed to help the player understand the

Figure 2. A logic challenge in form of an abstract schematic drawing of the plant's layout

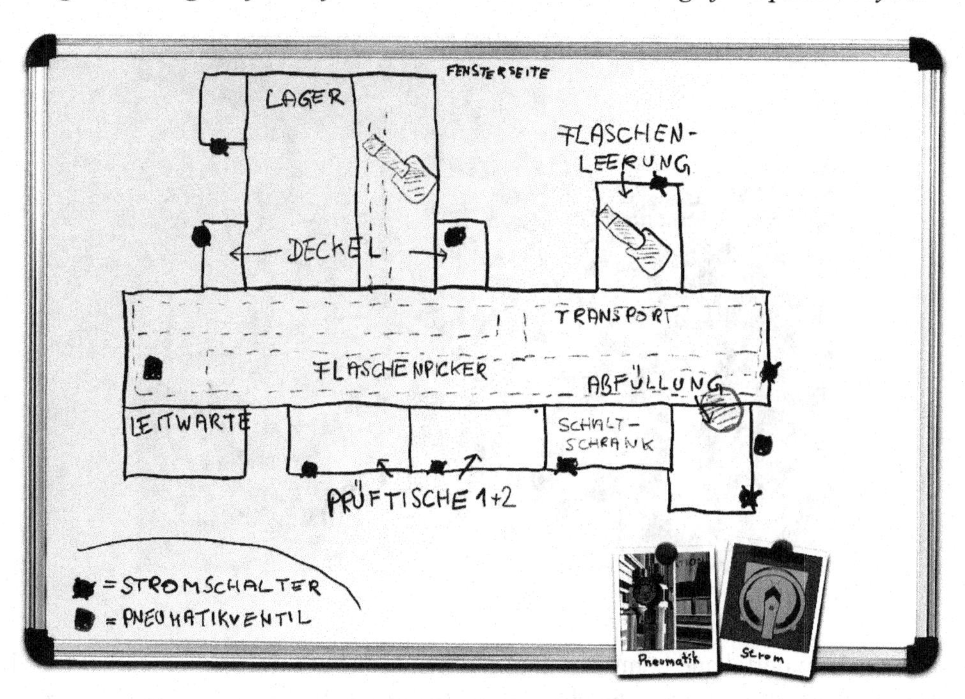

cause and affects of his actions and thereby the structure of the plant. There were no active, just passive *obstacles* (time pressure, inactive terminals etc.), because the real plant did not have any active obstacles either. Given that the serious game was a three-dimensional game with a first person perspective (Figure 3), there were many *degrees of freedom*. However, vertical movement was restricted to switch between walking and crouching (in order to inspect lower parts of the plant). Also, the player could not climb onto conveyor belts etc., since this was strictly prohibited within the real plant. *Game progression* was linear in order to have maximum control over the player's learning progress. *Multiplayer support* was omitted for the same reason. Since the challenges mainly consisted of puzzles, there were was no need to use *resources* in any form (except for time, which was restricted within boss battles, see above).

To some, these examples may sound trivial. But in the light of many serious games suffering from uninspired (educational) game design, our characterization of gameplay provides some handy and practical starting points for serious game design.

Game Analysis

In this section we will analyze a game according to our definition of gameplay. We want to get to the core of the gameplay to see what the essential parts are.

Portal

We picked the game Portal from Valve Software (2007). In the game, the player uses a gun, called the portal gun, to shoot either a blue or an orange shot. When a shot hits a suitable place in the level, e.g. on the walls, on the ceiling or on the floor, it opens up a portal in the same color. There can only be one portal of each color active at the same time, shooting the same color again removes the old portal and sets a new one. The portals are connected to each other, meaning that the player

Figure 3. First person perspective within the serious game

can go through the blue portal and come out the orange one and vice versa. In this way, the player can create new ways through the level. When going through a portal the game maintains the player's speed, momentum and direction. The player has to navigate through a series of increasingly complex levels and use the interconnected portals to reach and scale places otherwise inaccessible to him.

Challenges

There are two dominant challenges, *logic and physics*. The objective is always to find a way through the course to the next exit. The player has to understand the layout and the features of the course and deduce where and in which order he has to place the portals so that he can make his way to the exit. As the complexity goes up, this requires thinking a few steps ahead since he can only create one portal of a certain color at once.

The reasoning about where to put the portals is strongly connected with the understanding of the game's underlying physics system. Creating the portals at the right places implies using their movement preserving properties to catapult the player through the level. The solution often depends on building up momentum for a long jump by repeatedly jumping through portals or creating portals on an angled surface so that the resulting jumping curve will let the player reach the desired part of the level.

Methods

Portal uses inactive and active *obstacles*. As part of the obstacle course, the player is often confronted with walls that are too high or trenches that are too wide for him to jump over. The player has to navigate over moving platforms or use revolving walls to reach the next part of the level.

The player controls the *avatar* directly from a first person perspective. He can look around freely but movement is restricted to walking on the ground and jumping. The latter is augmented by using the physics system when using the portals to build speed and momentum, but still restricted to the physical simulation of the trajectory.

The only additional *ability* that the player gains throughout the game is the portal gun, giving him the ability to create portals at his will. The portal gun never runs out but only supports one portal of each color at a time. That means while the ability to create portals is unrestricted, the amount of active portals at any given time is severely constrained.

The only other important *resource* in the game is the avatar's life energy. There is no energy display during the game, but as long as the avatar doesn't get hit too often in a small time frame, his energy will not run out. If it does, however, Portal simply resets the game to a point prior to the avatar's death and lets the player try again. There is no limit as to how often the player can try a certain section in the game or how long it takes him to do it.

The *game progresses* linearly, with only one possible exit that leads to the next section. The *event progression* during the game happens in real-time. The *overview* of the playing field is incomplete and limited to the first-person view of the avatar. The only *opponents* in the game are fixed turrets that shoot at the avatar when he is in sight.

Essential Gameplay

As we can see, Portal's methods are very focused. There are not a lot of abilities or resources to handle. Most of the game's elements influence the player's ability to solve the main challenge of getting through the obstacle course. The limited overview from the first-person perspective is a design choice that puts the burden on the player to build a map of the level in his head first and then finding a way to the exit. The player can never see all the information at once and has to rely on the strategy he puts together piece by piece while exploring the level. This makes it so much more satisfying when a solution is found, because while everyone can acquire all the information he needs, it is still an achievement to put the things together in one's head.

Limiting the usage of portals is another design choice that makes finding the solution to a level satisfying. It lowers the amount of possibilities a player has to consider when constructing a solution: If more portals are required than are possible to create than the solution is wrong. Now the player only has to worry about the order and position of the two allowed portals. That means he has a better chance of understanding why a solution is correct or not.

Lastly, by setting portals the game world itself becomes part of the abilities of the player to some extent. Every ceiling becomes a possible shortcut, every ledge a possibility to gain momentum from falling or every angled surface a possible catapult. The fact the player can use the game world to defeat its obstacles and rise above its and the avatar's limitations is a rewarding experience.

We argue that these are the essential parts of the gameplay mechanics in Portal. We have shown how to extract these essential gameplay elements by methodically classifying challenges and methods and evaluating which combinations have the strongest impact on the gameplay. We think that this approach is suited for gameplay analysis and characterization.

CONCLUSION AND OUTLOOK

In this article, we looked at video games from the perspective of a scientist who wants to create games for a specific purpose. We looked at theories about games and found out that the defining characteristic of a game is its gameplay, or shell, because the other aspects are, for the most part, interchangeable. The gameplay is what defines the interactive elements of the game and is directly related to the things a player *does*, rather than what he perceives.

We then defined gameplay as a combination of challenges and methods. Challenges are the underlying concepts of a game, the fundamental cause for its existence. Methods are building blocks for interaction that define how the player experiences and approaches the challenges. Good gameplay is the creation of enjoyable tension between being under pressure from a challenge and understanding how to deal with it given the methods afforded by the game.

We presented collections of challenges and methods as part of a toolkit for the synthesis and analysis of gameplay in games. We see this definition and our collections as the foundation for studying the interactivity of video games. This has consequences for the research of cognitive effects of video games and game-based learning. We can enumerate the possible combinations of challenges and methods and use the resulting gameplay mechanics to research what their effects on the player are. With this we can study how to design games that support the learning process. Does it help to present the same challenge with two different methods to show how such a change allows for easier understanding of the problem domain?

Of course, defining the bare gameplay mechanics only assists up to a certain point in designing a game. A game designer would not invent another Portal if he simply threw the above mechanics together and called it a game. A game designer still has to think about how the challenges and methods manifest themselves in the actual game.

In this article we did not handle the creative process that translates the abstract methods to elements in the game. While we argue that a game can be characterized on the basis of its gameplay this does not mean we think that this is all there is to it. The creation of an overarching context for the game's elements is equally important. The background story provides the context for the decisions that went into the translation from gameplay mechanics to game elements: Obstacles can be represented by very different structures, such as hills, trenches or blocks. The artwork and the animation bring the elements to life and allow a deeper immersion. The music ties the whole game together and the sound effects underline the player's actions and achievements in conquering the challenges.

However, we think that with our definition of gameplay mechanics and our examples of synthesizing and analyzing games according to this definition we took an important step towards researching games based on their interactivity, the characteristic that distinguishes video games from books and movies. We believe that if a method from an existing game was exchanged with another method, a new game that plays differently would be attained. That is why we consider the definitions in this article as essential parts of the gameplay of a game.

We intend to use the definitions in this article to create new gameplay experiences as well as help the research in understanding games as an interactive medium in the future.

REFERENCES

Adams, E., & Rollings, A. (2006). *Fundamentals of game design*. Upper Saddle River, NJ: Prentice Hall.

Bethesda Game Studios. (2008). *Fallout 3* [PC game]. Rockville, Maryland: Bethesda Softworks.

Bethesda Softworks. (2006). *The elder scrolls IV: Oblivion* [PC game]. Novato, CA: 2K Games.

Blizzard Entertainment. (1998). *StarCraft* [PC game]. Irvine, CA: Blizzard Entertainment.

Bungie (2001). *Halo: Combat Evolved* [Xbox game]. Redmond, WA: Microsoft Game Studios.

Firaxis (2010). *Civilization V* [PC game]. Novato, CA: 2K Games.

Gee, J. P. (2007). *Good video games + good learning: Collected essays on video games, learning, and literacy*. New York, NY: Peter Lang.

Hebisch, E., & Jeong, Y. (2009). *Umsetzung von Spielmechaniken auf einer kugelförmigen Spielwelt.* Unpublished diploma thesis. University of Koblenz-Landau, Germany.

Hunicke, R., LeBlanc, M., & Zubek, R. (2004). MDA: A formal approach to game design and game research. *Discovery, 83*(3), 4.

Irem (1987). *R-Type* [Arcade game]. Hakusan, Japan: Irem.

Japan Studio. (2006). *Loco Roco* [Playstation Portable game]. Tokyo, Japan: Sony Computer Entertainment.

Juul, J. (2005). *Half-real: Video games between real rules and fictional worlds*. Cambridge, MA: MIT Press.

KCET. (1997). *Castlevania: Symphony of the Night* [PlayStation game]. Tokyo, Japan: Konami Corporation.

Lindley, C. A. (2005). The semiotics of time structure in ludic space as a foundation for analysis and design. *Game Studies*, 5. Retrieved from http://www.gamestudies.org/0501/lindley/

Maxis (1989). *SimCity* [PC game]. Emeryville, CA: Maxis.

Mäyrä, F. (2008). *An introduction to game studies: Games in culture*. Los Angeles, CA: SAGE Publishing.

Moretti, M., & Dondi, C. (2003). *Guide to quality criteria of learning games*. Bologna, Italy: SIG-GLUE.

Nintendo (1985). *Super Mario Bros* [NES game]. Kyoto, Japan: Nintendo.

Pajitnov, A. (1984). *Tetris* [Elektronika 60 game].

Parallax Software. (1995). *Descent* [PC game]. Beverly Hills, CA: Interplay Productions.

Quantic Dream. (2010). *Heavy Rain* [PS3 game]. Tokyo, Japan: Sony Computer Entertainment.

Rilling, S., & Wechselberger, U. (2011). A framework to meet didactical requirements for serious game design. *The Visual Computer*, *27*(4), 287–297. doi:10.1007/s00371-011-0550-6

Sellers, M. (2006). Designing the experience of interactive play . In Vorderer, P., & Bryant, J. (Eds.), *Playing Video Games: Motives, Responses, and Consequences* (pp. 9–22). New York, NY: Routledge.

Valve Software (2007). *Portal* [PC game]. Bellevue, WA: Valve.

Vorderer, P., Hartmann, T., & Klimmt, C. (2003). Explaining the enjoyment of playing video games: The role of competition. In *Proceedings of the 2nd International Conference on Entertainment Computing* (pp. 1-9). Pittsburgh, PA.

Wechselberger, U. (2009). Teaching me softly: Experiences and reflections on informal educational game design. *Transactions on Edutainment*, *II*, 90–104. doi:10.1007/978-3-642-03270-7_7

Wechselberger, U. (2011). A serious game compared to a traditional training. *Academic Exchange Quarterly*, *5*, 58–63.

ADDITIONAL READING

Aarseth, E. (2003). Playing research: Methodological approaches to game analysis. *Proceedings of the Digital Arts and Culture Conference, 2003*. Retrieved from http://hypertext.rmit.edu.au/dac/papers/Aarseth.pdf.

Bates, B. (2004). *Game design*. Boston, MA: Thomson Publishing.

Glassner, A. S. (2004). *Interactive storytelling: Techniques for 21st century fiction*. Wellesley: AK Peters.

Grau, O. (2007). *MediaArtHistories*. Cambridge, MA: MIT Press.

Holopainen, J. (2011). *Foundations of gameplay*. Doctoral Dissertation. Blekinge Institute of Technology, Sweden.

Lindley, C. A. (2005). The semiotics of time structure in ludic space as a foundation for analysis and design. *Game Studies*, 5. Retrieved from http://www.gamestudies.org/0501/lindley/

Pivec, M. (Ed.). (2006). *Affective and emotional aspects of human-computer interaction*. Amsterdam, The Netherlands: IOS Press.

Raessens, J., & Goldstein, J. H. (2005). *Handbook of computer game studies*. Cambridge, MA: MIT Press.

Ritterfeld, U. (2009). *Serious games: Mechanisms and effects*. London, UK: Routledge.

Schell, J. (2009). *The art of game design: A book of lenses*. Amsterdam, The Netherlands: Morgan Kaufmann.

Sicart, M. (2008). Defining game mechanics. *Game Studies*, 8. Retrieved April, 11, 2011, from http://gamestudies.org/0802/articles/sicart

Vorderer, P., & Bryant, J. (Eds.). (2006). *Playing video games: Motives, responses, and consequences*. New York, NY: Routledge.

Wechselberger, U. (2009). Teaching me softly: Experiences and reflections on informal educational game design. *Transactions on Edutainment*, *II*, 90–104. doi:10.1007/978-3-642-03270-7_7

KEY TERMS AND DEFINITIONS

Challenges (gameplay): Challenges are the basic underlying concepts a player has to deal with in a game.

Game: A game is a rule-based system with a variable and quantifiable outcome, where different outcomes are assigned different values, the player exerts effort in order to influence the outcome, the player feels emotionally attached to the outcome, and the consequences of the activity are optional and negotiable. (Juul, 2005, p. 36)

Gameplay: Gameplay is a combination of challenges and methods. Gameplay is achieved when a player approaches a challenge by means of the methods a game offers to him.

Methods (gameplay): Methods are the building blocks of the interaction in a game and describe how a player accesses the game and influences the outcome.

ENDNOTE

1 If the world is the game world, we'd rather speak of a recall challenge than trivia challenge.

Chapter 15
From a Game Story to a Real 2D Game

Chong-wei Xu
Kennesaw State University, USA

Daniel N. Xu
University of Wisconsin at Milwaukee, USA

ABSTRACT

How does one design and implement a 2D game? Specifically, how does one teach students how to develop a game from an idea or a game story? A technical guide has been developed, which includes the modeling principle, the generic global software structure framework, and the incremental development strategy. The technical guide is introduced via the development of the example game Othello from a standalone version to a single player version and then to a networked version. The technical guide not only has been applied for gaming but also for implementing teaching tools with dynamic behaviors.

INTRODUCTION

Games are an integration of Humanities, Mathematics, Physics, Artificial Intelligence, Animation, and graphical programming, just to name a few. Meanwhile, games are also industrial products. Many educational specialists have indicated that gaming is a good tool for education (Schollmeyer, J., 2006). Not only that, students gain real world industrial skills while they are learning. The variety of game genres plus the unlimited ways for implementing each game greatly encourages a student's imagination and problem solving ability. In order to educate students not only as programmers but also as architects, a technical guide for developing games has been derived. This technical guide will help guide students to model, design, and implement games following certain important principles. This chapter intends to introduce this technical guide by taking an easily described game Othello as an example.

DOI: 10.4018/978-1-4666-1634-9.ch015

This technical guide includes a model and a generic global software structure framework. The model is an abstraction of the common features of different genres of games, as shown in Figure 1. The abstraction emphasizes the separation of the GUI (Graphical User Interface) from the logic as well as the synchronization of the two units based on data passing. This model promotes the analysis of a game and the synthesis of the system, which is especially important when developing a networked game, where the data passed along networks for synchronizing the GUI and the logic is the key to its success. As Figure 1 illustrates, the model says that a game has a special GUI for displaying the scenes of the game. The GUI consists of so-called sprites, which are components animated independently. Another unit, the game logic unit, controls the animations and GUI. The movement of the sprites is controlled externally by those playing the game. The movements may cause collisions among sprites, which might further trigger a new piece of logic to change the structure and format of the sprites in the GUI. Even though different games may have different stories, sets of sprites, control logic, and numbers of players, the same model can be used to represent the majority of games.

Figure 1 indicates that the GUI of a game consists of GUI components, which are the sprites. The logic of a game refers to a set of states combined with certain units of artificial intelligence. All sprites should be held by data structures and the game logic should be defined by a state diagram. These two units are synchronized through the communication mechanism via the passing of the input states and the output states. These input states and output states are associated with the players' external actions and game's internal functions.

Figure 1 also reveals the fact that although games exist in hundreds of thousands of various forms, they share some common principles, which include certain specific OOP (Object-Oriented Programming) principles and general software principles. Following these principles, a global software structure called the three-layer framework for gaming has been derived and will be included in the technical guide to help students in designing games. This structure and framework will be detailed in the next section.

The GUI of a game forms a main game screen. Meanwhile, several other screens are associated with the main game screen as illustrated in Figure 2. Every screen has its own special role, which is suggested by its name. These screens are part

Figure 1. An abstract model of games

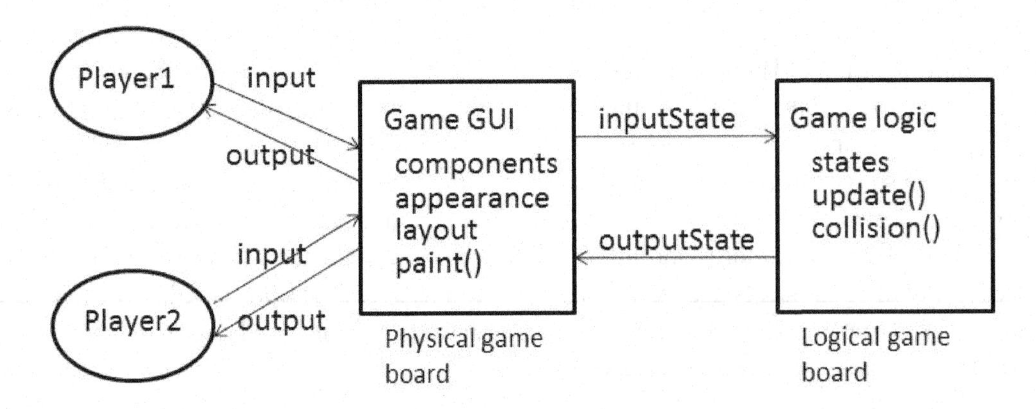

Figure 2. Multiple screens associated with a game

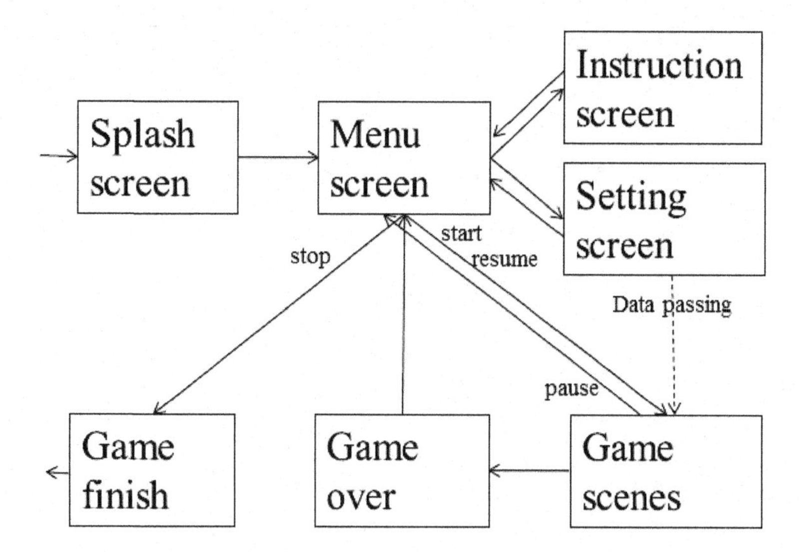

of the global software structure. However, this is beyond the scope of this chapter.

THE DERIVATION OF A GLOBAL SOFTWARE STRUCTURE OF GAMES

The global software structure of games is the realization of the model shown in Figure 1. Just like the supplies an artist needs to paint an oil painting, the GUI of a game needs a frame, a canvas, and a colourful paintbrush. In Java programming language APIs, the class JFrame is one of the top containers and can be used to define the outer size and the frame format of the game GUI. The class JPanel is selected as the canvas due to the rich power of the class. A JPanel class not only has ability to be a sub-container to contain and layout a variety of components and to implement event handlers for accepting the players' input actions but also it supports the Graphics object just like a colourful paintbrush for painting components. By extending the animation mechanism "Thread" or implementing the interface "Runnable", the JPanel implements the method run() for realizing a game loop to control the animation of all sprites and carries all game level methods, such as startGame(), pauseGame(), resumeGame(), and the like. Consequently, the game class Main.java that extends the JFrame and implements the main() method, which forms the first layer. And the class with the name of GamePanel.java that extends the JPanel forms the second layer. The class GamePanel.java has the responsibility to initialize all sprites in the game and to communicate with all sprites, thus becoming the control center for all sprites.

Box 1. Updating sprites: first method

```
public void updateComponent() {
  sprite1.updateSprite();
  sprite2.updateSprite();
}
```

Box 2. Updating sprites: second method

```
public void updateComponent() {
  for (int i = 0; i < arrayList.size(); i++) {
    anObj = arrayList.get(i);
    If ((anObj instanceof Circle) {// if the object is the type of Circle
      anObj.updateSprite();
    } else if (anObj instanceof Rectangle) {// if the object is the type of Rectangle
      ...
    }
  }
}
```

Box 3. Updating sprites: third method

```
public void updateComponent() {
  for (int i = 0; i < arrayList.size(); i++) {
    arrayList.get(i).updateSprite();
  }
}
```

All sprites are formed in the third layer of the global software structure. According to the definition, an individual sprite is an independently animated component. For a generalization, static components could also be counted as sprites with an animation speed of zero. This generalization uniforms the treatment to all sprites. The simplest animation format is the position animation, which is caused by a updateSprite() method, which updates the coordinates of a sprite according to a specified motion. Then another method paintSprite() repaints the sprite in the new position. That is, every individual sprite needs to have these two methods for implementing its animation. On the game level, the class GamePanel.java should have corresponding methods to invoke all updateSprite() and paintSprite() methods defined in every sprite as the pseudo code illustrates in Box 1.

But, what if a game has 10 or even 100 sprites? Using the above method would create a long list of sprites to update. One approach to simplify this code is to use a linear data structure, such as an ArrayList, to hold all sprites and a single "for" loop to access all these sprites. However, the prerequisite of using a linear data structure is that all elements have to be of the same type. How do we make different kinds of sprites be the same type? The solution is to use object oriented programming's inheritance. Assuming that all sprites are all geometrical shapes, then a superclass Sprite2D.java may be defined to convert all sprites to be the same type of Sprite2D. Correspondingly, the linear data structure can be defined as ArrayList<Sprite2D> to hold all these sprites. Unfortunately, a new problem is created. Which is, when an object is fetched from the linear data structure, the real type of the object should be identified by using the "instanceof" operator in order to call the method updateSprite() defined for that object. So, the code turns back into a long list for dealing with every sprite individually.

Fortunately, using abstract classes can solve this problem. That is, the class Sprite2D.java can be redefined as an abstract class AbsSprite2D.java, which declares two abstract methods updateSprite() and paintSprite(). We correspondingly redefine the linear data structure to be ArrayList<AbsSprite2D>. The abstract class separates the specification from the implementation and postpones the implementation of the two

Figure 3. A generic global software structure of games

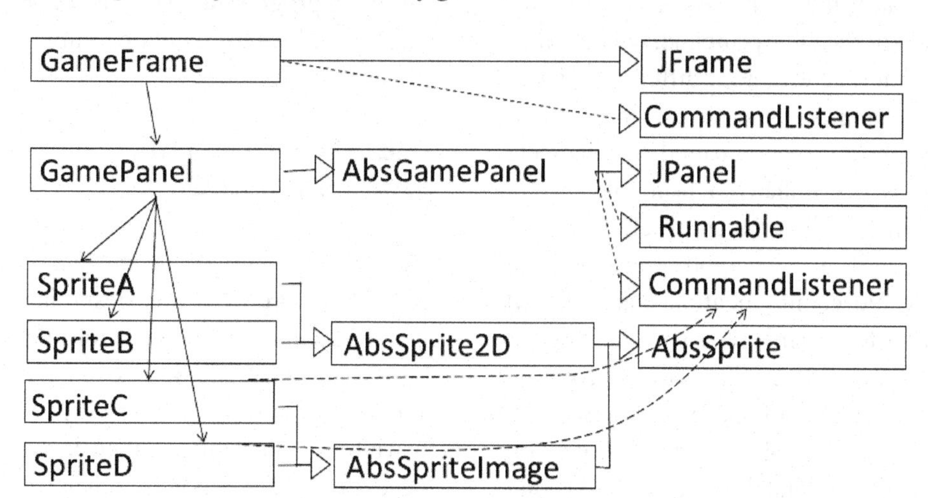

abstract methods until needed. Thus, the code shown above is simplified as expected. In other words, the long list is eliminated.

This dynamic binding allows the application of polymorphism for the coding and also allows all sprites in the game to form one single inheritance hierarchy by using multiple levels of abstract classes to form multiple branches for different kinds of sprites. The root class of the entire inheritance hierarchy is AbsSprite.java, which contains the common features of all sprites, such as Boolean properties "active" and "visible" and abstract methods, such as updateSprite() and paintSprite(). The property "active" is associated with the method updateSprite() for indicating whether the sprite can be updated or not; similarly, the property "visible" is associated with the method paintSprite() for indicating whether to paint the sprite or not. Other abstract classes, such as AbsSprite2D, form a branch of the inheritance hierarchy for all geometrical shapes. Examples are AbsSpriteImage, which forms a branch for all image sprites and AbsSprite3D, which forms a branch for all 3D sprites.

At this point, all classes and their relationships are merged into a global software structure as shown in Figure 3.

Using the same inheritance hierarchy with abstract classes, the class GamePanel.java in the second layer is also further separated into two classes, an abstract class AbsGamePanel.java and a concrete class GamePanel.java. The AbsGamePanel.java class implements the animation mechanism and declares abstract methods initSprites() and announceTermination(), which are implemented by the concrete class GamePanel.java.

THE VALUE OF THE GLOBAL SOFTWARE STRUCTURE OF GAMES

The total number of sprites in a game could be 10, 100, or more and will be added into a game during different stages of the game's development. This is especially true when the incremental development strategy is employed, and a game is developed step-by-step with sprites being added incrementally. This style of implementation requires the software architecture of the game to satisfy the so-called open-closed principle (OCP) (Meyer, 1998), which says, "Software entities should be open for extension but closed for modification." Due to the fact that the inheritance hierarchy can be extended by inserting or deleting any subclasses whenever the software needs, it meets the requirement of the "open" aspect of the OCP principle. On the other hand, the abstract classes among different levels of the inheritance hierarchy extract common shared features of a kind and won't be modified; it meets the "closed" aspect of the OCP principle.

In addition, the three-layer structure clearly arranges all units into an appropriate position with all possible communication channels; all sprites belong to a same type so that they all can be loaded into a linear data structure with a single "for" loop, and all sprites only need one thread to perform one game loop. Altogether, the structure of the first layer and the second layer is very stable and reusable for any game, and the structure of the third layer is very scalable and maintainable. Therefore, the three-layer framework allows developers to concentrate on the design and implementation of sprites without having to worry about the architecture.

Next, consider the two major properties of every sprite, namely the properties of "active" and "visible", as two "variations". The "variation principle," which says "Don't mix variations" and "Don't spread variations" (Shalloway & Trott, 2001) is employed to further separate the update function from the paint function. Thus, a sprite class that originally implements both display and motion functions can be further split into two classes such that the motion function will be encapsulated into an independent class (Xu, 2007), (Xu, 2008b). To support this split in the global software structure an interface named as AnIntMotion.java (an-interface-motion) is introduced to declare a set of abstract methods for implementing the motion patterns of animations. This separation supports the possibility for creating a motion library in advance. Whenever a sprite needs a certain motion pattern, it can search the library for the pattern and merge the existing motion with the sprite, allowing its behaviour to be changed dynamically. The motion library not only eases programming but also greatly increases the reusability and maintainability.

THE STORY OF THE GAME OTHELLO

Under the umbrella of the abstract model and the global software structure, designing and implementing a game solely depends on the game's story. Taking the game Othello as an example, the basic story of the game is that two players, distinguished by white and black pieces, take turns placing their pieces on an 8 x 8 board. Whenever two white pieces appear on opposite ends of a sequence of black pieces in a horizontal, vertical, or diagonal direction, the sequence of black pieces is flipped over to be white pieces; the same logic is true for the inverse. This continues until it's not possible to place any additional pieces on the board. The player who gains the most pieces on the board wins the game.

Embedding the Logic in a State Diagram

According to the story, the logic of the game Othello can be depicted as a state diagram, otherwise called Finite State Machine, as shown in Figure 4. The state diagram is a refinement of the logic unit in the abstract model shown in Figure 1. The state diagram not only defines the set of states but also implies the components that are needed for making the GUI unit. The state diagram could be further refined in the cyclic development process later on if necessary.

DESIGNING AND IMPLEMENTING OTHELLO

The process of design and implementation can be started based on the global software structure and

Figure 4. The state diagram of the game Othello

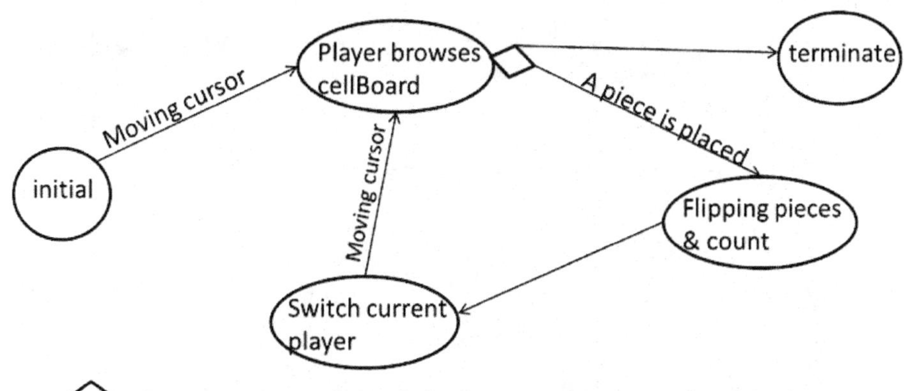

the state diagram. It is impractical to build up a complicated software system, such as a game, in one step. Therefore, the incremental development strategy should be applied, which is a process used to develop a complete game step-by-step with components added incrementally. Thus, the Java component model—JavaBeans—is a suitable choice as the enabling technology. Fortunately, the steps for building up a game have been given in the state diagram shown in Figure 4, which divides the entire logic into a collection of individual states. Every state can be designed and implemented independently and linked together through a set of events. In other words, the state diagram acts as a blueprint for building up sophisticated software. Meanwhile, the global structure of games shown in Figure 3 physically supports this strategy due to the flexibility and extensibility of the inheritance hierarchy in the third layer.

Initializing the Game

Based on the state diagram, the development of the game Othello can be started. The initial state, labelled "initial" in Figure 4, can be further divided into sub-states "initBoard" and "initGame". The sub-state "initBoard" deals with a board made of 8x8 cells. This consists of two sprite classes, CellBoard.java and CellSprite.java, which are added into the sprite inheritance hierarchy. Since both the cellboard and the cell are rectangles, an abstract class AbsSprite2D.java can be defined to abstract the common geometrical properties of the board and the cell sprites and is inserted into the inheritance hierarchy under the root class AbsSprite.java. The geometrical properties include location coordinates, width, height, color, and so forth. These two sprite classes directly extend the class AbsSprite2D.java, supplying data of their geometrical properties and override the two abstract methods updateSprite() and paintSprite() defined in the class AbsSprite.java. Actually, the cellBoard sprite consists of a 2D array of 8 x 8 cell sprites, which builds up the "has" relationship between them. After finishing the definition of these two sprites, the method initSprites() defined in the GamePanel.java class instantiates the two objects and adds them into the linear data structure. Depending on their roles, their updateSprite() methods contain the appropriate code for updating their appearances.

The sub-state "initGame" initializes the 2 black pieces and 2 white pieces into their pre-defined initial positions as shown in Figure 5. To implement the pieces, a PieceSprite.java class should be added. In the first stage, all pieces are graphi-

Figure 5. The implemented "initial" state

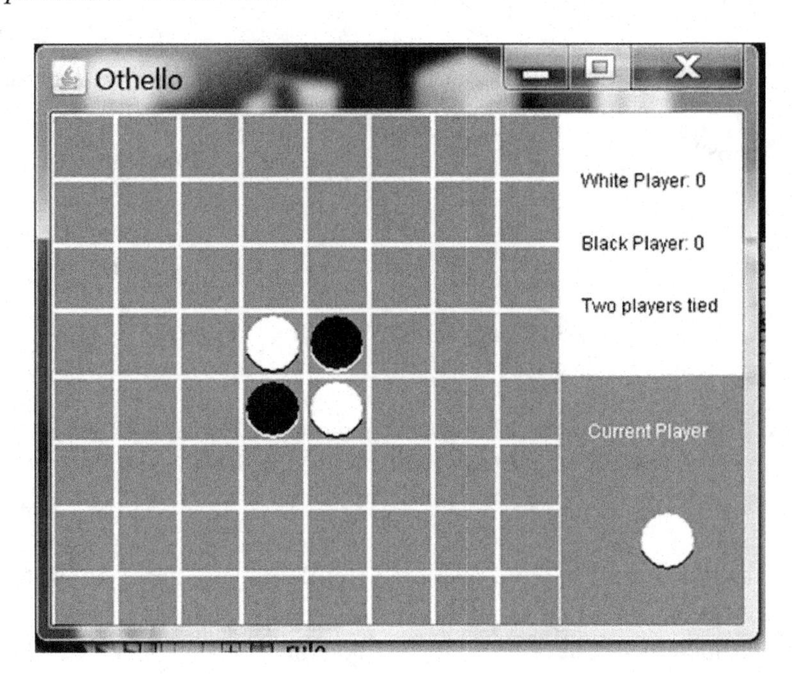

cal circles. Thus, the class PieceSprite.java still extends the abstract class AbsSprite2D.java with physical geometrical properties and overrides the two abstract methods. Later, the pieces will be changed to images. They can still extend the AbsSprite2D.java class or extend a new abstract class AbsSpriteImage.java. In order to control the pieces, a new board sprite, named PieceBoard.java, should be added. The pieceBoard consists of an 8x8 array of pieceSprites, which have the same coordinates as the cellSprites in the cellBoard. Following the completion of the pieceBoard, two black and two white pieces are coded onto the pre-defined coordinates on the pieceBoard.

At this moment, an important distinction should be made. In the previous step, the board that has been created is a cellBoard, which statically displays 8x8 cells on the background. However, all pieces have dynamic behaviours, including adding a new piece and flipping over other player's pieces. For handling these dynamic actions, the pieceBoard is necessary. At the same time, the class PieceBoard.java has a method addPiece(), which allows a player to place a new piece into a position in the pieceBoard indicated by a pair of coordinates (row, column). At the same time, the paintSprite() method, which is defined in the class PieceBoard.java, paints a piece that overlaps with the corresponding cell with the same coordinates. This cell is painted by the method paintSprite() defined in the class CellBoard.java.

For the entire game, two other boards are attached to the cellBoard. One is called the playerBoard, and the other is named the scoreBoard. The playerBoard displays either a white piece or a black piece for indicating the current player; the scoreBoard displays the players' scores. After completing the initial state, the game GUI is shown as in Figure 5 and the global software structure is depicted as in Figure 6.

External Actions

After the initial start-up of the game, as depicted in the state diagram in Figure 4, players may now start playing. The first external action of a

Box 4. The updateSprite function

```
public void updateSprite() {
  if the cell is index valid
    change the cell to use the special color;
    reset the old cell to use the background color;
}
```

Figure 6. The UML diagram of Othello after finishing the "initial" state

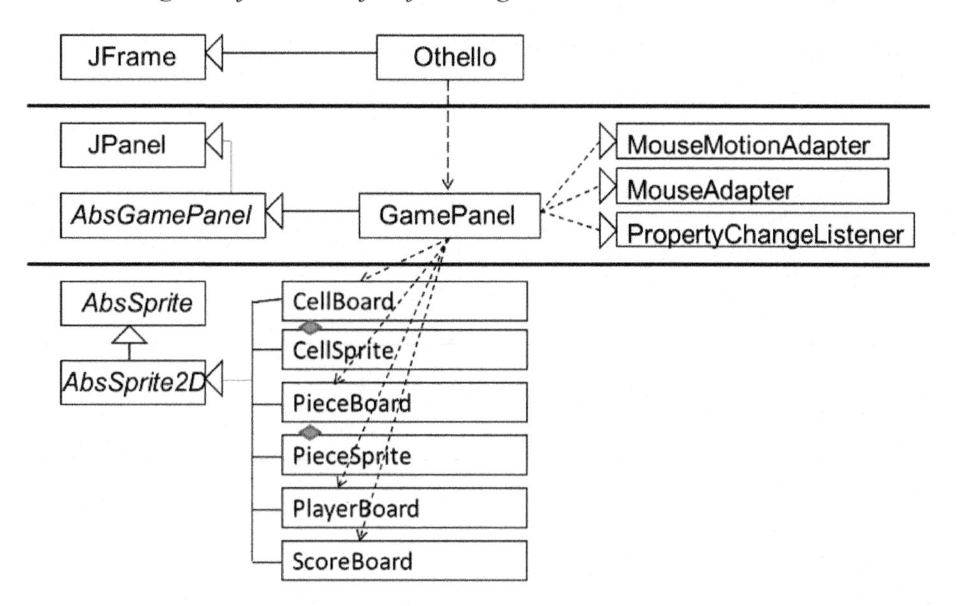

player is to move the mouse and enter the state of "Browse the cellBoard". To better indicate the current cell a player is in, the cellBoard displays a different color on the current cell. The mechanism for receiving the player's external actions depends on which input device is adopted for the game. For the game Othello the mouse is chosen as the input device. If the game is implemented for mobile devices like a cell phone, the input device would be the keys since cell phones have no mouse. No matter which device is used, when the player leaves an "old" cell and enters a "new" cell while browsing the cellBoard, the game follows the pseudo code below to display the special color for the new cell being entered and replace the color for the old cell with the background color. Before changing the color of the current cell, the cell should be verified as "index valid", which means the current cell should have a pair of valid indices for row and column.

Obviously, this is only a mechanism, which has no semantic meaning since the purpose of browsing the cellBoard is to look for those cells where a piece can be placed. These cells are called possible (or valid) playing cells. Which cell is a possible playing cell depends on the definition of a set of flipping rules. Consequently, the real browsing algorithm should be finalized after finishing the implementation of flipping rules. Similarly, the click action at this stage is also a mechanism, which only displays a piece on the cell that the player clicks on.

Box 5.Class RuleBase

```
public class RuleBase {
  private ArrayList<AbsRule> rules;

  public void initRules() {
    create the ArrayList rules;
    instantiate objects of all flipping rules and insert them into the ArrayList rules;
  }

  // implement two methods on the rule level for accessing all of the flipping rules
  public boolean isRuleValid() {
    for every flipping rule in the ArrayList rules
      if the cell is valid for this rule // the cell has a matching cell
        set ruleValid true;
    return ruleValid;
  }
  public void flipPiece() {
    for every individual rule in the ArrayList of rules
      set the flag flipFlag true for the rule;
      if the cell is valid for this rule
        flip pieces using this rule;
      set the flag flipFlag false to terminate the flip action;
  }
```

Box 6. Class AbsRule

```
public abstract class AbsRule {
  protected Color curPieceColor;
  protected PieceSprite[][] pieceBoard;
  private boolean flipFlag = false;

  public abstract boolean isValid(int newRow, int newCol);
  public abstract void flipPiece();
  ...
}
```

Rules for Flipping Pieces

The third state "Flipping pieces & count" shown in the state diagram is ready to be studied and implemented. Whether a sequence of pieces can be flipped over depends on a set of flipping rules. Once a new white piece is placed on the pieceBoard, the set of rules will search for another matching white piece along the eight directions left, right, up, down, upRight, upLeft, downRight, and downLeft. Thus, eight rules are needed. Each rule performs two functions: (1) searching for a matching piece, which is a piece that belongs to the current player on the opposite end of a sequence of opponent's pieces along any one of the eight directions, and (2) flipping over the sequence of the other player's pieces between the two matched current color pieces. The only difference among these eight rules is the search direction.

Based on these observations, the software structure of this rule unit is similar to that of the game framework. That is, a linear data structure can be used to hold the objects of all eight rules. Every time a player places a new piece on the pieceBoard, a "for" loop will invoke every rule in the linear data structure to perform the two functions listed above. Therefore, the class RuleBase.java defines the linear data structure, instantiates

Box 7. CellBoard's updateSprite method

```
public void updateSprite() {
  if the mouse enters an index valid cell
    call the method fireCheckRules() to invoke the propertyChangeListener defined in gamePanel to
      check whether the cell is rule valid;
    if the cell is a rule valid playing cell
      change its color to the current player's color;
      set false to terminate the rule validating action;
    change the "old" cell to the background color;
}
```

the objects of all rules, and adds them into the linear data structure as the pseudo code described in Box 5.

Meanwhile, an abstract class AbsRule.java is defined as the superclass with the declaration of two abstract methods isValid() and flipPiece(). The eight concrete rule classes extend the class AbsRule.java for overriding the two abstract methods.

Once a sequence of pieces has been flipped over, a counting function counts the number of pieces for the two colors on the pieceBoard and sets a Boolean variable true or false based on the condition associated with whether the game play can be continued or should be terminated (the diamond symbol in Figure 4).

Browsing the cellBoard

After implementing the flipping rules, the browsing algorithm and the clicking algorithm are able to be finalized. The browsing algorithm is based on the mechanism that has been developed for displaying a special color when a player moves the mouse and enters a current cell. Now, it can be improved to only display the possible playing cells using the flipping rules. In other words, browsing the cellBoard is implemented in two steps. Before the flipping rules had been set up, the browsing cellBoard algorithm filled a current cell with the special color. After the flipping rules were completed, the browsing cellBoard algorithm only fills valid playing cells with the current player's color via the following algorithm.

1. The MouseMotionAdapter is used to implement the mouseMoved() method. This method is associated with the gamePanel to accept the input of a player moving the mouse into the current cell. Then, the pair of the coordinates (row, column) of the current cell is transferred to the cellBoard sprite as the "inputState" from the GUI unit to the logic unit as shown in Figure 1.
2. The updateSprite() method defined in the cellBoard is invoked once in every game cycle. The updateSprite() method calls isValidMove() to get the pair of (row, column) data from the passed in "inputState" and to validate the coordinates to ensure they refer to a valid pair of indices.
3. If the method isValidMove() returns true, the updateSprite() method of the cellBoard calls the method fireCheckRules() to pass the coordinates of the current cell to all flipping rules to validate that is a valid playing cell. The validation result is carried out by the method getCheckValid(). The returning value, true or false, is the so-called "outputState" indicated in Figure 1, which is transferred from the logic unit to the GUI unit. If it returns true, then the cell is filled with the current player's color by using the mechanism that consists of the two methods

Box 8. PieceBoard's updateSprite method

```
public void updateSprite() {
  if the clicked cell is a valid playing cell
    get its row and column values
    add a piece at the cell position (row, column) with current piece color;
    call the method fireClickDone() to fire a property change to inform gamePanel;
}
```

enterNewCell() and exitOldCell(). This indication will help the player find a set of possible playing cells. These color switches are the "output" indicated in Figure 1.

4. Else if the method getCheckValid() returns false, which means the current cell is not a valid playing cell, then do nothing and go to step 1. Wait for the current player's next action.

The algorithm above is implemented in the method updateSprite() of the class CellBoard.java as shown in Box 7.

The browsing process completes a chain from the "input" to the "inputState", and then from the "outputState" to the "output". The identification of the values of the "inputState" and the "outputState" is especially important for the development of networked game later since they are the data that should be transferred through the network.

The Algorithm for Clicking the Selected Cell in pieceBoard

The goal of the player is to flip over a number of pieces that belong to the other player by placing a piece of his/her own. At any time, there is a set of possible playing cells that can be used for this purpose. After the game is implemented, the browsing algorithm described above displays the current player's color in the set of possible playing cells. The player applies his/her strategy to select one of the possible playing cells to place a piece in, which is the second action of the current player. This clicking action is described by the algorithm below.

1. For catching the player's click, the MouseAdapter is used to implement the method mouseClicked(), which delivers the coordinates (row, column) of the clicked cell to the pieceBoard at the third layer.
2. Whenever the pieceBoard receives the coordinates of the cell that was clicked and its method updateSprite() is invoked in the game loop, it calls the rules to validate whether the cell is a valid playing cell again. It calls the existing method again as that is easier than storing the valid status during browsing and then searching the status during clicking.
3. If the cell is a valid paying cell and the cell does not hold any piece (is null) when it is clicked, the player clicks the pieceBoard, the method addPiece() adds a piece to the pieceBoard at the selected cell coordinates, and the method paintSprite() displays the piece on the board.
4. The pieceBoard immediately calls the method fireClickDone() to pass the "clickDone" information to the gamePanel for further processing.
5. Otherwise, if the clicked cell is not a valid playing cell, do nothing and go to step 1.

This algorithm also goes over the "input" and the "output" chain and is implemented as the pseudo code in the class PieceBoard.java (Box 8).

Box 9. Handling a player clicking on a cell

```
class ClickChange implements PropertyChangeListener {
  public void propertyChange(PropertyChangeEvent evt) {
    if the fire property change operation is "clickDone"
      call the method flipPiece();
      count pieces for each player;
      if current piece color is WHITE
        set the current piece color as BLACK;
      else if current piece color is BLACK
        set the current piece color as WHITE;
  }
}
```

A Special Communication Mechanism

Two special methods, namely the fireCheckValid() and the fireClickDone(), appear in the code listed above for the implementation of algorithms for browsing the cellBoard and clicking the pieceBoard. They deal with a special communication mechanism, which is very important for the communications between two independent sprites.

As mentioned previously, the method fireCheckValid() is used to validate whether a cell is a valid playing cell by passing the coordinates of the cell from the cellBoard sprite to all of the flipping rules. Both the cellBoard and the flipping rules are sprites that are built up on the third layer. There are no relationships that have been set up between any of the sprites. That is, sprites do not have any direct communication links that can pass data to each other. The only possible path is from the source sprite that is sending data out to the gamePanel on the second layer, and then from the gamePanel to reach the target sprite. To make this special functioning communication path, a special PropertyChangeListener should be employed. The usage of this listener can be found in standard Java books, so it will not be detailed here.

Switching the Current Player

The last state that has not been touched in the state diagram is "Switching current player", which changes the current player from the white player to the black player or from the black player to the white player. It is a global control, so it is implemented in the gamePanel with an inner class of ClickChange. This class summarizes the major functions that should be performed after a player clicks on one of valid playing cells. These major functions include flipPiece(), countPieces(), and turning the playing authority over to the other player.

Completing the Game

When all functions are working correctly, some features need to be enhanced. The most important enhancements include changing the pieces from drawn circles to six images that are animated to simulate a flipping action. Also, providing sounds for some major actions, such as moving the mouse and flipping over the sequence of pieces is required.

At this point, it is clear that the design and implementation of the game Othello has followed the state diagram in Figure 4, based on the guidance of the three-layer framework in Figure 3 with the incremental development strategy for

Figure 7. The complete software structure of the game Othello

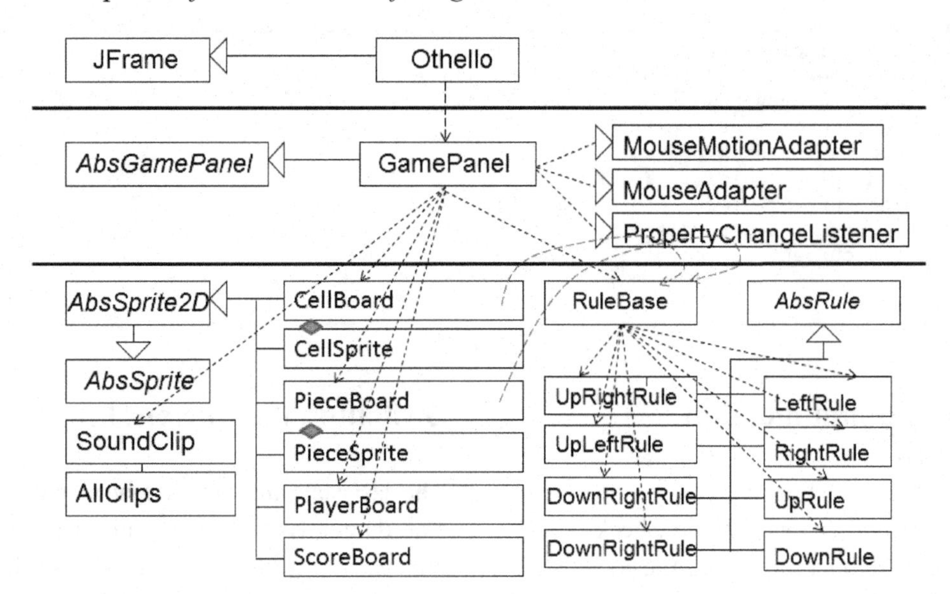

gradually inserting sprites into the inheritance hierarchy at different development stages. When the game is completed, its software structure can be summarized as the UML diagram shown in Figure 7. This process can be applied to the design and implementation of other games with a great degree of reusability and maintainability.

CONSTRUCTING A SINGLE-PLAYER VERSION OF OTHELLO

Othello is usually played by two human players. It can be converted into a single player game in case a player cannot find a human opponent. For that, the single human player needs to play the game against a computer player. Consequently, a new sprite, ComputerPlayerSprite.java, is added. For a human player, all actions are caught by human's eyes and controlled by human's brain. Now, the computer player needs to rely on code to perform all these actions as the human player does. The code should mimic a human's behaviours for playing the game, which can be achieved by merging units of artificial intelligence with the logic of the game.

The major human player's behaviours include browsing the cellBoard and clicking on the pieceBoard. The intelligence lies on the player's strategy for selecting one of the possible playing cells to click the game. No matter which strategy is used, the main goal is to ensure a maximum "earning" and a minimum "losing". Thus, the computer player should mimic this human's intelligence.

Normally, after the human player finishes his/her turn the game sends a signal to the object of the computer player through a Boolean property CpTurn, which is defined in the ComputerPlayerSprite.java class. As long as the method updateSprite() of the class is invoked in a game loop, it catches the true value of the variable CpTurn and calls the method allPossibleValidMoves() to collect all possible playing cells and then selects one of them to click on. All of the possible algorithms designed for this action can be divided into the three categories of "first", "random", and "best" algorithms.

Box 10. The First algorithm

```
public void allPossibleValidMoves() {
  get the pieceBoard;
  set the findFirst false;
  for every row
    for every column
      if the current cell is null with no piece
        if this cell is a valid playing cell
          add the cell into the ArryList with the name of validMoves;
          set findFirst true;
      if findFirst is true, break the inner "for" loop;
    if findFirst is true, break the outer "for" loop;
}
```

The First Algorithm

The "first" algorithm means that the computer player always selects the first valid playing cell that it finds to play the game. In other words,

1. The algorithm starts its search for a valid playing cell from the cell with coordinates (0, 0).
2. Whenever a cell is proved to be the "first" valid playing cell, the algorithm stops its search.
3. The computer player mimics a human player to "click" on the "first" valid playing cell by using its method updateSprite().

In other words, the computer player always takes the "first" valid playing cell to place its piece and all other valid playing cells are not searched. Therefore, its allPossibleValidMoves() method is as simple as the pseudo code described in Box 10.

Obviously, this algorithm saves on search time, but it is a blind algorithm without any intelligence, and it does not mimic the real behaviour of a human player. The biggest drawback is that the computer player always selects the first possible playing cell so the human player can easily estimate which cell the computer player will play. Thus, the human player can easily defeat the computer player. Therefore, the "first" algorithm is an unfair algorithm.

The Random Algorithm

The "random" algorithm collects all valid playing cells into a linear data structure and then generates a random index to pick one of those cells. Therefore, this algorithm requires a linear data structure and performs the following functions.

1. Whenever the computer player takes a turn, it goes over every cell from the cell (0, 0) until the cell (MAXCELL-1, MAXCELL-1) on the cell board to get the coordinates of (row, column), and invokes the method fireRuleCheck() to validate whether the cell is a valid playing cell.
2. If it is a valid playing cell, its coordinates are stored in the linear data structure validMoves.
3. After the computer player collects all valid playing cells, it generates a random number as an index used to select a valid cell. The computer player places a piece on the selected cell.

In comparison with the "first" algorithm, the "random" algorithm takes a longer time to search for all valid playing cells. However, the behaviour of this algorithm is closer to a real simulation of a

Box 11. Updating the computer player's sprite and finding the best move

```
public void updateSprite() {
  if CpTurn is true //computer player takes turn
    collect all valid playing cells into the ArrayList validMoves; at the same time, corresponding to every
      valid playing cell, count the number of pieces that could be flipped over and store them in
      ArrayList totalNumFlipAry with the same index as the valid playing cell;
    find the best playing cell by calling the method findBestMove(validMoves, totalNumFlipAry) below;
    click the best playing cell on the pieceBoard;
  set the variable CpTurn as false;
}
public Point findBestMove(ArrayList<Point> validMoves, ArrayList totalNumFlipAry) {
  find the maximum number and its index from the ArrayList totalNumFlipAry;
  return (the playing cell from the ArrayList validMoves indicated by the index that comes from the
    process above);
}
```

human player than the "first" algorithm. In addition, the human player won't be able to estimate the random behaviour of the computer player, which solves the unfairness problem embedded in the "first" algorithm. Therefore, the "random" algorithm is a practical algorithm with better performance. However, it still is a mechanism without intelligence.

The Best Algorithm – Flipping Over the Maximum Number of Pieces

The "best" algorithm collects all of the valid playing cells and then selects the one that will give the "best" playing result. There are many kinds of criteria for judging what the "best" result could be. The simplest one may be "flip over the maximum number of pieces". For that goal, the algorithm will do the following.

1. Whenever the computer player takes a turn, it collects all valid playing cells into the linear data structure validMoves.
2. During this process, the computer player also collects the total number of pieces that would be flipped over by all eight flipping rules. The value is stored in the other linear data structure totalNumFlipAry with the index that corresponds to the valid playing cell stored in the linear data structure of validMoves.
3. After that, the computer player goes through the linear data structure totalNumFlipAry to find the item that holds the maximum number of possible flipped pieces. It uses that index to find the corresponding cell coordinates from the validMoves data structure to select a cell.
4. The computer player places its piece on the selected cell.

This algorithm is coded in the method updateSprite() defined in the class ComputerPlayerSprite.java.

Intuitively, the "best" algorithm will take a longer time than both the "first" and the "random" algorithms. It not only needs to find all valid playing cells but also needs to collect the number of pieces that can possibly to be flipped over for every valid playing cell. It then takes more time to compare all the possible playing outcomes for determining the best choice. In addition, the algorithm requires more memory for its two linear data structures.

Actually, the "best" algorithm does not necessarily give the best playing outcome since flipping over the maximum number of pieces at this step could end up with the maximum number of its

Box 12. Updated version of the findBestMove function

```
public Point findBestMove(ArrayList<Point> validMoves,
    ArrayList<Integer> totalNumFlipAry) {
  // look for a valid cell that is closest to edges
  for every valid playing cell in the ArrayList validMoves
    get the coordinates of ith valid cell from the arrayList validMoves;
    calculate the distances from the ith cell to the four edges;
    compare the distances for finding the closest distance in x-axis direction;
    compare the distances for finding the closest distance in y-axis direction;
    compare the closest distance in the x and y directions for finding the closest distance to
      one of four edges;
    compare the new closest with the first element in the ArrayList allClosest
      if the new closest is closer to an edge than the first element, replace the first element
      else store the new closest in ArrayList allClosest

  if multiple valid cells are equally close to one of the edges
    select the one that will flip over maximum number of pieces as the best playing cell
}
```

own pieces to be flipped over by the opponent at the immediate next step. That is, when the human player takes a turn, he/she could obtain the best "revenge" result.

The Best Algorithm: Occupying Cells Along the Edges of the Board

Another "best" strategy could be more useful than the "flip the maximum number of pieces" method for winning the game. The new strategy is to occupy cells along the four edges of the cell board as early as possible. That is, the step 3 in the best algorithm described in the previous sub-section could be modified as follows.

3. After the computer player finishes the collection of all valid playing cells and corresponding number of possible pieces which could be flipped over,
   a. it goes through the first linear data structure validMoves to find the valid cell that is closest to one of the four edges of the cell board;
   b. if only one valid cell is the closest, take this cell as the selected cell;
   c. if multiple valid cells have the same degree of closeness to either one of four edges of the cell board, it goes to the second linear data structure totalNumFlipAry to select the cell that may flip over the "maximum" number of pieces. The selected cell is a closest cell with the maximum possible number of pieces to be flipped.

This modified step affects the implementation of the method findBestMove() as shown in Box 12.

This "best" algorithm gives a better performance for the computer player. However, if the human player also applies the same "best" tactic, then both players put forth their best efforts to occupy the cells closest to four edges. This can lead to the game ending in a stalemate. In other words, one player may not have any valid moves left due to the fact that the player cannot find a matching piece even though many cells are still available for placing pieces. To solve this problem, the game needs to build up a mechanism to allow any player to give up his/her turn to the opponent.

Figure 8. A traditional architecture model of networked games

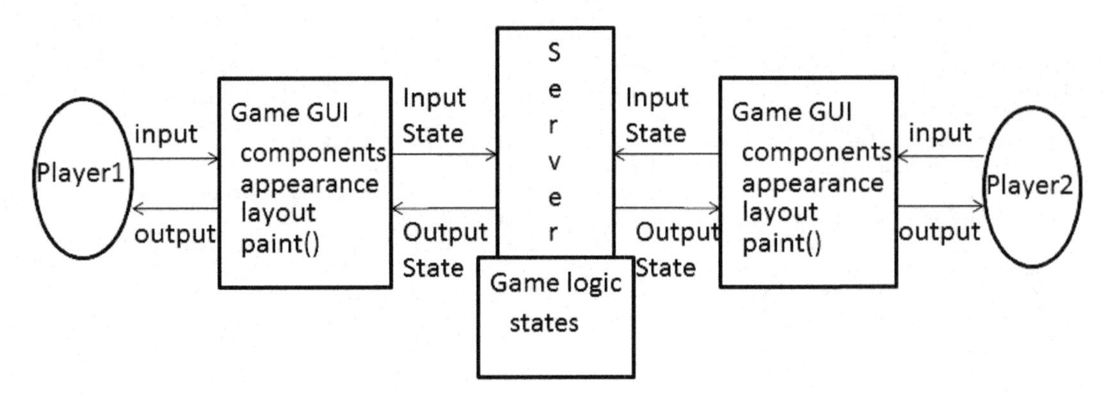

The Best Algorithm: Looking Ahead

According to the traditional principles of Artificial Intelligence, the best strategy for playing this chess style of game is to select a best play by looking several steps ahead, using either a depth-first or breadth-first searching algorithm before making a decision. For the game Othello, the simplified algorithm of this kind is to look up the next step. Very often, when player1 flips over some pieces of player2, player2 can immediately flip over pieces of player1. In order to avoid this kind of "revenge", player1 should look up the next step, so that his/her pieces won't be able to be flipped over again on player2's turn. Therefore, the algorithm can be further improved by applying the "best algorithm with maximum number of pieces to be flipped" twice. This means that after the computer player collects two sets of data and stores them in the two linear data structures, the computer player applies the "best" algorithm the first time to find the "best gain" result. And then the "best" algorithm is applied for a second time to mimic the human player's "best" algorithm for a "best revenge" result. Certainly, the difference between the "best gain" and the "best revenge" is the "net gain" of the computer player. Applying this strategy, the computer player can find out all of the "net gain" results corresponding to every possible playing cell. After finishing this process, the computer player picks up the playing cell with the maximum "net gain" as its selected cell. Clearly, this algorithm is time and memory extensive because every possible playing cell needs to apply the "best" algorithm twice and the game has to remember the new circumstance made by the first application of the "best" algorithm for supporting the second simulated execution of the "best" algorithm.

In practice, we could use the "look ahead" strategy instead of the "twice" algorithm described above. To accomplish this, a set of "safety rules" for reducing the degree of "revenge" would be defined. The safety rules may include "selecting the playing cell if it is the last empty cell along the direction for flipping pieces". This rule means that after the computer player places a piece into this empty cell, the human player cannot flip over the computer's pieces during the following step, since all pieces are the same color along the direction. This rule implies that the opponent has no possibility to find a matching piece for his/her "revenge". These rules are more heuristic oriented and more suitable for human player to use since they are not easily coded. Due to the time limitations, the "look ahead" and these "safety rules" have not been implemented yet.

DEVELOPING OTHELLO AS A NETWORKED GAME

Even though the game Othello can be played by a single human player with our implementation, it is still a two-player game in nature. Either version is suitable to be developed as a networked game. Here, only the two-player version will be taken as an example. A networked Othello game means that two players play the game using two different computers so that each player has his or her own game GUI. Whenever one player places a piece on his game GUI, it not only impacts his/her own GUI but also causes his/her opponent's GUI to display the same scene. That is, a player on one side will synchronously control the other player's GUI by sending the control parameters through a network. Besides the required network, a server is also necessary. This server plays the role of the mediator to transfer data and to synchronize actions between the two players.

Two Possible Architectures of Networked Games

The traditional approach for implementing a networked game is to insert a third computer, called a server, which directly communicates with two GUIs, called clients, to coordinate their actions as shown in Figure 8 (Fan, Ries, & Tenitchi, 1996). The two games connect to the server and through the server to reach each other. The two client sides only have the GUIs for accepting the players' inputs and displaying outputs for the players. The server holds the game logic for communicating with the two clients and synchronizes the players' actions to ensure and enforce that only one player is active at a time. Although every client is implemented as a Thread so that they run in parallel, the actual process in the background remains strictly sequential. Based on the game logic held at the server, the game state data is transferred by the server for making decisions (e.g., deciding the winner), and some of the state data are further transferred to the other side for keeping the state of both clients the same. This implementation inherently couples client and server units tightly and allows a limited number of players, thus rendering the software architecture unattractive from the viewpoints of reusability and maintenance. In addition, this architecture requires the game logic to be removed from the original standalone game body to the server, which requires a great deal of code modifications. Comparing Figure 3 with Figure 8, a networked game can be viewed as a merge of two standalone games with an overlap of two logic units by using the network as their linkage.

In order to reduce the burden of code modifications and to overcome problems such as the tightly coupled architecture discussed above, a new architecture shown in Figure 9 was adapted (Xu, 2008a). This new architecture keeps both GUI and logic on two clients as in the original game and leaves the server as a pure communication center with very weak controller functionality. Consequently, the two clients are almost the same as two copies of the original standalone game with some input and output states needing to be transferred to the server. This new architecture allows an original standalone game to be further developed as a networked game more quickly than the adoption of the previous architecture.

Web Services

The traditional enabling technology for developing networked games is socket programming (Fan, Ries, & Tenitchi, 1996), (Morrison, 2005). The Java programming language provides a Server Socket for implementing the server side and provides a Socket for implementing the client side. Currently, Web services (Deitel, & Deitel, 2008) have successfully exhibited the practical significance of the distributed computational model. Since Web services are standard software architecture, these externally available distributed

Figure 9. A new architecture model of networked games

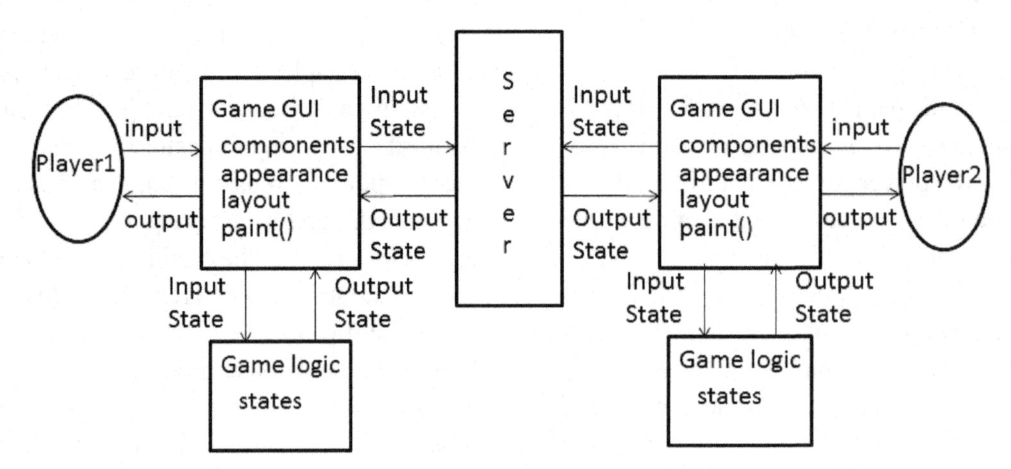

application components can now be employed to integrate computer applications that are written in different languages and run on different platforms. Making software services available for clients connecting from anywhere at any time, web services play important roles in the field of computing, including those of grid and cloud computing.

A web service was originally defined by the W3C as "a software system designed to support interoperable machine-to-machine interaction over a network". Web services have generated an immense amount of interest in recent times in terms of their applicability in social interactions including gaming scenarios (Rubel, 2008), (Ellis, 2008). Networked games need to be interactive and they are easily supported by Web services (Xu, Lei, & Xu, 2010). Thus, Web services are employed again to be the enabling technology for developing the networked Othello game.

A Web service has a set of web methods. These web methods can be invoked by clients of that Web service. However, a server that uses Web services causes a special issue. A server that is implemented by using sockets can send data out to its clients actively, but a server that is made of a Web service is a passive component. Therefore, a client needs two different approaches for communicating data with the server. When a client-to-server communication is needed, the client can send a request as a function call with or without parameters to invoke the related Web methods defined in the server. And the invoked methods immediately return the computational results as a response to the client. However, when server-to-client data sending is needed, the server cannot actively do so. The client needs to continuously probe the server to check whether the data that the client expected to be at the server has arrived or not. If yes, another client-to-server communication is issued in order to get the data from the server because the server cannot actively send the data to a client.

Determining Who is Player1 and Who is Player2

Based on the discussion above, the decisions about the architecture of the networked game and the enabling technology of the server have been selected for the development of the networked Othello game. According to the architecture as shown in Figure 9, the available standalone game is the basis of two clients; the major task is to build up the server and the communication mechanism from the two clients to the server.

The game Othello is a two-player game where the players alternate turns. In the case of the standalone version, two players may have a face-to-face discussion for determining who will be the first player and which color they will play with. However, for playing a networked game, this decision should be made by the server based on which client connects to the server first. The convention is that the first player will use white pieces and the second player will use black pieces. Therefore, the first step for turning the original Othello game into a client in the networked version is to have a decision maker for identifying who is player1 and who is player2. This decision maker is a simple Web method called connect(), which resides on the Web service.

- **Player1 connects to the server.** A player who would like to play the networked Othello game starts running a networked version of the game, which will go through the communication channel to invoke the web method connect() defined in the server. This web method increases the value of a variable named playerNumber by one to keep track of the number of connections from clients. The value of the playerNumber is set to 1 and then the client gets the value to indicate this player as player1.
- **Player2 connects to the server.** Next, the entire system waits for the player2's connection. Once player2 activates his/her networked game, similarly the game invokes the web method connect() defined in the Web service to get a new value of 2 for the variable playerNumber. Thus, this player becomes player2.

Consequently, the server needs a Web method connect() to receive connections from the two clients. But how does the client connect to the server? It is supported by the Web service technology. Whenever the original game Othello is assigned to be a client of the Web service, the client can define a variable "port" that plays the role of the end point of the communication channel between the client and the server. It is this port that supports the client's ability to talk with the server.

A New Class CommInterface.java

A new class is needed to define the variable "port" that starts the connection from the client to the server and performs all other required functions for communications. The class is named CommInterface.java (Communication Interface). When designing and implementing this class, very special attention should be paid because this class is instantiated as an object by each client and serves the needs of two clients at the same time. In other words, one class should take care of both player1 and player2, be able to identify player1 and player2 in order to guarantee the take-turn feature, and can loop forever with pause and resume abilities.

To satisfy all of the requirements listed above, the class CommInterface.java extends Thread to support the run() method for making loops and allow some "wait" methods to pause and resume certain functions. Clearly, the first task of the class is to connect to the server and be able to identify player1 and player2. The class works with the Web service to make the connection and identification as follows. First, the Web service implements the Web method connect() with the code in Box 13 and waits for the clients.

Next, the CommInterface.java class provides a method connectToServer() that gets the "port" and uses the statement port.connect() to connect to the server. After connecting, it immediately starts its run() method as the code in Box 14 shows.

The run() method, which is described by the following pseudo code, calls another Web method with the statement port.getPlayerStatus() to find out if the client is player1 or player2. Then, the different players execute different portions of the code.

The pseudo code indicates that after the two players have been identified as player1 or player2, the loop while(continueToPlay) starts, and the two players execute the same three methods in a different order. Player1 starts the game first. Thus, the code starts "wait for player1's action" to pause the loop and wait for player1's browsing and clicking. A t the same time, player2 is paused by the method "wait to receive data from player1". Once player1 clicks a cell, the method "wait for player1's action" resumes the loop and goes to the method "send the coordinates of the clicked cell to server by player1". The coordinates sent to the server from player1 are received by player2, since player2 is in the "wait to receive data from player1" state. The coordinates that are received by player2's commInterface are delivered to the client GUI of player2 so the scene on player2's GUI can be updated. After player1 completes his/her turn, player2 takes a turn. This causes player2 to execute the method "wait for player2's action". This causes player1 to enter the method of "wait to receive data from player2", which starts player2's turn. These two sequences of actions form a "sending-receiving loop", which enforces a sequential data transfer between the two clients until the game is terminated. Thus, the function of the commInterface can be depicted as shown in Figure 10.

Constructing a Web Service as the Server for the Networked Othello Game

Based on the description above, the major variables and methods of the Web service include

- An int variable playerNum with its getter and setter methods getPlayerNum() and setPlayerNum() for keeping track of the ID of players that have connected to the server. These two methods are not web methods but are internally used.
- connect(): a Web method to be invoked by games' connections, which increases the value of the variable playerNum by one for each connection.
- getPlayerStatus(): it returns the value of the playerNum to a client when it queries.
- An int variable playerArrival with its getter getPlayerArrival() and its setter setPlayerArrival() for indicating which player just sent data to the server.
- The int variables row and column with their getters getRow() and getColumn().
- setCell(): A major method with the parameters row, column, and player, which are called by a client to pass the coordinates of the playing cell to the server so that their values will be further transferred to the other client.

Thus, the server becomes a data carrier and a data exchange center. Two clients go through their own communication interface commInterface to send data to or receive data from the server.

Box 13. Implementing the connect function

```
@WebMethod(operationName = "connect")
@Oneway
public void connect() {
  setPlayerNum(getPlayerNum() + 1);
}
```

Box 14. The connectToServer method

```
private void connectToServer() {
  try {
    service = new itsWS_Service();
    port = service.getItsWSPort();
    port.connect();
  } catch (Exception ex) {
  }
  start(); // run()
}
```

Box 15. The run method

```
public void run() {
  try {
  int player = calling the web method port.getPlayerStatus() that returns the value of 1 or 2
  if (player == PLAYER1) { // PLAYER1 is a constant of 1
    play with WHITE piece;
    start its play;
  } else if (player == PLAYER2) { // PLAYER2 is a constant of 2
    play with BLACK piece;
  }

  while (continueToPlay) {
    if (player == PLAYER1) {
      wait for player1's action (browsing and clicking)
      send the coordinates of the clicked cell to server by player1;
      wait to receive data from player2;
    } else if (player == PLAYER2) {
      wait to receive data from player1;
      wait for player2's action (browsing and clicking);
      send the coordinates of the clicked cell to server by player2;
    }
  }
}
```

The Interface GamePanelInterface.java

Once the commInterface of player2 receives the coordinates of the playing cell from player1 through the server, the coordinates should be transferred to player2's GUI to change its scene according to the game logic; the same thing is repeated when player1 receives data from player2. The coordinates should be sent to the gamePanel of the game since all of the controlling functions are defined in the GamePanel.java class. For sending the coordinates, a Java interface GamePanelInterface.java is defined, which is inserted between the commInterface and the gamePanel as shown in Figure 11.

Figure 10. The interactions between the clients and the server in the networked game Othello

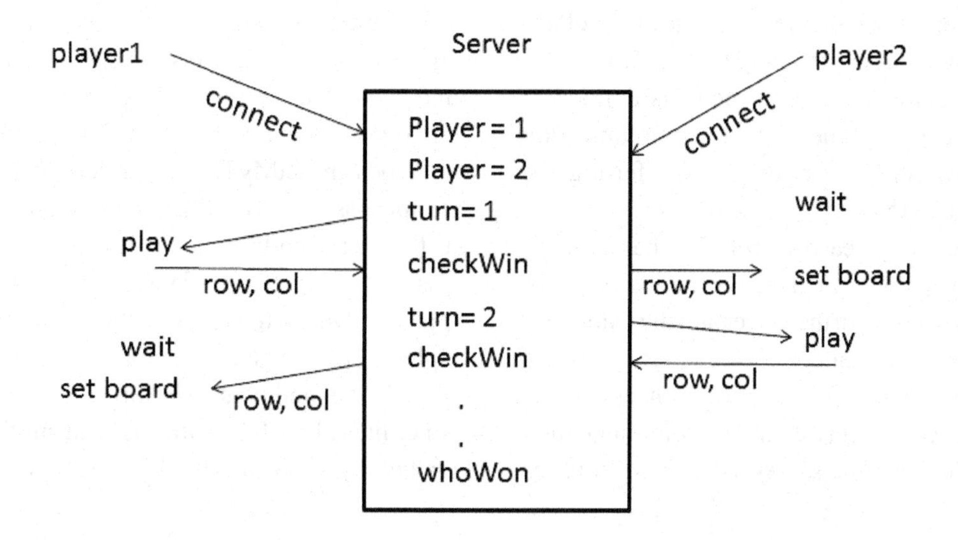

Figure 11. The real architecture of the networked game Othello

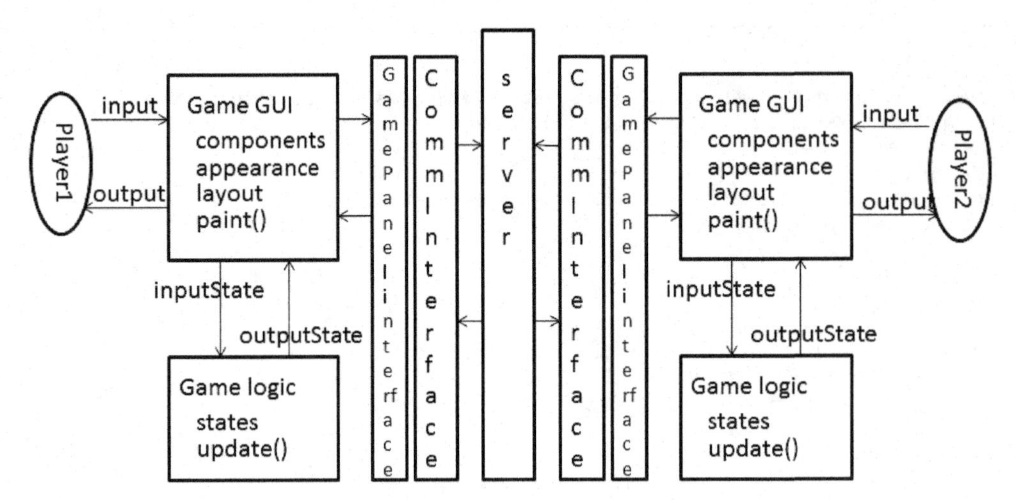

There are two reasons for using a Java interface here. One reason is that the standalone game and the server could be developed using different programming languages and/or different platforms. For example, the standalone game may be developed in JavaFX, but the server is implemented in Java. The interface becomes the "isolator" and the "translator" between the different programming languages. The second reason is that a definition of a Java interface can easily be reused and adapted in different games. Actually, the interface GamePanelInterface.java used in this networked Othello game is the same as that used in another networked game for Connect4 (Xu, Lei, & Xu, 2010). The interface GamePanelInterface.java is implemented by the class GamePanel.java so that all methods defined in the interface are called by the communication mechanism commInterface for transferring the received data to the game client.

Consequently, each client side has two interfaces. The commInterface is the interface for communicating with the server and the gamePanelInterface is the interface for communicating with the gamePanel. These two interfaces play the role of "glue" to form a complete communication channel between the server and client. In terms of these two interfaces, the server and the client form a loosely coupled architecture as depicted in Figure 11. Thus, it is clear that the different standalone games can be implemented as two clients without being impacted by adding the communication channel; only the contents of the server and these two interfaces will be affected by the different communication issues in different games. This design greatly simplifies the adoption of standalone games as networked ones.

The Implementation of GamePanelInterface.java

For the networked game Othello, the interface GamePanelInterface.java only declares three methods. The corresponding concrete methods are overridden in GamePanel.java. Two of these methods are setMyToken() and setMyTurn() that simply pass the two values generated by the server to the corresponding clients. That is, the server will assign a value of 1 to player1 and a value of 2 to player2. Due to the fact that there is a convention "player1 plays white piece and player2 plays black piece", the method setMyTurn() is not necessary since the value of myTurn has been implied by the value of myToken. The third method setOther()

means to set the coordinates received from the source client to the target client. In other words, when player1 sends the coordinates to the server and player2's commInterface probes the server and receives the coordinates, the values of coordinates will be transferred to the player2's game, which sets the player2's GUI to appear the same as player1's GUI.

Important Changes for the Original Standalone Version of the Game

The original standalone game is a desktop game. Since it now involves data communications, the game itself needs to be modified slightly.

The playerBoard needs to add an additional token. The original standalone Othello game has only one GUI. The GUI includes a playerBoard for displaying a piece token that indicates who the current player is. However, the networked version of the game has two GUIs. It requires the playerBoard to include two tokens. Besides the token for current player, it also needs another token to indicate which GUI is played by which player. For that, a new token called "You are player" is added into the same playerBoard. The new added token is a static indicator with different but fixed color in each GUI. However, the token for the current player on each game GUI has the same color and is dynamically switched whenever each player takes a turn. Thus, the game client that has the same color on these two tokens in the playerBoard takes the turn and the other game that has two tokens with different colors has to wait.

A Boolean variable allowPlay is added. As mentioned above, two tokens are shown in the playerBoard. Only the client that has two tokens with the same color can play the game. A Boolean variable allowPlay is added to catch this check and to prevent the inactive side from playing the game.

Each client side should have the same current piece color. Even though the GUIs are for two players, each GUI should have exactly the same current piece color. The exception is that the two "You are player" icons in the playerBoard have different piece colors.

A new Sprite, named triggerSprite, is added. Once player1 places his/her piece on the pieceBoard, it causes the flipping of some of player2's pieces. At the same time, the coordinates of the playing cell should be sent to player2's game on the remote side, which is expected to have the same effects on player2's board. That is, some code in the player2's game should mimic the behaviours of player1 to place a player1's piece on the same cell. However, who is going to play this role? From the implementation of the single player version, where a class ComputerPlayerSprite.java mimics a human player for browsing and clicking, here a new sprite, called triggerSprite, is added for performing these similar functions. The new triggerSprite is embedded in player2's game to be a representative of player1. When the coordinates sent by player1's game arrives at player2's game, the triggerSprite uses the coordinates to place player1's piece on player2's board. This duplicates the behaviours of player1 on the remote client. Player2 also has its representative embedded in player1's game. Consequently, the code of class TriggerSprite.java is very similar to the code of class ComputerPlayerSprite.java from the single player version of the Othello game.

Only the coordinates of a valid playing cell can be sent to the remote side. If the current player arbitrarily clicks a cell that is not a valid playing cell, the coordinates of the invalid cell cannot be sent to the other side. The reason is that every current player can only send one pair of coordinates to the remote side since the "sending-receiving loop" synchronizes the sequential data transferring. That is, each player can only send one message out when he/she takes a turn. Therefore, the coordinates should be approved by the game

Figure 12. Sending the browsing messages from player1 to player2

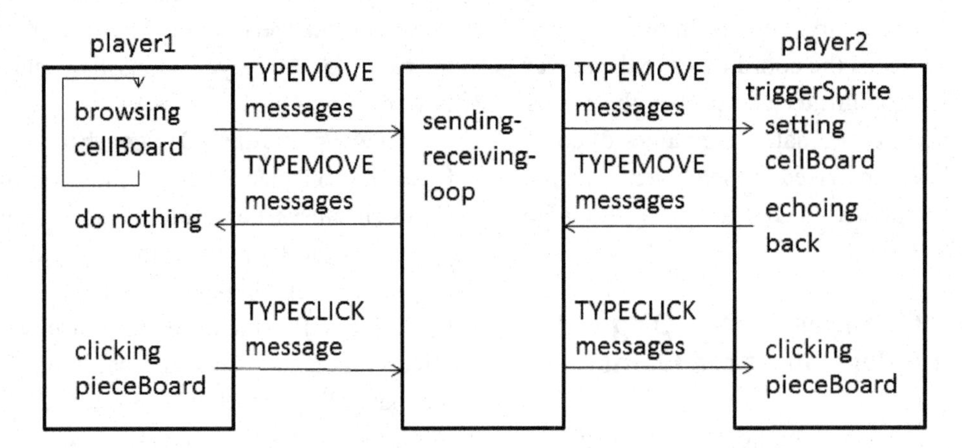

rules to be valid before sending out. Otherwise, the coordinates should not be sent.

Sending Mouse Data to the Remote Client

So far, the networked game Othello only sends one kind of message, which contains the coordinates of the cell clicked by a player. In the standalone version, the GUI also displays the current player's color when the player moves his/her mouse over one of the valid cells before selecting a cell via click. This moving information hasn't been sent to the other side yet since this message is different from the clicking message. To send this message to the remote side so the other player can also see the current player's browsing actions, a new version of the networked Othello is developed.

The transfer of the mouse actions introduces more challenges than only sending mouse clicks because the movement of the mouse is a sequence of random actions instead of a single click. The communication mechanism can only send one pair of coordinates for one player as discussed above. Therefore, the transmission of the moving data requires two big changes in the communication channel. One of them is the need to define two different kinds of messages, which are defined as the two constants TYPEMOVE and TYPECLICK, in class Consts.java. The sender needs to mark the sending message with the correct type and the receiver needs to distinguish the different kinds of data and give the appropriate treatment. Another challenge is determining how to use the single "sending-receiving loop" for transferring a sequence of messages.

The current function of the "sending-receiving loop" can be described as that player1 sends one message to player2 then player2 takes a turn to send one message to player1. That is, each client evenly sends one message only. However, sending the move action data implies that player1 sends a sequence of moving messages to player2 continuously until player1 sends a single clicking message, and then player2 takes a turn to do the same. In addition, this sequence has no fixed length. Obviously, this behaviour destroys the "even sending" condition of the "sending-receiving loop". In order to keep the "even sending" style of the "sending-receiving loop", a heuristic algorithm is adopted.

As discussed above, every client game has a triggerSprite as a "representative" of the other side to mimic the behaviours of the opponent. The heuristic algorithm takes advantage of this "representative" and asks it to perform the following action: when a remote client, say player2, receives a TYPEMOVE message from player1, in addition to showing player1's mouse move action on its GUI, triggerSprite also needs to echo back a TYPEMOVE message with a pair of coordinates (-1, -1) to its sender—player1. This algorithm, maintains the "even sending" one-to-one message used in the "sending-receiving loop", thus allowing for any player to send a sequence of moving messages with an unknown length. As a result, the click message is sent only one per turn, whenever the receiver sees that the message type is TYPECLICK, the triggerSprite still performs the original function to mimicking the click action sent by the remote client and switching the current player. In other words, the click action indicates the end of the current player's turn. The next player will start sending the sequence of TYPEMOVE messages and end its message sending with the single TYPECLICK message as well. The heuristic algorithm is depicted in Figure 12 when player1 repeatedly sends a sequence of TYPEMOVE messages to player2 until player1 clicks the pieceBoard and sends one TYPECLICK message to player2.

Implementation of the Heuristic Algorithm

1. In the previous version of the networked Othello game, the class GamePanel.java uses two inner classes to implement two listeners. One of them is the inner class TheMouseMotionAdapter that extends the class MouseMotionAdapter and implements the method mouseMoved() for catching the mouse movements. Now, the mouse movements need to be transferred to the other client, thus, the method mouseMoved() invokes a new method sentToCommInterface(msgType, row, column) to invoke the methods setMsgType(), setRowToServer(), and setColToServer() defined in the commInterface for setting the message type TYPEMOVE with the coordinates (row, column) to the server. The msgType is a new parameter that should be added in the commInterface and the server. Meanwhile, the Boolean variable "waiting" defined in the commInterface is also set as false to stop the waiting for the player's action and start the method sendMove().
2. When player1's moving data is sent to its commInterface, the method sendMove() invokes the communication end-point "port" to send the mouse move data, TYPEMOVE, and the player's ID to the server by calling the web method setCell() defined in the server. The web method setCell() also needs to add a new parameter referring to the message type.
3. As long as player1's moving data arrives on the server, the commInterface of the player2's game probes the moving data sent by player1, invokes the method receiveMove() for getting the values of the message type and the pair of coordinates (row, column) from the "port", and immediately calls the method toGamePanelSetOther() for passing the data to its game through the setOther() method defined in the interface GamePanelInterface.java.
4. Due to the class GamePanel.java implementing the interface GamePanelInterface.java, the moving data sent by player1 reaches the player2's gamePanel, which sends the data to the object triggerSprite.
5. The object triggerSprite's method updateSprite() is routinely invoked by the game loop implemented within the game Thread. If updateSprite() finds the message type is TYPEMOVE, it sets the cell according to the received coordinates for displaying the

received moving data on the cellBoard. At the same time, it invokes the method gamePanel.sendToCommInterface() to echo a pair of (-1, -1) back to the other player. The other player recognizes the pair of (-1, -1) is not a real coordinates so that it does nothing.

These steps start with player1 sending mouse data to player2 until player2 echoes the data back to player1 forming a cycle. The sensitivity and different speeds of the computers involved may affect this cyclic process such that sometimes the valid cells shown on the two game GUIs may appear to be shaking. This phenomenon will be less severe when the current player moves the mouse slowly. Whenever player1 selects a valid cell and clicks on it, the message TYPECLICK will be sent to player2 via the same route of communication mechanism and two game GUIs will flip over the same group of pieces.

CONCLUSION AND FUTURE WORK

This chapter describes the development of the game Othello to demonstrate different data structures and algorithms for implementing a game from its story to a standalone game, a single player version, and a networked game. The modeling principle, the global software structure framework, and some heuristic strategies are reusable for the development of other games, as well as teaching tools with dynamic behaviors.

The development of the standalone game illustrates the potential of the three-layer framework, the reusability of the global software structure, and the beauty of incremental development strategy. Implementing a game step-by-step according to its state machine and adding sprites one-by-one in different developing stages promote an interactive learning approach and has been proved to be a successful style of gaming, especially for beginners and students. The applications of important features of OOP, including abstract classes, interfaces, inheritance hierarchy, and polymorphism for gaming enrich the game contents and ease the understanding of these techniques.

The development of the single player version introduces the fundamental concepts of AI and the adoption of heuristic algorithms, which further opens circumstances of discoveries and excitements.

The networked game version encourages the design and employment of new architectures and new technologies. The new architecture and enabling technology improve the traditional tightly coupled architecture to be a loosely coupled architecture and increases the reusability of software units.

Certainly, imaginations have not been exhausted. The networked game Othello could be further developed to not only allow two players for playing the game but also allow more people to join the game as game judges or observers. That is, when the third, fourth, etc. person connects to the server, they will have a game GUI that shows all actions of the two players – including the possibility that these people could become advisers for the two players. The server implemented by Web services supports these possibilities. In addition to the simple two-player take-turn game like Othello, games like Snooker that need to transfer a complicated sequence of moving data through the communication channel is another more attractive candidate to be involved as one of future works.

REFERENCES

Deitel, P. J., & Deitel, H. M. (2008). *Internet & World Wide Web How to program* (4th ed.). Upper Saddle River, NJ: Prentice-Hall.

Ellis, S. (2008). The future is interactive, not online. Retrieved June 18, 2010, from http://thenewmarketing.com/blogs/steve_ellis/archive/2008/03/17/5467.aspx.

Fan, J., Ries, E., & Tenitchi, C. (1996). *Black art of java game programming*. Waite Group Press.

Meyer, B. (1998). *Object oriented software construction*. Upper Saddle River, NJ: Prentice Hall.

Morrison, M. (2005). *Beginning Mobile phone Game Programming*. Indianapolis, IN: Sams.

Rubel, S. (2008). The future is web services, not web sites. Retrieved from http://www.micropersuasion.com/2008/03/the-future-is-w.html

Schollmeyer, J. (2006). Games get serious. *The Bulletin of the Atomic Scientists*, *62*(4), 34–39. doi:10.2968/062004010

Shalloway, A., & Trott, J. (2001). *Design patterns explained – A new perspective on object-oriented design*. Reading, MA: Addison-Wesley.

Sharp, C. (2003). Business integration for games: An introduction to online games and e-business infrastructure. Retrieved from http://www.ibm.com/developerworks/webservices/library/ws-intgame/.

Xu, C.-W. (2007). A Hybrid Gaming Framework and Its Applications. *The International Technology, Education and Development Conference 2007 (INTED2007), Valencia, Spain*, March 7-9, 2007, pages 30000_0001.pdf.

Xu, C.-W. (2008a). A new communication framework for networked mobile games. *Journal of Software Engineering and Applications*, *1*(1), 20–25. doi:10.4236/jsea.2008.11004

Xu, C.-W. (2008b). Teaching OOP and COP technologies via gaming. In Ferdig, E. R. (Ed.), *Handbook of Research on Effective Electronic Gaming in Education*. Hershey, PA: IGI Global. doi:10.4018/978-1-59904-808-6.ch029

Xu, C.-W., Lei, H., & Xu, D. (2010). Networked games based on web services. *GSTF International Journal on Computing*, *1*(1), 170–175. doi:10.5176/2010-2283_1.1.28

ADDITIONAL READING

Bierre, K., & Phelps, A. (2004). The use of MUPPETS in an introductory java programming course. *SIGITE'04.* Salt Lake City, UT.

Brackeen, D., Barker, B., & Vanhelsuwe, L. (2004). *Developing games in java*. Berkely, CA: Peach Pit Press-New Riders Game Series.

Davison, A. (2005). *Killer game programming in java*. Sebastopol, CA: O'Reilly Media.

Doherty, D., & Leinecker, R. (2000). *JavaBeans Unleashed*. Sams.

Eberly, D. H. (2000). *3D game engine design*. San Francisco, CA: Morgan Kaufmann.

Flynt, J. P., & Salem, O. (2005). *Software Engineering For Game Developers*. Thomson Course Technology.

Hamer, C. (2004). *J2ME games with MIDP2*. New York, NY: Apress, Inc.

Hao, W-D., & Khurana, A. (2010). Video game design method for novice. *GSTF International Journal on Computing, 1* (1).

Jia, X. (2002). *Object-Oriented Software Development using Java, 2/e*. Addison-Wesley.

Johnson, R. E., & Foote, B. (1988). Designing reusable classes. *Journal of Object-Oriented Programming, 1*(2), 22–35.

Liang, D. (2008). *Introduction to Java Programming, 7/e*. Prentice-Hall.

Mayo, M. (2007). Games for science and engineering education. *CACM, 50*(7).

Monson-Haefel, R. (2004). *J2EE web services*. Reading, MA: Addison-Wesley.

Papazoglou, M. P. (2008). *Web services: Principles and technology*. Upper Saddle River, NJ: Prentice-Hall.

Penton, R. (2003). *Data structures for game programmers*. Portland, OR: Premier Press.

Phelps, A., Bierre, K., & Parks, D. (2003). MUPPETS: Multi-user programming pedagogy for enhancing traditional study. *CITC4'03*. Lafayette, IN.

Pree, W. (1995). *Design patterns for object-oriented software development*. Reading, MA: Addison-Wesley, ACM Press Series.

Pree, W. (1997). Component-based software development – A new paradigm in software engineering? *Software – Concepts and Tools, 18*(4), 169-174.

Rabin, S. (Ed.). (2002). *AI Game Programming Wisdom*. Hingham, MA: Charles River Media, Inc.

Szyperski, C., Gruntz, D., & Murer, S. (2002). *Component software: Beyond object-oriented programming* (2nd ed.). Reading, MA: Addison-Wesley.

Wolz, U., Barnes, T., Parberry, I., & Wick, M. (2006). Digital gaming as a vehicle for learning. *Proceedings of the thirty-Seventh SIGCSE technical Symposium on Computer Science Education*. Houston, TX.

Zyda, M. (2006). *Educating the next generation of game developers. Computer*. IEEE.

Zyda, M. (2007). Creating a science of games. *Communications of the ACM, 50*(7).

KEY TERMS AND DEFINITIONS

Component-Oriented Programming: Component-Oriented Programming (COP) enables programs to be constructed from pre-built software components, which are reusable, self-contained entities. These components should follow certain pre-defined standards, including interfaces, connections, versioning, and deployment to make themselves ready to use whenever from wherever.

Framework: A software framework is a semi-finished software architecture for an application domain, which can be adapted to the needs and requirements of a concrete application in the domain. Software frameworks consist of frozen spots and hot spots. On the one hand, frozen spots define the overall architecture of a software system, that is to say its basic components and the relationships between them. On the other hand, hot spots represent those parts where the programmers using the framework add their own code to add the functionality specific to their own project.

Game Genres: Games have a variety of different kinds including action games, role-playing games, adventure games, strategy games, simulation games, sport games, fighting games, casual games, educational games, puzzle games, and online games.

Object-Oriented Programming: Object-Oriented Programming (OOP) involves programming using objects. An object encapsulates attributes (data) and activities (operations) in a single entity so that many techniques, such as inheritance, abstract, interface, polymorphism, and so on can be applied to increase the software's reusability and maintainability.

Software Maintainability: Software maintainability includes modifiability and adaptability. Software should be easily modified, extended, and adapted for fixing errors and for satisfying new needs, new requirements, and new environments.

Software Reusability: Software exists in different forms throughout the software engineering process. The requirements specification, the architectural design, and the source code are all software in different formats. Software reusability includes the reuse of any software artefacts in various formats. The most intuitive reuse is in the reality of "plug-and-play" just like the hardware counterpart.

State Diagram: A state diagram is an informal term referring to a Finite State Machine (FSM). A game, including its sprites, can be defined by a set of states. At any given time, the game will be at a specific state. When an event happens, the specific state will be changed to another. A graphical diagram that describes the states and their changes caused by events is called a FSM.

Web Services: A Web service is a self-describing, self-contained software module available via a network for completing tasks, solving problems, or conducting transactions on behalf of a user or application. Web services constitute a distributed computer infrastructure to virtually form a single logical system. They are widely used in business domain.

Section 5
Serious Games

Chapter 16
Music Tutor Using Tower Defense Strategy

Golam Ashraf
National University of Singapore, Singapore

Ho Kok Wei Daniel
National University of Singapore, Singapore

Kong Choong Yee
National University of Singapore, Singapore

Nur Aiysha Plemping
National University of Singapore, Singapore

Ou Guo Zheng
National University of Singapore, Singapore

Teo Chee Kern
National University of Singapore, Singapore

ABSTRACT

Vivace is an online musical tower defense game using the tree-of-life metaphor, created using the Unity3D game engine. The game integrates basic music theory with the tower defense mechanic to motivate inspired learning. Vivace is different from most existing musical games, as it integrates notes and chords pedagogy in a puzzle-centric metaphor, as opposed to action/rhythm. This ensures that players undergo active learning by constantly applying and reiterating these musical concepts when tackling enemies and bosses through different levels. This article describes the procedural generation algorithms and game balancing strategies used to implement the game.

INTRODUCTION

Vivace uses the metaphor of a growing tree sapling that is battling pollutants and foreign organisms from reaching its "heart". If the heart of the tree dies, then the tree itself withers. Flowers act as towers, protecting the heart of the tree from enemies. Each flower represents a different musical note which is played when the flowers fire defensive volleys at nearby enemies. These musical notes attack and play cyclically, starting from the lowest point of the tree, allowing players to create a tune based on the type and location of planted flowers. The height of the tree is divided into regular intervals. Flowers grown in the same interval will be played at the same time, thus enabling the creation of chords.

Enemy agents serve as music teachers in disguise, as they motivate the player to place appropriate musical notes and build chords through the different game levels. Boss enemy agents do more damage to plants that do not contain correct

DOI: 10.4018/978-1-4666-1634-9.ch016

Figure 1. Screen shot of Vivace tower defense game

chords. Correct chords allow players to complete each level in an easier manner. The design blends well with casual games on a social web platform, so we allow players to share their completed game levels with other players. Since the flowers are musical notes, and the game state for any completed level plays out as a musical phrase, players can share their musical trees for peer rating. In addition to the rating system, we hope to emphasize community sharing, where players can vote for better songs and gain credits. Hence the motivational framework for the game is not only limited in-game, but also outside of it, in the form of song ratings and related bonus points that can be cashed out as in-game power-ups. (Danc, 2007)

In this article, we describe the motivation behind our game, the key aspects of the game design, mechanics and algorithms, design limitations and future extensions. The game can be played online at http://games.comp.nus.edu.sg/chordtutor/

WHY MUSIC AND TOWER DEFENSE?

Our preliminary brainstorming sessions were geared towards finding an appropriate game genre for teaching music to beginners. Amongst the many possibilities, we considered using board games and turn-based strategy games, because these lend well to cognitive exercises. For the look and feel, we wanted something that would calm the mind and allow players to concentrate on the music pedagogy. Thus simple art aesthetics combined with plant nurturing became a natural choice. Most music games revolve around rhythm and timed matching of existing musical phrases. The fast-paced action usually diminishes the underlying pedagogical goals. Existing educational games on music theory lean towards traditional classroom styled lessons, which defeats the fun-potential of game design.

A music learning game naturally leads towards content creation and sharing. Seeing the popularity of *social* games and sandbox music composition games such as the *Isle of Tunes* (Isle of Tune, 2010), we decided to include a system in which players would be able to share and rate songs composed by other players. We felt that this would inculcate a sense of community sharing among players; that they are learning music together.

In addition to the music creation mechanics, we wanted to make the educational use of our game more prominent. To do this, we added a boss agent which acts as a "teacher", teaching music theory to players in a guided manner. References were made to texts on music theory to help us design these lessons better.

DESIGN CHALLENGES

There were several challenges that we needed to address while creating this game. Firstly, we needed to create a gameplay that would integrate both the musical and tower defence elements seamlessly such that it would not be awkward for players used to one or both of these genres.

Secondly, we ran the risk of making the game too difficult and detailed that only a person with a musical background would be able to play it. Therefore, we made a conscious effort to restrict the mechanics and theory to beginners.

The third problem we faced was the creation of a procedurally generated tree which would form the playing field for the entire game. Although the tree generated was in 2D, there were issues with performance and how realistic the tree looked and how it contributed to the game mechanic itself.

The final problem lay in the issue of boss encounters and the complexity of boss Artificial Intelligence (AI). We had to balance game difficulty versus the pedagogical goals. Since enemy agents primarily attack players, we had to devise a teach-through-attack design that poses both a challenge as well as a learning experience for the player. Furthermore, the boss AI needs to adapt to different player abilities and choices, in order to fulfil its role as an effective teacher.

GAME DESIGN

Metaphors Affecting Game Design

During the design phase, we focused our energies on two areas that would need the most attention in the game – the musical and the social element of our game. Waves of enemies hover in from outer space, towards the home base located at the main trunk of the tree, and flowers grow out of branches protect the base. Figure 2 summarizes the main action items players need to perform to play this game.

Let us describe some music-related issues affecting our design choices.

1. There was a wide choice of musical elements that could be used in our game. Had we tried to integrate every element into the mechanics of the game, we ran the risk of overloading players and increasing the complexity of the

Figure 2. Summary of game design elements

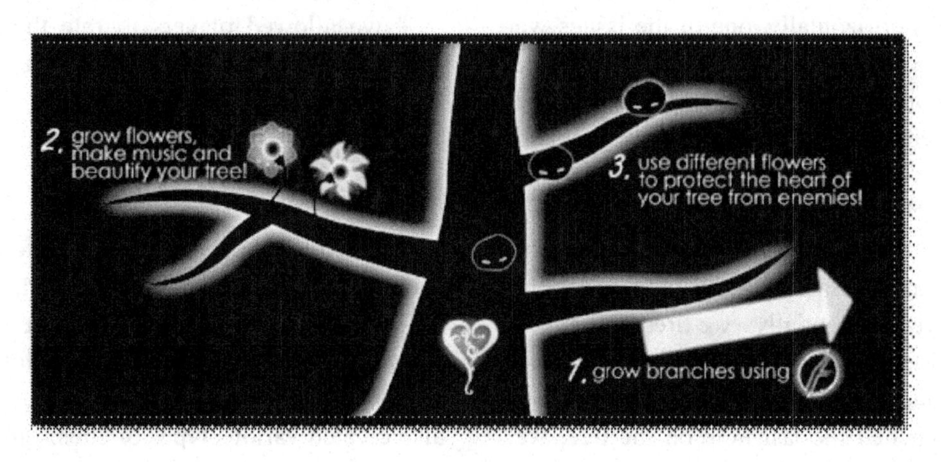

game to a point which was beyond the scope of the project. As such, we focused mainly on chord creation and musical melodies as they provided us with the most variation in our gameplay.

2. Each flower on the tree represents a tower that periodically fires a projectile, which represents one of the seven notes in a musical octave. This allowed for chord formation by grouping three towers firing simultaneous notes, if they were grown in the same interval. We rewarded players who correctly grouped towers to create the major family chords, by creating stronger and more powerful projectiles from these clusters of flowers that formed chords.
3. Since musical compositions consist of notes that span several octaves, we encourage players to upgrade existing flowers to fire notes in higher octaves, by increasing its resilience and making it harder to destroy by incoming enemy mobs.
4. Since the player had to remember each note within the octave accurately, it was important that the visual design of each flower was representative of the note the tower represented and the type of projectiles that each flower was able to shoot. For example, for the note "C", we mapped it to a "cold" type of flower, which would be able to shoot icicles at enemy mobs.
5. Since the tree was grown vertically and the flowers horizontally, one of the issues we faced was how to ensure that the firing of the projectiles by the flower would create a coherent melody. As such, we mapped the vertical area in the game into intervals and each interval represented one beat in a song. In our game, we fixed the tempo of the beats. The same way that notes in a score is played sequentially - we fired projectiles from the flowers sequentially starting from the bottom of the tree to the top. When we had reached the last note on the tree, we replayed the whole sequence again from the bottom, looping the melody till the game ends.
6. Many educational games focused on questions and puzzles that provide straightforward answers, which do not exercise a player's ability to create their own individual solutions. To avoid this, we designed the enemy and boss AIs to subtly steer players towards building certain combinations or certain towers. The AI would make it more likely for irrelevant or "unchorded" towers to be destroyed by monsters and bosses. In addition to these subtle hints from the enemy monsters and bosses, we also provided more direct forms of feedback in the form of tooltips describing the different types of enemies as well as an interactive fairy that acted as a guide for the players as they progressed through the game.
7. Even though we try to limit the types of chords and music created by players, we still give them enough room to exercise their creative abilities in creating new pieces of music using the game framework that we have created. Apart from the limitation of what constitutes a musically correct chord, we did not limit the players to building flowers in a certain melody or rhythm. Since musical aesthetics is subjective, we gave players the option of sharing the music they created with other players. Through sharing, we allowed players to rate the melodies that other players had created as a form of feedback for the composer.

Level and Enemy Design

To create a more engaging tower defence game, each tower had a unique projectile that it could use. These unique abilities could be combined when these towers are used in a chord formation to create new and more powerful combinations. We also created various types of enemy mobs, each with its own set of strengths and weaknesses. For example, short range enemies have a high amount of health while faster enemies have lower health but inflict more damage.

By scripting the types of enemies that spawn at different levels, we can subtly hint to players the combinations that they can build at particular levels. Boss monsters are a different type of enemy monster that is much stronger and more difficult to destroy than the normal monsters. These bosses add a sense of urgency to the process of finding a musically sound method of overcoming the level and have been programmed to punish players for wrong chords by destroying not one flower of the errant chord, but all. Before the boss proceeds to do so, he will give the player suggestions for change and if the player does not do anything about the chord, the boss will then destroy the offending chord. Despite this, the game is dynamic enough to pass a level even if a player does not follow the ideal combination of flower strengths against monster weaknesses, so long as the chords are musically sound.

Game Rewards

Rewards are an integral aspect of game design and in our own game in various forms. The first way we rewarded players was through player scores and progression through the game level. The interactive fairy gives players words of encouragement as they progress and congratulates them at a notable success during their game.

Players are also given rewards through in-game achievements that they earn for completing certain feats in the game. These achievements earn them a badge of recognition that other players can see. Rating the songs that other players post also gives players points that they can use on bonus items that boost their abilities while playing the game.

Design Limitations

Unlike most tower defence games where there is an element of navigation through a set of obstacles in the world, our game has no obstacles and enemy monsters and flower projectiles can pass through the branches of the tree. Currently, the tree grows vertically and most enemy monsters will swarm to the heart of the tree from the top of the screen. This meant that if the player places flowers at the top of the tree, the notes at the bottom become redundant since most of the combat takes place at the top of the tree. We overcame this problem by programming different targeting behaviours for the enemy monsters.

Another limitation of Vivace was that players did not have an option to change the duration or volume of the sounds generated by the flowers, making their compositions one dimensional without any variations in note length and tone. It was also difficult to add in background ambient music as it could interfere with the generated notes, creating noise rather than music as was initially intended.

Visual Design and Interactivity

An uncluttered visual scheme was chosen for this game (see Figure 3). Plant motifs were also used extensively to match the organic and natural form provided by the procedurally generated tree and to provide the game with a sense of fluidity and naturalism. Brighter colours and more ostentatious designs were used for the flowers to reinforce a sense of uniqueness for each flower for easier identification by players. Flowers were also designed to visually fit the type of prototype they used to shoot at enemy monsters.

Interaction with the players was further enhanced through various means of feedback and in-game input. Simple yet descriptive icons are used to describe the state of an enemy agent (*muted-attack, slow, reduced-armor*) when hit by a flower projectile. These icons were designed to

Figure 3. Rich user interface with contextual tooltips, hints and mini-map navigation

make it easier for the player to ascertain the enemy state and to plan the placing of their towers in an appropriate manner. As shown in Figure 3, a mini-map allows extended player navigation and early visualization of enemy swarms, to facilitate long compositions and strategic planning. It also shows an encounter between an enemy agent and a defending flower (*note C*).

IMPLEMENTATION

As mentioned earlier, we used the metaphor of a tree sapling. The player needs to add branches, leaves and flowers to this sapling using energy collected from rays of sunlight to help it grow. All these parts of the tree help to defend its heart against invading enemy monsters that take the form of insects and bacteria that want to infect the heart. Once a branch connects the tree to a ray of sunlight, a leaf can be planted near the light to harness energy points. These energy points serve as the expendable cash to plant and upgrade flowers. We had implemented 15 different levels with unique lesson objectives and progressive difficulty. We will now describe algorithms and implementation details of procedurally generated tree, the boss monster (teacher) AI, tower-defence mechanics and the game balancing algorithm.

Procedural Tree Generation

A typical tree consists of branches growing both upwards and outwards. Since branches come in different shapes and sizes, there is a need to control its growth and appearance, yet allow for enough variation for them to appear natural. Curve equations allowed us to generate the skeleton of the branch, as well as to extend it to simulate branch growth. We also needed to consider how to manage branch hierarchy and the branching factor, as these would affect flower placement and the performance of the game. We limited the maximum number of branches in the tree and the maximum branching factor for each branch. We then applied a pseudo-random recursive L-system tree generation algorithm after setting the main trunk shape and branching limits. The growth of each child branch is evenly spaced out, and alternately spawned (to cover both left and right sides) on its parent branch. All initial branches are grown at specified time offsets with respect to their parent, to give an even and fluid growth animation effect at the start of each level.

Listing 1. Pseudo code for growing specified number of branches

Function GrowBranches(current branch, growth rate, number of sub branches on this branch)
 Grow current branch;
 Stop if current branch is fully-grown or maximum number of branches hit;
 If tree trunk growth triggered by player, plant multiple branches on trunk;
 else plant multiple branches on current branch ;
 Recursively call GrowBranches for all child branches;

Listing 2. Pseudo code for creating a branch

Function CreateSideBranch(start point, end point, parent branch, user generated or computer generated)
 If user clicks on a valid parent branch, create a new branch object on the clicked area;
 else if computer generated branch
 Create a new branch object based on pre-calculated start points and end points;
 Add a new empty mesh to the branch object;
 Specify materials for the branch;
 Set the branch orientation based on start and end points of branch;
 Add 1st vertex for mesh calculation;
 Shift the user-clicked point to nearest parent branch vertex (prevent artifacts);
 Retrieve start points, end points and the distance of the branch for the Bezier curve;

We chose quadratic Bezier curves over B-Splines as they are faster to compute and provide decent curvature. The only caveat is that we have to additionally ensure that *c1* continuity is achieved at the junctions of new branches. Each branch is modelled with 4 control points, as shown in Figure 4. A typical branch in the tree has 3 different components: orientation, curvature and length. The orientation of the branch is determined by a vector which passes through the 1st and 4th control points of a Bezier curve. The curvature of the branch is determined by the 2nd and 3rd control points respectively (see Figure 4). We approximated the branch length with the distance between the 1st and 4th control points. Once the orientation, curvature and length have been determined, the growth of the branch can be animated by varying

Figure 4. Using quadratic Beziers to model branches

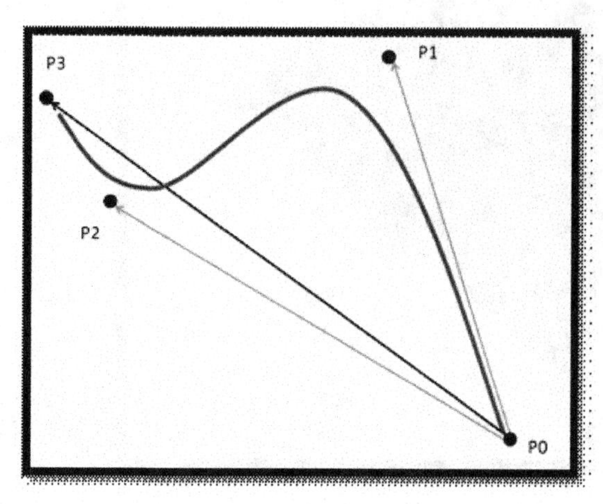

the *t* interpolation parameter in the Bezier curve from 0 (no growth) to 1 (fully grown).

Listing 3. Pseudo code for the generation of control points for a Bezier curve for the branch

Function UpdateBranchPoints(start point, end point, length of branch)
 Retrieve orientation vector of branch;
 Randomize sign +/- for p1;
 Vary distance p1 from p0;
 Rotate tangent on parent branch at growth point p0 by about 45 degrees to minimize artifacts;
 Freely rotate p1 by a small angle, about p0, to calculate p2;
Function GrowOneBranch(current branch, growth rate)
 If the branch is not fully grown, increase the growth timer value;
 Calculate position of next branch vertex by using the normalized growth timer as an input for Bezier interpolation of the branch's 4 control points;
 Set branch mesh colliders at specified time intervals;

A branch mesh can be thought of as a string of quads that gradually grows smaller towards the tip. We first discretized the smooth quadratic Bezier curve that defines the skeleton of the branch into 300 to 500 segments. For every discretized poly-segment (B_0B_1 in Figure 5), we then created the mesh quad by creating two cross-sectional vertices at each end (Q_0Q_1 and Q_2Q_3 in Figure 5). Branching discontinuities at joints may give away the illusion of a unified organic structure. We removed this problem by firstly constraining the branch's 2nd control point such that it maintains a range of angles around 45° to the parent branch-segment and secondly constraining the starting width of the base of the branch to a percentage of the width of the parent branch-segment.

Listing 4.: Pseudo code for adjusting branch width

To adjust branch's current max width while growing,

Figure 5. Branch mesh quad, where B_0B_1 represents a skeletal segment (branch root on B_0 side), and Q_0-Q_3 are the generated mesh vertices

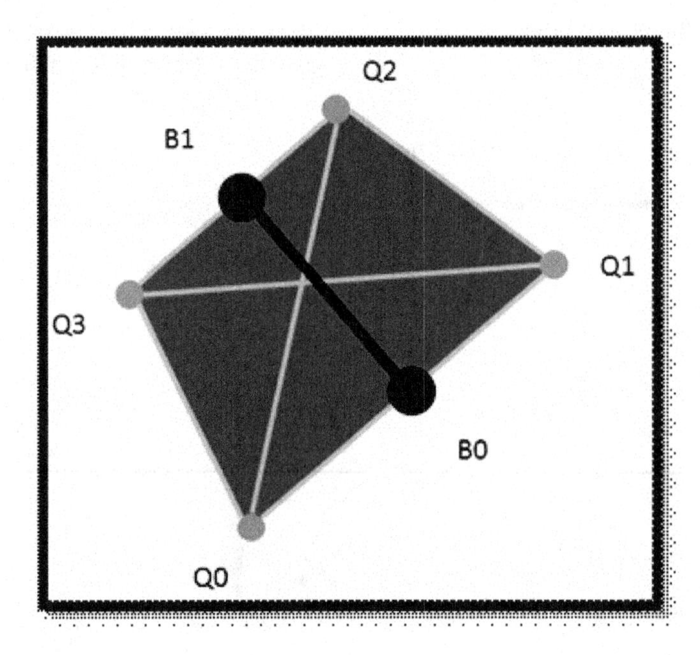

Find difference in branch max and min width;
Derive difference as a fraction of current branch vertex count;
Subtract result from current branch's max width;
To adjust branch's current min width while growing,
Find difference in branch max and min width;
Derive difference as a fraction of (current branch vertex count + 1);
Subtract result from current branch's min width;

Players were restricted to placing objects on branches only. Thus, we needed accurate mesh collision instead of ill-fitting bounding boxes. Each branch mesh could contain up to 1200 triangles, and a given tree can contain anywhere between 5 to 30 branch segments (including the trunk). This could entail a collision checking complexity of about 6000 to 36000 triangles per planting operation. In practice, we reduced the collision mesh complexity by sub-sampling the above-mentioned discretized branch skeleton, by a factor of 2 to 4 (based on the curvature). We also accelerated the query by culling all branches that do not span the horizontal slice (beat interval) enveloping the mouse click. This culling can be simply done against the Axis Aligned Bounding Boxes of all the branch segments, and a scaled offset into the skeletal list of the main trunk.

Listing 5. Pseudo code for creation and sub sampling of branch mesh

Specify start and end points of quad;
Derive quad vertices from start and end points;
Create triangle mesh using quad vertices;
As branch vertices increase during branch growth, derive larger quad vertices between branch vertices at specified intervals; (e.g. Branch vertices = 4, interval = 2, use 1st and 4th vertex as start and end points of the larger quad)
Create a larger triangle mesh based on the new quad vertices for sub sampling;
Recalculate bounding box for both meshes;
Select sub sampled mesh for use after branch is grown;

Finite State Machine for Boss Agent

Figure 6 shows the finite state machine used to control the boss enemy's behaviour. Each state, governs a distinct behaviour as described below. States were implemented as classes in the boss script to allow for proper encapsulation of trigger variables, as described below:

1. ***Idle* state:** This controls the boss' hovering and movement behaviour from point to point.
2. ***Attack* state:** This state encompasses the decision making module used by the boss to decide on an attack type.
3. ***Single Note Attack* State:** This state is responsible for a swoop attack aimed at a specific note in a wrong chord combination. Although this attack does not allow players to 'redeem' themselves, it is less damaging than the interval attack.
4. ***Single Interval Attack* State:** Implements a "chew-flower" attack spanning an entire beat interval. The sub states are split as follows:
   a. The boss decides on a target chord combination which is wrong. It obtains positional offsets from the leftmost and rightmost towers in that interval.
   b. The boss then moves to the pre-charge position, calculated from the leftmost tower.
   c. Waiting before charging: This is the sub state in which the player is given a window of opportunity to correct his chord combination. The boss will then charge if time exceeds limit and chord combinations are still wrong.
   d. Boss flinches if player corrects his chord combination.

Figure 6. Finite state machine for boss (teacher)

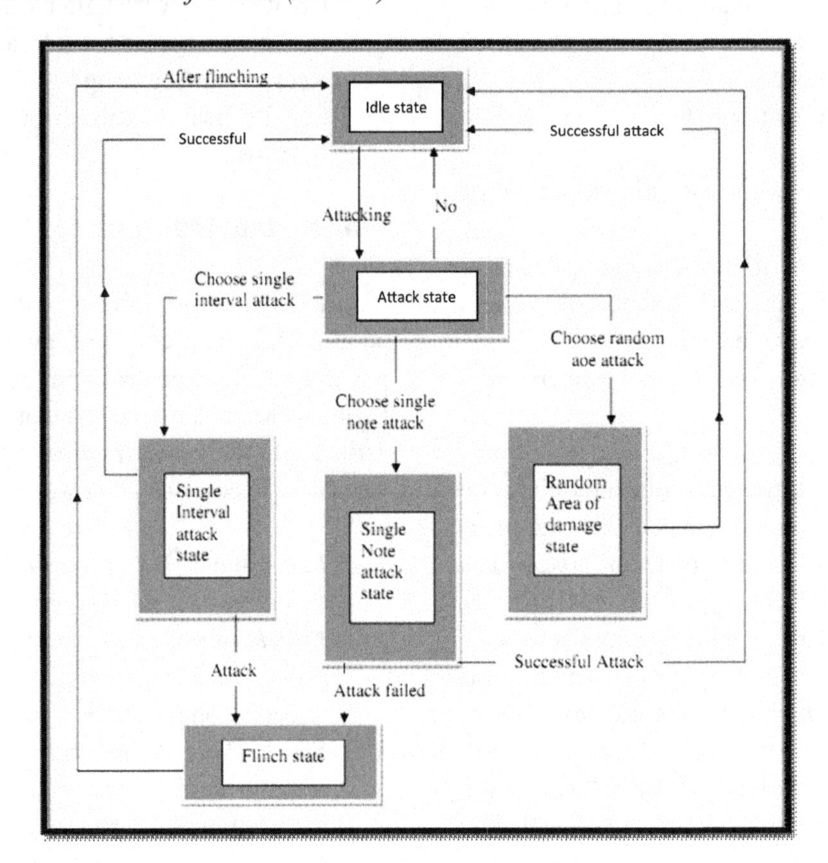

5. ***Flinch* State:** Implements boss' stunned behaviour after the player successfully corrects a chord combination.

We have mentioned earlier that the boss' primary goal is to play the role of a tutor and examiner, with a greater emphasis on musical theory. The boss-behavior was designed in such a way that it is clear to player what is being taught to them. This is done by encoding different behaviors for different player-mistakes, together with a speech bubble that includes a text description of the mistake. For example, when the player misplaces a note in an interval that creates a bad chord, the boss will say: "That is a wrong note!" A *Single Note Attack* follows this exclamation. If the player pieces together some notes that do not meet the basic requirements of a chord, the boss will activate a *Single Interval Attack*, and have a speech bubble saying, "What sort of chord is this?" We also added orbs circling around the boss to indicate the most probable chord we want the player to create in order to appease the boss. In addition, we implemented a preparation time for the *Single Interval Attack*. This allows the player to correct his mistake by learning from the circling orb hints for the correct chord notes. If the player manages to execute the correction, the boss enters a *flinch* state and stops its attack on the tree momentarily. Thus the boss agent serves as teacher-guide, one that informs and punishes students if hints are not taken. It is perhaps not an ideal role model of a teacher, but we feel it is permissive given the context of the game. The last thing we want to tell our players is that the boss enemy is actually the music-teacher, though in reality, we see players scrambling to rectify their mistakes, especially after they have lost an entire set of towers to a *Single Interval Attack.*

Tower Defense Mechanics

As shown in Figure 7, we have 7 different notes (*C, D, E, F, G, A, B*), 7 different chords (*C, Dm, Em, F, G, Am, Bdim*), and 7 different enemy types. These different elements were spread out over 15 different levels, so we paid careful attention to their attributes and risk-reward tension to ensure an enjoyable game experience. Table 1 lists important swarm mechanics variables, and six increasingly challenging instances of the type-B swarm agent (see 2nd left, bottom row in Figure 7).

As illustrated in Table 1, each enemy type is separated into six grades of difficulty. In this example, enemy species B is a mid-range attacker, where the range of attack increases and attack behavior changes to *"Straight for the heart"*, as the player progresses to the advanced levels. Attributes *"Destination Range"* and *"Retreating distance after attacking"* serve to alter its movement patterns before and after an attack.

Table 2 illustrates an aggregation of *"Attack Damage"* values across all enemy types, for different difficulty grades. It was overwhelming for a designer to manually refine all these values, and exhaustively play-test the game over 15 levels to suit different player styles. As such, we created a balancing mechanism to reduce amount of play testing we had to do before the release of the game. Before we describe the balancing mechanism, let us quickly cover the different attributes for the towers as well.

For the sake of brevity, we did not describe the attributes of chord combos and higher-octave power-ups here. As seen in Table 3, each flower is designed to serve a different purpose. For example, flower *C* serves as a frost tower, designed to slowdown enemies. Flower *A* is a short-ranged, area-damage tower, designed more for defense. Flower *F* deals damage over time (DoT) on targeted enemies. It is an effective weapon against harder to hit targets, e.g. enemies with high speed, or non-linear trajectories. These attack designs are heavily inspired by effects and custom games created in Warcraft 3 (Warcraft 3, 2010).

Game Balancing Algorithm

Let us now describe the core of the balancing algorithm. The player's solution to a tower-defense game is a complex process, where the success of future waves depends on the remaining health of older towers and strategic profit reaped in the past waves. We fixed some design variables, so that we could automatically organize the enemy waves, estimate the number and type of flowers

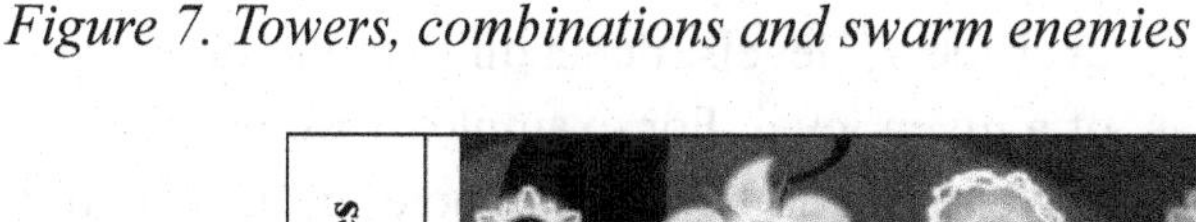

Figure 7. Towers, combinations and swarm enemies

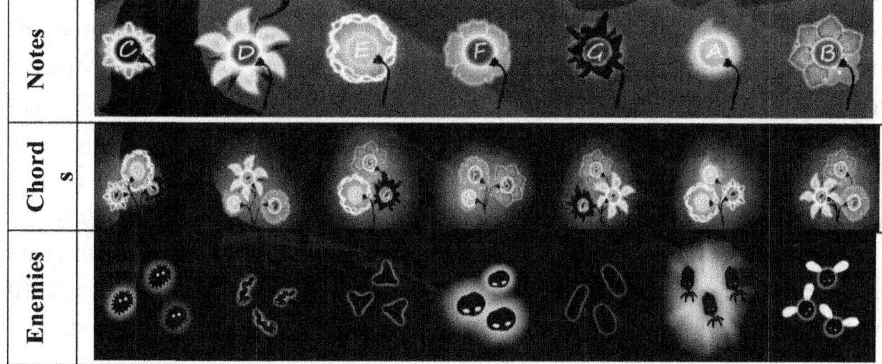

Table 1. Attribute values representing increasing difficulty (Grades 1→6) for type B enemies

	Grade 1	Grade 2	Grade 3	Grade 4	Grade 5	Grade 6
Armor	0	0.4	0.8	1.2	1.6	2
Attack Damage	2.8	3.84	4.88	5.92	6.92	8
Attack Range	4	4.2	4.4	4.6	4.8	5
Attack Cooldown	2	1.9	1.8	1.7	1.5	1.4
Acceleration	0.1	0.1	0.11	0.11	0.13	0.13
Movement Speed	0.26	0.47	0.68	0.89	1.09	1.3
Health	70	80	90	100	110	120
Straight For the heart	FALSE	FALSE	FALSE	FALSE	TRUE	TRUE
Post-attack retreat distance	6	6	6	6	6	6
Destination Range Sensitivity	2.5	2.5	2.5	2.5	2.5	2.5
Size	1	1.12	1.24	1.36	1.48	1.6

Table 2. Graded Attack Damage values across all enemy types

	Grade 1	Grade 2	Grade 3	Grade 4	Grade 5	Grade 6
A (melee)	2.8	3.84	4.88	5.92	6.92	8
B (swarm)	2.8	3.84	4.88	5.92	6.92	8
C (swarm)	2.8	3.84	4.88	5.92	6.92	8
D (tank)	2.8	3.24	3.68	4.12	4.56	5
E (swarm)	2.8	3.84	4.88	5.92	6.92	8
F (tank)	2.8	3.24	3.68	4.12	4.56	5
G (range)	5	7	9	11	13	15

needed to counter these waves, tweak puzzle design choices or agent attribute values should the enemies win at any stage, and finally to appropriately setup the light energy resources to control the game difficulty (once the best solution and its cost are known).

We started with intuitive values for the simplest and most difficult grades for all the attributes in each enemy type, guided by the agent's expected behavior; e.g. speed = slow | fast, damage = low | high, target = focused | nearest-neighbor, etc. These values were then linearly or exponentially interpolated over the intermediary grades 2 to 5.

We then arbitrarily fixed the type of enemy agents over the 15 levels, depending on the objectives of a given level. For example, if we wanted the player to learn the member notes of the *C* chord (notes *C, E, G*), we needed to put up suitable enemies that can stand up to longer range, high damage towers denoted by these member notes. So we chose from a mix of enemies that were fast moving, have non-linear trajectories, or have higher resilience. Once we crafted a set of enemy types for every level, we approached the balancing problem using the following steps:

Table 3. Attributes for flowers firing different notes on the C Major scale

	C	D	E	F	G	A	B
Health	100	100	100	100	100	100	100
Shot Type	Single Target	Single Target	Single Target	Single Target	Area	Area	Random
Damage	12	15	4	9	11	3.5	11
Attack Range	13	10	10	10	15	10	15
Projectile Lifespan	5	1	4	2	2	0.1	5
Projectile Launch Speed	5	10	-	8	5	-	-
Tracking Factor	0.1	0.7	1	1	1	0.2	-
Health Regeneration rate	0.5	0.5	0.5	0.5	0.5	0.5	0.0003
Gravity	0.4	2	0	1	5	0	0
Is homing	TRUE	FALSE	TRUE	TRUE	FALSE	FALSE	FALSE
Aditional Damage	Movement impairment	Armor reduction	-	DoT	Splash Damage	-	Attack cooldown

1. We fixed the number and difficulty grade of enemy agents, as well as the maximum number of agents per wave.
2. We constructed a heuristic that measured the overall strength of each flower and enemy agent, as a function of its health, motion, attack, and defense attributes.
3. We computed the wave sequence, by randomly constructing N agent-subsets from the fixed pool in step (1), and then sorting these waves in increasing order of strength, using the heuristic function in step (2). A wave-queue stores this sequence, ensuring that the weakest wave appears first, and the challenge increases with successive waves.
4. We popped the first wave from the queue, and aggregated the overall health and time afforded by the constituent enemy agents. The problem now was to find a configuration of towers at minimum cost that can defeat the incumbent wave. The cost is a weighted function of resource requirements (light-energy costs of building/upgrading units) and remaining aggregated health of the flowers. Constraints can be put on the maximum number of flowers or maximum light energy available for a given wave. The solution was obtained with an A* Search or Dynamic Programming, where every state is a start-end simulation run of the enemies versus the flowers. Branching actions are ignored unless there are no more free spots to plant flowers. Flower planting starts from the highest available branch, and proceeds downwards.
5. If we have a successful outcome computation for the first wave, we can pop the next wave, and repeat the search process, except that we start with the previous wave's flower instances, and have an additional choice of selling them.
6. If we encounter a failure-to-defend scenario after step (4), we could employ these strategies, in the following order of priority:
   a. Discard the strongest member from the recent wave and re-perform the search
   b. Relax the current resource constraints
   c. Designer reconfigures the pool of enemies, and restarts the search from step (1)
   d. Tweak one of the attack/defense/motion/health parameters of a related enemy (make weaker), or flower (make stronger)

7. The final and important piece in this balancing mechanism was to set up the locations and values of the light-energy resources. Note that we now know the resource requirements for the best solution, at every wave. If we fix the number and locations of the light-resource centers, we can easily compute the best cost for constructing branches and planting harvesting leaf units. As long as there are enough resources to build branches to the light energy sources, we can arbitrarily distribute the total energy resources required every wave, over the fixed light-resource centers. To adjust the game difficulty, we add on a variable extra percentage of the computed resource values, where the variable is tied to player performance (e.g. only 5% extra for experts, 60% extra for novices).

Listing 6. Pseudo code for enemy generation per wave

```
//Construct histogram of tower in
stage
towerSetup = RetrieveTowerSetup();
histogram = ConstructHistogramOfTower
s(towerSetup);
//evaluate strength of each enemy
type, based on strength/weakness of
each tower type and their numbers
wave = RetriveWaveSetup(wave_i);
enemyScore[] = EvaluateScoreOfEachE
nemyTypesBasedOnTowerSetup(wave_i,
histogram);
if(sum(enemyScore) > threshold)
        Balance(setup);
        LogGame(); //For game bal-
ancing by designer
SendWave(wave);
```

Software Methodology

Since the development of the game was limited by a time span of only 6 weeks, the team applied an agile software development process. From start to end, the game underwent 3 major milestones. As the development team comprised out of students, we adopted a modified SCRUM methodology. As students, we could not work on the game daily and assigned group working days when we would physically meet to work. During non-working days, only preparation work for the next development day way done. As such, most progress would happen when the team is working together physically.

During each physical meeting, we worked on our assigned tasks for the session and at the end of the day, report our progress. In addition, we listed out objectives that needed to be prepared by next meeting. This approach allowed the team to coordinate our progress effectively and reduce dependencies among each programmer. This was especially important due to Unity3D component based structure and inability to version control prefabs and scenes. Meeting up physically allowed the team to notify other team members about the

Table 4. Software milestones

Milestone 1 (Week 2)	• Basic music flower input • 1 type of enemy • Towers attack the same way • Basic tree growing mechanism
Milestone 2 (Week 3)	• Optimisation to tree growing mechanism • Tree growing mechanism can produce curls and additional branches • Towers have unique attacks • Replace all placeholder graphics • Different types of Unique enemies • 1 Boss type with simple boss AI
Milestone 3 (Week 6)	• Tree can grow taller and bigger • Towers have combined attacks • Enemies balancing • Online creation sharing platform • Achievements on online platform • Game balancing AI • New type of boss with a FSM model AI

resources they were using to prevent conflicts in file usage.

After each milestone, the team carried out user testing on a small group of testers for comments and improvements. These comments heavily aided the development of the user interface and the balancing of the game mechanics.

FUTURE DIRECTIONS

We have received some informal feedback from play-testers who do not know about music theory and play games. Most of them said that they concentrated more on memorizing chord combinations for the sake of winning the game, rather than learning anything new in music. We were surprised that despite excluding the action elements of rhythm and pattern games, our testers still felt this way. Clearly, this is a challenge that requires us to go back to the drawing board in order to better align the game with musical pedagogical goals. Perhaps we need a stronger musical context to complement the rather abstract tree-nurturing metaphor. Also, currently we have addressed notes and chord sounds of the piano. The addition of percussive, wind and string instruments will add interesting challenges to the design and balancing problems.

CONCLUSION

We have proposed the design and implementation of a piano notes and musical theory tutor game. We have attempted to closely align the game mechanics to concepts in music. For example, we tied in the rate of firing weapons to the beat of the composition, and bullets are equivalent to musical notes. We consciously chose a design genre different from rhythm and pattern games, and describe our motivations and design choices. We have described technical components required by the design, namely: 1) procedural tree generation and flower planting; 2) boss AI; 3) tower-defence mechanics; 4) game balancing algorithm.

The game can be viewed as a teaching and peer-rated content sharing platform. We hope to enrich the game mechanics and musical context, to serve as an entertaining and useful tool for music education.

REFERENCES

Cook, D. (2007). Celestial music. *Lost Garden*. Retrieved from http://www.lostgarden.com/2007/09/celestial-music.html

Isle of Tune. (2010). Retrieved from http://isleoftune.com/

Warcraft 3 (2002). Retrieved from http://us.blizzard.com/en-us/games/war3/

Wei, D. H. K., Yee, K. C., Guozheng, O., Kern, T. C., & Plemping, N. A. (2011). Vivace: Chord tutor. *SoC Games Portal: National University of Singapore*. Retrieved from http://games.comp.nus.edu.sg/chordtutor/

ADDITIONAL READING

Collins, K. (2008). *Game sound: An introduction to the history, theory, and practice of video gamemusic and sound design*. Cambridge, MA: MIT Press.

Divnich, J. (2008). *The divnich tapes: Music & Rhythm Game 'Fade" on the Way Out.* San Francisco, CA: Gamasutra. Retrieved from http://www.gamasutra.com/view/news/112140/The_Divnich_Tapes_Music__Rhythm_Game_Fad_On_The_Way_Out.php

Fox, B. (2005). *Game interface design*. Boston, MA: Thomson Course Technology PTR.

Hoffmann, L. (2009). Learning through games. *Communications of the ACM, 52*(8), 21–22. doi:10.1145/1536616.1536624

Michael, D., & Chen, S. (2006). *Serious games: Games that educate, train and inform*. Boston, MA: Thomson Course Technology PTR.

Pichlmair, M., & Fares, K. (2007). Levels of sound: On the principles of interactivity in music video games. *Proceedings of the digital games research association 2007 conference* (pp. 424-430). Tokyo, Japan: DIGRA Digital Games Research Association.

Velden, J. v. (2011). *Pon pon pata pon sounds of power and the power of sound.* Retrieved from http://www.scribd.com/doc/59601689/Pon-Pon-Pata-Pon-Sounds-of-Power-and-the-Power-of-Sound

KEY TERMS AND DEFINITIONS

Boss: The master enemy agent in a computer game that is usually quite difficult to defeat.

Cooldown: A period of time before a feature can be re-activated.

Chord: Two or more notes in any harmonic set that is heard simultaneously.

Octave: Interval between two identical notes, typically separated by the number of notes in the scale.

Procedural Generation: Using rules, template data and computer code to generate content.

Sandbox: Systems allowing players to freely explore all game features without any win-loss objective.

Tempo: Pace of a given piece of musical composition.

Tower Defense: A strategy game where the player builds static defenses against incoming enemies.

Chapter 17
Low Cost Immersive VR Solutions for Serious Gaming

Damitha Sandaruwan
University of Colombo School of Computing, Sri Lanka

Nihal Kodikara
University of Colombo School of Computing, Sri Lanka

Chamath Keppitiyagama
University of Colombo School of Computing, Sri Lanka

Rexy Rosa
University of Colombo, Sri Lanka

Kapila Dias
University of Colombo School of Computing, Sri Lanka

Ranjith Senadheera
Sri Lanka Navy, Sri Lanka

Kanchana Manamperi
Sri Lanka Navy, Sri Lanka

ABSTRACT

Games are used for other purposes than providing entertainment. This chapter is particularly interested in serious games, also known as simulators, with immersive virtual reality environments that are used for training and teaching purposes. These simulators have very stringent requirements and as a result, they are expensive to build. However, the authors managed to develop a ship handling simulator for the Sri Lanka Navy, at a cost of less than $20,000, which is an order of magnitude less costly than the cheapest available ship handling simulators. The cost of the simulator was kept at a minimum by using Commodity-Off-The-Shelf (COTS) hardware, Free and Open Source Software (FOSS), and also by adopting a development strategy which kept the client involved in the complete life cycle of the development. The availability of the required manpower at a very low cost in Sri Lanka was also beneficial.

INTRODUCTION

Computer based games are usually associated with entertainment. However, there is a particular genre of games called serious games that do not focus on providing entertainment but focuses on solving real world problems (VSTEP, 2010), (Ilan Papini, 2010). Ship handling simulators, firing simulators, and flight simulators that are used for training are some of the examples of serious games (FlightSafety International Inc, 2010) (Simulator Systems International (SSI), 2010) (Transas Marine Limited, 2010) (Oceanic

DOI: 10.4018/978-1-4666-1634-9.ch017

Consulting Corporation, 2008). Although these simulators are not usually referred to as games they use the same technology and design concepts used in games.

Serious games mentioned above have more stringent requirements that are not usually expected of games for entertainment. These games model real world physics, and responses to external events are expected in real time. Therefore, they require higher computational power and high end components, in contrast to games that are purely entertainment. In addition, a higher level of expertise is required to model the real world physics as accurately as possible. Certain serious games such as the ship handling simulators and flight simulators used for training purposes also require the users to be in an immersive virtual reality environment when they interact with the game. This leads to the high cost of some serious games (FlightSafety International Inc, 2010), (Simulator Systems International (SSI), 2010) (Transas Marine Limited, 2010) (Oceanic Consulting Corporation, 2008).

The performance of computer hardware has improved tremendously over the past decade and the cost has also gone down. It is now possible to use commodity-off-the-shelf (COTS) hardware to model real world physics in real time with the accuracy required for serious games. In addition, there are free and open source software (FOSS) packages that could be used to build physics simulators and rendering engines to create the virtual environments. Accordingly, low cost serious games equal to commercial high end simulators can be developed using off-the-shelf hardware and open source software components.

This chapter focuses on the design challenges in developing immersive virtual reality solutions for serious games using COTS hardware and FOSS. These challenges are more or less common to most serious games and we attempt to keep the discussion generic. The term serious game is very broad; it is not possible to address the design issues related to the broad spectrum of games covered by this term. In this chapter, we limit our focus on serious games used for training in handling vehicles, such as ships, flights, and cars in immersive virtual environments.

To ease the discussion, we use the example of building a ship handling simulator called "*Vidusayura*" to satisfy several training requirements of the Sri Lanka Navy (SL navy) as shown in Figure 1, it was built using the COTS hardware and FOSS were used to develop it.

Sri Lanka Navy

The Sri Lankan Navy (SLN) is the naval arm of the Sri Lankan Armed Forces. They have a training division to train the crew for their ships. One

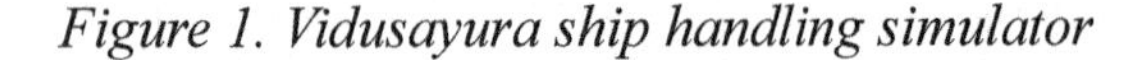

Figure 1. Vidusayura ship handling simulator

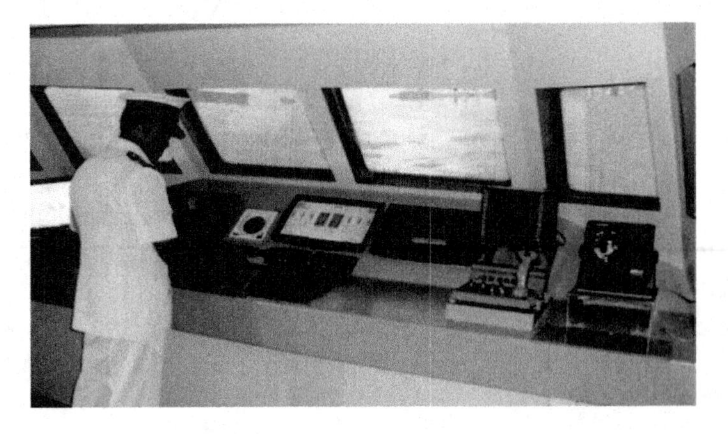

of the tasks of the training division is to train the bridge crew for the medium scale ships the SL navy owns. They have qualified teaching staff to take on this task. The training has two main components; maritime class room lectures and ship maneuvering training in real ships. The crew is given lessons on bridge procedures, regulations for preventing collisions at sea, and observing the basic principles of keeping a navigational watch etc. SLN utilize ships from their fleet for training and this proves to be extremely costly. They are aware of the potential of ship handling simulators. Therefore, they have decided to explore the possibility of a custom made, low cost ship handling simulator.

WHAT IS REAL-TIME SIMULATION?

A simulation is a representation of the operation or features of a system through the use or operation of another (Philippe Venne, 2010). In general, that is the concept behind a simulation and computer based simulation is also based on the same concept. There are two kinds of computer based simulations, real-time simulations and non real-time simulations (Wikipedia, 2011). Throughout this chapter, we focus on real-time simulations which respond to inputs at the same rate as the actual operation.

There are two kinds of time setups in simulations, fixed time-steps and variable time-steps. In the fixed time-step method, the simulation considers discrete equal time segments and time moves forward. These time segments are very small and response rate of the simulation is less than or equal to the number of discrete time segments per second. At the end of each discreet time segment the simulation system considers the status of all variables and solves equations to obtain the next state of the dynamic bodies included in the simulation. In the variable time-step method, simulations do not use equal size discrete time segments. The size of the discreet time segment is dynamic. However, fixed time-steps method is more suitable for real-time simulations (Philippe Venne, 2010).

In the fixed time step method, the real time required to solve equations and obtain the next state of all dynamic bodies may be equal, shorter or longer than the simulation time. Figure 2 illustrates all three possibilities of the time step variations.

In real-time simulations, the simulation time and real clock time are synchronized at the end of each fixed time segment. However, most of the time complexity of the mathematical equations in the simulation has to be simplified for real-time simulations.

DEVELOPMENT STRATEGIES

As mentioned in the introduction, serious games can be used for a broad range of applications, e.g. military, government, educational, corporate and health care (Susi, et al., 2007). Serious games permit learners to undertake tasks and experience situations which otherwise would be impossible

Figure 2. Variation of the simulation time and real clock time

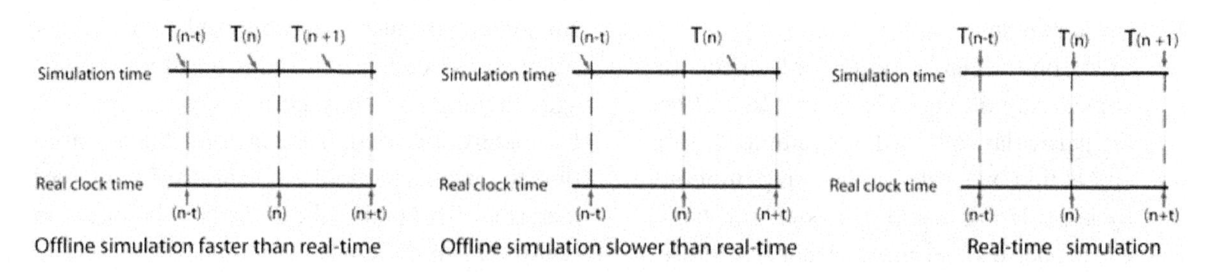

in the real world due to factors such as cost, time, logistical problems and safety (Corti, 2006). A serious game with an immersive VR solution provides a more realistic environment to the users, and it replaces the existing real world scenario to a certain extent. For example, ship simulators can be used to replace the considerable amount of conventional maritime class room lectures and substantial hours of ship maneuvering trainings with real ships. Similar examples can be found in other areas such as health care, medical and military. However, serious gaming applications cannot be used to replace the entire real world teaching or learning process. It can replace only a segment of real teaching or learning process. Obviously, it is not possible to simulate all possible scenarios in the real world. Therefore, it is very important to identify the possible and most important activities to be simulated in the virtual environment.

Ultimately, the physical and behavioral realism (Yin, et al., 2010) of the entire virtual environment should be at a sufficient level to obtain an immersive feeling. Behavioral realism can be improved by increasing the accuracy of the motion prediction algorithms and realism of the objects or scenarios in the rendered visuals. Physical realism can be obtained by interfacing the VR solution with real equipment and providing a more realistic physical environment to the user to interact with the VR solution. This gives greater ecological validity (Rizzo, et al., 2006) to the VR solution, and it is a very important factor in any virtual reality application (Rizzo, et al., 2006).

Throughout real-time simulation development there are conflicting goals to be achieved.

Behavioral realism:

- Predicting accurate motions of objects by considering all possible factors that affect the particular activity. For example, in ship simulation we can predict ship motions by considering a ship's physical/mechanical properties and most responsive ocean wave, but the real world scenario is more complex. The ocean consists of complex wave patterns, and wind and sea currents are also present. We can predict a ship's motion by considering all possible factors that affects the ship motion.
- Simulate all dynamic objects in the virtual environment with realistic physics.

Focusing on the above factors, we can achieve a higher level of behavioral realism, but it requires more computational time. This leads to non real-time motion predictions.

Physical realism:

- Provide a realistic environment with greater ecological validity for the user to interact with the VR solution. For example, in ship simulation we can interface a real ship throttle and a wheel to handle the ship and physically construct the ship bridge similar to a real world ship bridge.
- Use more realistic 3D objects and simulate real world effects. For example, in ship simulation we can model realistic naval vessels, moving or fixed targets, cultural objects and scenes of the navigation areas with high quality maps and materials.
- Enhance the user perception by using a tiled seamless multi-display vision system.

Incorporating more realistic physics and increasing the accuracy of the motion prediction requires more complex equations, and it takes extra time to solve them. Modeling more realistic 3D objects requires more polygons, and it directly affects the rendering time. More realistic maps and materials cause a similar problem. Cylindrical seamless screens with edge blending and image stitching can enhance the physical realism. Ultimately, due to all these factors, it takes more time to render a particular scenario and to project the visual. This makes the entire simulation a non real-time simulation.

DEVELOPMENT TECHNIQUES

Identify Simulation Requirements

At the initial stage of the development of a simulator it is important to consider the following factors.

1. Study the existing real word scenario, and identify all possible activities.
2. Identify the activities to be simulated to enhance the quality of the existing teaching and learning process.
3. Identify the activities to be simulated to reduce the cost of the existing teaching and learning process.
4. Identify new activities which are not possible in the real world but possible in virtual world and helpful to reduce the cost or enhance the quality of the existing teaching and learning
5. Identify activities that cannot be simulated at all.

According to our requirements, we focus on the enhancement of the quality or the reduction of the cost of existing teaching learning process and sometimes the focus should be on both aspects. An extensive feasibility study should be carried out to identify all possible activities to be simulated in the VR solution. Above mentioned factors are illustrated in Figure 3.

Based on the results of the feasibility study it is possible to prioritize the activities to be simulated in the virtual reality environment. Prioritizing the activities to be simulated is essential before the design and development phases of the VR solution. Subsequently, the system architecture can be finalized during the iterative development process; all necessary activities will be simulated according to their priority level.

At the initial stage of the *Vidusayura* project, conventional maritime education process was investigated. In local context, maritime education consists of maritime class room lectures and ship maneuvering training with real ships. In a conventional classroom, a ship's motions can only

Figure 3. Activities to be identified in the initial stage

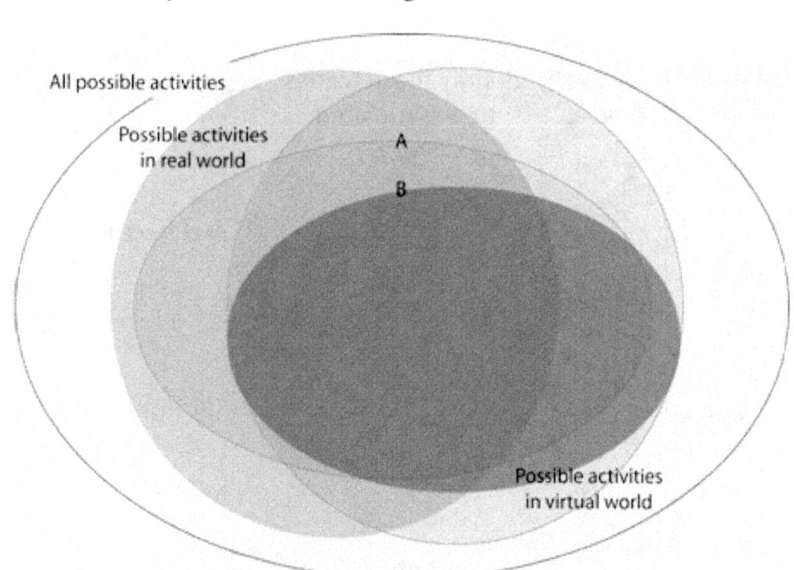

Table 1. Sample activates identified for simulation

Activities to be simulated	*Reduced the cost*	*Enhance the quality*	*Priority*
Simulation medium class ship	√	√	1.0
Simulate throttle			1.1
Simulate rudder			1.1
Simulate RPM meter			1.2
Simulate haze, fog		√	8.0
Simulate cultural objects		√	5.0

be explained one at time, but practically several motions can occur at the same time. Consequently, it is not easy to understand the phenomena. However, the ship simulation system can be used to demonstrate complex ship motions. Conventional ship maneuvering training involves hundreds of sea hours. However, the immersive ship simulation system can be utilized for training activities such as bridge procedures, regulations for Preventing Collisions at Sea, observing the basic principles of keeping a navigational watch and more. In conventional ship maneuvering training, it is not possible to create real scenarios such as terrorist attacks, rapid environmental changes, etc. Table 1 gives some of the training activities at SL navy identified for simulation.

At the end of the initial feasibility study, the simulation development life cycle was agreed upon, and the entire simulation development life cycle (SDLC) aligns with the classical software development life cycle as illustrated in Figure 4. The client (SL Navy) was also directly involved in each development phase. This is a unique feature that is not usually seen in traditional software development life cycle.

Once the activities to be simulated are identified the next step, according to the diagram, is to come up with a suitable architecture for the simulator. This study and the study of other available ship handling simulators lead to the develop-

Figure 4. Simulation Development Life Cycle

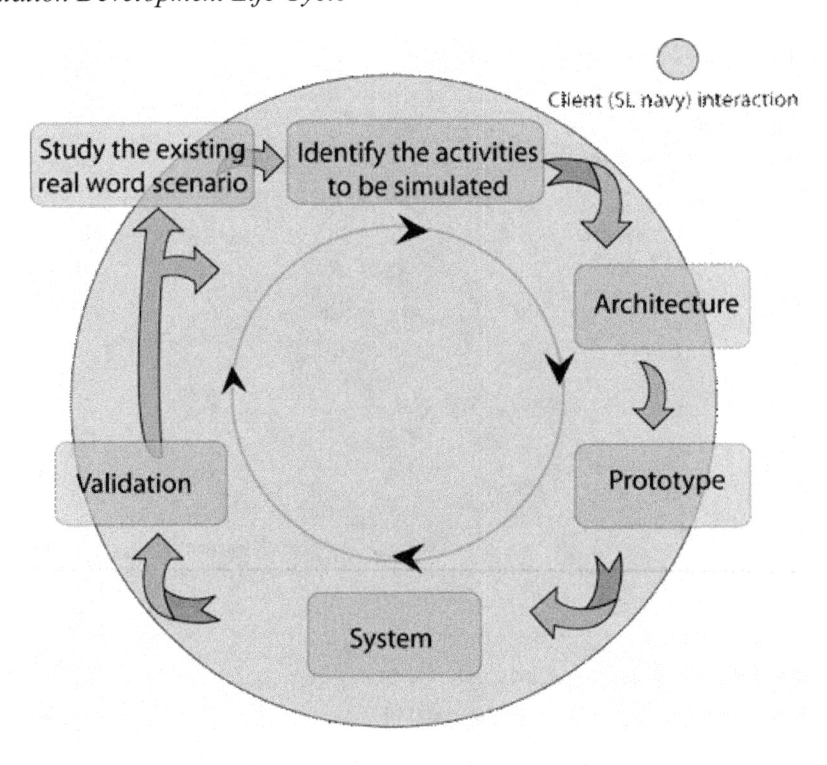

ment of a suitable architecture for the ship handling simulator.

The Architecture of the VR Solution

The immersive environment consists of mathematical computation modules, databases with the vehicle's physical and mechanical data/environment conditions, visual rendering engines and a seamless display system. The mathematical computation modules consider user interactions, vehicle's physical/ mechanical conditions and environmental conditions to predict real-time vehicle motions and the states of all other dynamic bodies in the virtual environment. The visual rendering engine considers the predicted states of all dynamic bodies to generate the relevant scenery. Finally, the user can see the generated visual through a seamless display system. The entire process is real-time. Simply, it responds in real-time to user interactions and other variations. The higher level structure of an immersive virtual environment is illustrated in Figure 5. This structure is complied with most commonly used vehicle simulators such as car, flight and ship simulators.

Vidusayura software architecture perspectives are based on distributed independent modules which are run in separate workstations. Distributing the computational work load among separate workstations is very effective and an important factor in this type of real-time simulations. However, the communication protocols for each and every module were clearly defined in the design phase. In Vidusayura, VR solution real-time ship motion predictions are done in separate workstations and visual rendering is done separately in multiple workstations as illustrated in Figure 6.

By using the same strategy, all modules were developed separately. Module programmers developed and tested their modules independently. They used simulated inputs for their modules for the development and testing. That was possible because the relevant protocols were clearly defined prior to the development process.

Physics-Based Modeling

The position of a particle in the real world at time t can be described as a vector x (t), which describes the translational motion of the particle with respect to its initial position. Ships, cars or flights are rigid bodies and they are more complicated. In addition to translational motions, there are rotational motions. Those rotational motions can be described as a vector R (t). To locate a rigid body such as a ship in the real world, both vectors, translational x (t) and rotational motions R (t), should be considered. A rigid body can form six degrees of freedom complex motions as shown in Figure 7.

Figure 5. The higher level structure of an immersive virtual environment

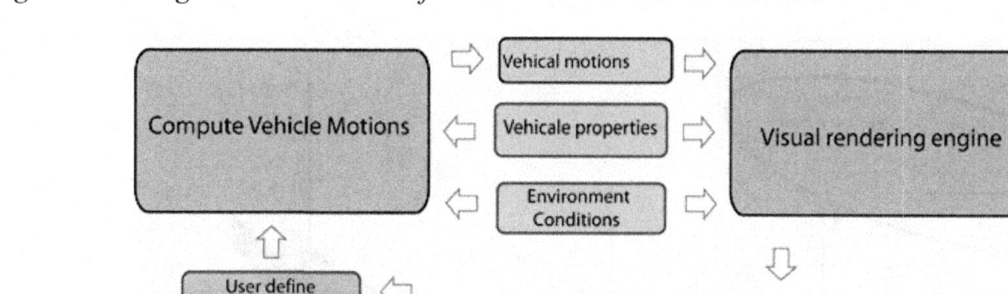

Figure 6. Distributed independent modules

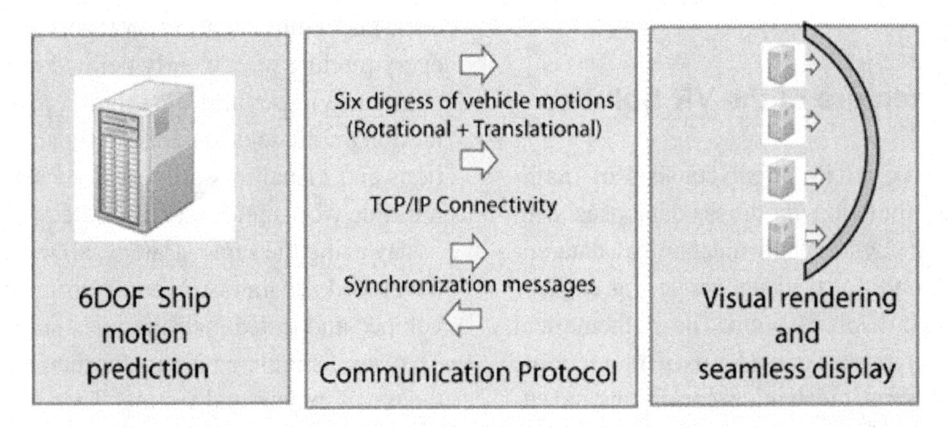

Figure 7. Six degrees of freedom rigid body motions

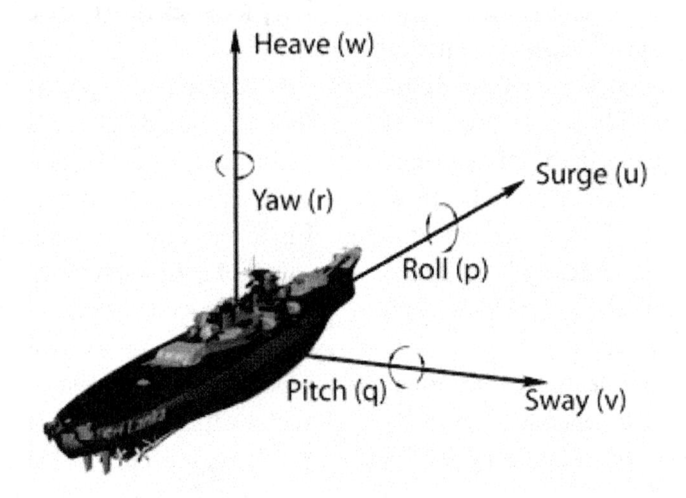

Figure 8. Reference frames and coordinate systems

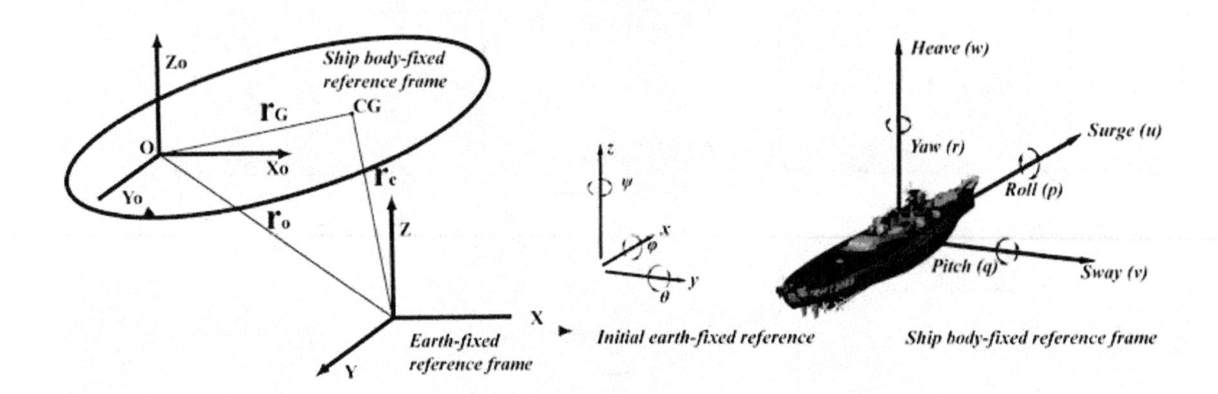

A rigid body occupies a volume of space and has a particular shape. The shape of a rigid body can be defined in terms of a fixed and unchanging space called body space. The motion prediction algorithm converts the given body space descriptions into world space. Two reference frames are defined as shown in Figure 8.

Figure 9 illustrates the initial earth-fixed reference frame with $[x\ y\ z]^T$ coordinate system and the ship body-fixed reference frame with $[x_o\ y_o\ z_o]^T$ coordinate system (Fossen, 1996). In order to simplify, the ship-body-fixed axes coincide with the principal axes of inertia. The origin of the ship body-fixed frame coincides with the center of gravity of the ship ($r_{G=}$ $[0\ 0\ 0]^T$) (Fossen, 1996).

The magnitudes of the position, orientation, forces, moments, linear velocities and angular velocities are respectively denoted by $[x\ y\ z]^T$, $[\psi\ \theta\ \varphi]^T$, $[X\ Y\ Z]^T$, $[K\ M\ N]^T$, $[u\ v\ w]^T$ and $[p\ q\ r]^T$ as shown by Table 2. The position-orientation vector in the XY plane is expressed as $\eta = [x\ y\ \psi]^T$ and the linear-angular velocity vector is expressed as $v = [u\ v\ r]^T$. The rate change of Position-orientation vector is expressed as $\dot{\eta} = R(\psi)v$ (Perez & Blanke, 2002) where,

$$R(\psi) = \begin{bmatrix} \text{Cos}\,\psi & -\text{Sin}\,\psi & 0 \\ \text{Sin}\,\psi & \text{Cos}\,\psi & 0 \\ 0 & 0 & 1 \end{bmatrix}. \quad (1)$$

More generic and simple equations of surge motion can be expressed as follows.

$$M\dot{u} = T_e - R_u u\left|u\right| \quad (2)$$

In Equation (2) M, R_u and T_e denote the vehicle's mass, resistance coefficient and single driving force (effective thrust) that is transmitted from the engine. The term $R_u u\left|u\right|$ can be regard-

Table 2. Six possible degrees of freedom ship motion (Society of Naval Architects and Marine Engineers, 1950)

Degrees of freedom	Forces Moments	Linear/ Angular Velocities	Positions Euler Angles
Surge	X	u	x
Sway	Y	v	y
Heave	Z	w	z
Yaw	N	r	ψ
Pitch	M	q	θ
Roll	K	p	φ

Figure 9. Vertically projected ship body and height fields variation of the ocean surface

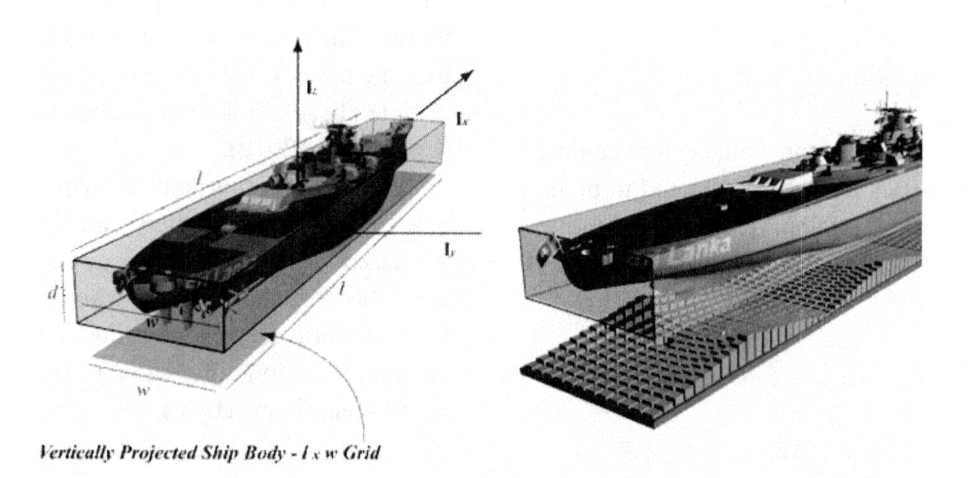

ed as resistance forces, which depend on the instantaneous dynamics of the vehicle. There are various types of equations to predict six degrees of freedom motions depending on the characteristic of the vehicle. There are motion prediction equations with large number of model parameters. Deriving equations for real-time simulation of a vehicle is challenging because there should be a way to evaluate these parameters. Consider the following examples.

Surge Motion of a Ship

The surge motion of a ship is described by simplification of the non-linear speed equation as follows (Fossen, 1996).

$$M\dot{u} = T_e - X_{u|u|}u\left|u\right| - (M + X_{vr})vr \quad (3)$$

In Equation (3), M and T_e respectively denote the ship's mass and single driving force (effective propeller thrust) that is transmitted from the ship engine by her propellers. The two terms $X_{u|u|}u\left|u\right|$ and $\left(M + X_{vr}\right)vr$ can be regarded as damping forces, which depend on the instantaneous dynamics of the vessel. $\left(M + X_{vr}\right)vr$ represents excess drag force due to combined yaw and sway motion. $X_{u|u|}u\left|u\right|$ corresponds to the quadratic resistance force at the forward speed u.

Forward Motion of a Car

If we have to model a car, a much more complex set of parameters has to be estimated to predict just the forward motion. The forward motion of a car can be described as follows (Popp & Schiehlen, 2010).

Vehicle body

$$m\ddot{x} = -mg\sin\propto - W_L - F_{xV} - F_{xH},$$

$$0 = mg\cos\propto - F_L - F_{zV} - F_{zH},$$

$$0 = -M_L + M_V + M_H - (h - r)\left(F_{xV} + F_{xH}\right) + l_V F_V - l_H F_{zH}.$$

Front axle

$$m_V\ddot{x} = -m_V g\sin\propto + F_{xV} + T_V,$$

$$0 = F_{zV} - N_V,$$

$$I_V Æ_V = M_V - rT_V - e_V N_V.$$

Rear axle

$$m_H\ddot{x} = -m_H g\sin\pm + F_{xH} + T_H,$$

$$0 = F_{zH} - N_H,$$

$$I_H Æ_H = M_H - rT_H - e_H N_H.$$

In the above equations, the mass of the vehicle body is denoted by m, the mass of both wheels at each axle is m_V, m_H, and I_V, I_H are the corresponding moments of inertia. α denotes angle of the car and the ground, and all other symbols represent the dynamic properties related to active forces. There are nine equations available for the nine unknowns. The dynamics of the car are completely determined by these equations (Popp & Schiehlen, 2010).

All of these motion prediction equations consist of several model parameters, and these equations are based on some assumptions and simplifications. However, all of these equations represent the real world scenario, but due to assumptions and simplifications, they cannot represent the real world scenario exactly as it is.

The parameters in the above equations should be evaluated by experiments or we have to assume reasonable values based on a scientific process. In serious games, accurate evaluation of the model parameters and simplification of the real-time motion prediction equations are very important because it directly affects the realistic behavior of the vehicle. Deriving equations for real-time simulation of a vehicle is challenging because there are two major issues:

1. There should be a proper scientific method to evaluate all model parameters with respect to the particular vehicle.
2. Accurate and realistic motion prediction equations require more computational power. Over simplification of real-time motion prediction equations leads to an inaccurate model.

In Vidusayura, we developed real-time algorithms to predict six degrees of freedom ship motions. There are several model parameters in surge, sway and yaw motion predictions. However, it is possible to evaluate these parameters.

Surge Motion

All model parameters related to surge the motion $M\dot{u} = T_e - X_{u|u|}u\left|u\right| - (M + X_{vr})vr$ can be evaluated. When the ship travels a straight course and at a constant speed u, $\left(M + X_{vr}\right)vr = 0$ and $T_e - X_{u|u|}u\left|u\right| = 0$. Then, if we know T_e T_e and u then it is possible to calculate the $X_{u|u|}$. The force of the propeller is related to its rate of revolutions n, the diameter D and the thrust coefficient K_t and it is given by $T_e = K_t n^2 D^4$ (K J Rawson). X_{vr} = 0.33 M (Perez & Blanke, 2002).

Yaw Motion

The yaw motion of the ship is described by using the first order Nomoto model, and the transfer function from the rudder angle δ r is equivalent to the time domain differential equation as follows (Fossen, 1996) (Tzeng & Chen, 1997, 1997).

$$\tau\ \dot{r} + r = K\delta \tag{4}$$

K is the steady-state gain of the system, and Ais the time constant. K and Acan be calculated by using Maneuvering Tests (Journée & Pinkster, 2002, 2002). This model behaves more accurately within the linear region of a ship's steering function up to rudder angles of approximately 35 degrees. (K J Rawson) (Fossen, 1996)

Sway Motion

Sway motion is calculated based on the transfer function model with constant parameters. It is equivalent to the time domain differential equation given below (Fossen, 1996) (Tzeng & Chen, 1997, 1997).

$$\tau_v \dot{v} + v = K_v(\tau\ \dot{r} + r) \tag{5}$$

K_v is the steady-state gain in sway, and $Ä_v$ is the time constant in sway. The time constant and the steady state constant have been assumed to be $K_v = 0.5L$,l- length of the ship and $Ä_v = 0.3$ Ä (Barauskis & Friis-Hansen, 2007) (Fossen, 1996).

Wave Model

We use the multivariable ocean wave model introduced by Ching-Tang (Chou & Fu, 2007) to model the sea surface and determine the height field of the sea surface. In that model,

$$h(x, y, t) = \sum_{i=1}^{n} A_i sink_i[(xcos,_i + ysin,_i) - \acute{E}_i t + \AE_i] \qquad (6)$$

The above function represents the water surface height on the Z axis direction. A is the wave amplitude, k is the wave number and this number is defined as $2\pi / \lambda$ where λ the wave length. ω is the pulsation which is defined as the $2\pi f$ by the frequency f. A, k, and f are time (t) dependent variables. θ is the angle between the X axis and the direction of the wave. φ is the initial phase which can be selected randomly between 0 - 2π.

Heave, Pitch, and Roll

The ship's position orientation vector in the XY-horizontal plane is calculated with respect to time by solving the surge, sway and yaw equations. That means that we know x, y, u, v and ψ in respect to time. Then according to Archimedes' principle, (Nakayama & Boucher, 1998) we assume the translational motion (heave) and rotational motions (pitch, and roll) are generated by the swellness of water under the ship. It can be calculated by using the height variations of the sea surface. We assume that the shape of the ship is cuboid as illustrated in Figure 8 and the ship body is vertically projected onto the sea surface to get the $l \times w$ bounding box (w is the width of the ship). It is divided in to $1m \times 1m$ cells for convenience as illustrated in Figure 9.

We evaluate the height fields at the center points of each $1m \times 1m$ cell, and we assume that the ship is not actually present when the height field is calculated. We can use Ching-Tang's wave model (Chou & Fu, 2007) to calculate the height fields. We assume that the projected bounding box and its points move with the ship. Then, at any given time we can calculate the height fields according to the ship's orientation and the wave propagation. We can obtain the forces and moments to generate the heave, pitch and roll motions by calculating the height fields for the overall bounding box, calculating the difference of height fields between the front and rear halves of the bounding box and calculating the difference of height fields between the port and starboard halves of the bounding box.

In this case, if we divide the $l \times w$ bounding box into $1cm \times 1cm$ cells instead of $1m \times 1m$ cells then it needs a huge computational power to predict the heave, pitch and roll motions. If we assume the real ship's shape, instead of a box, and $1mm \times 1mm$ cells then it takes several hours to predict heave, pitch and roll motions.

Implementation of the Real-time Motion Prediction Algorithms

There are three major techniques to implement motion prediction algorithms. Depending on the requirements, we can select the most appropriate technique.

1. Derive motion prediction algorithms and the other basic rigid body dynamics are implemented based on first principles of the physics.
2. Use free and open source physics dynamic engine and implement basic rigid body dynamics in the virtual environment and the derived motion prediction algorithms.
3. Combine the above two techniques.

In other words, a set of differential equations are solved in real-time to predict the ship motion. There are pre defined differential equation solvers such as Dormand-Prince, Runge-Kutta solver (Ashino, Nagase & Vaillancourt, 2000). These solvers can be used to implement motion prediction algorithms from the scratch.

There are free and open source physics dynamic engines (Game Physics Simulation, 2011) (NVIDIA Corporation, 2011) (Smith, 2011). They have different features and functionalities (Boeing & Bräunl, 2007) (PhysX, 2009). Most open source physics engines support basic rigid body dynamics (Boeing & Bräunl, 2007). They are usually optimized for specific applications. For example, you can try to simulate 6DOF ship motion with respect to wave conditions and ships physical and mechanical properties with available physics engines, but natively most physics engines do not support such scenarios. However, they support basic vehicle dynamics. The following table shows a comparison of seven open source physics engines.

The above details are based on a research article published in 2007 (Boeing & Bräunl, 2007) and physics engine APIs change with the time. The latest physics engines have the capability to utilize the features available in latest CPUs and GPCs. Those physics engines are more suitable for entertainment activities such as computer games (Boeing & Bräunl, 2007). Natively, they do not support six degrees of freedom ship motions in real-time with respect to ship's throttle, wheel (rudder) and environment conditions such as wave frequency and wave height.

In Vidusayura, depending on the requirements a third motion prediction algorithm implementation technique was used. Real-time motion prediction algorithms were implemented by using C++ and PhysX. Free and open source physics dynamic engine was incorporated into the OGRE3D visual rendering engine to simulate basic rigid body dynamics in the virtual environment. The detailed discussion of the real-time visual rendering is given in the following section.

During this process, current states or behavior of all dynamic bodies in the virtual environment are determined. The time taken for this process depends on the available computational capability, efficiency of differential equation solvers and complexity of the equations.

3D Modeling and Real-Time Visual Rendering

The quality of a real-time rendering is determined by realism of the modeled 3D objects and visually attractive real world effects such as shadows, reflections and shading. There are two major activities behind the real-time visual rendering.

1. Determine the current states or behavior of all dynamic bodies in the virtual environment
2. Render all bodies in the particular region of the virtual environment

The process of determining of current states or behavior of all dynamic bodies in the virtual environment was discussed in the previous section. Simplified versions of 3D meshes with low polygon count simplify the calculation of collisions with other meshes. The time taken for

Table 3. A comparison of selected open source physics engines

Physics Engine	Cost	Platform support Win32/Linux	Basic rigid body dynamic support	Generic (6D) support	Basic vehicle dynamic support	6DOF ship motion support*
PhysX	Free	Yes	Yes	Yes	Yes	No
Bullet	Free	Yes	Yes	Yes	Yes	No
JigLib	Free	Yes	Yes	Yes	Yes	No
Newton	Free	Yes	Yes	Yes	Yes	No
ODE	Free	Yes	Yes	No	No	No
Tokamak	Free	Yes	Yes	Yes	No	No
True Axis	Free	Yes	Yes	Yes	Yes	No

* Native support for 6DOF ship motion with respect to wave condition, ships physical and mechanical properties

the rendering process depends on the available computational capability (CPU/GPU) and the complexity of the 3D models.

There are free and open source rendering engines (OGRE 3D, 2010) (Crystal Space Team, 2011) that have different features and functionalities. However, most rendering engines have a particular target application to which they are optimized, and they support different constraints, e.g. VDrift (VDrift.net, 2010) rendering engine optimized to car simulations.

Initially, the SL navy required simulating ships in the deep sea, inner harbor and outer harbor. It was necessary to incorporate real world geographical sceneries with cultural objects, moving/fixed targets, several environmental conditions and a wide range of visibility and illumination effects such as daytime, dusk and night. Available rendering engines were evaluated based on the above requirements and the result is given in below Table 4.

After the evaluation of existing rendering engines, because of the plugins available for water rendering, Ogre 3D (Object-Oriented Graphics Rendering Engine) was selected to the rendering process. There are several physics engines which are compatible with Ogre 3D such as PhysX and ODE (NVIDIA Corporation, 2011) (Smith, 2011). Due to its flexibility and greater community support, PhysX (NVIDIA Corporation, 2011) was selected and integrated with Ogre 3D for basic rigid body dynamics in the virtual environment.

The polygon count of the scenery is very important. The complexity of the scenery depends on the number of polygons. In most of the virtual reality applications, the user navigates the virtual environment. Depending on the application, the user moves closer to the objects and observes information from the entire object. Similarly, the user can observe the object from several kilometers away, or the user observes only a section of the object. Based on this concept, all objects in the virtual environment can be classified. In ship simulation, ship moves within the sea area. Consequently, the objects in the shoreline and above are less important, but objects in the harbor such as other ships and the pier are very important. These objects should be modeled with more details. Depending on the importance for the navigation lessons, the polygonal count of the 3D objects can be reduced. There are three techniques to reduce the polygon count.

Table 4. Analysis of selected rendering engines (based on Vidusayura requirements)

	Rendering Engines				
Requirements	*3D Game Studio*	*Crystal Space*	*Ogre 3D*	*Reality Engine*	*Torque*
Cost	*Free/Commercial*	*free*	*free*	*free*	*Commercial*
Platform support Win32/Linux	*Windows only*	*yes*	*yes*	*yes*	*yes*
3d mesh support	*yes*	*yes*	*yes*	*yes*	*yes*
C++ support	*yes*	*yes*	*yes*	*yes*	*yes*
Collision detection	*yes*	*yes*	*No*	*yes*	*yes*
Render Outdoor environment	*yes*	*yes*	*yes*	*yes*	*yes*
*water rendering capability**	*No*	*No*	*yes*	*No*	*No*

* **Native support to render high quality pretty water scenes with** water depth effects, smooth transitions, foam effects, caustics, underwater god rays

1. Simplify the shape of the object and reduce the polygon count
2. Replace the entire object with a billboard or replace part of the object with an image and reduce the polygon count
3. Combine above two techniques and reduce the polygon count

Initially, the SL navy wanted to simulate their ships in local harbor. The SL navy provided navigational maps of the selected harbor area. We used Google earth (Google, 2011) images and navigational maps to finalize the shore line as shown in the Figure 10.

A sequence of digital pictures was taken from the sea while keeping the same distance from the shoreline. The same process was repeated several times to get different image sequences with different distances from the shore line. Various moving and fixed targets and cultural objects were observed in the sea around the selected harbor. The relative sizes of the observed objects were recorded according to a selected earth fixed object. Digital pictures were taken from different distances and textures of cultural objects were captured. Blender and 3D studio max (Autodesk, 2012) (Blender Foundation, 2010) were used to create 3D mesh models such as naval vessels, moving or fixed targets, cultural objects and scenes of navigation areas. During this process, the main focus was to model more realistic models with low polygons and entire shore line was modeled with billboards. Ogre Scene Exporter (OgreMax, 2010) was used to export the modeled 3D Scenes to Ogre 3D compatible format. The outcome was at a quite satisfactory level as shown in the Figure 11.

Enhance the User Perception and Ecological Validity

Seamless images and videos are regularly used in various virtual reality applications such as autonomous navigation and virtual walkthroughs (Liu, 2006) (Ferreira, 1999). *Alexandre et al.* (1999) presented a multiple-display visualization system that can greatly enhance user perception for maritime virtual reality applications. The common approach to provide multiple synchronized views uses a powerful centralized processing unit to support the rendering process on all screens. *Xiuwen Liu et al.* proposed the Multi-Projector Tiled Display System for Marine simulators (Liu, 2006). In their approach, they developed a low-cost multi projector seamless tiled display system using commodity hardware for the virtual reality based marine simulators. Based on the above approaches, we developed the distributed architecture that sup-

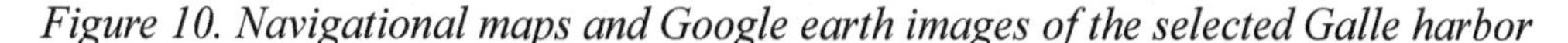

Figure 10. Navigational maps and Google earth images of the selected Galle harbor

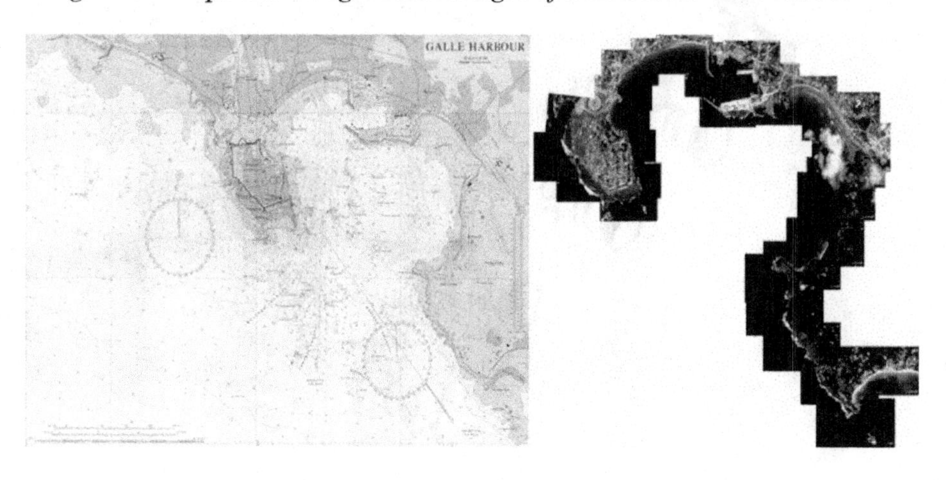

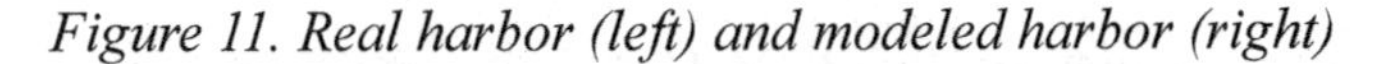

Figure 11. Real harbor (left) and modeled harbor (right)

ports a flexible and reliable visualization system while giving the users a sensation of immersion with low-end graphics workstations.

In Vidusayura, seamless multi display system is based on the client-server architecture. It supports real-time six degrees of freedom autonomous navigation system with 300^0 field of view. The server computer sends the navigational instructions (latitude, longitude, altitude, roll, pitch, and yaw) to the six client computers as shown in the Figure 12.

In each client computer, the same virtual environment is loaded, and they get the position and the orientation values from the parental node. Each virtual camera inherits its position and orientation from the parental node while maintaining 60 degrees with respect to the adjacent virtual cameras. To create a tiled 300° field of view (FOV), each virtual camera occupies 60^0 angle of view (Junker, 2006). Navigational instructions (latitude, longitude, altitude, roll, pitch, and yaw) are sent to the virtual cameras from the master computer over the network and multi projector seamless tiled display system was constructed using three multimedia projectors with 2500 *ANSI lumens.*

Figure 12. Structure of the vision system

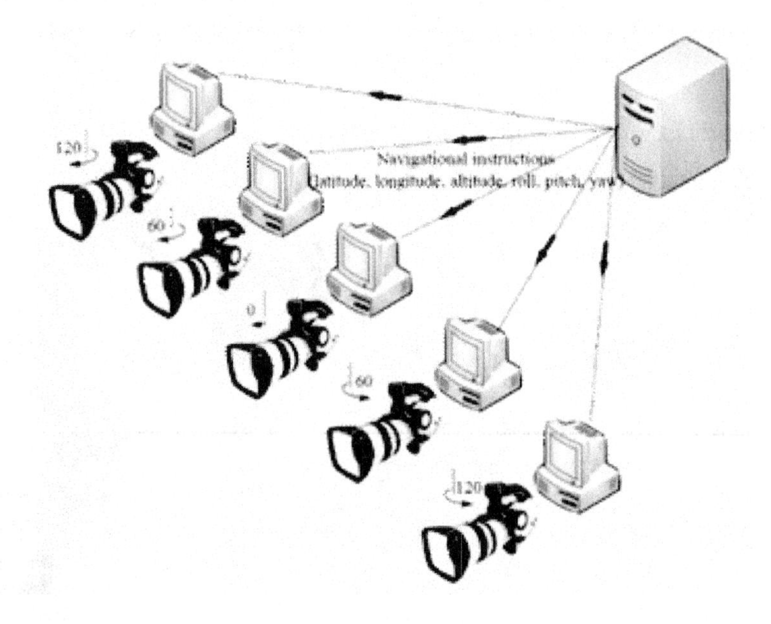

However, the major development challenge in user perception enhancement can be represented as follows. Consider simulation of real world scenery.

A. Qualitative and quantitative properties of the real world scenery.
B. Qualitative and quantitative properties of the scenery which are generated by the virtual cameras.
C. Qualitative and quantitative properties of the scenery projected to large screens by using the commodity multimedia projectors.

For more realistic immersive VR solution with a tiled display system, we have to satisfy the following condition:

Qualitative and quantitative properties of the real world scenery should be identical to the qualitative and quantitative properties of the projected scenery on to large screens. (A Ξ C)

To satisfy the **A Ξ C,** we can modify the virtual camera, program the GPU to generate visual transformations and use available functionalities in the multimedia projectors/display panels. This process requires activities such as geometry distortion adjustment and edge blending.

Ecological validity of the virtual environment is very important (Rizzo, et al., 2006). It helps to enhance the immersive feeling within the virtual environment. To enhance the ecological validity it is necessary to provide more realistic physical environment to the user which is quite identical to real world situations.

Vidusayuara software solution was interfaced with real ship bridge equipment. Discarded real equipment was integrated into the bridge. In this case, the real throttle and wheel were interfaced to provide a more realistic virtual environment and a strong sense of immersion. A large number of people including naval experts, sailors, navy officers, national and international researchers, game lovers and general public used this immersive environment as shown in Figure 1.

VALIDATION TECHNIQUES

The entire simulation system can be validated under different criteria. As we discussed before, real-time simulation is a representation of a real world scenario with assumptions and simplifications. Validating the simulated results against its own assumptions and simplifications is one validation technique. Consider a particular real world scenario which was identified at the initial stage as a possible activity to simulate in the virtual environment. Then, record all the state changes of the dynamic bodies in that particular scenario within a given time frame and simulate it in the virtual environment. Comparison of the real world scenario and the simulated scenario provide another kind of validation to the VR solution. User testing is the other technique to validate an immersive VR solution. In this technique, the VR solution is validated based on the user experience. Some of these validation techniques are quantitative while the others are qualitative. Quantitative validation methods can be used to validate activities such as accuracy of the real-time motion predictions and qualitative validation methods can be used to validate activities such as user perception enhancement and ecological validity of the entire VR solution.

The Vidusayura validation process was divided into three segments

1. Validate against non real-time and more accurate model: Ignore assumptions, limitations, and constraints of the algorithms and predict accurate ship motions without considering the time constraint and then predict the real-time ship motions with assumptions, limitations and constraints. Comparison of both results gives the deviation of the ship

motion predictions due to the assumptions and constraints.

2. Real world sea trials: Consider real world sea trial and simulate a similar sea trial with the VR solution. Comparison of the real sea trials and the simulation results gives overall deviation.
3. User tests: Compare the real world user perception and the user perception of the immersive VR solution with experienced real world users. This comparison can be used to identify accuracy of the real-time ship motion algorithms, defects of the seamless display system and perception enhanced techniques.

We focused mainly on the second and third validation techniques and obtained satisfactory outcomes (Damitha Sandaruwan, 2010). We used first technique to identify several deviations in our computational ship model. In our computational ship model, we assume that the shape of the ship is identical to a cuboid for mathematical convenience as illustrated in Figure 13.

We simulated above illustrated hull shapes (actual hull shape and simplified hull shape) under similar conditions by using accurate non real-time application and simulation results were compared. For example, we obtained deviated results for heave pitch and roll response amplitude operators.

Ship motion prediction algorithms were validated with respect to the benchmark sea trials of the "Esso Ossaka Tanker" (Auke Visser's International Esso Tankers site, 2011). Esso Osaka is a tanker with an approximately box type geometric shape. It is an oil tanker that has been used for extensive maneuvering studies in the open literature (International Towing Tank Conference, 2002). Main particulars for the Esso Osaka are Length (between perpendiculars) -325m, Beam -53m, draft -21.73m, Displacement- 4 319,400 tones. The simulated and the actual trajectories are given in Figure 14.

CONCLUSION

We discussed the challenges in developing a low cost virtual reality solution for serious games. The material presented in this chapter is mainly based on the experience that we gained in developing a ship-handling simulator, Vidusayura, for the Sri Lanka Navy. We used commodity off the shelf hardware and free and open source software to build that simulator. The hardware cost has come down over the years, and there is a wide selection of required open source software available to make this task feasible. Vidusayura was built using the commonly available desktop PCs, commodity projectors, and the open source rendering engine OGRE.

Figure 13. Real shape and simplified shape of the ship

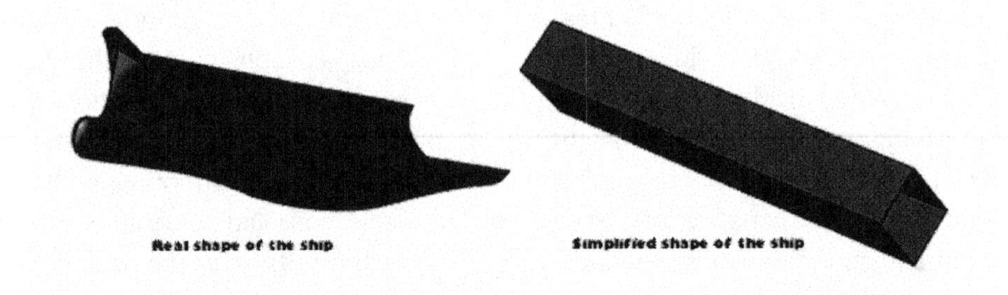

Figure 14. Validation against the "Esso Ossaka" sea trials

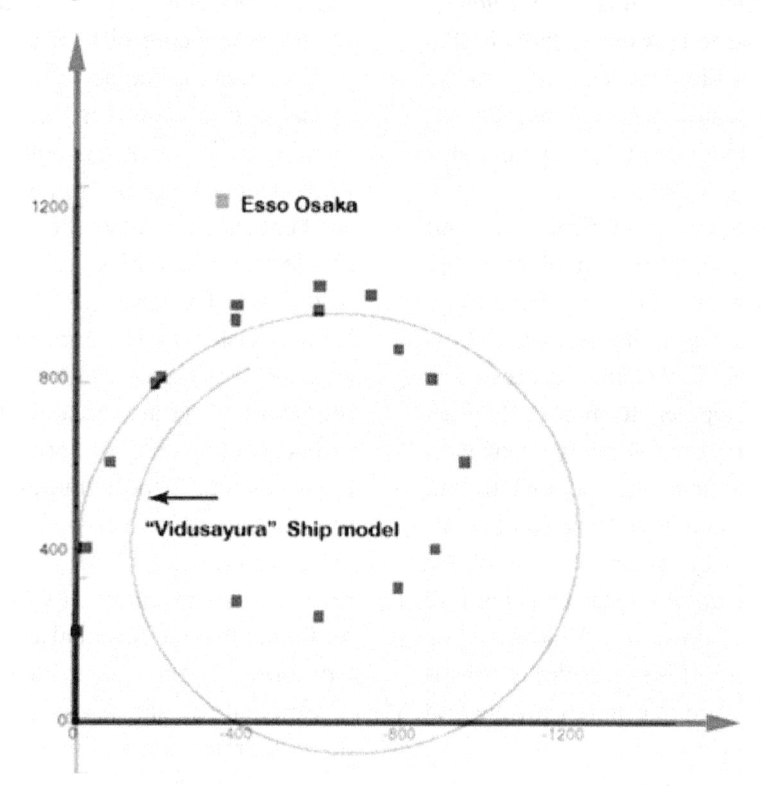

In addition to using the above mentioned hardware and software, we also got the client, Sri Lanka Navy, involved in the development cycle to bring the cost down. Timely client feedback simplified the development process, and the client also contributed by providing discarded equipments from their ships to build a bridge to improve the physical realism.

We noted that the techniques and the software and hardware that we used for Vidusayura can be

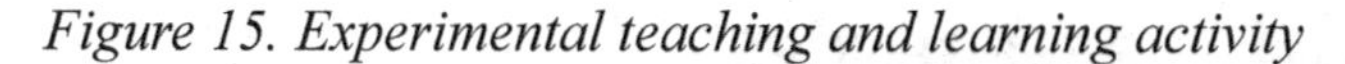
Figure 15. Experimental teaching and learning activity

used to build other vehicle simulators. We generalized our experience and presented them in this chapter, but we used Vidusayura as an example whenever such an example was required. We are now in the process of developing a car simulator based on the experience that we gained.

Vidusayura still requires more refinements and extensions. Since the client was involved in the development process from the very beginning, the client also has acquired the necessary skills required to continue the development process. The dynamics of the development team have changed with the Sri Lanka Navy's active involvement with the software development and model building process. The team is now in the process of refining and extending the behavioral realism of the Vidusayura. Various experimental teaching and learning activities are done with Vidusayura as shown in the Figure 15. It is successful in substituting some of real world maritime teaching and learning activities.

In Sri Lanka, the human resource cost for software development is moderate. The total development cost inclusive of human resource and equipment costs was less than 20000USD and total time duration was less than 24 month. However, a PC based simulation system may be acquired for $100,000 (USD) and full mission simulator with greater behavioural and physical realism may easily cost $2,000,000 (USD) (Robson, 2004). This is a huge cost reduction and due to the enhancement of technological capability of commodity-off-the-shelf (COTS) hardware, the gap between the required behavioural realism and feasible behavioural realism was gradually reduced.

FUTURE RESEARCH DIRECTIONS

Our main focus is the reuse of the same architecture and the technologies used in the ship simulator to develop a low cost ground vehicle simulator to impart safe driving techniques to all road users. It allows drivers to experience dangerous driving situations in a controlled learning environment.

Investigating the accuracy level of an entire virtual environment for expected teaching and learning activities is extremely important. This leads to identifying the minimum level of physical and behavioural realism for the expected teaching and learning activities and enhance the quality of training. Extensive research was carried out for user task analysis, user interface evaluations and user experience evaluations in virtual environments. There are research studies which lead to determining the user presence in the virtual environment. All these studies consider a specific component and have carried out research to solve varied research problems. After considering past research, our research was focused to ensure a minimum level of physical and behavioural realism for training and assessment objectives.

Another research focus is to enhance the Vidusayura real-time motion predictions to predict complex ship motions such as motion prediction in narrow water channel. Narrow channel effects should be simulated accurately along with the six degree of freedom ship motions. This will enable effective real-time navigation training using the ship simulator in navigations within a harbor or passing another ship.

REFERENCES

Ashino, R., Nagase, M., & Vaillancourt, R. (2000). Behind and beyond the matlab ode. Retrieved from http://citeseerx.ist.psu.edu/viewdoc/download?doi=10.1.1.99.5871&rep=rep1&type=pdf

Auke Visser's International Esso Tankers site. (2011). *Esso Osaka - (1973-1985)*. Retrieved from http://www.aukevisser.nl/inter/id427.htm

Autodesk. (2012). The autodesk 3D studio max documentations. Retrieved from http://usa.autodesk.com/

Barauskis, G., & Friis-Hansen, P. (2007). Description of inputs for the fast-time numerical navigator. *Safety at Sea.*

Bélanger, J., & Paquin, J.-N. (2010). The what, where and why of real-time simulation. *PES IEEE, general meeting.* Minneapolis, MN.

Blender Foundation. (2010) Open source 3D content creation suite. Retrieved from http://www.blender.org/

Boeing, A., & Bräunl, T. (2007). *Evaluation of real-time physics simulation systems. GRAPHITE 2007* (pp. 281–288). Perth, Australia: ACM Press.

Chou, C. T., & Fu, L. C. (2007). Ships on real-time rendering dynamic ocean applied in 6-dof platform motion simulator. *CACS International Conference.* Taichun, Taiwan.

Corti, K. (2006). Games based learning: A serious business application. *PIXELearning Limited.*

Crystal Space Team. (2011). *Crystal Space Real-time 3D Graphics.* Retrieved from http://www.crystalspace3d.org/main/Main_Page

Ferreira, A. G. Cerqueira, R., Filho, W.C., & Gattass, M. (1999). Multiple display viewing architecture for virtual environments over heterogeneous networks. *IBGRAPI '99 Proceedings of the XII Brazilian Symposium on Computer Graphics and Image Processing* (pp. 83 - 92). Washington, DC.: IEEE Computer Society

FlightSafety International Inc. (2010). *Flight-Safety Simulation.* Retrieved from http://www.flightsafety.com/fs_service_simulation_systems.php

Fossen, T. I. (1996). *Guidance and control of ocean vehicles.* Chichester, UK: John Wiley.

Game Physics Simulation. (2011). *Bullet Open Source Physics Library.* Retrieved from http://bulletphysics.org/wordpress/

Google. (2011). Retrieved from Google Earth: http://www.google.com/earth/index.html

International Towing Tank Conference (2002). Esso osaka specialist committee - Final report and recommendations to the 23rd ittc. *Proceedings of 23rd ITTC.* Venice, Italy.

Journée, J. M. J., & Pinkster, J. J. (2002). *Introduction in ship hydromechanics.* The Netherlands: Delft University of Technology.

Junker, G. (2006). *Pro OGRE 3D programming.* New York, NY: Apress.

Liu, C. X. (2006). Construct low-cost multi-projector tiled display system for marine simulator. *16th International Conference on Artificial Reality and Telexistence (ICAT'06)* (pp. 688-693). Hangzhou, China.

Nakayama, Y., & Boucher, R. F. (1998). *Introduction to Fluid Mechanics.* Oxford, UK: Butterworth-Heinemann.

NVIDIA Corporation. (2011). *NVIDIA physx.* Retrieved from http://developer.nvidia.com/technologies/physx

Oceanic Consulting Corporation. (2008). *Centre for Marine Simulation.* Retrieved from http://www.oceaniccorp.com/FacilityDetails.asp?id=7

OGRE 3D (2010). *Object-oriented graphics rendering engine.* Torus Knot Software Ltd. Retrieved from http://www.ogre3d.org/

OgreMax. (2010). The ogremax documentations. Retrieved from http://www.ogremax.com/

Papini, I. (2010). *Virtual sailor.* Retrieved from http://www.hangsim.com/vs/

Perez, T., & Blanke, M. (2002). *Mathematical Ship Modeling for Control Applications. Technical Report.* Australia: The University of Newcastle.

PhysX (2009). Popular physics engines comparison physx, havok and ode. Retrieved from http://physxinfo.com/articles/?page_id=154

Popp, K., & Schiehlen, W. (2010). *Ground vehicle dynamics*. Berlin, Germany: Springer-Verlag.

Rawson, K. J., & Tupper, E. C. (2001). *Basic ship theory* (*Vol. 1*). Oxford, UK: Butterworth-Heinemann.

Rizzo, A.A., Bowerly,T., Buckwalter, G., Klimchuck, D., Mitura, R., & Parsons, T.D. (2006). A virtual reality scenario for all seasons the virtual classroom. *CME 3*.

Robson, C. S. (2004). Review of Business.

Sandaruwan, D. (2010). A six degrees of freedom ship simulation system for maritime education. *The International Journal on Advances in ICT for Emerging Regions*, *3*, 34–47.

Simulator Systems International (SSI). (2010). *Vehicle simulation*. Retrieved from http://simulatorsystems.com/vehiclesimulators/

Smith, R. (2011). Open dynamics engine. Retrieved from http://www.ode.org/

Society of Naval Architects and Marine Engineers. (1950). *Nomenclature for treating the motion of a submerged body through a fluid*. Jersey City, NJ: The Society of Naval Architects and Marine Engineers.

Susi, T., Johannesson, M., & Backlund, P. (2007). *Serious Games - An Overview. Technical Report HS- IKI -TR-07-001*. Sweden: University of Skövde.

Transas Marine Limited. (2010). *Transas simulation products*. Transas Marine International. Retrieved from http://www.transas.com/products/simulators/

Tzeng, C.-Y., & Chen, J.-F. (1997). Fundamental properties of linear ship steering dynamic models. *Journal of Marine Science and Technology*, *7*(2), 79–88.

VDrift.net. (2010). Retrieved from http://vdrift.net/

VSTEP. (2010). *Ship simulator extremes*. Retrieved from http://www.shipsim.com/

Wikipedia. (2011). *Real-time computing*. Retrieved from http://en.wikipedia.org/wiki/Real-time_computing

Yin, Y., Sun, X., Zhang, X., Liu, X., Ren, H., Zhang, X., & Jin, Y. (2010). Application of virtual reality in marine search and rescue simulator. *The International Journal of Virtual Reality*, *9*(3), 19–26.

ADDITIONAL READING

Aldrich, C. (2009). *The Complete Guide to Simulations and Serious Games*. San Francisco, CA: Pfeiffer- John Wiley & Sons.

Bourg, D. M. (2002). *Physics for game developers*. Sebastopol, CA: O'Reilly Media, Inc.

Millington, I. (2007). *Game physics engine development*. Academic Press.

KEY TERMS AND DEFINITIONS

Development Challenges: Challenges to be faced in the development stage of the Virtual reality application.

Immersive VR Solution: Virtual reality application which provide closer experience to the real world scenario.

Low Cost VR Solution: Virtual reality applications which are developed by using Commodity-Off-The-Shelf (COTS) hardware, Free and Open Source Software (FOSS).

Motion prediction: Predict the motion of vehicles such as ship, car and flight in a virtual reality application.

Real-Time Vehicle Motion: Motions of a simulated vehicle which responses in real-time to driver interactions.

Serious Games: Games which are used for more purposes than providing entertainment.

Simulation Requirements: I identified scenarios to be simulated in the Virtual reality application.

Compilation of References

Abstract and Sealed Classes and Class Members. (2011) In *MSDN Library*. Retrieved from http://msdn.microsoft.com/en-us/library/ms173150%28v=VS.100%29.aspx

Adams, E., & Rollings, A. (2006). *Fundamentals of game design*. Upper Saddle River, NJ: Prentice Hall.

Albrecht, T. (2009a). The latency elephant. *Seven Degrees of Freedom*. Retrieved from http://seven-degrees-of-freedom.blogspot.com/2009/10/latency-elephant.html

Albrecht, T. (2009b). Pitfalls of Object Oriented Programming. *Seven Degrees of Freedom*. Retrieved from http://seven- degrees-of-freedom.blogspot.com/2009/12/pitfalls-of-object-oriented-programming.html

Andrade, G. D., Santana, H. P., Furtado, A. W. B., Leitão, A. R. G. A., & Ramalho, G. L. (2004) Online adaptation of computer games agents: A reinforcement learning approach. *1st Brazilian Symposium on Computer Games and Digital Entertainment (SBGames2004)*.

Anitescu, M., & Potra, F. A. (1996). Formulating dynamic multi-rigid-body contact problems with friction as solvable linear complementarity problems. *Computer*, *93*, 1–21.

Arkin, R. (1998). *Behavior-based robotics*. Cambridge, MA: MIT Press.

Arrow, K. J. (1951). *Social Choice and Individual Values*. New York, NY: John Wiley & Sons.

Ashino, R., Nagase, M., & Vaillancourt, R. (2000). Behind and beyond the matlab ode. Retrieved from http://citeseerx.ist.psu.edu/viewdoc/download?doi=10.1.1.99.5871&rep=rep1&type=pdf

Assiotis, M., & Tzanov, V. (2006). A distributed architecture for MMORPG. *NetGames '06: Proceedings of the 5th ACM SIGCOMM workshop on Network and system support for games* (pp. 4). Singapore: ACM.

Auke Visser's International Esso Tankers site. (2011). *Esso Osaka - (1973-1985)*. Retrieved from http://www.aukevisser.nl/inter/id427.htm

Autodesk. (2012). The autodesk 3D studio max documentations. Retrieved from http://usa.autodesk.com/

Auto-Implemented Properties. (2011) In *MSDN Library*. Retrieved August 10, 2011 from, http://msdn.microsoft.com/en-us/library/bb384054.aspx

Bäck, T. (1996). *Evolutionary algorithms in theory and practice*. New York, NY: Oxford University Press.

Baekkelund, C. (2006). A new look at learning and games. In Rabin, S. (Ed.), *AI Game Programming Wisdom 3* (pp. 687–692). Hingham, MA: Charles River Media.

Baker, R. S. J. D., Corbett, A. T., et al. (2006). *Adapting to When Students Game an Intelligent Tutoring System*. Proceedings from the *8th International Conference on Intelligent Tutoring Systems*. Jhongli, Taiwan

Bakkes, S., Spronck, P., & Postma, E. O. (2004). Team: The team-oriented evolutionary adaptability mechanism. In Rauterberg, M. (Ed.), *ICEC, ser* (*Vol. 3166*, pp. 273–282). Lecture Notes in Computer Science New York, NY: Springer.

Balch, T., & Hybinette, M. (2000). Social potentials for scalable multi-robot formations. In *Proceedings of the IEEE International Conference on Robotics and Automation* (pp. 73-80). Piscataway, NJ: IEEE.

Balch, T., & Arkin, R. C. (1998). Behavior-based formation control for multi-robot teams. *IEEE Transactions on Robotics and Automation, 14*(6), 926–939. doi:10.1109/70.736776

Baraff, D. (1994). Fast contact force computation for non-penetrating rigid bodies. *Proceedings of the 21st annual conference on Computer graphics and interactive techniques - SIGGRAPH '94*, 23-34. New York, NY: ACM Press. doi: 10.1145/192161.192168

Baraff, D. (1999). Physically based modeling course notes. *ACM SIGGRAPH*, (2-3).

Baraff, D. (1989). Analytical methods for dynamic simulation of non-penetrating rigid bodies. *Computer, 23*(3), 223–232.

Barauskis, G., & Friis-Hansen, P. (2007). Description of inputs for the fast-time numerical navigator. *Safety at Sea.*

Bartholdi, J. J. III, Tovey, C. A., & Trick, M. A. (1989). The computational difficulty of manipulating an election. *Social Choice and Welfare, 6*, 227–241. doi:10.1007/BF00295861

Beigbeder, T., Coughlan, R., Lusher, C., Plunkett, J., Agu, E., & Claypool, M. (2004). The effects of loss and latency on user performance in unreal tournament. *NetGames '03: Proceedings of the 2nd ACM SIGCOMM Workshop on Network and System Support for Games*. Portland, OR.

Bélanger, J., & Paquin, J.-N. (2010). The what, where and why of real-time simulation. *PES IEEE, general meeting.* Minneapolis, MN.

Bergen, G. (1999). A fast and robust gjk implementation for collision detection of convex objects. Retrieved from http://www.win.tue.nl/~gino/solid/jgt98convex.pdf

Bergen, G. (2001). Proximity queries and penetration depth computation on 3d game objects. Presented at *Game Developers Conference, 2001*. San Jose, CA. Retrieved from http://www.win.tue.nl/~gino/solid/gdc2001depth.pdf

Bergen, G. (2004). *Collision detection in interactive 3d environments*. San Francisco, CA: Morgan Kaufmann Publishers.

Bethesda Game Studios. (2008). *Fallout 3* [PC game]. Rockville, Maryland: Bethesda Softworks.

Bethesda Softworks. (2006). *The elder scrolls IV: Oblivion* [PC game]. Novato, CA: 2K Games.

Bezerra, C. E., Cecin, F. R., & Geyer, C. F. R. (2008). A3: A novel interest management algorithm for distributed simulations of MMOGs. Proceedings from the *2008 12th IEEE/ACM International Symposium on Distributed Simulation and Real-Time Applications.* Vancouver, Canada.

Bezerra, C. E., & Geyer, C. F. (2009). A load balancing scheme for massively multiplayer online games. *Multimedia Tools and Applications, 45*(1-3), 263–289. doi:10.1007/s11042-009-0302-z

Bharambe, A., Pang, J., & Seshan, S. (2006). Colyseus: A distributed architecture for online multiplayer games. *NSDI'06: Proceedings of the 3rd Conference on 3rd Symposium on Networked Systems Design & Implementation* (pp. 12-12). San Jose, CA: USENIX Association.

BigWorld Technology. (2011). Retrieved from http://www.bigworldtech.com

BinSubaih,, A., & Maddock, S. C. (2008). Game portability using a service-oriented approach. *International Journal of Computer Games Technology, 7*. doi:doi:10.1155/2008/378485

Blender Foundation. (2010) Open source 3D content creation suite. Retrieved from http://www.blender.org/

Blizzard Entertainment. (1998). *StarCraft* [PC game]. Irvine, CA: Blizzard Entertainment.

Blizzard Entertainment. (2009). StarCraft Brood War. Version 1.16.1.

Blizzard Entertainment. (2011). Blizzard Entertainment. Retrieved from http://us.blizzard.com/en-us/.

Boeing, A., & Bräunl, T. (2007). *Evaluation of real-time physics simulation systems. GRAPHITE 2007* (pp. 281–288). Perth, Australia: ACM Press.

Bolstad, W. M. (2007). *Introduction to Bayesian Statistics*. Wiley-Interscience. doi:10.1002/9780470181188

Booth, T. L. (1967). *Sequential machines and automata theory*. New York, NY: John Wiley & Sons, Inc.

Border, K. C., & Jordan, J. S. (1983). Straightforward elections, unanimity and phantom voters. *The Review of Economic Studies, 50*(1), 153–170. doi:10.2307/2296962

Borkar, S., & Chien, A. A. (2011). The future of microprocessors. *Communications of the ACM, 54*(5), 67–77. doi:10.1145/1941487.1941507

Bourke, P. (1998). *Determining whether or not a polygon (2D) has its vertices ordered clockwise or counterclockwise*. Retrieved from http://paulbourke.net/geometry/clockwise/index.html

Boutros, D. (2006). A detailed cross-examination of yesterday and today's best selling platform games. *Gamasutra*. Retrieved from http:///www.gamasutra.com/view/feature/1851/a_detailed_crossexamination_of_.php

Bungie (2001). *Halo: Combat Evolved* [Xbox game]. Redmond, WA: Microsoft Game Studios.

Byl, P. B. (2004). *Programming believable characters for computer games. Game Development* (1st ed.). Hingham, MA: Charles River Media.

Cai, W., Xavier, P., Turner, S. J., & Lee, B.-S. (2002). A scalable architecture for supporting interactive games on the internet. *Proceedings of the 16th Workshop on Parallel and Distributed Simulation* (pp. 60-67). Washington, D.C.

Cameron, S. (1997). Enhancing GJK: Computing minimum and penetration distances between convex polyhedra. Proceedings from *IEEE International Conference on Robotics and Automation.* Albuquerque, NM.

Cao, A., Chintamani, K. K., Pandya, A. K., & Ellis, R. D. (2009). NASA TLX: Software for assessing subjective mental workload. *Behavior Research Methods, 41*(1), 113–117. doi:10.3758/BRM.41.1.113

Catto, E., & Park, M. (2005). Iterative Dynamics with Temporal Coherence, 1-24. Retrieved from http://erwincoumans.com/ftp/pub/test/physics/papers/IterativeDynamics.pdf

Champandard, A. J. (2007). Behavior trees for next-gen game ai. *Game Developers Conference 2007*. Retrieved from http://aigamedev.com/insiders/presentation/behavior-trees/

Champandard, A. J. (2011). This year in game ai: Analysis, trends from 2010 and predictions for 2011. *AiGameDev.com*. Retrieved from http://aigamedev.com/open/editorial/2010-retrospective/

Champandard, A. J. (2008). Getting started with decision making and control systems. In Rabin, S. (Ed.), *AI Game Programming Wisdom 4* (pp. 257–264). Boston, MA: Course Technology.

Chanel, G., & Rebetez, C. Bétrancourt. M., & Pun, P. (2008). *Boredom, engagement and anxiety as indicators for adaptation to difficulty in games.* New York, NY: ACM Press

Chen, J. (2006). *Flow in games*. Unpublished Master's Thesis. School of Cinematic Arts Los Angeles, University of Southern California, California.

Cho, B. H., Jung, S. H., Seong, Y. R., & Oh, H. R. (2006). Exploiting intelligence in fighting action games using neural networks. *IEICE – Transactions on Information Systems. E (Norwalk, Conn.), 89-D*(3), 1249–1256.

Chou, C. T., & Fu, L. C. (2007). Ships on real-time rendering dynamic ocean applied in 6-dof platform motion simulator. *CACS International Conference.* Taichun, Taiwan.

Cohen, T. (2010). A Dynamic Component Architecture for High Performance Gameplay. *Game Developers Conference Canada*. Retrieved from http://www.insomniacgames.com/a-dynamic-component-architecture-for-high-performance-gameplay/

Compton, K., & Mateas, M. (2006). Procedural level design for platform games. Paper presented at the *Second Artificial Intelligence and Interactive Digital Entertainment International Conference (AIIDE).* Marina del Rey, CA.

Conitzer, V., & Sandholm, T. (2003). *Universal voting protocol tweaks to make manipulation hard.* Paper presented at the 18th International Joint Conference on Artificial Intelligence, Acapulco, Mexico.

Cook, D. (2007). Celestial music. *Lost Garden*. Retrieved from http://www.lostgarden.com/2007/09/celestial-music.html

Corti, K. (2006). Games based learning: A serious business application. *PIXELearning Limited.*

Cottle, R. W., Pang, J. S., & Stone, R. E. (1992). *The linear complimentarity problem*. New York, NY: Academic Press.

Coulom, R. (2008). *Whole-history rating: A bayesian rating system for players of time-varying strength. Conference on Computers and Games*, Beijing, China

Cranor, L. F. (1996). Declared-strategy voting: An instrument for group decision-making. Doctoral Dissertation. Washington University, St. Louis.

Cranor, L. F., & Cytron, R. K. (1996).*Towards an information-neutral voting scheme that does not leave too much to chance.* Paper presented at the Midwest Political Science Association Annual Meeting, Chicago, Illinois.

Crash, C. (2011). Retrieved from http://www.cometcrash.com/

Crystal Space Team. (2011). *Crystal Space Real-time 3D Graphics*. Retrieved from http://www.crystalspace3d.org/main/Main_Page

Csikszentmihalyi, M. (1990). *Flow: The psychology of optimal experience*. New York, NY: Harper Perennial.

Dangauthier, P., Herbrich, R., Minka, R., & Graepel, T. (2007). Trueskill through time: Revisinting the history of chess. [Cambridge, MA: MIT Press.]. *Advances in Neural Information Processing Systems, 20*, 337–344.

Dawson, C. (2002). Formations. In Rabin, S. (Ed.), *AI game programming wisdom*. Hingham, MA: Charles River Media.

de Vleeschauwer, B., van den Bossche, B., Verdickt, T., de Turck, F., Dhoedt, B., & Demeester, P. (2005). Dynamic microcell assignment for massively multiplayer online gaming. Paper presented at the *Proceedings of 4th ACM SIGCOMM Workshop on Network and System Support for Games.* New York, NY.

Defenders, C. (2008). Retrieved from http://www.crystaldefenders.jp/na/index.html

Deitel, P. J., & Deitel, H. M. (2008). *Internet & World Wide Web How to program* (4th ed.). Upper Saddle River, NJ: Prentice-Hall.

Deneubourg, J. L., Pasteels, J. M., & Verhaeghe, J. C. (1983). Probabilistic behaviour in ants: A strategy of errors? *Journal of Theoretical Biology, 105*, 259–271. doi:10.1016/S0022-5193(83)80007-1

Derenick, J. C., & Spletzer, J. R. (2007). Convex optimization strategies for coordinating large-scale robot formations. *IEEE Transactions on Robotics, 23*(6), 1252–1259. doi:10.1109/TRO.2007.909833

Doherty, D., & O'Riordan, C. (2008). Effects of communication on the evolution of squad behaviours. In Darken, C., & Mateas, M. (Eds.), *AIIDE*. Paso Alto, CA: The AAAI Press.

Dongarra, J. (2009). Free Matrix Library List. Retrieved from http://www.netlib.org/utk/people/JackDongarra/la-sw.html

Eberly, D. (2008). Intersection of Convex Objects: The Method of Separating Axes. Retrieved from www.geometrictools.com/Documentation/MethodOfSeparatingAxes.pdf

Eberly, D. (2008a). Boolean operations on intervals and axis-aligned rectangles. Retrieved from www.geometrictools.com/Documentation/BooleanIntervalRectangle.pdf

Eberly, D. (2008b). Dynamic Collision Detecting using Oriented Bounding Boxes. Retrieved from www.geometrictools.com/Documentation/DynamicCollisionDetection.pdf

Eberly, D. H. (2004). *Game Physics*. San Francisco, CA: Morgan Kaufmann.

Elkind, E., & Lipmaa, H. (2005). *Hybrid voting protocols and hardness of manipulation.* Paper presented at the 16th Annual International Symposium on Algorithms and Computation (ISAAC), Sanya, Hainan, China.

Ellis, S. (2008). The future is interactive, not online. Retrieved June 18, 2010, from http://thenewmarketing.com/blogs/steve_ellis/archive/2008/03/17/5467.aspx.

Emergent Game Technologies. (2007). Retrieved from http://www.emergent.net

Ericson, C. (2004, August). The gilbert-johnson-keerthi (gjk) algorithm. Presented at *SIGGRAPH 2004*, Los Angeles, CA. Sony Computer Entertainment America. Retrieved from http://realtimecollisiondetection.net/pubs/

Ericson, C. (2005). *Real-time collision detection: Christer Ericson.* Amsterdam [etc.: Morgan Kaufmann.
Jovanoski, D. (2008). Bachelor seminar: The Gilbert -Johnson -Keerthi (GJK) algorithm. Retrieved from reference.kfupm.edu.sa/content/b/a/bachelor_seminar__the_80305.pdf

Erleben, K. (2004). Stable, robust, and versatile multibody dynamics animation. Unpublished doctoral dissertation. University of Copenhagen, Denmark. Retrieved from http://www2.imm.dtu.dk/visiondag/VD05/graphical/slides/kenny.pdf

Esparcia-Alcázar, A. I., Martínez-García, A. I., Mora, A. M., Merelo, J. J., & García-Sánchez, P. (2010). Controlling bots in a first person shooter game using genetic algorithms. In *IEEE Congress on Evolutionary Computation* (pp.1-8). Washington, DC: IEEE Press.

Fan, J., Ries, E., & Tenitchi, C. (1996). *Black art of java game programming.* Waite Group Press.

Ferreira, A. G. Cerqueira, R., Filho, W.C., & Gattass, M. (1999). Multiple display viewing architecture for virtual environments over heterogeneous networks. *IBGRAPI '99 Proceedings of the XII Brazilian Symposium on Computer Graphics and Image Processing* (pp. 83 - 92). Washington, DC.: IEEE Computer Society

Firaxis (2010). *Civilization V* [PC game]. Novato, CA: 2K Games.

FlightSafety International Inc. (2010). *FlightSafety Simulation.* Retrieved from http://www.flightsafety.com/fs_service_simulation_systems.php

Fossen, T. I. (1996). *Guidance and control of ocean vehicles.* Chichester, UK: John Wiley.

Fredslund, J., & Mataric, M. J. (2002). A general algorithm for robot formations using local sensing and minimal communication. *IEEE Transactions on Robotics and Automation, 18*(5), 837–846. doi:10.1109/TRA.2002.803458

Fu, D., & Houlette, R. (2004). The ultimate guide to fsms in games. In Rabin, S. (Ed.), *AI Game Programming Wisdom 2* (pp. 283–302). Hingham, MA: Charles River Media.

Funge, J. (2004). *Artificial intelligence for computer games.* Wellesley, MA: A K Peters.

Game Physics Simulation. (2011). *Bullet Open Source Physics Library.* Retrieved from http://bulletphysics.org/wordpress/

Game, A. I. Conference (2010). *Game/AI Conference.* Retrieved from http://gameaiconf.com/

Games, E. A. (2010). Battlefield: Bad Company 2. Retrieved October 05, 2011, from http://www.battlefield-badcompany2.com.

Gee, J. P. (2007). *Good video games + good learning: Collected essays on video games, learning, and literacy.* New York, NY: Peter Lang.

Gibbard, A. (1973). Manipulation of voting schemes: A general result. *Econometrica: Journal of the Econometric Society, 41*(3), 587–601. doi:10.2307/1914083

Gilleade, K. M., & Dix, A. (2004). Using frustration in the design of adaptive videogames. Proceedings from the 2004 *ACM SIGCHI International Conference on Advances in computer entertainment technology* (pp. 228-232). Singapore. Association for Computing Machinery.

Goetschalckx, R., Missura, O., Hoey, J., & Gartner, T. (2010). *Games with Dynamic Difficulty Adjustment using POMDPs.* International Conference on Machine Learning, Haifa, Israel.

Goldberg, D. E. (1989). *Genetic Algorithms in search, optimization and machine learning.* Reading, MA: Addison-Wesley.

Goldberg, D. E., Korb, B., & Deb, K. (1989). Messy genetic algorithms: Motivation, analysis, and first results. *Complex Systems, 3*(5), 493–530.

Google. (2011). Retrieved from Google Earth: http://www.google.com/earth/index.html

Gorman, B., & Humphrys, M. (2005). Towards integrated imitation of strategic planning and motion modelling in interactive computer games. In *Proc. 3rd ACM Annual International Conference in Computer Game Design and Technology (GDTW 05)*(pp 92-99).

Gregory, J. (2009). *Game Engine Architecture.* Wellelsley, MA: AK Peters.

Guendelman, E., Bridson, R., & Fedkiw, R. (2003). Non-convex rigid bodies with stacking. *ACM Transactions on Graphics*, *22*(3), 871. doi:10.1145/882262.882358

Hager, W. W. (1988). *Applied Numerical Linear Algebra*. Upper Saddle River, NJ: Prentice Hall.

Hampel, T., Bopp, T., & Hinn, R. (2006). A peer-to-peer architecture for massive multiplayer online games. Paper presented at the *Proceedings of 5th ACM SIGCOMM workshop on Network and System Support for Games*. Singapore.

Hartley, T., Mehdi, Q., & Gough, N. (2004). Applying Markov decision processes to 2d real time games. In Medhi, Q., Gough, N., Natkin, S., & Al-Dabass, D. (Eds.) *CGAIDE 2004 5th International Conference on Intelligent Games and Simulation* (pp. 55-59).

Havok. (1998). Havok Inc. Retrieved from http://www.havok.com

Hebisch, E., & Jeong, Y. (2009). *Umsetzung von Spielmechaniken auf einer kugelförmigen Spielwelt*. Unpublished diploma thesis. University of Koblenz-Landau, Germany.

Hecker, C. (1998). *Rigid body dynamics*. Retrieved from http://chrishecker.com/Rigid_Body_Dynamics

Hecker, C. (2009). My liner notes for spore/spore behavior tree docs. *Chris Hecker's Website*. Retrieved July from http://www.chrishecker.com/My_Liner_Notes_for_Spore/Spore_Behavior_Tree_Docs

Hedegaard, R. (n.d.). Convex combination. MathWorld--A Wolfram Web Resource created by Eric W. Weisstein. Retrieved from http://mathworld.wolfram.com/ConvexCombination.html

Heijden, M., van der Bakkes, S., & Spronck, P. (2008). Dynamic formations in real-time strategy games. In *2008 IEEE Symposium On Computational Intelligence and Games* (pp. 47-54). doi: 10.1109/CIG.2008.5035620

Heinermann, A. (2010). BWTA. Retrieved from http://code.google.com/p/bwta/wiki/BWTA.

Heinermann, A. (2011). BWAPIManual. Retrieved from http://code.google.com/p/bwapi/wiki/BWAPIManual?tm=6.

Heinermann, A. (2011). BWSAL Overview. Retrieved from http://code.google.com/p/bwsal/.

Hennessy, J. L., & Patterson, D. A. (2007). *Computer architecture: A Quantitative Approach* (4th ed.). San Francisco, CA: Morgan Kaufmann.

Hesprich, D. (1998). QuakeC Reference Manual. Retrieved from http://pages.cs.wisc.edu/~jeremyp/quake/quakec/quakec.pdf

Higgins, D. (2002a). Pathfinding design architecture. In Rabin, S. (Ed.), *AI Game Programming Wisdom* (pp. 122–132). Hingham, MA: Charles River Media.

Higgins, D. (2002b). How to achieve lightning fast a. In Rabin, S. (Ed.), *AI Game Programming Wisdom* (pp. 133–145). Hingham, MA: Charles River Media.

Hunicke, R. (2005). The case for dynamic difficulty adjustment in *games*. Proceedings from the 2005 *ACM SIGCHI International Conference on Advances in computer entertainment technology* (pp. 429-433). Valencia, Spain: Association for Computing Machinery.

Hunicke, R., & Chapman, V. (2004). AI for Dynamic Difficulty Adjustment in Games. Proceedings from the *Challenges in Game AI Workshop, Nineteenth National Conference on Artificial Intelligence.*

Hunicke, R., LeBlanc, M., & Zubek, R. (2004). MDA: A formal approach to game design and game research. Paper presented at *AAAI Workshop Challenges in Game Artificial Intelligence*. San Jose, CA.

Hunicke, R., LeBlanc, M., & Zubek, R. (2004). MDA: A formal approach to game design and game research. *Discovery*, *83*(3), 4.

International Towing Tank Conference (2002). Esso osaka specialist committee - Final report and recommendations to the 23rd ittc. *Proceedings of 23rd ITTC*. Venice, Italy.

Irem (1987). *R-Type* [Arcade game]. Hakusan, Japan: Irem.

Isla, D. (2005). Handling complexity in the halo 2 ai. *Proceedings of the Game Developers Conference 2005*. Retrieved from http://www.gamasutra.com/gdc2005/features/20050311/isla_01.shtml

Isle of Tune. (2010). Retrieved from http://isleoftune.com/

Japan Studio. (2006). *Loco Roco* [Playstation Portable game]. Tokyo, Japan: Sony Computer Entertainment.

Jerez, J., & Suero, A. (2003). *Newton*. Retrieved from http://www.newtondynamics.com

Johnson, G. (2006). Smoothing a navigation mesh path. In Rabin, S. (Ed.), *AI Game Programming Wisdom 3* (pp. 129–139). Charles River Media.

Journée, J. M. J., & Pinkster, J. J. (2002). *Introduction in ship hydromechanics*. The Netherlands: Delft University of Technology.

Junker, G. (2006). *Pro OGRE 3D programming*. New York, NY: Apress.

Juul, J. (2005). *Half-real: Video games between real rules and fictional worlds*. Cambridge, MA: MIT Press.

Juul, J. (2009). *Fear of failing? The many meanings of difficulty in video games. The Video Game Theory Reader 2*. New York, NY: Routledge.

Kabus, P., Terpstra, W. W., Cilia, M., & Buchmann, A. P. (2005). *Addressing cheating in distributed MMOGs*. Paper presented at the *Proceedings of 4th ACM SIGCOMM Workshop on Network and System Support for Games*. Hawthorne, NY.

Kacic-Alesic, Z., Nordenstam, M., & Bullock, D. (2003). A practical dynamics system. In *Proceedings of the 2003 ACM SIGGRAPH Eurographics Symposium on Computer animation* (pp. 7-16). Aire-la-Ville, Switzerland: Eurographics Association.

KCET. (1997). *Castlevania: Symphony of the Night* [PlayStation game]. Tokyo, Japan: Konami Corporation.

Knafla, B. (2011a). Introduction to behavior trees. *Altdevblogaday Series About Data-Oriented Behavior Trees*. Retrieved from http://altdevblogaday.com/2011/02/24/introduction-to-behavior-trees/

Knafla, B. (2011b). Shocker: Naive object-oriented behavior tree isn't data-oriented. *AltDevBlogADay Series About Data-Oriented Behavior Trees*. Retrieved from http://altdevblogaday.com/2011/03/10/shocker-naive-object-oriented-behavior-tree-isnt-data-oriented/

Knafla, B. (2011c). Data-oriented streams spring behavior trees. *AltDevBlogADay Series About Data-Oriented Behavior Trees*. Retrieved from http://altdevblogaday.com/2011/04/24/data-oriented-streams-spring-behavior-trees/

Knafla, B. (2011d). Data-Oriented Behavior Tree Overview. *AltDevBlogADay Series About Data-Oriented Behavior Trees*. Retrieved from http://altdevblogaday.com/2011/07/09/data-oriented-behavior-tree-overview/

Koster, R. (2004). *A Theory of Fun for Game Design*. Phoenix, AZ: Paraglyph Press.

Koza, J. R. (1992). *Genetic Programming: On the programming of computers by means of natural selection*. Cambridge, MA: MIT Press.

Kwok, M., & Yeung, G. (2005). Characterization of user behavior in a multi-player online game. *ACE '05: Proceedings of the 2005 ACM SIGCHI International Conference on Advances in Fomputer Entertainment Ttechnology* (pp. 69-74). Valencia, Spain: ACM.

Laasonen, J. (2008). *Muodostelmien hallinta reaaliaikastrategiapeleissä*. Master's thesis, University of Turku, Finland.

Laird, J. (2001). It knows what you're going to do: Adding anticipation to a quakebot. Proceedings from *Fifth International Conference on Autonomous Agents* (pp. 385-392). Retrieved November 21, 2010 from http://ai.eecs.umich.edu/people /laird/ papers/Agents01.pdf.

Laird, J. E. (2001). Using a computer game to develop advanced AI. *Computer*, *34*(7), 70–75. doi:10.1109/2.933506

LeBlanc, M. (2000). *Formal Design Tools: Emergent Complexity, Emergent Narrative*. Game Developers Conference.

LeGrand, R. (2008). Computational aspects of approval voting and declared-strategy voting. Doctoral Dissertation. Washington University, St. Louis.

LeGrand, R., & Cytron, R. K. (2008). *Approval-rating systems that never reward insincerity*. Paper presented at the 2nd International Workshop on Computational Social Choice (COMSOC), Liverpool, England.

Lewis, A. M. (1997). High precision formation control of mobile robots using virtual structures. *Autonomous Robots*, *4*(4), 387–403. doi:10.1023/A:1008814708459

Linden, L. (2002). Strategic and tactical reasoning with waypoints. In Rabin, S. (Ed.), *AI Game Programming Wisdom* (pp. 211–220). Hingham, MA: Charles River Media.

Lindley, C. A. (2005). The semiotics of time structure in ludic space as a foundation for analysis and design. *Game Studies*, 5. Retrieved from http://www.gamestudies.org/0501/lindley/

Lindley, C. A., & C. C. Sennersten (2008). *Game play schemas: From player analysis to adaptive game mechanics* (pp. 1-7). International Journal of Computer Games Technology.

Littlejohn, S. W. (1989). *Theories of Human Communication*. Belmont, CA: Wadsworth Publishing Company.

Liu, C. X. (2006). Construct low-cost multi-projector tiled display system for marine simulator. *16th International Conference on Artificial Reality and Telexistence (ICAT'06)* (pp. 688-693). Hangzhou, China.

Llopis, N. (2009). Data-oriented design (Or why you might be shooting yourself in the foot with oop). *Games From Within*. Retrieved from http://gamesfromwithin.com/data-oriented-design

Llopis, N. (2011a). Data-oriented design now and in the future. *Games From Within*. Retrieved from http://gamesfromwithin.com/data-oriented-design-now-and-in-the-future

Llopis, N. (2011b). High-performance programming with data-oriented design. In Lengyel, E. (Ed.), *Game Engine Gems 2* (pp. 251-261). Natick, MA: A K Peters.

Lorch, W. (2000). An introduction to graph algorithms. Retrieved September 10, 2010, from http://www.cs.auckland.ac.nz/~ute/220ft/graphalg/node21.html.

Lötstedt, P. (1984). Numerical simulation of time-dependent contact friction problems in rigid body mechanics. *Society for Industrial and Applied Mathematics: Journals on Scientific Computing*, *5*(2), 24.

Macedonia, M. R., Zyda, M. J., Pratt, D. R., Barham, P. T., & Zeswitz, S. (1994). NPSNET: A network software architecture for large-scale virtual environments. *Presence (Cambridge, Mass.)*, *3*(4), 265–287.

Manslow, J. (2006). Practical Algorithms for In-Game Learning. In Rabin, S. (Ed.), *AI Game Programming Wisdom 3* (pp. 599–616). Hingham, MA: Charles River Media.

Manslow, P. (2002). Learning and adaptation. In Rabin, S. (Ed.), *AI Game Programming Wisdom* (pp. 557–566). Hingham, MA: Charles River Media.

MasterOfChaos. (2011). Chaoslauncher. Version 0.5.3. Retrieved from http://wiki.teamliquid.net/starcraft/Chaoslauncher.

Maxis (1989). *SimCity* [PC game]. Emeryville, CA: Maxis.

Mäyrä, F. (2008). *An introduction to game studies: Games in culture*. Los Angeles, CA: SAGE Publishing.

McGlinchey, S. (2003). Learning of ai players from game observation data. In Medhi, Q., Gough, N., & Natikin, S. (Eds.), *GAME-ON 2003 4th International Conference on Intelligent Games and Simulation* (pp. 106-110).

Meiländer, D., Ploss, A., Glinka, F., & Gorlatch, S. (2011). Software development for real-time online interactive applications on clouds. *Frontiers in Artificial Intelligence and Applications*, *231*. doi:doi:10.3233/978-1-60750-831-1-81

Meyer, B. (1998). *Object oriented software construction*. Upper Saddle River, NJ: Prentice Hall.

Michalewicz, Z. (1996). *Genetic algorithms + data structures = Evolution programs* (3rd ed.). Berlin, Germany: Springer Verlag.

Microsoft. (2011). Microsoft Visual Studio. Retrieved from http://www.microsoft.com/visualstudio/en-us.

Millington, I. (2006). *Artificial Intelligence for Games*. San Francisco, CA: Morgan Kaufmann.

Millington, I., & Funge, J. (2009). *Artificial intelligence for games* (2nd ed.). San Francisco, CA: Morgan Kauffman.

Montoya, R., Mora, A., & Merelo, J. J. (2002). Evolución nativa de personajes de juegos de ordenador. In E. Alba, F. Fernández, J. A. Gómez, F. Herrera, J. I. Hidalgo, J.-J. Merelo-Guervós, and J. M. Sánchez, (Eds.) *Actas del Primer Congreso Español de Algoritmos Evolutivos*, (pp.212-219) AEB′ 02, Universidad de Extremadura.

Moon, S. (2011). Retrieved http://www.fluffylogic.net/games/savage-moon/

Moretti, M., & Dondi, C. (2003). *Guide to quality criteria of learning games*. Bologna, Italy: SIG-GLUE.

Morrison, M. (2005). *Beginning Mobile phone Game Programming*. Indianapolis, IN: Sams.

Moulin, H. (1980, January). On strategy-proofness and single peakedness. *Public Choice*, *35*(4), 437–455. doi:10.1007/BF00128122

Müller-Iden, J. (2007). Replication-based scalable parallelization of virtual environments. Doctoral dissertation. Retreived from http://pvs.uni-muenster.de/jmueller/jmi_thesis.pdf. Universität Münster, Germany.

Muratori, C. (2006). *Implementing GJK* [MP4]. Molly Rocket Nebula. Retrieved from http://mollyrocket.com/849

Muratori, C. (n.d.). *Implementing GJK*. Retrieved from http://mollyrocket.com/849

Nae, V., Prodan, R., & Fahringer, T. (2008). Neural network-based load prediction for highly dynamic distributed online games. Paper presented at *The 14th International Euro-Par Conference European Conference on Parallel and Distributed Computing*. Las Palmas de Gran Canaria, Spain.

Naffin, D. J., & Sukhatme, G. S. (2004). Negotiated formations. In *Proceedings of the International Conference on Intelligent Autonomous Systems (IAS)*. In F. Groen, N. Amato, A. Bonarini, E. Yoshida, & B. Kröse (Eds.), *Proceedings of the 8th International Conference on Intelligent Autonomous Systems* (pp. 181-190).

Nakayama, Y., & Boucher, R. F. (1998). *Introduction to Fluid Mechanics*. Oxford, UK: Butterworth-Heinemann.

Namee, B. M. (2004). *Proactive persistent agents: Using situational intelligence to create support characters in character-centric computer games*. Doctoral Dissertation. Trinity College of Dublin, Ireland.

Ng, B., Si, A., Lau, R. W. H., & Li, F. W. B. (2002). A multi-server architecture for distributed virtual walk-through. Paper presented at the *Proceedings of the ACM symposium on Virtual Reality Software and Technology*. Hong Kong, China.

Nguyen, H. (2007). *GPU gems 3*. Reading, MA: Addison-Wesley.

Nintendo (1985). *Super Mario Bros* [NES game]. Kyoto, Japan: Nintendo.

Nisan, N. (2007). Introduction to Mechanism Design (for Computer Scientists). In Nisan, N., Roughgarden, T., Tardos, E., & Vazirani, V. V. (Eds.), *Algorithmic Game Theory* (pp. 209–241). Cambridge, UK: Cambridge University Press. doi:10.1017/CBO9780511800481.011

Nuclex framework. (2011). Retrieved from http://nuclex-framework.codeplex.com/

NVIDIA Corporation. (2011). *NVIDIA physx*. Retrieved from http://developer.nvidia.com/technologies/physx

NVIDIA. (2011). *PhysX*. Retrieved from http://www.nvidia.com/object/physx_new.html

Oceanic Consulting Corporation. (2008). *Centre for Marine Simulation*. Retrieved from http://www.oceaniccorp.com/FacilityDetails.asp?id=7

OGRE 3D (2010). *Object-oriented graphics rendering engine*.Torus Knot Software Ltd. Retrieved from http://www.ogre3d.org/

OGRE. (2011). Retrieved from http://www.ogre3d.org/

OgreMax. (2010). The ogremax documentations. Retrieved from http://www.ogremax.com/

Ögren, P., Fiorelli, E., & Leonard, N. E. (2002). Formations with a mission: Stable coordination of vehicle group maneuvers. In D.S. Gilliam & J. Rosenthal (Eds.), *Electronic Proceedings of the 15th International Symposium on Mathematical Theory of Networks and Systems*. Retrieved May 27, 2011, from http://www.nd.edu/~mtns/papers/4615_3.pdf

Override. (2011) In *MSDN Library*. Retrieved from http://msdn.microsoft.com/en-us/library/ebca9ah3%28v=VS.100%29.aspx

Paanakker, F. (2008). Risk-adverse pathfinding using influence maps. In Rabin, S. (Ed.), *AI Game Programming Wisdom 4* (pp. 173–178). Hingham, MA: Charles River Media.

Pajitnov, A. (1984). *Tetris* [Elektronika 60 game].

Pantel, L., & Wolf, L. C. (2002). On the impact of delay on real-time multiplayer games. *NOSSDAV '02: Proceedings of the 12th International Workshop on Network and Operating Systems Support for Digital Audio and Video* (pp. 23-29). Miami, Florida, USA: ACM.

Papini, I. (2010). *Virtual sailor*. Retrieved from http://www.hangsim.com/vs/

Parallax Software. (1995). *Descent* [PC game]. Beverly Hills, CA: Interplay Productions.

Patterson, D. A. (2004). Latency lags bandwidth. *Communications of the ACM*, *47*(10), 71–75. doi:10.1145/1022594.1022596

Pelland, S. (2009). *A Wii Bit of History*. Retrieved from www.Gamasutra.com

Perez, T., & Blanke, M. (2002). *Mathematical Ship Modeling for Control Applications. Technical Report*. Australia: The University of Newcastle.

PhysX (2009). Popular physics engines comparison physx, havok and ode. Retrieved from http://physxinfo.com/articles/?page_id=154

Pillosu, R., Jack, M., Kirst, K., Mohr, M., & Zielinski, M. (2009). Coordinating agents with behavior trees. *Paris Game AI Conference 2009*. Retrieved from http://staff.science.uva.nl/~aldersho/GameProgramming/Papers/Coordinating_Agents_with_Behaviour_Trees.pdf

Pinter, M. (2002). Realistic turning between waypoints. In Rabin, S. (Ed.), *AI Game Programming Wisdom* (pp. 186–192). Hingham, MA: Charles River Media.

PixelJunk Monsters. (2011). Retrieved from http://pixeljunk.jp/

Ploss, A. (2011). *On Efficient Dynamic Communication for Real-Time Online Interactive Applications in Heterogeneous Environments.* Doctoral dissertation. University of Müenster, Germany.

Ploss, A., Glinka, F., Gorlatch, S., & Müller-Iden, J. (2007). Towards a high-level design approach for multi-server online games. *GAMEON'2007: Proceedings of the 8th International Conference on Intelligent Games and Simulation* (pp. 10-17). Bologna, Italy.

Pogamut 2. (n.d.). Retrieved from http://artemis.ms.mff.cuni.cz/pogamut/tiki-index.php

Popp, K., & Schiehlen, W. (2010). *Ground vehicle dynamics*. Berlin, Germany: Springer-Verlag.

Pottinger, D. C. (1999a). Coordinated unit movement. *Gamasutra*, January 22, 1999. Retrieved May 27, 2011, from http://www.gamasutra.com/features/19990122/movement_01.htm

Pottinger, D. C. (1999b). Implementing coordinated movement. Gamasutra, January 29, 1999. Retrieved May 27, 2011, from http://www.gamasutra.com/features/19990129/implementing_01.htm

Priesterjahn, S., Goebels, A., & Weimer, A. (2005). Stigmergetic communication for cooperative agent routing in virtual environments. *In International Conference on Artificial Intelligence and the Simulation of Behaviour* (AISB'05).

Priesterjahn, S., Kramer, O., Weimer, A., & Goebels, A. (2006). Evolution of human-competitive agents in modern computer games. In *IEEE Congress on Computational Intelligence 2006, CEC'06* (pp. 777-784). Vancouver, Canada.

Procaccia, A. D., & Rosenschein, J. S. (2007). Junta distributions and the average-case complexity of manipulating elections. *Journal of Artificial Intelligence Research*, *28*, 157–181.

Puustinen, I., & Pasanen, T. A. (2006). Game theoretic methods for action games. In T. Honkela, T. Raiko, J. Kortela & H. Valpola (Eds.), *Proceedings of the Ninth Scandinavian Conference on Artificial Intelligence (SCAI 2006)* (pp. 183-188). Espoo, Finland: Finnish Artificial Intelligence Society.

Quantic Dream. (2010). *Heavy Rain* [PS3 game]. Tokyo, Japan: Sony Computer Entertainment.

Quazal. (2011). Quazal Net-Z. Retrieved from http://www.quazal.com

Rabin, S. (2000). A* aesthetic optimizations. In DeLoura, M. (Ed.), *Game programing gems*. Hingham, MA: Charles River Media.

Rabin, S. (2002). *AI Game Programming Wisdom*. Hingham, Massachusetts: Charles River Media.

Rakkarsoft. (2011).RakNet. Retrieved from http://www.rakkarsoft.com

Rawson, K. J., & Tupper, E. C. (2001). *Basic ship theory (Vol. 1)*. Oxford, UK: Butterworth-Heinemann.

Raydium Game Engine. (2011). Retrieved August 10, 2011 from, http://radium.org

Reynolds, C. W. (1999). Steering behaviors for autonomous characters. In *Proceedings of the Game Developers Conference* (pp. 763-782). San Francisco, CA: Miller Freeman Game Group.

Reynolds, C. W. (1987). Flocks, herds, and schools: A distributed behavioral model. *Computer Graphics*, *21*(4), 25–34. doi:10.1145/37402.37406

Rilling, S., & Wechselberger, U. (2011). A framework to meet didactical requirements for serious game design. *The Visual Computer*, *27*(4), 287–297. doi:10.1007/s00371-011-0550-6

Rizzo, A.A., Bowerly,T., Buckwalter, G., Klimchuck, D., Mitura, R., & Parsons, T.D. (2006). A virtual reality scenario for all seasons the virtual classroom. *CME 3*.

Robson, C. S. (2004). Review of Business.

Rolling, A., & Adams, E. (2003). *Andrew Rollings and Ernest Adams on Game Design*. Berkely, CA: Pearson Education- New Riders Game series.

Rosedale, P., & Ondrejka, C. (2003). *Enabling player-created online worlds with grid computing and streaming*. Retrieved from http://www.gamasutra.com/resource_guide/20030916/rosedale_01.shtml

Rubel, S. (2008). The future is web services, not web sites. Retrieved from http://www.micropersuasion.com/2008/03/the-future-is-w.html

Salen, K., & Zimmerman, E. (2003). *Rules of play: Game design fundamental*. Cambridge, MA: MIT Press.

Saltsman, A. (2009). *Game changers: Dynamic difficulty*. Retrieved from http://www.gamasutra.com/blogs/AdamSaltsman/20090507/1340/Game_Changers_Dynamic_Difficulty.php

Sandaruwan, D. (2010). A six degrees of freedom ship simulation system for maritime education. *The International Journal on Advances in ICT for Emerging Regions*, *3*, 34–47.

Satterthwaite, M. A. (1975). Strategy-proofness and arrow's conditions: Existence and correspondence theorems for voting procedures and social welfare functions. *Journal of Economic Theory*, *10*(2), 187–217. doi:10.1016/0022-0531(75)90050-2

Schell, J. (2008). *The art of game design - A book of lenses*. Waltham, MA: Morgan Kaufman.

Schollmeyer, J. (2006). Games get serious. *The Bulletin of the Atomic Scientists*, *62*(4), 34–39. doi:10.2968/062004010

Schrum, J., & Miikkulainen, R. (2009). Evolving multi-modal behavior in NPCs. In *IEEE Symposium on Computational Intelligence and Games, CIG 2009*, (pp. 325-332). Milan, Italy.

Schwab, B. (2004). *AI game engine programming. Game Development Series*. Hingham, MA: Charles River Media.

Sellers, M. (2006). Designing the experience of interactive play. In Vorderer, P., & Bryant, J. (Eds.), *Playing Video Games: Motives, Responses, and Consequences* (pp. 9–22). New York, NY: Routledge.

Shabana, A. (1994). *Computational dynamics*. Chichester, UK: John Wiley & Sons, Inc.

Shalloway, A., & Trott, J. (2001). *Design patterns explained – A new perspective on object-oriented design*. Reading, MA: Addison-Wesley.

Sharp, C. (2003). Business integration for games: An introduction to online games and e-business infrastructure. Retrieved from http://www.ibm.com/developerworks/webservices/library/ws-intgame/.

Simulator Systems International (SSI). (2010). *Vehicle simulation*. Retrieved from http://simulatorsystems.com/vehiclesimulators/

Small, R., & Bates-Congdon, C. (2009). Agent Smith: Towards an evolutionary rule-based agent for interactive dynamic games. In *IEEE Congress Evolutionary Computation, CEC '09,* (pp. 660-666). Trondheim, Norway.

Smed, J., & Hakonen, H. (2006). *Algorithms and networking for computer games*. Chichester, UK: John Wiley & Sons. doi:10.1002/0470029757

Smith, R. (2004). *Open dynamics engine*. Retrieved from http://www.ode.org

Society of Naval Architects and Marine Engineers. (1950). *Nomenclature for treating the motion of a submerged body through a fluid*. Jersey City, NJ: The Society of Naval Architects and Marine Engineers.

Soni, B., & Hingston, P. (2008). Bots trained to play like a human are more fun. In *IEEE International Joint Conference on Neural Networks, IJCNN 2008, (IEEE World Congress on Computational Intelligence)* (pp. 363-369). Hong Kong, China.

Sørensen, M. J. (2003). Artificial potential field approach to path tracking for a nonholonomic mobile robot. In *Proceedings of the 11th Mediterranean Conference on Control and Automation*. Rhodes, Greece.

Space, C., III. (2011). Retrieved from http://www.crystalspace3d.org/main/Main_Page

Spronck, P., Sprinkhuizen-Kuyper, I., & Postma, E. (2003). Online adaptation of game opponent ai in simulation and in practice. In Medhi, Q., Gough, N., & Natikin, S. (Eds.), *GAME-ON 2003 4th International Conference on Intelligent Games and Simulation* (pp. 93 - 100).

Sterren, W. (2001). Terrain Reasoning for 3D Action Games. Retrieved from http://www.gamasutra.com/features/20010912/sterren_01.htm

Stewart, D., & Trinkle, J. C. (1996). An implicit time-stepping scheme for rigid body dynamics with coulomb friction. *International Journal for Numerical Methods in Engineering, 39*, 2673–2691. doi:10.1002/(SICI)1097-0207(19960815)39:15<2673::AID-NME972>3.0.CO;2-I

Straatman, R., Sterren, W., & Beij, A. (2005). Killzone's AI: Dynamic procedural combat tactics. Retrieved from http://www.cgf-ai.com/docs/straatman_remco_killzone_ai.pdf

Straatman, R., Beij, A., & Sterren, W. (2006). Dynamic tactical position evaluation. In Rabin, S. (Ed.), *AI Game Programming Wisdom 3* (pp. 389–403). Hingham, MA: Charles River Media.

Sun Microsystems. (2009). I. Project Darkstar. Retrieved from http://www.projectdarkstar.com

Susi, T., Johannesson, M., & Backlund, P. (2007). *Serious Games - An Overview. Technical Report HS- IKI -TR-07-001*. Sweden: University of Skövde.

Sweetser, P., & Wyeth, P. (2005). GameFlow: A model for evaluating player enjoyment in games. *Computer Entertainment, 3*(3), 3–3. doi:10.1145/1077246.1077253

Team Liquid. (2010). Liquipedia: The StarCraft Encyclopedia. Retrieved from http://wiki.teamliquid.net/starcraft/Main_Page.

Teiner, M., Rojas, I., Goser, K., & Valenzuela, O. (2003). A hierarchical fuzzy steering controller for mobile robots. In *2003 International Conference Physics and Control. Proceedings (Cat. No.03EX708)* (pp. 7-12). doi: 10.1109/CIMSA.2003.1227193

Terdimann, P. (2007). Sweep-and-prune. Retrieved from www.codercorner.com/SAP.pdf

Thurau, C. (2006). Behavior Acquisition in Artificial Agents. Doctoral Dissertation. Bielefeld (Germany): Bielefeld University.

Thurau, C., Bauckhage, C., & Sagerer, G. (2004). Imitation learning at all levels of game-ai. In *Proc. Int. Conf. on Computer Games, Artificial Intelligence, Design and Education* (pp. 402–408).

Thurau, C., Bauckhage, C., & Sagerer, G. (2003). Combining self organizing maps and multilayer perceptrons to learn bot-behaviour for a commercial game. In Mehdi, Q. H., Gough, N. E., & Natkine, S. (Eds.), *GAME-ON* (pp. 119–123). EUROSIS.

Totilo, S. (2007). A higher standard – Game designer jonathan blow challenges super mario's gold coins, "unethical" mmo design and everything else you may hold dear about games. *MTV Networks.* Retrieved from http://multiplayerblog.mtv.com/2007/08/08/a-higher-standard-game-designer-jonathan-blow-challenges-super-marios-gold-coins-unethical-mmo-design-and-everything-else-you-may-hold-dear-about-video-games/

Tournament, U. (2004). *Ut2004*. Retrieved from http://www.unrealtournament2004.com/ut2004/modes.html.

Tozour, P. (2002a). First-person shooter ai architecture. In Rabin, S. (Ed.), *AI Game Programming Wisdom* (pp. 387–396). Hingham, MA: Charles River Media.

Tozour, P. (2002b). The evolution of game ai. In Rabin, S. (Ed.), *AI Game Programming Wisdom* (pp. 3–15). Hingham, MA: Charles River Media.

Transas Marine Limited. (2010). *Transas simulation products*. Transas Marine International. Retrieved from http://www.transas.com/products/simulators/

Tremblay, J., Bouchard, B., & Bouzouane, A. (2010). Adaptive game mechanics for learning purposes: Making serious games playable and fun. Proceedings from the *International Conference on Computer Supported Education: session "Gaming platforms for education and reeducation",* Valencia, Spain.

Tutorial, D. (2011) In *MSDN Library*. Retrieved from http://msdn.microsoft.com/en-us/library/aa288459%28v=vs.71%29.aspx

Tzeng, C.-Y., & Chen, J.-F. (1997). Fundamental properties of linear ship steering dynamic models. *Journal of Marine Science and Technology*, *7*(2), 79–88.

Um, S., Kim, T., & Choi, J. (2007). Dynamic difficulty controlling game system. IEEE International Conference on Consumer Electronics, 2007. ICCE 2007. Digest of Technical Papers. Las Vegas, Nevada.

Unreal (n.d.). Retrieved from http://www.unreal.com

Unreal Engine. (2011). Wikipedia: The free encyclopedia. Retrieved from http://en.wikipedia.org/wiki/Unreal_Engine

Unreal Script. (n.d.). Unreal Engine Site. Retrieved from http://www.unrealengine.com/features/unrealscript/

Unreal(2009). Wikipedia: The free encyclopedia (2009). Retrieved from http://en.wikipedia.org/wiki/Unreal

Valdes, R. (2004). The artificial intelligence of halo 2. *HowStuffWorks.com*. Retrieved from http://electronics.howstuffworks.com/halo2-ai.htm

Valve Software (2007). *Portal* [PC game]. Bellevue, WA: Valve.

Van Dongen, J. (2010). AI in swords and soldiers (part 1). *Joost's Dev Blog*. Retrieved from http://joostdevblog.blogspot.com/2010/12/ai-in-swords-soldiers-part-1.html

Van Dongen, J. (2011). AI in swords and soldiers (part 2). *Joost's Dev Blog*. Retrieved from http://joostdevblog.blogspot.com/2011/01/ai-in-swords-soldiers-part-2.html

VDrift.net. (2010). Retrieved from http://vdrift.net/

Verth, J. V., Brueggemann, V., Owen, J., & McMurry, P. (2000). Formation-based pathfinding with real-world vehicles. In *Proceedings of the Game Developers Conference*. Retrieved May 27, 2011, from http://citeseerx.ist.psu.edu/viewdoc/summary?doi=10.1.1.17.2031

Vondrak, M. (2006). *Crisis physics library.* Retrieved from http://crisis.sourceforge.net/.

Vorderer, P., Hartmann, T., & Klimmt, C. (2003). Explaining the enjoyment of playing video games: The role of competition. In *Proceedings of the 2nd International Conference on Entertainment Computing* (pp. 1-9). Pittsburgh, PA.

VSTEP. (2010). *Ship simulator extremes*. Retrieved from http://www.shipsim.com/

Vygotsky, L. S. (1976). Play and its role in the mental development of the child. *Journal of Russian & East European Psychology*, *5*(3), 6–18. doi:10.2753/RPO1061-040505036

Wang, X., & Devarajan, V. (2004). 2d structured mass-spring system parameter optimization based on axisymmetric bending for rigid cloth simulation. In *Proceedings of the 2004 ACM SIGGRAPH International Conference on Virtual Reality Continuum and its Applications in Industry Vrcai 04* (pp. 317-323). New York, NY: ACM Press.

Warcraft 3 (2002). Retrieved from http://us.blizzard.com/en-us/games/war3/

Wark, M. (2007). *Gamer theory*. Cambridge, MA: Harvard University Press.

Waveren, J., & Rothkrantz, L. J. M. (2001). Artificial Player For Quake III Arena. In Medhi, Q., Gough, N., Natkin, S., & Al-Dabass, D. (Eds.) *GAME-ON 2001 2nd International Conference on Intelligent Games and Simulation* (pp 48 – 55).

Wechselberger, U. (2009). Teaching me softly: Experiences and reflections on informal educational game design. *Transactions on Edutainment*, *II*, 90–104. doi:10.1007/978-3-642-03270-7_7

Wechselberger, U. (2011). A serious game compared to a traditional training. *Academic Exchange Quarterly*, *5*, 58–63.

Wei, D. H. K., Yee, K. C., Guozheng, O., Kern, T. C., & Plemping, N. A. (2011). Vivace: Chord tutor. *SoC Games Portal: National University of Singapore*. Retrieved from http://games.comp.nus.edu.sg/chordtutor/

Weisstein, E. (2007). *Gravitational potential energy.* Retrieved from http://scienceworld.wolfram.com/physics/GravitationalPotentialEnergy.html.

Weisstein, E. W. (n.d.) *Convex. MathWorld*--A Wolfram Web Resource created by Eric W. Weisstein. Retrieved from http://mathworld.wolfram.com/Convex.html

What Is a Game Loop? (2011) In *MSDN Library*. Retrieved August 10, 2011 from, http://msdn.microsoft.com/en-us/library/bb203873.aspx

Wiki, U. (2007). *Unreal unit*. Retrieved from http://wiki.beyondunreal.com/Legacy:Unreal_Unit.

Wikipedia. (2011). *Real-time computing*. Retrieved from http://en.wikipedia.org/wiki/Real-time_computing

Xu, C.-W. (2007). A Hybrid Gaming Framework and Its Applications. *The International Technology, Education and Development Conference 2007 (INTED2007), Valencia, Spain*, March 7-9, 2007, pages 30000_0001.pdf.

Xu, C.-W. (2008a). A new communication framework for networked mobile games. *Journal of Software Engineering and Applications*, *1*(1), 20–25. doi:10.4236/jsea.2008.11004

Xu, C.-W. (2008b). Teaching OOP and COP technologies via gaming. In Ferdig, E. R. (Ed.), *Handbook of Research on Effective Electronic Gaming in Education*. Hershey, PA: IGI Global. doi:10.4018/978-1-59904-808-6.ch029

Xu, C.-W., Lei, H., & Xu, D. (2010). Networked games based on web services. *GSTF International Journal on Computing*, *1*(1), 170–175. doi:10.5176/2010-2283_1.1.28

Yager, R. R., & Filev, D. P. (1994). *Essentials of Fuzzy Modeling and Control*. New York, NY: John Wiley & Sons.

Yin, Y., Sun, X., Zhang, X., Liu, X., Ren, H., Zhang, X., & Jin, Y. (2010). Application of virtual reality in marine search and rescue simulator. *The International Journal of Virtual Reality*, *9*(3), 19–26.

Young, B. J., Beard, R. W., & Kelsey, J. M. (2001). A control scheme for improving multi-vehicle formation maneuvers. In *Proceedings of the American Control Conference* (pp. 704-709).

Zadeh, L. A. (1965). Fuzzy sets. *Information and Control*, *8*(3), 338–353. doi:10.1016/S0019-9958(65)90241-X

Zanetti, S., & Rhalibi, A. E. (2004). Machine learning techniques for FPS in Q3. In *Proceedings of the 2004 ACM SIGCHI International Conference on Advances in computer entertainment technology,ACE '04* (pp. 239-244). New York, NY: ACM.

Zhou, L. (1991). Impossibility of strategy-proof mechanisms in economies with pure public goods. *The Review of Economic Studies*, *58*(1), 107–119. doi:10.2307/2298048

About the Contributors

Ashok Kumar is an Assistant Professor in the School of Computing and Informatics at the University of Louisiana at Lafayette. He has teaching experience on algorithmic and architectural aspects of game engine design along with other courses such as Computer Architecture, Operating Systems, and Embedded Systems. He has several years of academic and industrial experience with research and development, and he has published in a variety of areas including video game design, intelligent systems, sensor-enabled systems, and low power design. He obtained his bachelor's degree from the Indian Institute of Technology, BHU, India, and his master's and doctoral degrees from the University of Louisiana at Lafayette in Louisiana, USA.

Jim Etheredge received the M.S. degree in Computer Science from the University of Southwestern Louisiana in 1986 and the PhD. in computer science from the University of Southwestern Louisiana in 1989. He is currently an Associate Professor of Computer Science at the University of Louisiana at Lafayette in Louisiana and the coordinator for the Video Game Design and Development concentration of the undergraduate computer science curriculum. His research and teaching interests include video game design and development, artificial intelligence, multi-agent game systems, and database management systems.

Aaron Boudreaux is a PhD. student in computer science at the University of Louisiana at Lafayette. He received his bachelor's degree in computer science from Nicholls State University in 2004 and his master's from the University of Louisiana at Lafayette in 2007. He is currently working on completing his dissertation on the effects of performance control parameters on heterogeneous, autonomous agent teams in video games. Aaron has published around 10 papers in peer-reviewed conferences and journals on both artificial intelligence and computer science education topics. In addition to his research, Aaron is heavily involved with UL Lafayette's Video Game Design and Development curriculum.

* * *

Golam Ashraf received his PhD in Computer Engineering from NTU, Singapore. He teaches game development, animation and interactive media at School of Computing, National University of Singapore. His research interests are in real time graphics, computational aesthetics, multimedia analysis, and pedagogical game design.

William Bittle graduated from the University of South Carolina Upstate with a Bachelors of Science in Computer Science and is currently employed as a Web Application Developer. In this position he has expanded his knowledge in a variety of programming languages and problem solving techniques. Outside of work he enjoys game programming and the sharing of ideas through his personal blog. He has been interested in game programming since the beginning of his undergraduate degree and primarily focuses on the collision detection and physical modeling fields. With a passion for learning and teaching, he promotes simple and readable code in addition to algorithmic design. He is the creator of the open source collision detection and physics library, dyn4j, written in the Java programming language.

Bruno Bouchard is an associate professor and a scientist working at the Department of Mathematics and Computer Science of the University of Quebec at Chicoutimi (UQAC). He received a PhD in computer science from the University of Sherbrooke (Canada) in 2006. He then completed a postdoctoral fellowship at the University of Toronto in 2007. He co-founded in 2008, with the professor Abdenour Bouzouane, the LIARA laboratory, which possesses a cutting edge smart home infrastructure dedicated to research in the field of assistive technologies for Alzheimer people. His research is sponsored by the Natural Sciences and Engineering Research Council of Canada (NSERC), the Quebec Research Fund on Nature and Technologies (FQRNT), the Canadian Foundation for Innovation (CFI), Bell Canada, UQAC and its foundation. He contributed to application domains as varied as cognitive assistance, serious games for elders and activity recognition.

Abdenour Bouzouane is a professor at the University of Quebec at Chicoutimi since 1997. He received in 1993 a PhD in Computer Engineering from the Ecole Centrale de Lyon. His research is funded by NSERC and FQRNT of Canada and related to the domain of artificial intelligence, data mining and smarthome. He is co-founder of Ambient Intelligence Laboratory for the Recognition of activities (LIARA).

Alex J. Champandard has worked in industry as a senior AI programmer for many years, most notably for Rockstar Games. There he worked the motion technology that powers the R.A.G.E. engine (proprietary middleware used in all Rockstar titles internally, including the animation-intensive ROCKSTAR'S TABLE TENNIS) and was initially involved developing the artificial intelligence and gameplay for MAX PAYNE 3. Alex regularly consults with leading studios in Europe, most recently on the multiplayer bots for KILLZONE 2 at Guerrilla Games. With a strong academic background in artificial intelligence, Alex authored the book AI Game Development and often speaks about his research — in particular at the Paris Game AI Conference which he co-organizes with his wife Petra. He's Associate Editor for the IEEE Transactions on Computational Intelligence and AI in Games (TCIAIG), is a member of the Program Committee for the Artificial Intelligence in Interactive Digital Entertainment (AIIDE) conference, and part of the Organizing Committee of Computational Intelligence in Games (CIG) as Industry Liaison.

Ron K. Cytron is a professor of computer science and engineering at Washington University. His research interests include optimized middleware for embedded and real-time systems, fast searching of unstructured data, hardware/runtime support for object-oriented languages, and computational political science. Ron has over 100 publications and 5 patents. He has received the SIGPLAN Distinguished Service Award and is a co-recipient of SIGPLAN Programming Languages Achievement Award. He served as Editor-in-Chief of ACM Transactions on Programming Languages and Systems for 6 years. He participated in writing the Computer Science GRE Subject Test for 8 years and chaired the effort for 3 years. He is a Fellow of the ACM.

Luke Deshotels spent his childhood on a farm in Mamou, Louisiana. He received his B.S. in computer science at the University of Louisiana at Lafayette where he was recognized as the outstanding graduate of science. He has won a master's fellowship at the University of Louisiana and is pursuing his M.S. there under the supervision of Professor Maida in preparation for a PhD in Computer Science elsewhere. During the spring and summer of 2011 he worked with a team of undergraduates to construct the StarCraft AI that this chapter is based on. He is president of the ACM National Alpha Student Chapter and served for two years as an AmeriCorps team member. His main research interests are computational neuroscience and parallel processing.

Kapila Dias is a Senior Lecturer attached to the University of Colombo School Of Computing. He has obtained his B.Sc. from the University of Colombo, Post Graduate Diploma from University of Essex, UK., and Master of Philosophy from the University of Wales College of Cardiff, U.K. He is presently the Head of the Department of Department of Communication and Media Technologies. His research interests include Computer Aided Software Engineering and Multimedia in Education.

Antonio M. Mora García is currently working as a Research Associate at the University of Granada where he got his PhD in 2009. He has been involved in several Spanish funded research projects and published a number of papers in top-rated international journals and conferences. His areas of interest include Ant Colony Optimization, Multi-Objective Optimization and Genetic Algorithms and their applications to Pathfinding or Video Games problems among others.

Frank Glinka received his Computer Science degree from the University of Muenster in 2006 and is now a Research Associate at the department of computer science in the group of Prof. Sergei Gorlatch. He has worked as a Work Package Leader for the European research project edutain@grid covering the topic of real-time application services and is currently completing his PhD thesis titled "Developing Grid Middleware for a High-Level Programming of Real-Time Online Interactive Applications".

Sergei Gorlatch is Full Professor of Computer Science at the University of Muenster (Germany) since 2003. Earlier he was Associate Professor at the Technical University of Berlin, Assistant Professor at the University of Passau, and Humboldt Research Fellow at the Technical University of Munich. Professor Gorlatch has more than 100 peer reviewed publications in renowned international journals and conferences. He is often delivering invited talks at international conferences and serves at their programme committees. He was principal investigator in several international research and development projects in the field of parallel, distributed, Grid and Cloud computing, funded by the European Commission, as well as by German national bodies. Professor Gorlatch holds MSc degree from the State University of Kiev, PhD degree from the Institute of Cybernetics of Ukraine, and the Habilitation degree from the University of Passau (Germany).

Ou Guozheng is a final year undergrad in National University of Singapore, majoring in Communications and Media. He is one of the programmers in the development of this game. Guozheng has a range of interests ranging from the designing the game art to gameplay programming. Guozheng has also dwelled in special effects, storyboarding and camera works for films; 2 of his school projects came out as one of the best in that semester. Guozheng is currently a game designer and technical artist with a game company called Reversal Studios.

Thomas Hartley is a specialist in Artificial Intelligence, computer games design and digital media. He is a lecturer in computer games development at the University of Wolverhampton. He also coordinates, manages, and promotes the Institute of Gaming and Animation's course, student, and curriculum activities. He holds a BSc (Honors) in Computer Science and received his PhD from the University of Wolverhampton in 2010. His research interests include artificial intelligence, online learning, machine learning, virtual environments and computer games development.

Erik Hebisch was born in Köthen, Germany in 1983 and completed his Abitur at Pestalozzi-Gymnasium in Unna in 2002. In school, he would often doodle boardgames on quad-ruled paper and create little stories around them. During his Zivildienst in 2003 he discovered his interest in graphics programming and decided to study Computational Visualistics at the University of Koblenz-Landau in Koblenz. At university, he regularly got together with friends to play video games until the early morning and analyzed their virtues and shortcomings. He finished his diploma thesis on gameplay mechanics and their realization on a spherical game world in 2009. Since 2009, he works as junior researcher at the Fraunhofer Institute for Industrial Engineering IAO. His research topics include software engineering for multicore and cloud computing architectures.

Ho Jie Hui is a final year student from National University of Singapore's School of Computing who is graduated at the end of 2011 with a major in Communications and Media. Jie Hui was an intern at Singapore-MIT GAMBIT (Singapore) where she was part of the QA team. She is very interested in games, game design and human-computer interaction and would love the opportunity to study more about them if money was not a concern. Lerpz the Rocketeer is her first major game-research project.

Lim Kang Ming Kenny is a School of Computing graduate majoring in Communications and Media (with specialization in Games), at the National University of Singapore. He in an alumnus of the Singapore-MIT GAMBIT Summer Program 2010 – in which he undertook the role of a quality assurance lead for the game Afterland, that was later nominated as a finalist in the 2nd Independent Games Festival Student Competition held in GDC China 2010. Kenny was in charge of the conceptualization and implementation of the game design and level design for Lerpz the Rocketeer. Kenny also holds an intermediate proficiency in Japanese and is currently seeking for a job in the game industry.

Ben Kenwright is completing his Ph.D. in computer science at Newcastle University, researching synthetic human motions in highly-parallel execution environments. His current research interests lie in the areas of computer graphics, scientific visualization, immersive environments and human-computer interaction.

Chamath Keppitiyagama has obtained his B.Sc from the University of Colombo and M.Sc and Ph.D in Computer Science from the University of British Columbia (UBC) in Vancouver. He is currently working as a Senior Lecturer at the University of Colombo School Of Computing (UCSC). He is one of the coordinators of the undergraduate program at UCSC. He has research interests in Distributed Computing, Computer Networks, Operating Systems and Sensor Networks.

Teo Chee Kern recently graduated from the National University of Singapore in 2011 with a Bachelor of Computing majoring in Communication and Media. He is the network programmer that handles all the network storage, rating, and saving of the game state during the development of "Vivace". Moreover, he also actively participates in the design of the game as well as the algorithm. He has developed a strong interest in game development since high school. He has been developing his own mini games using various platform such as XNA framework, flash and also JAVA for android. He is currently a backend (network system) programmer in a game company named Reversal Studios.

Bjoern Knafla researches parallel programming for computer and video games with a strong focus on artificial intelligence (AI) in games. He laid the foundation for his work as a research associate in the programming languages / methodologies research group at the University of Kassel, Germany. During this time, he taught and supervised theses in the areas of game development, artificial intelligence, general-purpose computations on graphics processing units (GPGPU), and algorithms and data structures. These days, Bjoern discusses his findings and insights as an AltDevBlogADay.com author and as @bjoernknafla on twitter. He speaks at conferences and freelances as a parallelization + AI + gamedev consultant.

Nihal Kodikara is a Professor attached to the University of Colombo School of Computing (UCSC). He has obtained his B.Sc. from the University of Colombo and M.Sc and PhD in Computer Science from the University of Manchester, U.K. He is presently the Deputy Director of the UCSC. His research interests include Computer Graphics, Computer Vision and e-Goverance.

Jussi Laasonen is a post graduate student at the University of Turku in Finland. Currently he is working for his PhD thesis on cheating prevention in computer games and his research interests include machine learning and algorithms for computer games. He is also a software engineer at a Finnish communications services provider Elisa Corporation, where he is currently working in web application development projects utilizing agile methodologies.

Luo Lan recently graduated from the National University of Singapore in 2011 with a Bachelors in Computing (Communication & Media). She likes making and looking at web designs and has tutored students in IT techniques for 2D and web design, video production and visual effects at NUS Faculty of Social Sciences. She is currently a web developer at Techsailor Group Pte. Ltd.

Rob LeGrand is an Assistant Professor of Computer Science at Angelo State University. His current research interests include computational social choice, artificial intelligence (especially multiagent systems, decentralized decision-making and machine learning) and algorithmic game theory. He also has publications in the areas of static program analysis and programming languages for real-time systems. Rob earned his M.C.S. at Texas A&M University in 1999, with research in cryptography and algorithms, and his PhD at Washington University in St. Louis in 2008, in the Department of Computer Science and Engineering, with research in computational social choice. He holds two patents and is a member of the Association of Computing Machinery. He is originally from San Angelo, Texas.

Esther Luar is a student at the National University of Singapore (School of Computing), completing her Bachelor of Computing (Honours) in Communications and Media, as well as a minor in Sociology. Her primary interest lies in designing, regardless of medium. She sets up and maintains websites for various organizations and individuals, provides networking and technical support, designs logos and letterheads, transcribes lecture notes and other teaching materials, and helps to construct and maintain professional audit spreadsheet tools. She also takes an interest sociological studies, geology, botany and cooking, and hopes to be involved in the National Park Service in the future.

Kanchana Manamperi has obtained his BCSc from the University of Colombo School of Computing. He is a member of Modeling & Simulation research group at the University of Colombo School of Computing. He worked as a Software engineer in Virtusa Pvt.Ltd. He is currently working as a Research and Development officer at the Sri Lanka, Navy. His research interests include Computer Graphics, Computer Simulation and Computer gaming.

Dominik Meilaender received his degree in computer science from the University of Muenster in 2009. He has been a research associate and PhD student at the Department of Computer Science at Muenster in the group for parallel and distributed systems since September 2009. He is working in the area of distributed and Cloud systems and is actively involved in the EU Network of Excellence S-Cube. Dominik Mailaender has published 10 papers in reviewed international conferences and journals and has gained experience in cooperation with international companies within the EU-FP6 project edutain@grid.

Juan J. Merelo received the B.Sc. degree in Theoretical Physics and the PhD degree, both from the University of Granada, Spain, in 1988 and 1994. He has been Visiting Researcher at Santa Fe Institute, Santa Fe, NM, Politecnico Torino, Turin, Italy, RISC-Linz, Linz, Austria, the University of Southern California, Los Angeles, and Université Paris-V, Paris, France. He is currently Full Professor at the Computer Architecture and Technology Department, University of Granada. His main interests include neural networks, genetic algorithms, and artificial life.

Graham Morgan is a faculty member at Newcastle University and together with his team carries out fundamental research in distributed systems, protocols, algorithms and middleware. A major part of his work is in the area of high performance solutions for attaining realistic simulations for the video game industry. Working with some of the top studios in the video games' world Graham created and now runs an advanced masters programme in video game engineering. Many graduates from this programme have gone on to achieve high profile positions in the video games industry.

Nur Aiysha Plemping is a final year student in the National University of Singapore, majoring and Communications and Media. She is the artist in the development of the game "Vivace". She has a passion for art and illustration which was triggered after playing through Final Fantasy, and has never really died since. She has honed her abilities through various mediums from comics and board games to computer games and digital art. Her interest in making more immersive and beautiful games has led her to explore various disciplines such as user experience and interface design.

Alexander Ploss received his degree in Mathematics from the University of Muenster in 2007 and was a Research Associate in the department of computer science in the group for parallel and distributed systems at Muenster until july 2011. During this time, he worked in the international research project edutain@grid and received his doctoral degree in 2011 for his dissertation on "Efficient Dynamic Communication for Real-Time Online Interactive Applications in Heterogeneous Environments". He has published about 20 papers in peer-reviewed conferences and journals as well as book chapters on the scalability of interactive applications, e.g., massively multiplayer online games, with the focus on communication aspects.

Brandon Primeaux is currently pursuing his Masters Degree in Computer Science at the University of Louisiana at Lafayette. He received his Bachelors in Computer Science from the same university while pursuing his interests in video game development, with his primary interests being engine architecture. As a graduate student, he has taken an active role in not only furthering his own education, but passing along his knowledge to younger students coming through the undergraduate program.

Tim Roden is an Associate Professor of Computer Science at Angelo State University. His research interests include computer game development and mobile computing. He founded the Entertainment Computing Laboratory and Technology Innovation and Commercialization Laboratory at ASU. He currently teaches and supervises student research projects. Tim earned his PhD from the University of North Texas in 2005. In 2007 Tim established an innovative new curriculum in computer game development at ASU. The program received national recognition in 2010 when the Princeton Review named it one of North America's top 50 game design programs. Tim is also heavily involved in technology commercialization projects including current projects in secure mobile cloud computing and wearable medical devices.

Benjamin Rodrigue is a resident of southern Louisiana. He has had an interest in technology from a young age. Throughout his childhood, video games fascinated him. He did not consider learning how to create video games until college. His first experience in programming happened during his junior year of high school when he took Computer Science I. He decided to further his understanding of computer science in college. During his college years, he enriched his understanding of video games through the University of Louisiana at Lafayette's video game design and development concentration in Computer Science. He graduated from this University and applied to become a graduate student to further his work in this area.

Rexy Rosa is an Associate Professor attached to the University Of Colombo Faculty Of Science. He has obtained his B.Sc. from the University of Colombo and M.Sc and PhD from the University of Pittsburgh, USA. He is presently the Head of the Department of Physics University of Colombo. His research interests include Gamma Ray Spectroscopy and Physics Education.

Damitha Sandaruwan has obtained his B.Sc from University Of Colombo Faculty Of Science. He is currently doing his M.Phil research and working as a media officer at the University of Colombo School of Computing (UCSC). He is a member of Modeling & Simulation research group at the UCSC He has research interests in real-time simulations, Mathematical modeling and 3D graphics.

Ranjith Senadheera is a Rear Admiral in Sri Lanka navy. At present he is the head of Electrical, Electronics and Information Technology Departments. He has obtained his M.Sc in Defense Studies from Kothalawala Defense University, Sri Lanka. He has over thirty three years of naval service after joining the Naval and Maritime Academy of Tricomalee and gained engineering graduation from Indian Naval College of Engineering and Specialized in Electrical and Electronics from Indian Naval Ship Valsura, Jamnagar. He is a Charted Engineer of Institute of Engineers, Sri Lanka. His research interests are in the field of network centric surveillance and response model development.

Jouni Smed got his first computer at the age of thirteen – and ever since he has been keen in designing, programming and playing computer games. But there is more to computers than just games, so he got serious for a while and received his PhD in Computer Science and became an Adjunct Professor in the University of Turku, Finland. Jouni Smed has organized and taught game development on diverse topics ranging from game algorithms and networking in multiplayer games to game software construction, game design and interactive storytelling. He is also the co-founder of Game Tech&Arts Lab, which aims at bringing together technologically- and artistically-oriented students to work on the same game projects. Jouni Smed's research interests range from code tweaking to software processes and from simple puzzles to multisite game development. He lectures, he supervises students, and yes, he still likes to design, program and play computer games.

Jonathan Tremblay is a PhD student at McGill University. His Master's thesis topic was the implementation and design of different adaptive difficulty models. This chapter is based on his work with the LIARA lab, where video games are used to improve the cognitive functions of Alzheimer patients. Jonathan's fields of interest are AI, game and level design, dynamic difficulty and procedurally generated content. He is also an independent game developer. He strives to reach a wider audience of players and to create meaningful human interactions. He is currently working on a novel model for adaptive behaviour tree.

Ulrich Wechselberger received his degree in 2005 in educational sciences from the University of Koblenz-Landau/Germany. Afterward, he was involved in several research projects dealing with blended learning, e-portfolios, and notably game-based learning as a staff member of the Knowledge Media Institute and the Institute for Computational Visualistics at the University of Koblenz-Landau. In his doctoral dissertation and several publications he investigated the didactical and motivational possibilities and limitations of game-based learning. Another topic was the design of serious games from both theoretical and practical perspectives. In 2011, Dr. Wechselberger moved to the Center for Educational Research and Teacher Education at the University of Paderborn/Germany. His current research covers new media and instructional design in the field of teacher education.

Daniel Ho Kok Wei is a final year undergrad in National University of Singapore, majoring in Communications and Media (With Specialisation in Games). He is one of the programmers in the development of this game "Vivace". Daniel's passion for developing video games started since the age of fourteen. He had developed multiple projects on the Unity platform over the past few years, some of which have gained recognition. He and his team from GAMBIT Singapore-MIT Game Lab, Summer Programme 2011, worked on the game titled "Robotany". It has been nominated one of the sixth finalists of Independent Game Festival China 2011. Daniel is constantly pursuing knowledge that would help him produce better games for people to enjoy.

Chong-wei Xu is currently a Professor emeritus of Computer Science in Department of Computer Science at Kennesaw State University. He received his Master's in Computer Science from University of Wisconsin-Madison and his PhD in Computer Science from Michigan State University. His current research interests mainly include Internet and distributed/parallel system and gaming technologies. He has been awarded two NSF grants, two state-level Yamacraw grants, and two university-level initiative grants for research on Internet and parallel/distributed computing systems. He has been awarded eleven university and college supported research funds on gaming. He has published on these topics in journals such as GSTF International Journal on Computing, Journal of Software Engineering and Applications, Journal of Parallel and Distributed Computing, Computational Statistics and Data Analysis, Journal of Statistical Computation and Simulation, Computational Statistics Quarterly, IEEE Trans. on Software Engineering, and in many proceedings of international conferences.

Daniel Xu is currently a master's graduate student in Computer Science at University of Wisconsin-Milwaukee. He received a bachelor's degree in Computer Science (CS) and Electrical Engineering (EE) from Georgia Technology Institute. He holds a solid grasp of fundamentals in EE and CS, an aptitude of team working, a passion for learning and making, a zest for challenges and a goal for being a computer professional. He is interested in doing research in distributed systems, databases, and gaming. He has joined several research projects and published one journal article in GSTF International Journal on Computing and a proceeding paper on gaming.

Kong Choong Yee is a final year undergrad in National University of Singapore, majoring in Communications and Media (Specialisation in Games). He is the programmer behind the mechanic of tree generation in the development of "Vivace". His primary interest lies in computer games and special effects for videos. During the course of the degree, he has gained exposure in developing games in various platforms such as Flash and Unity, as well as making special effect videos with Adobe software. He is currently one of the game designers in a game company called Reversal Studios.

Index